CABLE SYSTEM TRANSIENTS

CABLE SYSTEM TRANSIENTS

THEORY, MODELING AND SIMULATION

Akihiro Ametani

Emeritus Professor, Doshisha University, Japan

Teruo Ohno

Manager, Tokyo Electric Power Company, Japan

Naoto Nagaoka

Professor, Doshisha University, Japan

IEEE

WILEY

This edition first published 2015

Registered office

John Wiley & Sons Singapore Pte. Ltd., 1 Fusionopolis Walk, #07-01 Solaris South Tower, Singapore 138628.

For details of our global editorial offices, for customer services and for information about how to apply for permission to reuse the copyright material in this book please see our website at www.wiley.com.

Library of Congress Cataloging-in-Publication Data

Ametani, Akihiro.
Cable system transients : theory, modeling and simulation / Akihiro Ametani, Teruo Ohno, Naoto Nagaoka.
pages cm
Includes bibliographical references and index.
ISBN 978-1-118-70212-3 (cloth)
1. Transients (Electricity)–Simulation methods. 2. Electric lines–Simulation methods. I. Ohno, Teruo. II. Nagaoka, Naoto. III. Title.
TK3226.A479 2015
621.387′84–dc23

2015007952

ISBN: 9781118702123

A catalogue record for this book is available from the British Library.

Typeset in 10/12 Times by SPi Global, Chennai, India
Printed and bound in Singapore by Markono Print Media Pte Ltd

1 2015

Contents

About the Authors

Akihiro Ametani received his PhD degree from the University of Manchester (UMIST), Manchester, UK, in 1973. He was with UMIST from 1971 to 1974, and with Bonneville Power Administration for the summers from 1976 to 1981, and developed the electromagnetic transients program. Since 1985, he has been a Professor at Doshisha University, Kyoto, Japan. In 1988, he was a Visiting Professor at the Catholic University of Leuven, Belgium. From April 1996 to March 1998, he was the Director of the Science and Engineering Institute, Doshisha University, and the Dean of the Library and Computer/Information Center from April 1998 to March 2001. He was the Chairperson of the Doshisha Council. He is a Chartered Engineer in the UK, a distinguished member of the International Council on Large Electric Systems, and a Life Fellow of the Institute of Electrical and Electronic Engineers. He was awarded D. Sc. from the University of Manchester in 2010.

Teruo Ohno received his BS degree from the University of Tokyo, Tokyo, Japan in Electrical Engineering in 1996, his MS degree from the Massachusetts Institute of Technology, Cambridge, USA in Electrical Engineering in 2005, and his PhD degree from Aalborg University, Aalborg, Denmark in Energy Technology in 2012. Since 1996 he has been with the Tokyo Electric Power Company, Inc., where he is currently involved in power system studies, in particular, on cable systems, generation interconnections, and protection relays. He was a secretary of Cigré WG C4.502, which focused on technical performance issues related to the application of long HVAC cables. He is a member of the Institute of Electrical and Electronics Engineers and the Institute of Electrical Engineers of Japan.

Naoto Nagaoka received his BS, MS, and PhD degrees from Doshisha University, Kyoto, Japan, in 1980, 1982, and 1993, respectively. In 1985, he joined Doshisha University, where since 1999 he has been a Professor. From April 2008 to March 2010, he was the Dean of the Student Admission Center, Doshisha University. From April 2010 to March 2012, he was the Director of both the Liaison Office and the Center of Intellectual Properties, Doshisha University. He is a member of the Institution of Engineering and Technology and the Institute of Electrical Engineers of Japan.

Preface

Power transmission by cable is widely used in densely populated areas. Recently off-shore windfarms have become quite common, especially in Europe, and a number of off-shore windfarms are under construction and planned. Thus, a number of submarine cables have been installed and constructed. Submarine cables are also commonly used to connect an island to a mainland. Further, in Denmark all the overhead lines above 100 kV are replaced by underground cables. Thus, transients in cable systems become a very important subject, especially in long and complex cable systems.

The most significant difference of a cable from an overhead line is that a single-phase cable is composed of multi-conductors, that is, a core and a metallic sheath (shield), while a single overhead line is a single conductor. Thus, a three-phase cable (single-core coaxial cable) becomes a six conductor system. When the three-phase cable is enclosed in a conducting pipe, it becomes a seven conductor system. Therefore, an analysis of cable voltages and currents necessitates a theory of multi-conductors.

Another significant difference is that a cable is, in most cases, buried underground. This results in the propagation velocity of the earth-return mode along the cable being far smaller than that of an overhead line, which is nearly the velocity of light in free space. Also, the propagation velocity between a core and a metallic sheath (called "coaxial mode") is determined by the relative permittivity ε_i of an insulator between the core and the sheath, which ranges from two to four, that is coaxial mode velocity $c_c = c_0/\sqrt{\varepsilon_i}$, where $c_0 \fallingdotseq 300\ \mathrm{m}/\mu\mathrm{s}$ (velocity of light).

There are various types and kinds of cables: (1) a power transmission cable, a communication cable and a control/single cable; (2) a directly buried or tunnel installed underground cable, a submarine cable and an overhead cable such as a gas-insulated bus; (3) a single-core coaxial (SC) cable, a multi-core cable, and a pipe-enclosed type (PT) cable; (4) circular or cylindrical, and flat-shaped cables; (5) normal-bonded and cross-bonded cables. This makes an analysis of cable voltages and currents far more complicated than that of an overhead line. As a matter of fact, the overhead line is categorized as just one of the cables, that is, a cable composed only of a core.

This book deals with transients in a power system cable. In Chapter 1, various cables manufactured and used in practice are described.

Chapter 2 explains the impedance and admittance formulas of typical cables, that is, an SC cable and a PT cable. Exact but complicated formulas for numerical calculations are described. Also simple but approximate formulas for a hand calculation are explained so readers understand the physical meaning of the formulas.

In Chapter 3, theories of wave propagation in various cables are described. Section 3.1 explains a basic theory to handle a multi-conductor system called "modal theory". Then, wave propagation characteristics, which are the basis of a transient analysis, are investigated for an SC cable, a PT cable and a cross-bonded cable.

Chapter 4 discusses cable modeling for transient simulations by using well-known EMTP (electromagnetic transients program). A method of calculating the sequence impedances of a cable system is explained by using a lumped PI-circuit model. As the most conventional modeling method for a transient analysis, Dommel's distributed line model is explained first. Then, Semlyen's and Marti's frequency-dependent line models are described. Also, frequency-dependent line models using vector fitting are explained.

In Chapter 5, transients in a single-phase cable are investigated based on experimental results and EMTP simulation results. Then, analytical calculations are carried out based on the theory explained in Chapter 3 so as to be able to understand the surge phenomena in a cable physically and theoretically.

Chapter 6 deals with field test results on various three-phase cables. A comparison with simulation results is carried out. Surge characteristics and the effect of various parameters are investigated based on the field test and simulations results. Also, EMTP simulation results by frequency-independent and -dependent line (cable) models prepared in the EMTP are discussed.

Chapter 7 explains abnormal transients in high voltage large cable systems where reactive power compensation is inherent. Because of a large capacitance due to a long cable and a large inductance of a shunt reactor, series and parallel resonance appears in the large cable systems. Also, system islanding, slow-front overvoltages, leading current interruption, zero missing phenomenon and cable discharge become significant problems. EMTP simulations are carried out, and the characteristics of the above mentioned transients are investigated based on the simulation result.

Chapter 8 describes transients in distributed generation systems where various cables are involved. Modeling of various components in a windfarm and a solar plant by the EMTP are explained. Handling EMTP simulations of transients in the windfarm and in the solar plant is explained, and the EMTP input data are described in detail.

Akihiro Ametani
March 2015

Acknowledgements

Most of the work in this book has been carried out at Doshisha University, Kyoto, Japan, and partially at the US Department of Energy, Bonneville Power Administration, Portland, OR, USA and at the Tokyo Electric Power Company, Tokyo, Japan. The authors are grateful for the provision of the necessary facilities and for the financial support they have received to carry out their work.

1

Various Cables Used in Practice

Teruo Ohno
Tokyo Electric Power Company, Tokyo, Japan

1.1 Introduction

Transmission system operators (TSOs) throughout the world have been seeing growing numbers of transmission line projects in recent years for different reasons including the increase of cross-border trade, renewable energy sources, smart grid projects, the replacement of aging facilities, and in some countries due to growing demand.

Until recently, TSOs have responded to these necessary transmission upgrades mostly by the introduction of overhead lines (OHLs). HVAC underground cable systems have been used, but their applications have been mainly limited to densely populated areas. As such, HVAC underground cable systems are limited both in length and number to date.

This tendency has been changing over the past 10 years as the service experience of HVAC, especially EHV AC, cable systems has become satisfactory [1]. The applications of HVAC cable systems are proposed more often in order to protect the landscape and also public health (e.g., EMF). Hence, HVAC cable systems recently planned or installed are longer than those installed previously.

For example, in Denmark, after receiving public and political pressures to underground its OHLs, Danish TSO, Energinet.dk, published a report on the future expansion and undergrounding of its transmission grid on the 3rd of April 2008 [2]. The report proposed and compared five principles (A–E in Figure 1.1). From the five principles, the Danish government has selected Principle C, as shown in Figure 1.2, in which all new 400 kV lines will basically be undergrounded.

A similar tendency can be observed on HVDC, especially EHV DC, submarine cable systems. The NorNed cable, which connects Norway and the Netherlands, and the BritNed cable, which connects the Netherlands and the UK, are symbolic examples of such a trend. These cable lines, mainly for cross-border trades, have a total length of 580 and 260 km, respectively.

Cable System Transients: Theory, Modeling and Simulation, First Edition.
Akihiro Ametani, Teruo Ohno and Naoto Nagaoka.

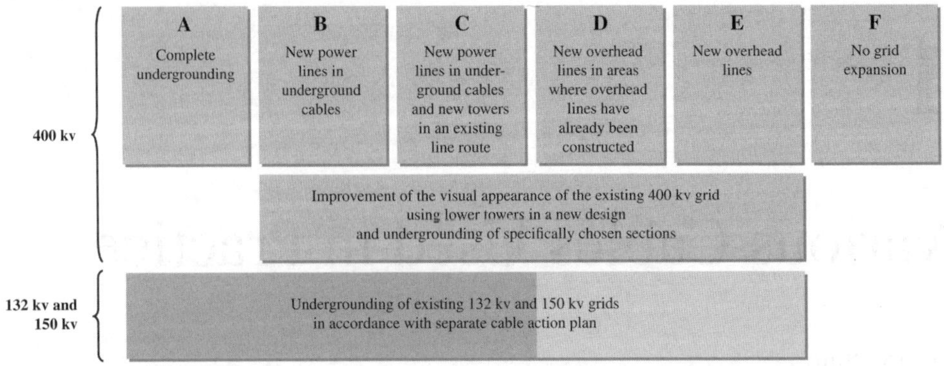

Figure 1.1 Five principles for the future grid expansion (from [2])

The scale of these projects is beyond the level many people expected at the beginning of this century.

As these cable projects increase, there is an increased need to study cable system transients. In particular, the introduction of long cable systems may cause peculiar phenomena, such as resonance overvoltages, which require careful attention. Severe temporary overvoltages in the power system with long cable systems which can be caused in specific network conditions or configurations have been reported [3–9].

Cable modeling for studies on cable system transients, as discussed in Chapter 4, requires the understanding of cable systems. This chapter first discusses the cable itself and then introduces the laying configuration and the sheath bonding, that is, the cable as the cable system. Various cables used in practice are explained in the following two sections – land cables in Section 1.2 and submarine cables in Section 1.3. Section 1.4 discusses the laying configuration including the sheath bonding. The main focus of this chapter is on how these physical characteristics of cable systems affect their electrical characteristics.

Land cables in Section 1.2 cover three major cable types, that is, XLPE (cross-linked polyethylene, PE) cables, SCOF/SCFF (self-contained oil-filled/self-contained fluid-filled) cables and HPOF/HPFF (high-pressure oil-filled/high-pressure fluid-filled) cables. The term "fluid-filled" is used to include both oil-filled cables and gas-filled cables, but most fluid-filled cables are oil-filled cables in actual installations. Even though XLPE cables are increasingly selected for new cable lines in many countries, SCOF cables and HPOF cables are still a popular choice in some countries. HPOF cables are selected, in particular, for the replacement of old HPOF cables since it is often possible to continue using their steel pipes even after the cable replacement.

The laying configuration and the sheath bonding affect cable system transients as the cable itself does. They need to be modeled correctly in order to obtain accurate impedance/admittance of the cable system or reasonable simulation results. Section 1.4 discusses different laying conditions and sheath bonding methods together with their impact on the cable system transients.

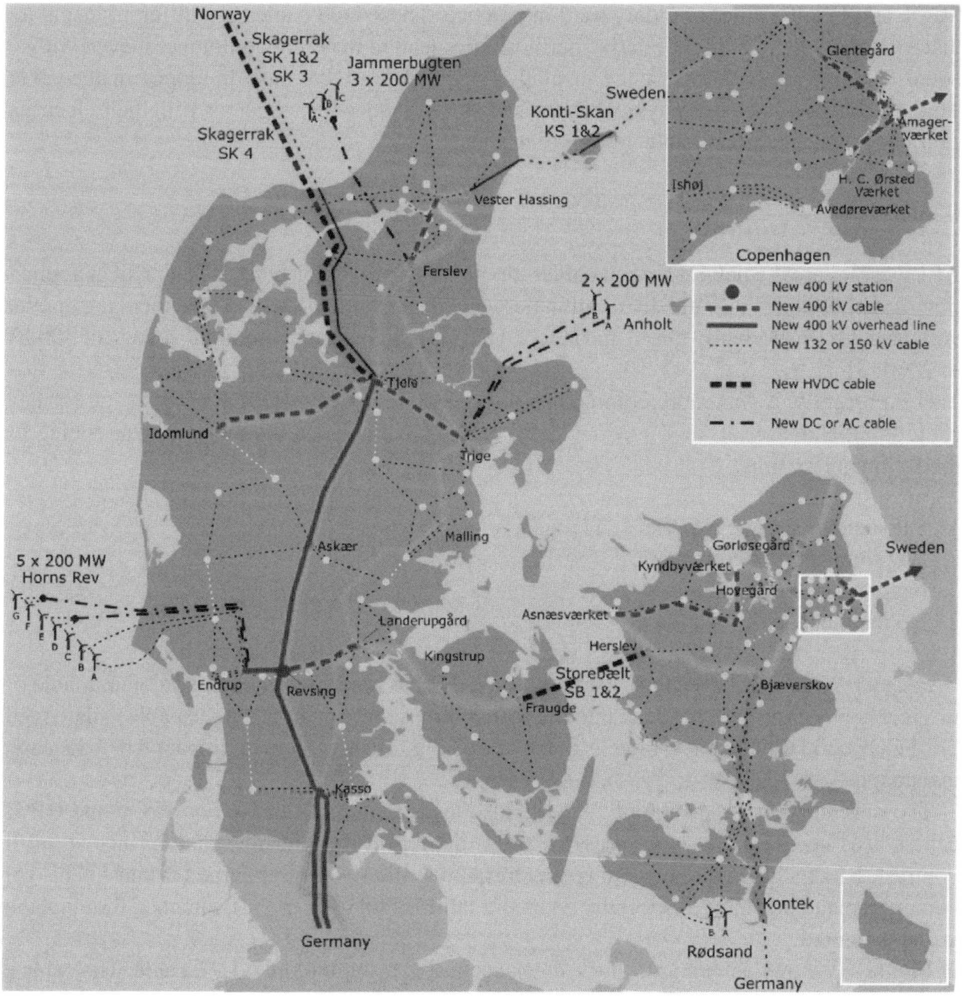

Figure 1.2 Grid expansion plan based on Principle C (from [2])

1.2 Land Cables

1.2.1 Introduction

One obvious difference of a cable from an OHL is that the outer surface of the cable is insulated from the core conductor. This leads to the relatively large admittance of cables compared with OHLs.

The difference in the admittance can be observed not only between cables and OHLs but also between different types of cables. This section explains three types of land cables – XLPE,

SCOF and HPOF cables – widely used in practice. These land cables use different insulating materials, and their different relative permittivities lead to different admittances, even with an equal length. The thickness of the insulation layer also affects the admittance of the cable. This section also discusses physical characteristics of land cables together with their electrical characteristics that affect cable system transients.

1.2.2 XLPE Cables

XLPE cables are the newest type of cables among the three major types. The practical application of XLPE cables into the distribution network started in the 1960s. Since then, the applied voltage has been gradually raised and reached the current maximum nominal voltage of 500 kV in 2000 [3], [10].

Until recently, it was more common to select SCOF cables or HPOF cables even for new cable lines. However, in recent years XLPE cables have become the most popular choice for the following reasons:

1. Satisfactory service experience [1]
2. Environmental effect
3. No pressure system
4. No maintenance

For item 2, since XLPE cables use cross-linked PE as the insulating material, flammable oil or greenhouse gases will not leak into the soil or atmosphere even when the cable is damaged. For SCOF and HPOF cables, there is a risk of causing fire or ecological impact when the cable is damaged.

The difference in the insulating material also leads to items 3 and 4. For SCOF and HPOF cables, it is necessary to apply pressure to insulating oil in order to maintain insulation. In contrast, XLPE cables do not require the pressure system including large oil tanks. This can become a major advantage especially when the cable is laid in an urban center and the available space is limited.

Figure 1.3 shows an example of the cross-section of a single-core XLPE cable. The cable is composed of the following layers:

1. Core conductor
2. Inner semiconducting layer
3. Insulation layer
4. Outer semiconducting layer
5. Metallic sheath
6. Outer cover

This configuration of layers is basically identical for single-core XLPE cables with recent technology, regardless of the cross-section size and the adopted material for each layer. The following describes each layer from inside to outside.

1.2.2.1 Core Conductor

The core conductor carries load currents of the cable and is made of copper or aluminum. The copper or aluminum wire is wound to form a stranded conductor as shown in Figure 1.4. Even

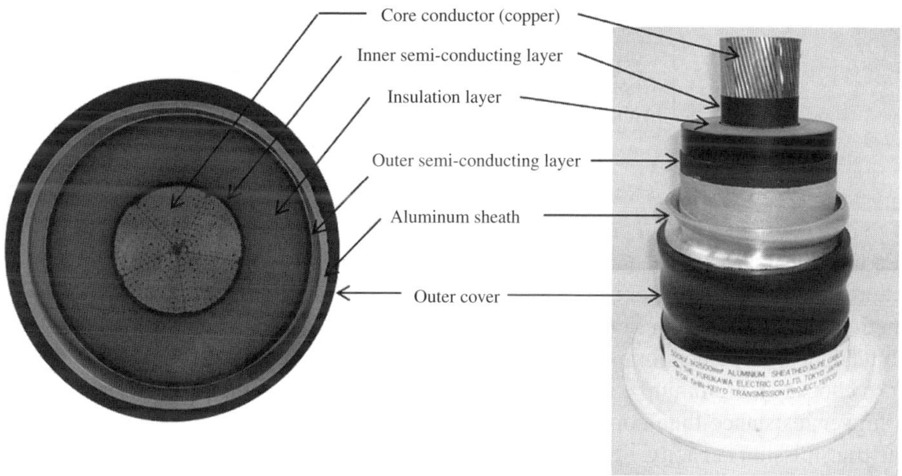

Figure 1.3 Single-core XLPE cable. Courtesy of VISCAS Corporation

Figure 1.4 Stranded conductor. Courtesy of VISCAS Corporation

though it is stranded, the core conductor is considered to be identical to the solid conductor when building a cable model for EMT studies.

The a.c. resistance of a solid conductor is larger than its d.c. resistance due to the skin effect and the proximity effect. The conversion from d.c. resistance to a.c. resistance is made using (1.1) [11].

$$R_{ac} = R_{dc}(1 + y_s + y_p) \tag{1.1}$$

where

R_{ac} = a.c. resistance of a solid conductor
R_{dc} = d.c. resistance of a solid conductor
y_s = skin effect factor
y_p = proximity effect factor

Table 1.1 Resistivity of core conductors

Core conductor	Resistivity at 20 °C (Ωm)
Copper	1.7241×10^{-8}
Aluminum	2.8264×10^{-8}

The d.c. resistance in (1.1) can be calculated from the resistivity of copper and aluminum given in Table 1.1 [11]. The temperature correction is not normally performed since the temperature of the core conductor can be lower than 20 °C for the cable energization. However, it is sometimes necessary to consider the temperature correction when matching simulation results with field measurements, for example for a forensic analysis. It is becoming increasingly popular to choose an aluminum conductor due to recent copper price increases. Since aluminum has higher resistance than copper, a larger cross-section is required in order to provide the same transmission capacity.

Because of the skin effect, the density of an a.c. current is not uniform throughout the cross-section. In fact, the density of an a.c. current is higher at the outer surface of the solid conductor, which reduces the effective area of the cross-section. The formula for calculating the skin effect factor is given in IEC 60287-1-1 [11]. Most EMT-type programs can calculate the skin effect factor, and it is common practice to consider the skin effect when building a cable model for EMT studies.

The proximity effect occurs between conductors of the other phases. Because of the proximity effect, the density of an a.c. current becomes higher at the perimeter close to conductors of the other phases. This also reduces the effective area of the cross-section, but its effect is much smaller than that of the skin effect. The formula for calculating the proximity effect factor is given in IEC 60287-1-1 [11]. Many EMT-type programs cannot calculate the proximity effect factor, and it is common practice not to consider the proximity effect between conductors when building a cable model for EMT studies.

Both the skin effect and the proximity effect are more significant for a cable with a larger cross-section. In order to reduce these effects, a segmental conductor as in Figure 1.5 is normally adopted for a cable with a large cross-section. The segmental conductor is also called the Milliken conductor. In the segmental conductor, each segment is insulated from another by a semiconducting tape or sometimes by an insulating tape. Recently, an enameled conductor has also been considered in order to virtually eliminate the skin effect and the proximity effect.

When EMT-type programs calculate the skin effect factor, they cannot consider a segmental conductor and overestimate the skin effect factor. As a result, they produce higher a.c. resistance than the actual. The same is true for an enameled conductor, which is introduced for the same reason as the segmental conductor. When modeling these types of conductors, it becomes necessary to calculate the skin effect factor outside EMT-type programs depending on the type of study being performed.

1.2.2.2 Inner Semiconducting Layer (Conductor Screen)

The inner semiconducting layer is applied around the core conductor in order to equalize the electric field strength at the outer surface of the core conductor. This layer is also referred to as the conductor screen. Since the resistivity of the semiconducting layer is much larger than the

Figure 1.5 Segmental conductor. Courtesy of VISCAS Corporation

core conductor, the current does not flow through the semiconducting layer. Therefore, it can be considered as a part of the insulation layer for calculating the series inductance of the cable.

In contrast, for calculating the capacitance of the cable, the inner semiconducting layer should be considered as a part of the core conductor. This is because the potential of the inner semiconducting layer becomes equal to the core conductor.

1.2.2.3 Insulation Layer

XLPE cables use cross-linked PE as the insulating material. In the planning stage, the relative permittivity of XLPE is normally set between 2.3 and 2.5 depending on the utility's practice. In the implementation study stage it is given by a cable manufacturer and after the installation it can be calculated from field measurements.

1.2.2.4 Outer Semiconducting Layer (Insulation Screen)

The outer semiconducting layer is applied between the insulation layer and the metallic sheath in order to equalize the electric field strength at the inner surface of the metallic sheath. This layer is also referred to as the insulation screen. Similarly to the inner semiconducting layer, the outer semiconducting layer should be treated as a part of the insulation layer for the inductance calculation and should be treated as a part of the insulation layer for the capacitance calculation.

When building a cable model in EMT-type programs, it is common practice to set the radius of each layer for the accurate calculation of the cable inductance. That is, the inner and outer semiconducting layers are treated as a part of the insulation layer. This treatment of the semiconducting layers makes the cable capacitance smaller than the actual value. In order to compensate for the error in the cable capacitance, the relative permittivity of the main insulation needs to be converted as [12]:

$$\varepsilon_{i1} = \varepsilon_{i0} \frac{\ln\left({R_3}/{R_2}\right)}{\ln\left({R_{so}}/{R_{si}}\right)} \tag{1.2}$$

where

ε_{i1} = effective relative permittivity of the main insulation
ε_{i0} = actual relative permittivity of the main insulation
R_2 = outer radius of the core conductor
R_3 = outer radius of the outer semiconducting layer
R_{si} = outer radius of the inner semiconducting layer
R_{so} = outer radius of the insulation layer

1.2.2.5 Metallic Sheath (Metallic Screen)

There are several types of metallic sheath. The most common types are the lead (alloy) sheath, aluminum tape sheath, and copper tape sheath. Figure 1.3 shows the corrugated aluminum sheath, but it is becoming more popular to choose the plain aluminum tape sheath. The lead sheath is manufactured by applying extruded lead alloy on the outer semiconducting layer. The aluminum tape sheath and the copper tape sheath are welded longitudinally to guarantee watertight construction in a radial direction. When water sealing in a longitudinal direction is necessary, the water swelling tape is placed inside these types of metallic sheath.

The lead sheath has a larger resistance than the aluminum tape sheath and the copper tape sheath. In addition, it is expensive, toxic, and heavy. In spite of these unfavorable characteristics, the lead sheath is still adopted due to its high resistance to corrosion especially when a cable is installed in a moist environment.

The copper wire sheath is often formed inside the lead sheath or the aluminum tape sheath in order to carry the required fault current and also to improve the conductivity of the metallic sheath. The copper wire sheath is wound to form a stranded conductor as shown in Figure 1.6. When building a cable model for EMT studies, it is modeled as a solid conductor. The resistivity of copper is modified so that the modeled sheath has identical resistance to the actual copper wire sheath [13]. The resistivity values of lead, copper, and aluminum are shown in Table 1.2.

1.2.2.6 Outer Cover

The outer cover is normally made of PE or polyvinyl chloride (PVC). PVC has better performance than PE in terms of fire resistance. PVC is non-flammable whereas PE is flammable. However, since PVC releases toxic hydrogen chloride gas when burned, PE is the more

Table 1.2 Resistivity of metallic sheaths

Metallic sheath	Resistivity at 20 °C (Ωm)
Lead or lead alloy	21.4×10^{-8}
Copper	1.7241×10^{-8}
Aluminum	2.84×10^{-8}

Figure 1.6 Copper wire sheath. Courtesy of VISCAS Corporation

preferred choice and is increasingly used by many utilities. Even though PVC has much larger permittivity than PE [14], it does not have a noticeable impact on cable system transients.

1.2.3 SCOF Cables

SCOF cables were preferred until recently over XLPE cables because of their reliability. They are less susceptible to a defect that can be introduced during a manufacturing process at a factory or a joint assembly process at a site. In addition, the defect can be found through routine maintenance and can be fixed before causing a cable failure and a subsequent blackout.

Due to the improvement in the manufacturing process of XLPE cables and the joint assembly process, SCOF cables have lost their popularity to XLPE cables. Currently, it is not common to choose SCOF cables for the construction of a new cable line or even for the replacement of an old SCOF cable line. However, SCOF cables are still in the majority in terms of installed cables. Therefore, cable system transients with SCOF cables will be studied for the safe operation of existing cable lines and for forensic studies.

The physical and electrical characteristics of SCOF cables are very similar to those of XLPE cables. Figure 1.7 shows the construction of a SCOF cable.

There are two main differences between SCOF cables and XLPE cables:

1. Hollow conductor
   The hollow conductor is used so that it can serve as an oil duct for insulation. In the setup of the cable model, the inner diameter of the conductor does not become zero for SCOF cables.
2. Insulating material
   The impregnated paper is used as the insulating material. The paper wrapped outside the hollow conductor is impregnated with insulating oil. Kraft paper has been used as the insulating material, but in recent years, it is common to choose PPLP (polypropylene laminated paper) because of its better dielectric characteristic. The relative permittivity and tanδ of the impregnated paper are shown in Table 1.3. Because of the larger relative permittivity, the charging capacity of SCOF cables is normally larger than that of XLPE cables.

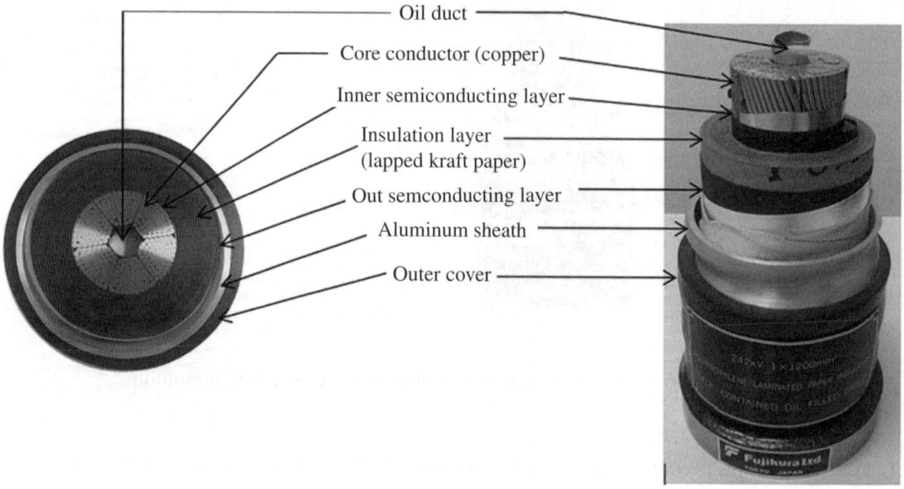

Figure 1.7 SCOF cable. Courtesy of VISCAS Corporation

Table 1.3 Typical values of relative permittivity and tanδ of the impregnated paper

	Relative permittivity	Tanδ (at 80 °C)
Kraft paper	3.2–3.5	0.17–0.2
PPLP	2.7–2.8	0.07–0.08

1.2.4 HPOF Cables

HPOF cables have a significantly different structure compared with XLPE and SCOF cables. Normally, cables of three phases are enclosed in one steel pipe as shown in Figure 1.8.

The cable in the steel pipe has only the conductor and the main insulation. The material and the construction of the conductor and the main insulation are basically the same as those of SCOF cables. One difference is that it is not necessary to use a hollow conductor in HPOF cables as the steel pipe serves as an oil duct.

The steel pipe works as the metallic sheath offering a path for the current return in case of cable faults and shielding for the intrusion of humidity. It is shared by three-phase cables enclosed in it. Because of the larger resistivity and diameter of the steel pipe, compared with the aluminum or copper tape, HPOF cables have a higher zero-sequence impedance (not earth-return, but pipe-return propagation mode), both in resistance and reactance, compared with XLPE or SCOF cables. It is typical for HPOF cables that the zero-sequence impedance is larger than the positive-sequence impedance, which is not the case with XLPE and SCOF cables. The relative permeability of the steel pipe is normally specified as one.

In addition to the proximity effect between conductors of the other phases, the proximity effect occurs between inner conductors and the pipe in the case of HPOF cables. The formula for calculating the impedance and admittance, taking this proximity effect into account, is

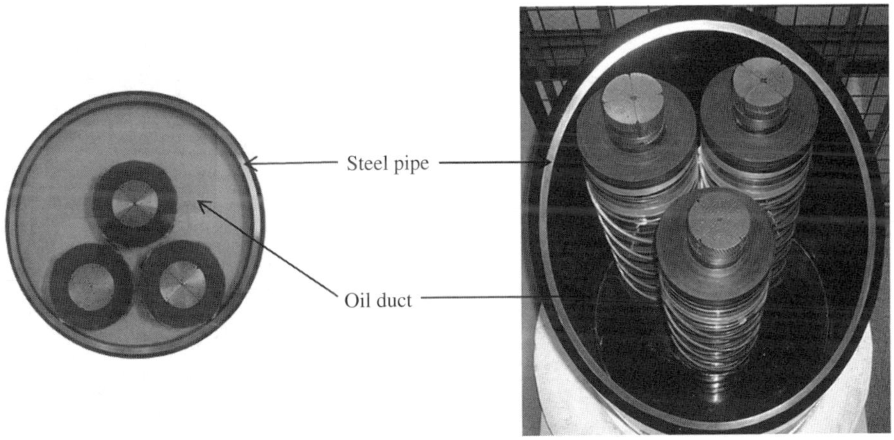

Figure 1.8 HPOF cable. Courtesy of VISCAS Corporation

given in Reference [15]. Many EMT-type programs can consider the proximity effect between inner conductors and the pipe of HPOF cables.

1.3 Submarine Cables

1.3.1 Introduction

Submarine cables are on average longer than land cables since they are often proposed for cross-border trades and for the integration of off-shore wind farms. Long HVAC submarine cables require shunt reactors to compensate for their charging capacity. When the length of a cable line is further increased, it becomes necessary to install shunt reactors at multiple points along the cable length in order to avoid the reduction of transmission capacity for active power or temporary overvoltages. The reactive power compensation incurs an additional cost for the cable project especially when it requires the installation of shunt reactors in the sea. Even though HVDC submarine cables need converter stations, the overall project cost becomes lower for the HVDC option at some cable line length.

1.3.2 HVAC Submarine Cables

XLPE cables are the most preferred option as HVAC submarine cables for the same reasons as land cables. HVAC submarine cables have a layer of armor outside the XLPE land cables as shown in Figure 1.9. The armor helps submarine cables to endure the tensile force they experience during installation. In addition, it can help to avoid a cable failure when a submarine cable is damaged in the sea by a third party.

The most common type of armor is steel wire. As is the case with the copper wire sheath, it is modeled as a solid conductor when building a cable model for EMT studies. The resistivity of steel is modified so that the modeled armor has identical resistance to the actual steel wire armor.

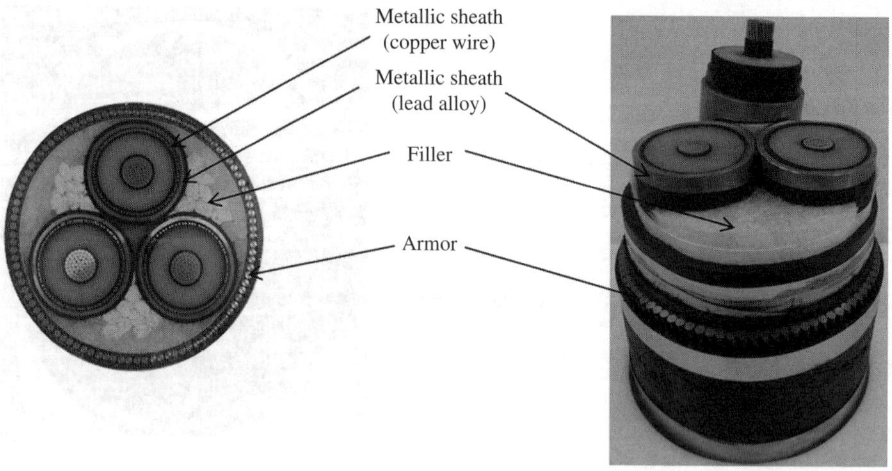

Figure 1.9 HVAC submarine cable. Courtesy of VISCAS Corporation

Some HVAC submarine cable has three-phase cables enclosed together in armor. In this case, the submarine cable is often modeled as a pipe-type cable, considering the armor as a pipe of the pipe-type cable.

1.3.3 HVDC Submarine Cables

Until recently, XLPE cables could not be adopted in a HVDC cable project with LCC–HVDC (line-commutated converter HVDC) technology. The space charge was formed in the main insulation for XLPE cables, and it could lead to an insulation breakdown when the voltage polarity was reversed. The reversal of the voltage polarity occurs with LCC–HVDC when the power flow is reversed. Therefore, SCOF cables have been selected for HVDC cable projects with LCC–HVDC.

The most common type of SCOF cables for HVDC submarine cables is MI (mass impregnated) cables. MI cables are used since the insulating oil cannot be supplied at cable terminations. Figure 1.10 shows the construction of MI cables. Unlike SCOF land cables, it is still common to use Kraft paper as the insulating material. However, it is expected that PPLP will soon become more common because of its better performance as an insulating and dielectric material.

An XLPE cable was first adopted in a HVDC cable project with LCC–HVDC in 2012 for the Hokkaido–Honshu HVDC link in Japan [16, 17]. The 250 kV XLPE cable has overcome the problem caused by the voltage polarity reversal by adding nanoparticles to the insulating material. It is expected that the successful operation of this cable will, in future, lead to the increased application of XLPE cables to HVDC cable projects with LCC–HVDC.

Other than the Hokkaido–Honshu HVDC link, XLPE cables are selected for HVDC cable projects with VSC–HVDC (voltage-source converter HVDC) technology. The VSC–HVDC reverses the power flow by the reversal of the current flow. The reversal of the voltage polarity does not occur with VSC–HVDC, which enables the selection of XLPE cables.

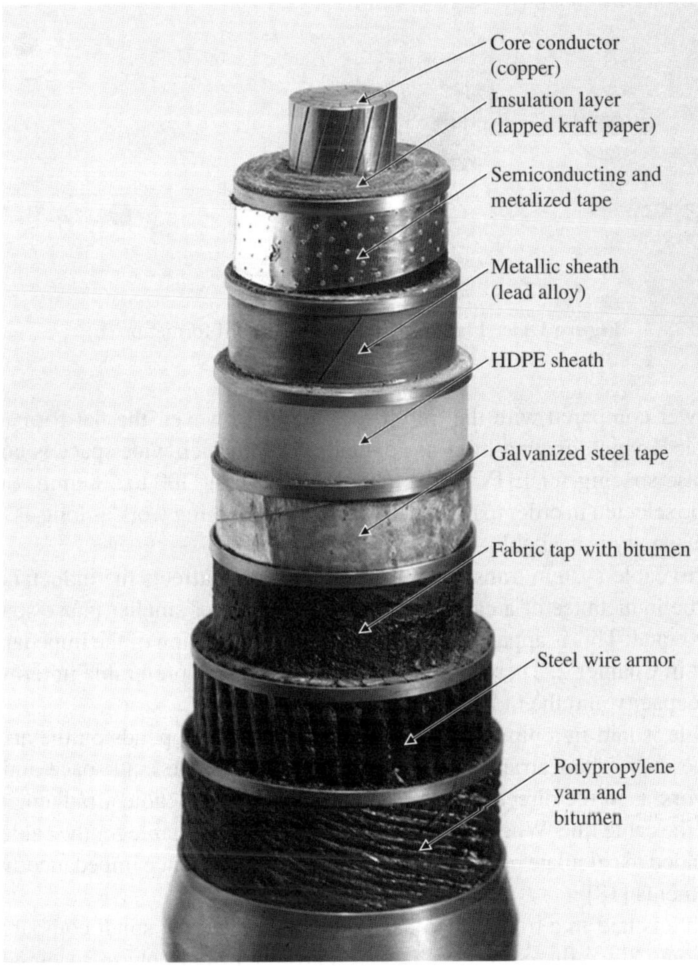

Figure 1.10 MI cable. Courtesy of Nexans

1.4 Laying Configurations

1.4.1 Burial Condition

Cables are, in most cases, buried in the following conditions:

- Directly buried
- Buried in a pipe or a duct
- Buried in a trough
- Buried in a tunnel

When cables are directly buried, they are laid in a flat formation or a trefoil formation as shown in Figure 1.11. The flat formation is often preferred as the required depth of a trench

Figure 1.11 Flat formation (a) and trefoil formation (b)

can be shallower compared with the trefoil formation. However, the flat formation requires wider land space, which may prohibit its application when such wide space is not available.

Typical phase spacing for EHV cables ranges from about 300 to 500 mm. Smaller phase spacing may be selected in order to reduce the cost for the digging work as long as the necessary transmission capacity is available.

In relation to cable system transients, this phase spacing affects the inductance of a cable line. In fact, the inductance of a cable line becomes larger for smaller phase spacing if other conditions are equal. This is apparent from the theoretical equation of the impedance of a cable line discussed in Chapter 2. Therefore, larger phase spacing is preferable in terms of both the transmission capacity and the inductance of a cable line.

When a cable is laid in a pipe or a duct, its phase spacing depends on the arrangement of the pipe or the duct. Their arrangement is affected by available land space and the cost for the digging work as in the directly buried cable. Obviously the cable arrangement affects the impedance of the cable line. When surplus pipes or ducts are available for the cable installation, it is recommended to calculate positive-sequence and zero-sequence impedances with different cable arrangements [18].

When a cable is laid in a trough, its phase spacing tends to be small compared with other laying conditions. Since three-phase cables are laid in a trough, phase spacing is limited by the size of the trough. When the trough is made of concrete, the cable is often modeled as a pipe-type cable, considering the trough as a pipe of the pipe-type cable.

When a cable is laid in a tunnel, some utilities decide to choose small phase spacing in order to save space in a tunnel and install more cables in it. Some cables are laid even next to each other with a cable of other phases like a triplex cable. This can be justified considering the high construction cost of a tunnel, especially around big cities where a tunnel needs to accommodate many cables.

Tunnel installed cables are normally modeled as a pipe-type cable considering the tunnel as a pipe of the pipe-type cable. Reinforced concrete is considered electrically conductive because of the steel and concrete.

1.4.2 Sheath Bonding

When considering the sheath bonding, two important factors are the sheath voltage and the sheath current. If only the former factor is considered, the most favorable sheath bonding is solid bonding. However, it causes higher sheath current, leading to a larger sheath circuit loss. Hence, it is only applied to submarine cables, which do not have joints in a cable line.

In contrast, if only the sheath current is considered, the most favorable sheath bonding is single-point bonding. The sheath current in the normal operating condition becomes zero when single-point bonding is adopted. However, it causes higher sheath voltage, which requires the installation of an earth continuity cable (ECC). In fact, single-point bonding is often adopted together with the cross bonding as discussed later in this section.

Cross bonding is applied considering both the sheath voltage and the sheath current. It can suppress the sheath voltage while limiting the sheath current. It is dominantly applied to a cable line with three or more cable sections.

1.4.2.1 Solid Bonding

An example of solid bonding is shown in Figure 1.12 even though this configuration is not applied to land cables due to the large sheath current. As the sheath circuit is grounded at every joint (earthing joint, EJ) and termination, the sheath voltage is suppressed to virtually zero at these points. Only the sheath voltage caused by the grounding resistance remains at joints and terminations. The value of the grounding resistance is normally around 5–10 Ω at joints and 0.1–1 Ω at terminations. The grounding resistance at terminations is lower since the grounding circuit can be connected to the substation grounding grid, except for the termination at the transition point between the underground cable (UGC) and the OHL in the mixed UGC/OHL line.

The solid bonding is applied to submarine cables, which do not have joints in a cable line. The metallic sheath of a submarine cable is grounded every 2–4 km in order to suppress the sheath voltage as shown in Figure 1.13. As a result, the sheath current of a submarine cable becomes larger than that of land cables. The metallic sheath of a submarine cable often has a large cross-section in order to reduce the sheath circuit loss.

1.4.2.2 Single-point Bonding

An example of single-point bonding is illustrated in Figure 1.14. Single-point bonding is applied to a short cable line with one or two cable sections. The sheath circuit on the left

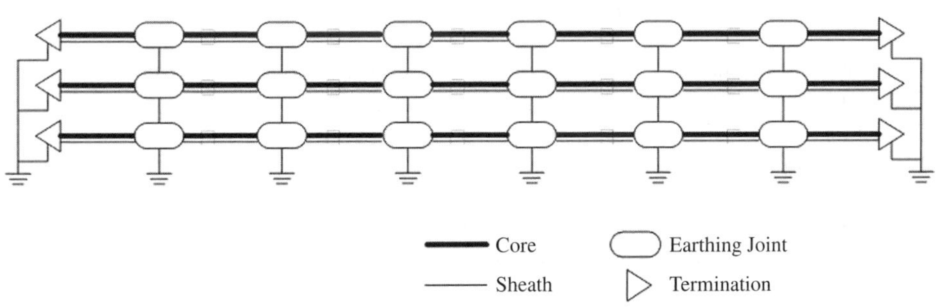

Figure 1.12 Solid bonding

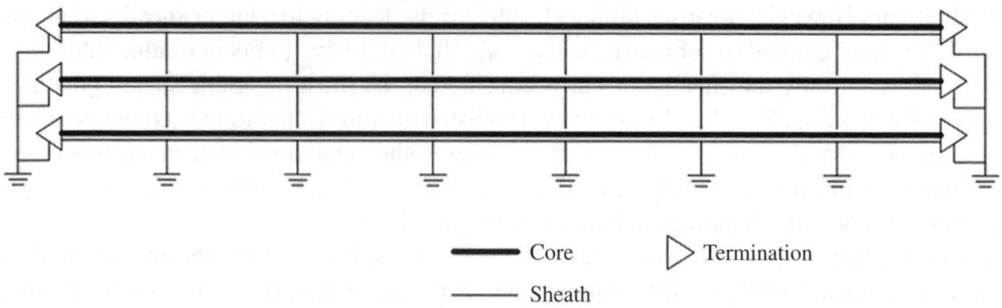

Figure 1.13 Solid bonding of a submarine cable

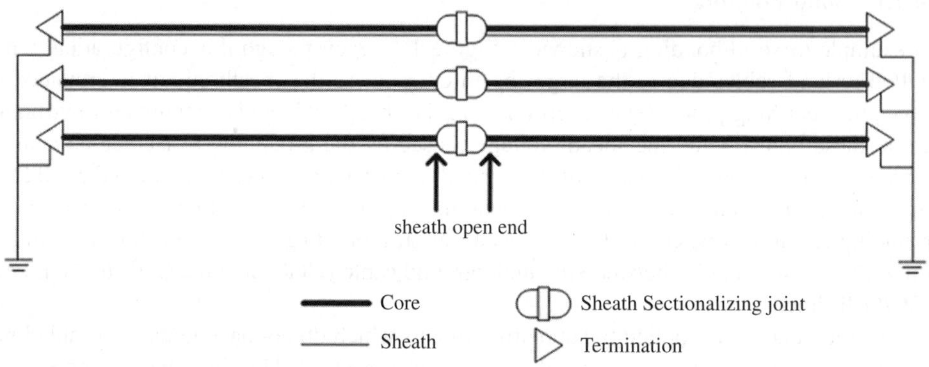

Figure 1.14 Single-point bonding (two sections)

side is insulated from the sheath circuit on the right side by sheath sectionalizing joints (SSJs). This creates open ends in the sheath circuit.

The magnitude of the continuous sheath voltage, which is induced by positive-sequence power flow in a core conductor in normal operating conditions, is equivalent to that in the case of cross bonding. The short-term sheath voltage at the sheath open end becomes an issue in the case of single-point bonding. The short-term sheath voltage under the following conditions is studied, and the earth continuity cable (ECC) and sheath voltage limiters (SVLs) are installed as a countermeasure as shown in Figure 1.15:

- Single-line-to-ground (SLG) faults (external to the targeted major section)
- Three-phase faults (external to the targeted major section)
- Switching surges
- Lightning surges (only for mixed UGC/OHL)

The power-frequency component of the short-term sheath voltage under SLG faults and three-phase faults is calculated using the formulas given in References [19–22]. In addition, it is sometimes necessary to study the transient component of the short-term sheath voltage, especially in order to evaluate the performance of SVLs. The study is performed using EMT-type programs.

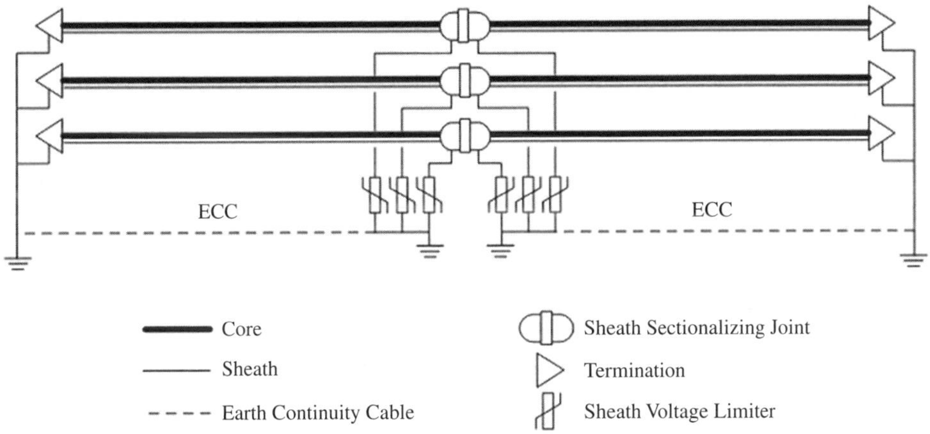

Figure 1.15 Single-point bonding (two sections) with ECC and SVLs

1.4.2.3 Cross Bonding

An example of cross bonding is illustrated in Figure 1.16. The cross bonding is applied to a cable line with three or more cable sections.

As in single-point bonding, the sheath circuit on the left side is insulated from the sheath circuit on the right side by SSJs. In the cross bonding, however, the sheath circuit on the left side is connected to the sheath circuit of a different phase cable on the right side as shown in Figure 1.16. For example, the sheath circuit expressed by the dotted line goes with the phase a cable in the first minor section, with the phase b cable in the second minor section, and with the phase c cable in the third minor section.

Thanks to this connection of the sheath circuit, the cross bonding can suppress the sheath voltage while limiting the sheath current. Assuming the following three conditions, the vector sum of the continuous sheath voltage of three minor sections becomes zero:

- Positive-sequence core current
- Trefoil formation (balanced circuit)
- Equal length of three minor sections

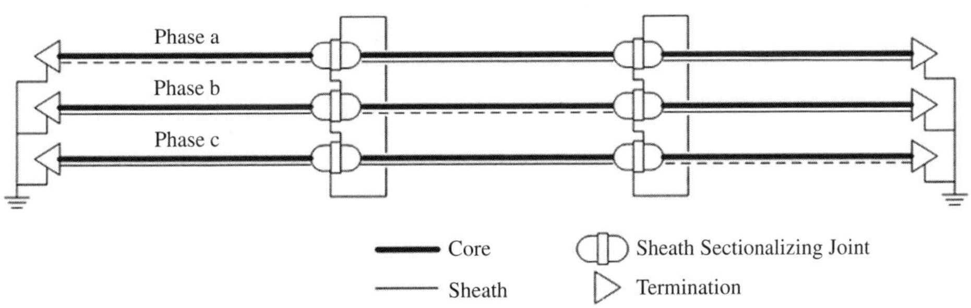

Figure 1.16 Cross bonding (three sections)

The continuous sheath voltage of three minor sections can be illustrated as in Figure 1.17. This means that the sheath current becomes zero if these conditions are satisfied. However, in an actual installation, these conditions are not completely satisfied, which causes an imbalance in the continuous sheath voltage of the three minor sections. The sheath current flows due to this imbalance, but it is much smaller than that in the solid bonding.

Even though the continuous sheath voltage is suppressed by the small sheath current and the balanced sheath voltage, the short-term sheath voltage cannot be suppressed by the cross bonding. As a countermeasure for the short-term sheath voltage, SVLs are installed at sheath sectionalizing joints as shown in Figure 1.18.

SVLs are normally arranged in a star formation, and the neutral point of SVLs is grounded as shown in Figure 1.18. The short-term sheath voltage can exceed the ratings of SVLs when the grounding resistance at a SSJ is high. In this case, ECC can be installed connecting the neutral points of SVLs to grounding wires at terminations and EJs. The installation of ECC

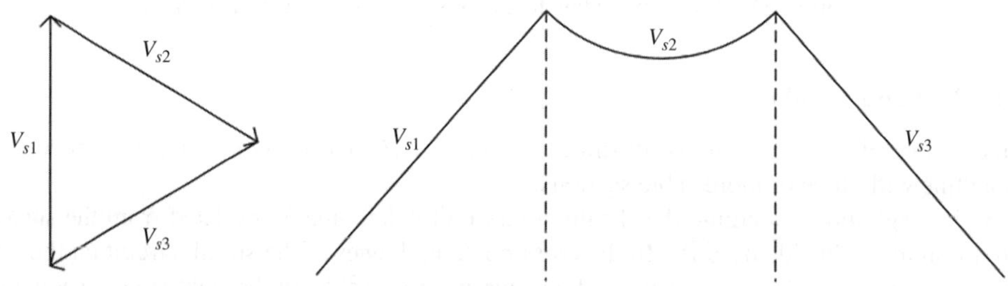

V_{s1}: sheath voltage of the first minor section, V_{s2}: sheath voltage of the second minor section and V_{s3}: sheath voltage of the third minor section

Figure 1.17 Continuous sheath voltage in a major section of a cross-bonded cable

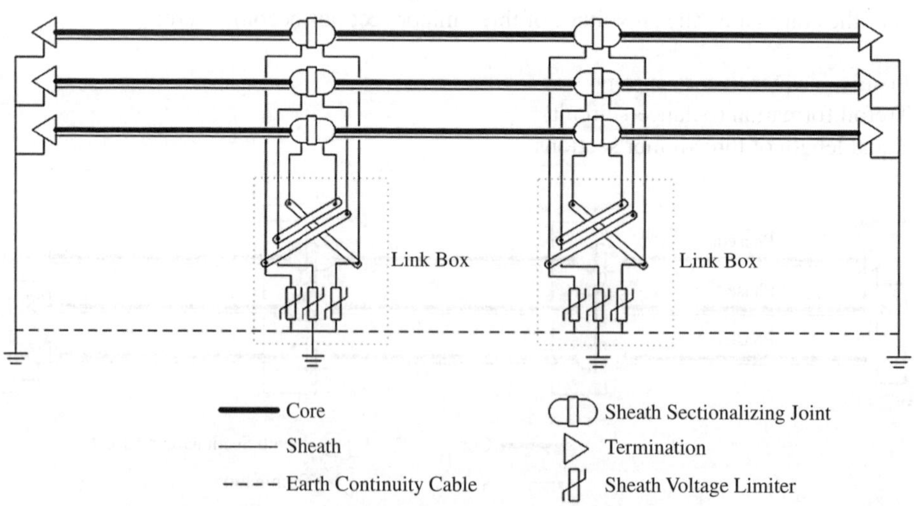

Figure 1.18 Cross bonding (three sections) with ECC and SVLs

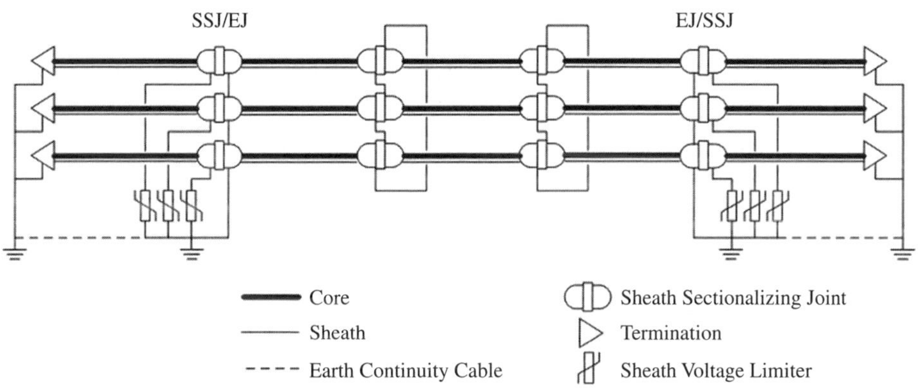

Figure 1.19 Combination of cross bonding and single-point bonding

lowers the effective grounding resistance at the SSJ. Other countermeasures include changing the neutral point from solidly grounded to ungrounded and changing the SVL connection from the star formation to the delta formation [21].

When the number of minor sections is not a multiple of three, one or two minor sections cannot become a part of the cross-bonding configuration. For example, if the number of minor sections is five, two minor sections cannot become a part of the cross-bonding configuration. In such a situation, single-point bonding is applied to the remaining two minor sections as shown in Figure 1.19. In the figure, SSJ/EJ and EJ/SSJ means that:

- The sheath circuit of the left side is insulated from the sheath circuit of the right side as in SSJ.
- The sheath circuit of the left side (EJ/SSJ) or the right side (SSJ/EJ) is grounded as in EJ.

As the single-point bonding is applied to these minor sections, it is important to confirm that the sheath voltage of these minor sections does not exceed the ratings of SVLs, joints, and the sheath.

References

[1] CIGRE WG B1.10 (2009) Update of Service Experience of HV Underground and Cable Systems. *CIGRE Technical Brochure* **379**.

[2] Energinet.dk (2008) Technical Report on the Future Expansion and Undergrounding of the Electricity Transmission Grid. https://selvbetjening.preprod.energinet.dk/NR/rdonlyres/CC966C3A-FE78-41D8-9DC7-6B455210B502/0/TechnicalReportSummary.pdf (accessed 27 November 2013).

[3] Momose, N., Suzuki, H., Tsuchiya, S., and Watanabe, T. (1998) Planning and development of 500 kV underground transmission system in Tokyo metropolitan area. *CIGRE Session*, 37-202.

[4] Colla, L., Lauria, S., and Gatta, F.M. (2007) Temporary overvoltages due to harmonic resonance in long EHV cables. International Conference on Power System Transients (IPST), Lyon, France.

[5] Rebolini, M., Colla, L., and Iliceto, F. (2008) 400 kV AC new submarine cable links between Sicily and the Italian mainland. Outline of project and special electrical studies. *CIGRE Session*, C4-116.

[6] Tokyo Electric Power Company (2009) Assessment of the Technical Issues relating to Significant Amounts of EHV Underground Cable in the All-island Electricity Transmission System. http://www.eirgrid.com/media/Tepco%20Report.pdf (accessed 27 November 2013).

[7] Akhmatov, V. (2006) Excessive over-voltage in long cables of large offshore windfarms. *Wind Engineering*, **30** (5), 375–83.
[8] Ohno, T. (2012) Dynamic study on the 400 kV 60 km Kyndbyværket–Asnæsværket line. PhD thesis. Aalborg University.
[9] CIGRE WG C4.502 (2013) Power System Technical Performance Issues Related to the Application of Long HVAC Cables. *CIGRE Technical Brochure* **556**.
[10] Kawamura, T., Kouno, T., Sasaki, S. *et al.* (2000) Principles and recent practices of insulation coordination in Japan. *CIGRE Session*, 33-109.
[11] IEC 60287-1-1 ed. 1.2 (2001) Electric cables – Calculation of the current rating – Part 1-1: Current rating equations (100 % load factor) and calculation of losses – General.
[12] Gustavsen, B. (2001) Panel session on data for modeling system transients. Insulated cables. Proceedings of the IEEE Power Engineering Society Winter Meeting, Columbus, OH, USA.
[13] CIGRE WG B1.30 (2013) Cable Systems Electrical Characteristics. *CIGRE Technical Brochure* **531**.
[14] Kersting, W.H. (2012) *Distribution System Modeling and Analysis*, CRC Press.
[15] Ametani, A. (1980) General formulation of impedance and admittance of cables. *IEEE Transactions on Apparatus and Systems*, **99** (3), 902–10.
[16] Watanabe, C., Itou, Y., Sasaki, H. *et al.* (2013) Practical application of +/-250 kV DC-XLPE cable for Hokkaido–Honshu HVDC link. *IEEJ Transactions on Power and Energy*, **134** (1), 64–75 (in Japanese).
[17] Murata, Y., Sakamaki, M., Abe, K. *et al.* (2013) Development of high-voltage DC-XLPE cable. *SEI Technical Review*, **182**, 48–55 (in Japanese).
[18] Nakanishi, H., Inoguchi, H., Hashimoto, I. *et al.* (1991) A study of zero-sequence current induced in a cable system. *IEEE Transactions on Power Delivery*, **6** (4), 1352–8.
[19] CIGRE Working Group 07 Study Committee 21 (1973) The design of specially bonded systems. *Electra* No. **28**.
[20] CIGRE Working Group 07 Study Committee 21 (1976) The design of specially bonded systems (part II). *Electra* No. **47**.
[21] CIGRE Working Group 07 Study Committee 21 (1990) Guide to the protection of specially bonded systems against sheath overvoltages. *Electra* No. **128**.
[22] CIGRE WG B1.18 (2005) Special Bonding of High Voltage Power Cables. *CIGRE Technical Brochure* **283**.

2

Impedance and Admittance Formulas

Akihiro Ametani
Graduate School of Electrical Engineering, Doshisha University, Kyoto, Japan

The impedance and admittance of a cable are expressed in a matrix form, because even a single-phase cable becomes a two-conductor system. The voltages and currents along the cable are related in the following matrix equations [1, 2].

$$d(V)/dx = -[Z] \cdot (I) \tag{2.1}$$

$$d(I)/dx = -[Y] \cdot (V) \tag{2.2}$$

where (V) and (I) are vectors of the voltages and currents at a distance x along the cable. $[Z]$ and $[Y]$ are square matrices of the impedance and the admittance.

In general, impedance and admittance matrices of a cable can be expressed in the following forms.

$$[Z] = [Z_i] + [Z_p] + [Z_c] + [Z_0] \tag{2.3}$$

$$[Y] = s \cdot [P]^{-1} \tag{2.4a}$$

$$[P] = [P_i] + [P_p] + [P_c] + [P_0] \tag{2.4b}$$

where $[P]$ is the potential coefficient matrix, and $s = j\omega$.

In the above equations, matrices with subscript "*i*" concern a single-core coaxial (SC) cable, and the matrices with subscripts "*p*" and "*c*" are related to a pipe enclosure of a pipe-type (PT) cable. The matrices with subscript "*o*" concern cable outer media, that is, air space and earth. When a cable has no pipe enclosure, there exists no matrix with subscripts "*p*" and "*c*".

Cable System Transients: Theory, Modeling and Simulation, First Edition.
Akihiro Ametani, Teruo Ohno and Naoto Nagaoka.

In the formulation presented here, the following assumptions are made.

1. The displacement currents and dielectric losses are negligible.
2. Each conducting medium of a cable has a constant permeability.
3. The pipe thickness is greater than the penetration depth of the pipe wall in a PT cable.

The details will be explained in the following sections.

2.1 Single-core Coaxial Cable (SC Cable)

2.1.1 Impedance

When an SC cable consists of a core, a sheath and an armor conductor as shown in Figure 2.1, the impedance is given in the following form based on the results of Appendix 2.A [3].

$$[Z] = [Z_i] + [Z_0] \tag{2.5}$$

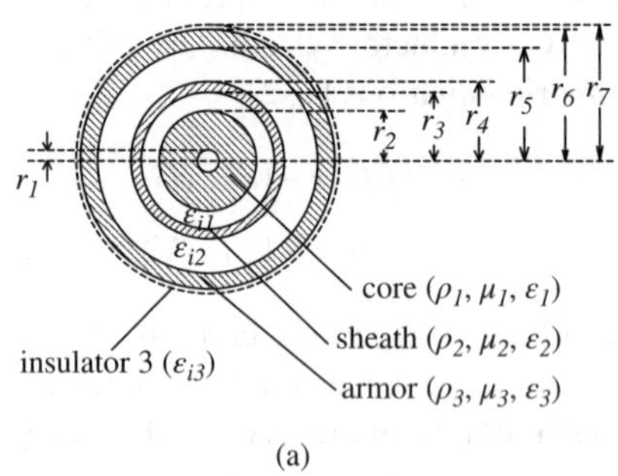

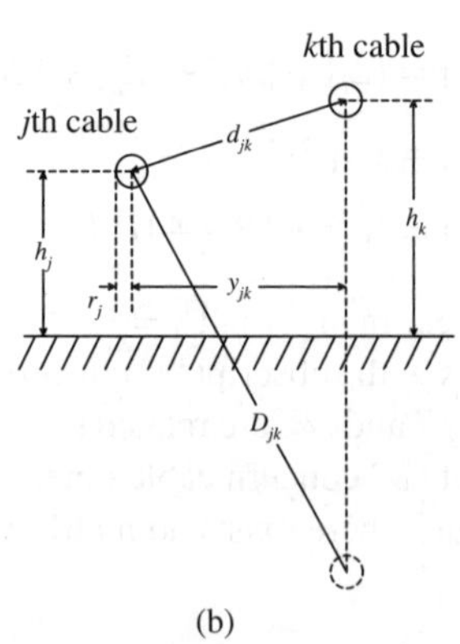

Figure 2.1 An SC cable system: (a) SC cable cross section; and (b) system configuration

where

$[Z_i]$ = SC cable internal impedance matrix

$$= \begin{bmatrix} [Z_{i1}] & [0] & \cdots & [0] \\ [0] & [Z_{i2}] & \cdots & [0] \\ \vdots & \vdots & \ddots & \vdots \\ [0] & [0] & \cdots & [Z_{in}] \end{bmatrix} \tag{2.6}$$

$[Z_0]$ = impedance matrix of the cable outer medium (earth-return impedance)

$$= \begin{bmatrix} [Z_{011}] & [Z_{012}] & \cdots & [Z_{01n}] \\ [Z_{012}] & [Z_{022}] & \cdots & [Z_{02n}] \\ \vdots & \vdots & \ddots & \vdots \\ [Z_{01n}] & [Z_{02n}] & \cdots & [Z_{0nn}] \end{bmatrix} \tag{2.7}$$

All the off-diagonal submatrices of $[Z_i]$ are zero.

A diagonal submatrix in (2.6) expresses the impedance matrix of the jth phase SC cable. When the SC cable consists of a core, a sheath and an armor, the impedance matrix is given by:

$$[Z_{ij}] = \begin{bmatrix} Z_{ccj} & Z_{csj} & Z_{caj} \\ Z_{csj} & Z_{ssj} & Z_{saj} \\ Z_{caj} & Z_{saj} & Z_{aaj} \end{bmatrix} \tag{2.8}$$

where

$$\begin{aligned}
Z_{ccj} &= \text{core self-impedance} \\
&= z_{cs} + z_{sa} + z_{a4} - 2z_{2m} - 2z_{3m} \\
Z_{ssj} &= \text{sheath self-impedance} \\
&= z_{sa} + z_{a4} - 2z_{3m} \\
Z_{aaj} &= \text{armor self-impedance} = z_{a4} \\
Z_{csj} &= \text{mutual impedance between the core and sheath} \\
&= z_{sa} + z_{a4} - z_{2m} - 2z_{3m} \\
Z_{caj} &= \text{mutual impedance between the core and armor} \\
&= z_{a4} - z_{3m} \\
Z_{saj} &= \text{mutual impedance between the sheath and armor} \\
&= z_{caj}
\end{aligned} \tag{2.9}$$

where

$$z_{cs} = z_{11} + z_{12} + z_{2i} \tag{2.10a}$$

$$z_{sa} = z_{20} + z_{23} + z_{3i} \tag{2.10b}$$

$$z_{a4} = z_{30} + z_{34} \tag{2.10c}$$

When the SC cable consists of a core and a sheath, the matrix of (2.8) is reduced to a 2×2 matrix.

$$[Z_{ij}] = \begin{bmatrix} Z_{ccj} & Z_{csj} \\ Z_{csj} & Z_{ssj} \end{bmatrix} \tag{2.11}$$

where

$$Z_{ccj} = z_{cs} + z_{s3} - 2z_{2m} \tag{2.12a}$$

$$Z_{ssj} = z_{s3} \tag{2.12b}$$

$$Z_{csj} = z_{s3} - z_{2m} \tag{2.12c}$$

$$z_{s3} = z_{20} + z_{23} \tag{2.12d}$$

If an SC cable consists only of a core, the submatrix is reduced to one element.

$$[Z_{ij}] = Z_{ccj} = z_{11} + z_{12} \tag{2.13}$$

The component impedances per unit length in the above equations are given in the following equations for an SC cable shown in Figure 2.1a [3–5].

- z_{11}: internal impedance of core outer surface

$$z_{11} = (s\mu_0\mu_1/2\pi) \cdot (1/x_2D_1) \cdot \{I_0(x_2) \cdot K_1(x_1) + K_0(x_2) \cdot I_1(x_1)\}$$

- z_{12}: core outer insulator impedance

$$z_{12} = (s\mu_0\mu_{i1}/2\pi) \cdot \ln(r_3/r_2)$$

- z_{2i}: internal impedance of sheath inner surface

$$z_{2i} = (s\mu_0\mu_2/2\pi) \cdot (1/x_3D_2) \cdot \{I_0(x_3) \cdot K_1(x_4) + K_0(x_3) \cdot I_1(x_4)\}$$

- z_{2m}: sheath mutual impedance

$$z_{2m} = \rho_2/2\pi r_3 r_4 D_2$$

- z_{20}: internal impedance of sheath outer surface

$$z_{20} = (s\mu_0\mu_2/2\pi) \cdot (1/x_4D_2) \cdot \{I_0(x_4) \cdot K_1(x_3) + K_0(x_4) \cdot I_1(x_3)\}$$

- z_{23}: sheath outer insulator impedance

$$z_{23} = (s\mu_0\mu_{i2}/2\pi) \cdot \ln(r_5/r_4)$$

- z_{3i}: internal impedance of armor inner surface

$$z_{3i} = (s\mu_0\mu_2/2\pi) \cdot (1/x_5D_3) \cdot \{I_0(x_5) \cdot K_1(x_6) + K_0(x_5) \cdot I_1(x_6)\}$$

- z_{3m}: armor mutual impedance

$$z_{3m} = \rho_3/2\pi r_5 r_6 D_3$$

- z_{30}: internal impedance of armor outer surface

$$z_{30} = (s\mu_0\mu_3/2\pi)\cdot(1/x_6D_3)\cdot\{I_0(x_6)\cdot K_1(x_5)+K_0(x_6)\cdot I_1(x_5)\}$$

- z_{34}: armor outer insulator impedance

$$z_{34} = (s\mu_0\mu_{i3}/2\pi)\cdot\ln(r_7/r_6)$$

where

$$D_1 = I_1(x_2)\cdot K_1(x_1) - I_1(x_1)\cdot K_1(x_2)$$
$$D_2 = I_1(x_4)\cdot K_1(x_3) - I_1(x_3)\cdot K_1(x_4)$$
$$D_3 = I_1(x_6)\cdot K_1(x_5) - I_1(x_5)\cdot K_1(x_6)$$
$$x_k = B_k\sqrt{s}, B_2 = r_2\sqrt{\mu_0\mu_1/\rho_1}$$
$$B_3 = r_3\sqrt{\mu_0\mu_2/\rho_2}, B_4 = r_4\sqrt{\mu_0\mu_2/\rho_2}$$
$$B_5 = r_5\sqrt{\mu_0\mu_3/\rho_3}, B_6 = r_6\sqrt{\mu_0\mu_3/\rho_3}$$

A submatrix of the earth return impedance $[Z_0]$ in (2.7) is given in the following form.

$$[Z_{0jk}] = \begin{bmatrix} Z_{0jk} & Z_{0jk} & Z_{0jk} \\ Z_{0jk} & Z_{0jk} & Z_{0jk} \\ Z_{0jk} & Z_{0jk} & Z_{0jk} \end{bmatrix} \quad (2.14)$$

When the SC cable consists of a core and a sheath, the above matrix is reduced to:

$$[Z_{0jk}] = \begin{bmatrix} Z_{0jk} & Z_{0jk} \\ Z_{0jk} & Z_{0jk} \end{bmatrix} \quad (2.15)$$

If the SC cable consists only of a core, the matrix includes only one element.

$$[Z_{0jk}] = Z_{0jk} \quad (2.16)$$

Z_{0jk} in (2.14), (2.15), and (2.16) is the earth-return impedance between the jth and kth cables. When a cable system is overhead, the impedance is given by Carson [6]. When a cable system is underground, the impedance given by Pollaczek [7] is used. If a cable is above a stratified earth, the earth-return impedance developed by Nakagawa *et al.* [8] can be used.

2.1.2 Potential Coefficient

The admittance matrix of a cable system is evaluated from the potential coefficient matrix as given in (2.4). In the SC cable case, $[P_p]$ and $[P_c]$ are zero, and when the cable system is underground, $[P_0]$ is also zero. Thus, based on the result of Appendix 2.B,

(a) Overhead cable

$$[P] = [P_i] + [P_0] \quad (2.17)$$

(b) Underground cable

$$[P] = [P_i] \quad (2.18)$$

where

$$[P_i] = \text{cable internal potential coefficient matrix}$$

$$= \begin{bmatrix} [P_{i1}] & 0 & \cdots & 0 \\ 0 & [P_{i2}] & & \vdots \\ \vdots & & \ddots & \vdots \\ 0 & \cdots & 0 & [P_{in}] \end{bmatrix} \tag{2.19}$$

$$[P_0] = \text{potential coefficient matrix of the system in air}$$

$$= \begin{bmatrix} [P_{011}] & [P_{012}] & \cdots & [P_{01n}] \\ [P_{012}] & [P_{022}] & \cdots & [P_{02n}] \\ \vdots & \vdots & \ddots & \vdots \\ [P_{01n}] & [P_{02n}] & \cdots & [P_{0nn}] \end{bmatrix} \tag{2.20}$$

All the off-diagonal submatrices of $[P_i]$ are zero. A diagonal submatrix expresses the potential coefficient matrix of an SC cable. When the SC cable consists of a core, a sheath and an armor as shown in Figure 2.1a, the diagonal submatrix is given in the following form.

(See Appendix 2.B [2, 3].)

$$[P_{ij}] = \begin{bmatrix} p_{cj} + p_{sj} + p_{aj} & p_{sj} + p_{aj} & p_{aj} \\ p_{sj} + p_{aj} & p_{sj} + p_{aj} & p_{aj} \\ p_{aj} & p_{aj} & p_{aj} \end{bmatrix} \tag{2.21}$$

where

$$p_{cj} = (1/2\pi\varepsilon_0\varepsilon_{i1}) \cdot \ln(\mathrm{r}_3/\mathrm{r}_2) \tag{2.22a}$$

$$p_{sj} = (1/2\pi\varepsilon_0\varepsilon_{i2}) \cdot \ln(\mathrm{r}_5/\mathrm{r}_4) \tag{2.22b}$$

$$p_{aj} = (1/2\pi\varepsilon_0\varepsilon_{i3}) \cdot \ln(\mathrm{r}_7/\mathrm{r}_6) \tag{2.22c}$$

When the cable consists of a core and a sheath, the above matrix is reduced to:

$$[P_{ij}] = \begin{bmatrix} p_{cj} + p_{sj} & p_{sj} \\ p_{sj} & p_{sj} \end{bmatrix} \tag{2.23}$$

If the cable consists only of a core, then $[P_{ij}]$ includes only one element.

$$[P_{ij}] = p_{cj} \tag{2.24}$$

The submatrices of $[P_0]$ are given in the following form.

$$[P_{0jk}] = \begin{bmatrix} p_{0jk} & p_{0jk} & p_{0jk} \\ p_{0jk} & p_{0jk} & p_{0jk} \\ p_{0jk} & p_{0jk} & p_{0jk} \end{bmatrix} \tag{2.25}$$

where p_{0jk} is the space pontential coefficient and is given for the case of Figure 2.1b by:

$$p_{0jj} = (1/2\pi\varepsilon_0) \cdot \ln(2h_j/r_{7j}) \tag{2.26a}$$

$$p_{0jk} = (1/2\pi\varepsilon_0) \cdot \ln(D_{jk}/d_{jk}) \tag{2.26b}$$

where

$d_{jk}^2 = (h_j - h_k)^2 + y_{jk}^2, D_{jk}^2 = (h_j + h_k)^2 + y_{jk}^2$
h_j, h_k = height of the jth and kth conductors, respectively
y_{jk} = separation distance between the jth and kth conductors

2.2 Pipe-enclosed Type Cable (PT Cable)

2.2.1 Impedance

The impedance matrix of a PT cable shown in Figure 2.2, where an inner conductor is assumed to be an SC cable, is given in the same manner as the SC cable case [2, 9, 10].

2.2.1.1 Pipe Thickness Assumed to be Infinite

$$[Z] = [Z_i] + [Z_p] \tag{2.27}$$

2.2.1.2 Pipe Thickness Being Finite

$$[Z] = [Z_i] + [Z_p] + [Z_c] + [Z_0] \tag{2.28}$$

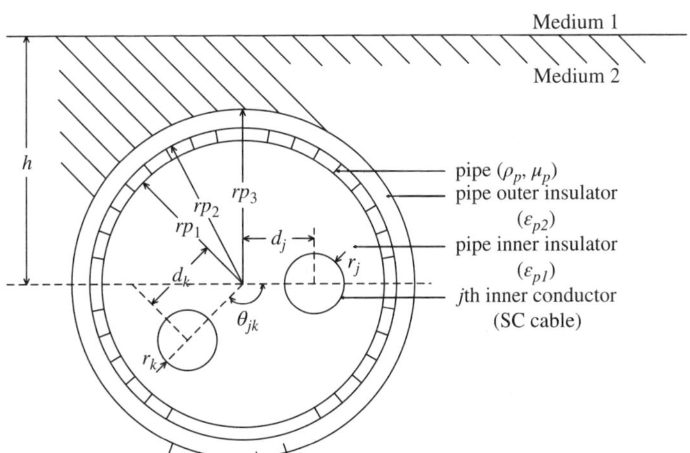

Figure 2.2 A PT cable

where

$[Z_i]$ = SC cable internal impedance matrix

$$= \begin{bmatrix} [Z_{i1}] & [0] & \cdots & [0] & 0 \\ [0] & [Z_{i2}] & \cdots & [0] & 0 \\ \vdots & \vdots & \ddots & \vdots & \vdots \\ [0] & [0] & \cdots & [Z_{in}] & 0 \\ 0 & 0 & \cdots & 0 & 0 \end{bmatrix} \tag{2.29}$$

$[Z_p]$ = pipe internal impedance matrix

$$= \begin{bmatrix} [Z_{p11}] & [Z_{p12}] & \cdots & [Z_{p1n}] & 0 \\ [Z_{p12}] & [Z_{p22}] & \cdots & [Z_{p2n}] & 0 \\ \vdots & \vdots & \ddots & \vdots & \vdots \\ [Z_{p1n}] & [Z_{p2n}] & \cdots & [Z_{pnn}] & 0 \\ 0 & 0 & \cdots & 0 & 0 \end{bmatrix} \tag{2.30}$$

$[Z_c]$ = mutual impedance matrix between pipe inner and outer surfaces

$$= \begin{bmatrix} [Z_{c1}] & [Z_{c1}] & \cdots & [Z_{c1}] & Z_{c2} \\ [Z_{c1}] & [Z_{c1}] & \cdots & [Z_{c1}] & Z_{c2} \\ \vdots & \vdots & \ddots & \vdots & \vdots \\ [Z_{c1}] & [Z_{c1}] & \cdots & [Z_{c1}] & Z_{c2} \\ Z_{c2} & Z_{c2} & \cdots & Z_{c2} & Z_{c3} \end{bmatrix} \tag{2.31}$$

$[Z_0]$ = earth-return impedance matrix

$$= \begin{bmatrix} [Z_0] & [Z_0] & \cdots & [Z_0] & Z_0 \\ [Z_0] & [Z_0] & \cdots & [Z_0] & Z_0 \\ \vdots & \vdots & \ddots & \vdots & \vdots \\ [Z_0] & [Z_0] & \cdots & [Z_0] & Z_0 \\ Z_0 & Z_0 & \cdots & Z_0 & Z_0 \end{bmatrix} \tag{2.32}$$

In (2.29) and (2.30), the last column and row correspond to the pipe conductor. Thus, these should be omitted when the pipe thickness is assumed to be infinite. A diagonal submatrix of $[Z_i]$, that is, (2.29), is given in (2.8). A submatrix of $[Z_p]$, (2.30), is given in the following form.

$$[Z_{pjk}] = \begin{bmatrix} Z_{pjk} & Z_{pjk} & Z_{pjk} \\ Z_{pjk} & Z_{pjk} & Z_{pjk} \\ Z_{pjk} & Z_{pjk} & Z_{pjk} \end{bmatrix} \tag{2.33}$$

When an inner conductor consists of a core and a sheath, (2.33) is reduced to a 2×2 matrix, and when the inner conductor consists only of a core, (2.33) is further reduced to a column matrix in the same manner as explained in the case of $[Z_i]$. [See (2.8), (2.11) and (2.13).] This is the same for all the other impedance and admittance matrices explained in this section.

Z_{pjk} in (2.33) is the impedance between the jth and kth inner conductors with respect to the pipe inner surface, and is given by [11–14]:

$$Z_{pjk} = (s\mu_0/2\pi) \cdot [\mu_p K_0(x_1)/\{x_1 K_1(x_1)\} + Q_{jk} + 2\mu_p \sum_{n=1}^{\infty} C_n/\{n(1+\mu_p) + x_1 K_{n-1}(x_1)/K_n(x_1)\}] \tag{2.34}$$

$$Q_{jj} = \ln[(r_{p1}/r_j) \cdot \{1 - (d_j/r_{p1})^2\}] \tag{2.35a}$$

$$Q_{jk} = \ln[r_{p1}/\sqrt{d_j^2 + d_k^2 - 2d_jd_k \cos\theta j_k}] - \sum_{n=1}^{\infty} C_n/n \tag{2.35b}$$

$$C_n = (d_jd_k/r_{p1}^2)^n \cdot \cos(n\theta_{jk}) \tag{2.35c}$$

and

$$x_1 = \beta_1\sqrt{s}, \beta_1 = r_{p1}\sqrt{\mu_0\mu_p/\rho_p} \tag{2.36}$$

A submatrix and the last row and column elements of $[Z_c]$ in (2.31) are given in the following form.

$$[Z_{c1}] = \begin{bmatrix} Z_{c1} & Z_{c1} & Z_{c1} \\ Z_{c1} & Z_{c1} & Z_{c1} \\ Z_{c1} & Z_{c1} & Z_{c1} \end{bmatrix} \tag{2.37a}$$

$$Z_{c1} = Z_{c3} - 2z_{pm} \tag{2.37b}$$

$$Z_{c2} = Z_{c3} - z_{pm} \tag{2.37c}$$

$$Z_{c3} = z_{p0} + z_{p3} \tag{2.37d}$$

where

$$z_{pm} = \rho_p/(2\pi r_{p1}r_{p2}D_p) \tag{2.38a}$$

$$z_{p0} = (s\mu_0\mu_p/2\pi x_2D_p) \cdot \{I_0(x_2) \cdot K_1(x_1) + K_0(x_2) \cdot I_1(x_1)\} \tag{2.38b}$$

$$z_{p3} = (s\mu_0/2\pi) \cdot \ln(r_{p3}/r_{p2}) \tag{2.38c}$$

and

$$D_p = I_1(x_2) \cdot K_1(x_1) - I_1(x_1) \cdot K_1(x_2) \tag{2.39a}$$

$$x_1 \ : \text{given in (2.36)} \tag{2.39b}$$

$$x_2 = \beta_2\sqrt{s} \tag{2.39c}$$

$$\beta_2 = r_{p2}\sqrt{\mu_0\mu_p/\rho_p} \tag{2.39d}$$

A diagonal submatrix of $[Z_0]$ in (2.32) is given by:

$$[Z_0] = \begin{bmatrix} Z_0 & Z_0 & Z_0 \\ Z_0 & Z_0 & Z_0 \\ Z_0 & Z_0 & Z_0 \end{bmatrix} \tag{2.40}$$

where Z_0 in the above matrix is the self earth-return impedance of the pipe.

2.2.2 *Potential Coefficient*

The potential coefficient matrix of a PT cable shown in Figure 2.2 is given in the following form [9, 10].

2.2.2.1 Pipe Thickness Assumed to be Infinite

$$[P] = [P_i] + [P_p] \tag{2.41}$$

2.2.2.2 Pipe Thickness Being Finite

(a) Underground cable

$$[P] = [P_i] + [P_p] + [P_c] \tag{2.42}$$

(b) Overhead cable

$$[P] = [P_i] + [P_p] + [P_c] + [P_0] \tag{2.43}$$

where

$[P_i]$ = SC cable internal potential coefficient matrix

$$= \begin{bmatrix} [P_{i1}] & [0] & \cdots & [0] & 0 \\ [0] & [P_{i2}] & \cdots & [0] & 0 \\ \vdots & \vdots & \ddots & \vdots & \vdots \\ [0] & [0] & \cdots & [P_{in}] & 0 \\ 0 & 0 & \cdots & 0 & 0 \end{bmatrix} \tag{2.44}$$

$[P_p]$ = pipe internal potential coefficient matrix

$$= \begin{bmatrix} [P_{p11}] & [P_{p12}] & \cdots & [P_{p1n}] & 0 \\ [P_{p12}] & [P_{p22}] & \cdots & [P_{p2n}] & 0 \\ \vdots & \vdots & \ddots & \vdots & \vdots \\ [P_{p1n}] & [P_{p2n}] & \cdots & [P_{pnn}] & 0 \\ 0 & 0 & \cdots & 0 & 0 \end{bmatrix} \tag{2.45}$$

$[P_c]$ = potential coefficient matrix between pipe inner and outer surfaces

$$= \begin{bmatrix} [P_c] & [P_c] & \cdots & [P_c] & P_c \\ [P_c] & [P_c] & \cdots & [P_c] & P_c \\ \vdots & \vdots & \ddots & \vdots & \vdots \\ [P_c] & [P_c] & \cdots & [P_c] & P_c \\ P_c & P_c & \cdots & P_c & P_c \end{bmatrix} \tag{2.46}$$

$[P_0]$ = potential coefficient matrix of the pipe in air

$$= \begin{bmatrix} [P_0] & [P_0] & \cdots & [P_0] & P_0 \\ [P_0] & [P_0] & \cdots & [P_0] & P_0 \\ \vdots & \vdots & \ddots & \vdots & \vdots \\ [P_0] & [P_0] & \cdots & [P_0] & P_0 \\ P_0 & P_0 & \cdots & P_0 & P_0 \end{bmatrix} \tag{2.47}$$

In (2.44) and (2.45), the last column and row corresponding to the pipe conductor, should be omitted when the pipe thickness is being assumed infinite. A diagonal submatrix of $[P_i]$ in (2.44) is given in (2.21). Submatrix $[P_{pjk}]$ of $[P_p]$ in (2.45) is given in the following form.

$$[P_{pjk}] = \begin{bmatrix} P_{pjk} & P_{pjk} & P_{pjk} \\ P_{pjk} & P_{pjk} & P_{pjk} \\ P_{pjk} & P_{pjk} & P_{pjk} \end{bmatrix} \tag{2.48}$$

P_{pjk} in the above equation is the potential coefficient between the jth and kth inner conductors with respect to the pipe inner surface, and is given in the following equation using Q of (2.35).

$$P_{pjj} = Q_{jj}/2\pi\varepsilon_{p1}\varepsilon_0, P_{pjk} = Q_{jk}/2\pi\varepsilon_{p1}\varepsilon_0 \quad (2.49)$$

A submatrix and the last column and row elements of $[P_c]$ in (2.46) are given by:

$$[P_c] = \begin{bmatrix} P_c & P_c & P_c \\ P_c & P_c & P_c \\ P_c & P_c & P_c \end{bmatrix} \quad (2.50a)$$

$$P_c = 1/2\pi\varepsilon_{p2}\varepsilon_0 \cdot \ln(r_{p3}/r_{p2}) \quad (2.50b)$$

A submatrix and the last row and column elements of the space potential coefficient matrix$[P_0]$ is given in the following form.

$$[P_0] = \begin{bmatrix} P_0 & P_0 & P_0 \\ P_0 & P_0 & P_0 \\ P_0 & P_0 & P_0 \end{bmatrix} \quad (2.51a)$$

$$P_0 = (1/2\pi\varepsilon_0) \cdot \ln(2h/r_{p3}) \quad (2.51b)$$

2.3 Arbitrary Cross-section Conductor

In the previous sections, only a circular shape conductor is described. In practice, a conductor cross section is not always circular but it could be trapezoidal or fan-shaped. No theoretical formula is available for calculating the impedances and admittances of a conductor with an arbitrary cross section. In this section, an approximate method to deal with such a conductor is explained [1, 15].

2.3.1 *Equivalent Cylindrical Conductor*

An arbitrary cross-section conductor in Figure 2.3a is approximated by a cylindrical conductor in Figure 2.3b of which the inner (r_i) and outer (r_0) radii are defined in the following equation. The derivation is given in Appendix 2.C.

$$r_0 = l/2\pi, r_i = \sqrt{(l/2\pi)^2 - S/\pi} \text{ (m)} \quad (2.52)$$

where S is the cross-sectional area of the original conductor (m^2) and l is the circumferential length (m).

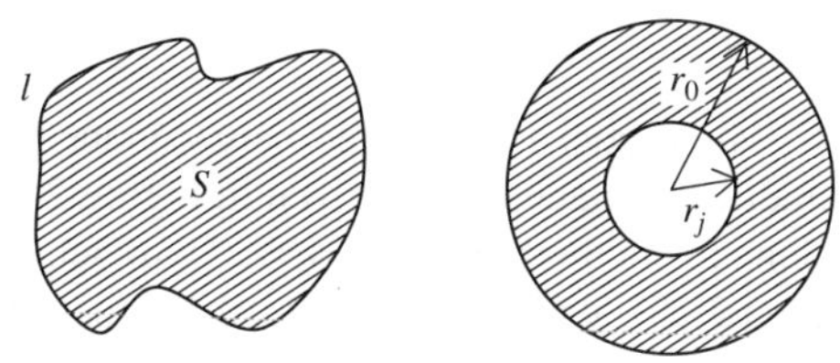

Figure 2.3 (a) Arbitrary cross-section conductor and (b) its equivalent cylindrical conductor

By applying the above, a multi-conductor system composed of arbitrary cross-section conductor is represented by a system composed of circular (cylindrical) conductors. Then, the impedance and admittance of the system are evaluated by using the formulas explained in the previous sections.

2.3.2 Examples

2.3.2.1 Cable A

A test circuit for a fan-shaped multi-conductor system and the cross-section of an actual cable as discussed in Reference [16] are illustrated in Figure 2.4a and b, respectively. The equivalent cable system obtained by the method in Section 2.3.1 is shown in Figure 2.4c. Since this equivalent cable becomes a PT cable with three power conductors and one neutral wire, it is possible to calculate its impedance using EMTP (electromagnetic transients program) Cable Constants [17]. The validity of the equivalent cylindrical conductors can be tested comparing the measured results with those calculated by the Cable Constants using the equivalent PT cable. The calculated and measured results are shown in Table 2.1. The absolute values of calculated impedances at frequency $f = 50$ Hz agree well with those of measured ones whether or not the neutral conductor is grounded. The measured phase angles differ from the calculated ones by a maximum 17.7%. At $f = 5$ kHz, the absolute vaule of calculated impedances agree well with the measured ones if the neutral conductor is not grounded, while the error is about 15% if it is grounded. At $f = 10$ kHz, the absolute values of calculated impedances differ from the measured ones by about 10% whether or not the neutral conductor is grounded. The phase angle error is about 3% if the neutral conductor is grounded, and is about 7% if it is not grounded.

2.3.2.2 Cable B

Figure 2.5a illustrates the cross-section of a 275 kV 2500 mm^2 CSZV cable which consists of five segment conductors [18]. It is represented by an equivalent cylindrical conductor as shown in Figure 2.5b. The outer circumferences of the five segment conductors act as a wire shield, but the effect of the wire shield on the core conductor impedances is not very strong in the frequency range below 100 Hz. Therefore, the effect of the wire shield is neglected.

The results calculated by the method are compared with the measured ones [18] as given in Table 2.2. The error by the approximation is 10% for the magnitude, and less than 5% for the phase angle.

2.3.2.3 Cable C

Figure 2.6 shows cable C which consists of three core conductors with fan-shaped cross section. Its equivalent cylindrical conductors are given in Figure 2.6b. The cable impedances calculated by the finite-element method (FEM) (H.W. Dommel, personal communication, 1989) are compared in Figure 2.7 with those obtained by the approximate method. It is assumed that there exists no leakage flux in the FEM, and the pipe is infinitely thick in the approximate method.

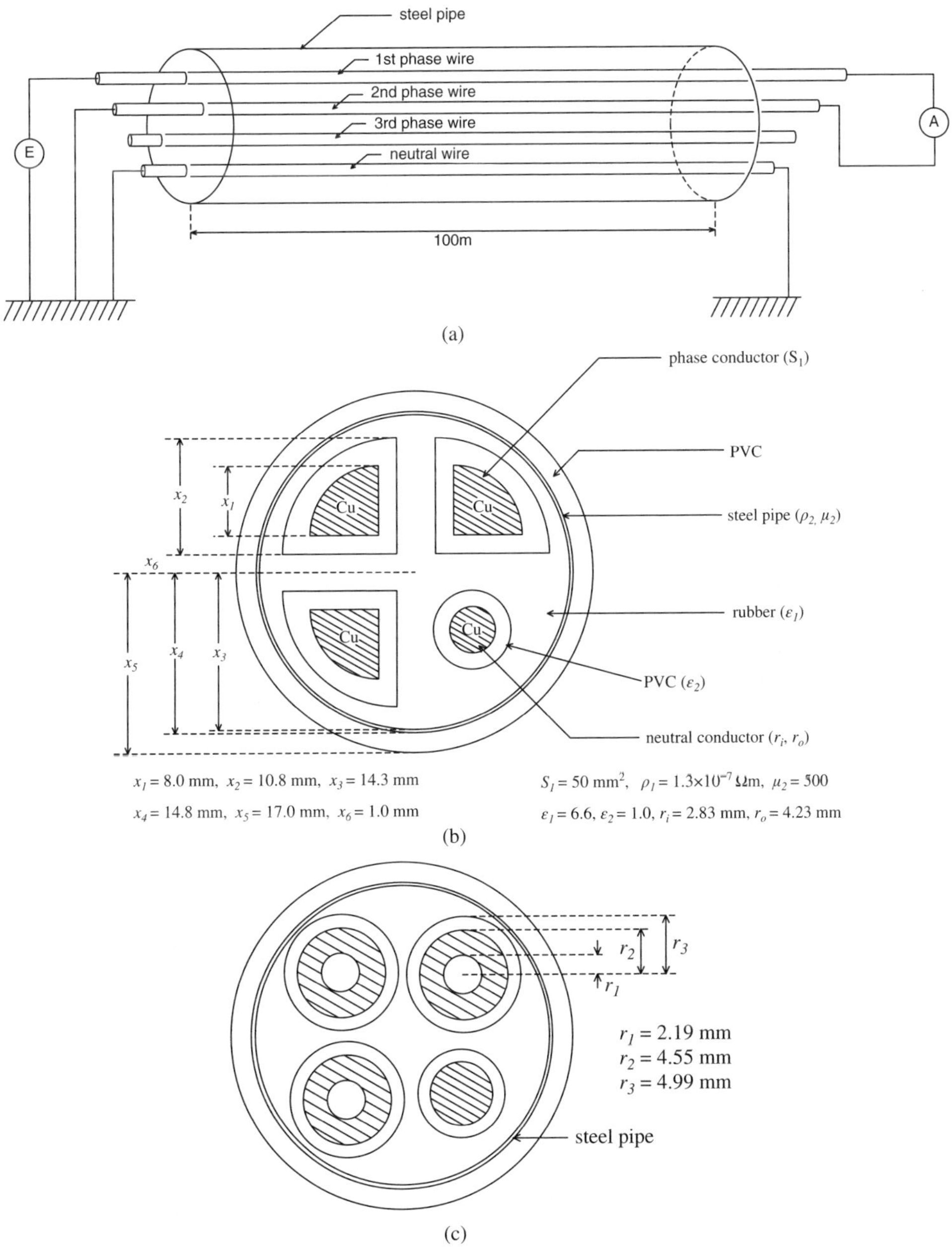

Figure 2.4 (a) Test circuit for cable A; (b) cross section of cable A; and (c) equivalent PT cable

Table 2.1 Measured and calculated impedances of cable A

Neutral grounding	Frequency f (Hz)	50	5000	100 000
	Meas. value Z (mΩ)	0.36	10.13	18.28
	θ (deg)	20.30	68.50	74.20
Grounded	Calc. value Z (mΩ)	0.35	8.62	16.38
	θ (deg)	16.70	79.20	76.40
	Error ΔZ (%)	−2.78	−14.91	−10.39
	Error $\Delta\theta$ (%)	−17.73	15.62	2.96
Isolated	Calc. value Z (mΩ)	0.35	10.41	16.15
	θ (deg)	19.70	79.80	79.30
	Error ΔZ (%)	−2.78	2.76	−11.65
	Error $\Delta\theta$ (%)	−2.96	16.50	6.87

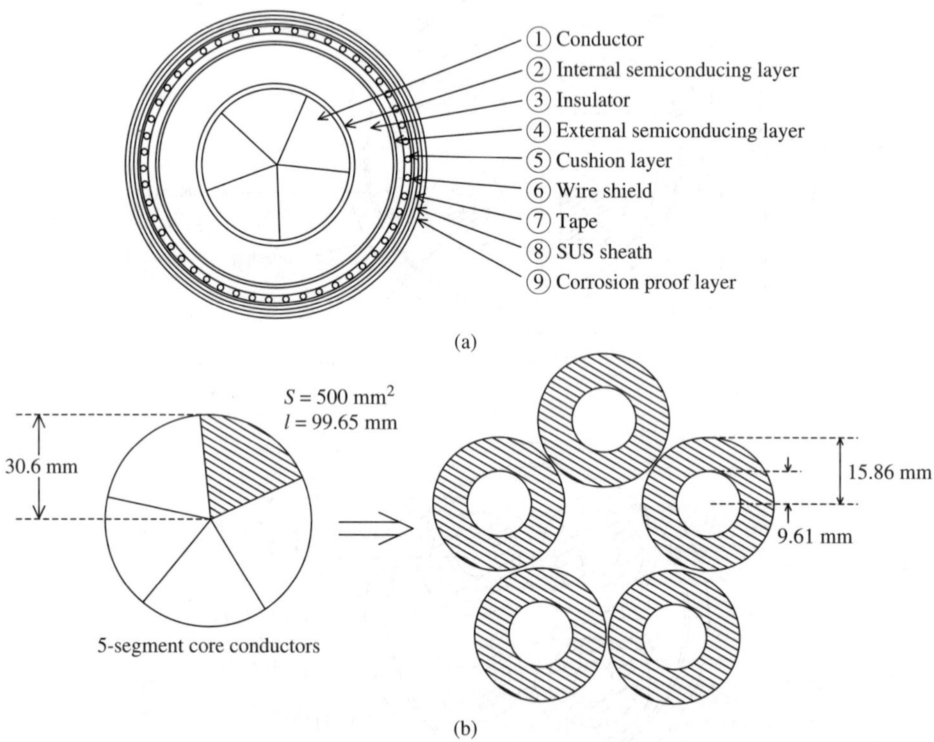

Figure 2.5 275 kV/2500 mm² CSZV cable: (a) cross section; and (b) equivalent cylindrical conductors

The results obtained by the approximate method agree well with the FEM. The calculation error of the approximate method is caused by the distance between the two conductors.

From the examples, it is confirmed that the approximate method explained in Section 2.3.1 shows a satisfactory accuracy in comparison with measured results and calculated results by the FEM.

Table 2.2 Measured and calculated impedance of cable B

Frequency f (Hz)	30	50	60	75	100
Meas. value Z (mΩ)	0.3293	0.5385	0.6420	0.7945	1.0430
θ (deg)	77.56	80.84	81.55	82.12	82.57
Calc. value Z (mΩ)	0.2933	0.4834	0.5789	0.7224	0.9616
θ (deg)	76.50	83.40	84.45	85.52	86.56
Error ΔZ (%)	−10.93	−10.23	−9.83	−9.07	−7.80
Error $\Delta\theta$ (%)	−1.37	3.17	3.56	4.14	4.83

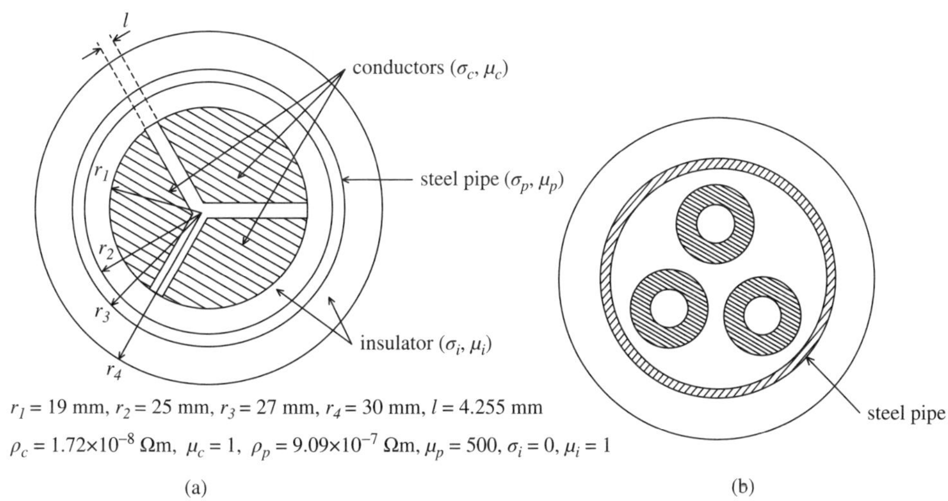

Figure 2.6 (a) Cross section of cable C; and (b) cross section of equivalent circular conductors

2.4 Semiconducting Layer Impedance

2.4.1 Derivation of Impedance

Figure 2.8 illustrates the cross section of a two-layered coaxial cylindrical conductor composed of medium 1 (resistivity ρ_1, permeability μ_1, inner and outer radii a and b) and medium 2 (ρ_2, μ_2, b', c) [19]. Medium 1 corresponds to a conductor and medium 2 to a semiconducting layer in a cable where the conductor has a semiconducting layer on its outer surface. The same is applicable to an outer semiconducting layer on the insulation. In the case of a semiconducting layer on the inner surface of a shield conductor (metallic sheath), medium 1 is the semiconducting layer and medium 2 is the shield conductor.

Appendix 2.D summarizes a derivation of an accurate impedance of the two-layered conductor in Figure 2.8 by solving Maxwell's equation. Since the resistivity of a cable conductor ($\rho_1 = 10^{-8}$Ωm) is far smaller than that of a semiconducting layer ($\rho_2 = 0.01$–10 Ωm [20, 21]), no current is assumed to flow between media 1 and 2 across the boundary assuming that the penetration depth d becomes greater than the thickness (c-b' = 1–5 mm) of medium 2, that is,

$$d = \sqrt{\rho_2/\omega\mu_2} = 35.6/\sqrt{f} \text{ to } 356\sqrt{f} \text{ (m)} > c - b' \tag{2.53}$$

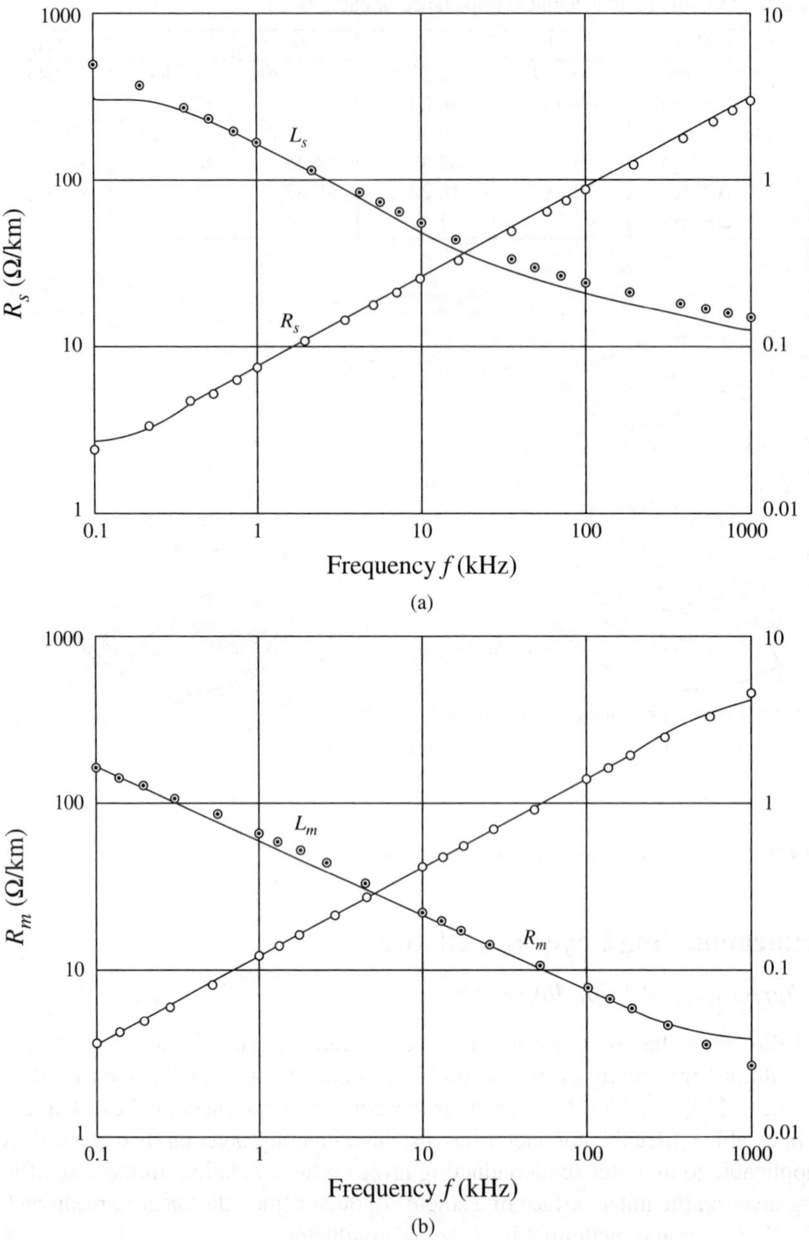

Figure 2.7 Calculated results by FEM (—) and approximate method (o) for cable C: (a) self-impedance $Z_s = R_s + j\omega L_s$; and (b) mutual impedance $Z_m = R_m + j\omega L_m$

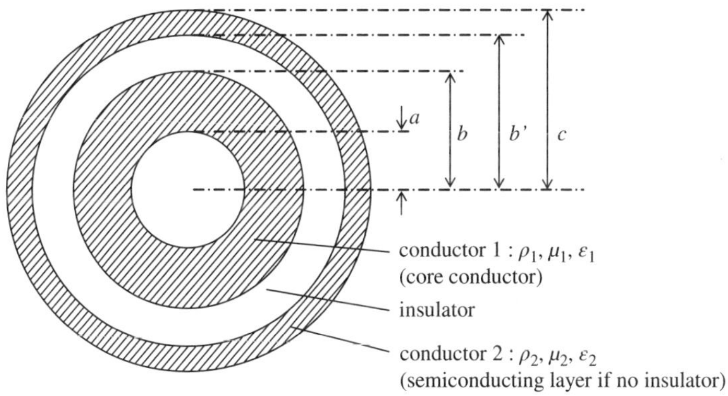

Figure 2.8 Cross section of a two-layered conductor

where $\mu_2 = \mu_0$ = permeability of medium 2 and f is the frequency of the current.

The above condition is almost always satisfied in a frequency range less than 1 GHz which is enough to deal with a wave propagation characteristic and a transient on a cable. This means that the impedance of the semiconducting layer is evaluated as an impedance of an independent (isolated) coaxial cylindrical conductor which has been given by Schelkunoff [5], and the total impedance of the two-layered (conductor/semiconducting layer) is obtained as the parallel impedance of the conductor and the semiconducting layer impedances [19].

The impedance of a cable consisting of inner conductor 1, outer conductor 2 and an insulator in Figure 2.8 is given in the following form [1–3].

$$Z_{11} = z_{10} + z_{12} + z_{2i} - z_{2m} + Z_{12} \tag{2.54a}$$

$$Z_{12} = Z_{22} - z_{2m} \tag{2.54b}$$

$$Z_{22} = z_{20} \tag{2.54c}$$

where

$$z_{10} = (m_1\rho_1/2\pi b)A/B \tag{2.55a}$$

$$z_{2i} = (m_2\rho_2/2\pi b')E/F \tag{2.55b}$$

$$z_{2m} = \rho_2/2\pi b'cF \tag{2.55c}$$

$$z20 = (m_2\rho_2/2\pi c)R/F \tag{2.55d}$$

$$z_{12} = j\omega(\mu_0/2\pi)\ln(b'/b) \tag{2.55e}$$

$$A = K_1(x_1)\cdot I_0(x_2) + I_1(x1)\cdot K_0(x_2)$$

$$B = K_1(x_1)\cdot I_1(x_2) - I_1(x_1)\cdot K_1(x_2)$$

$$E = I_0(x_3)\cdot K_1(x_4) + K_0(x_3)\cdot I_1(x_4)$$

$$F = K_1(x_3)\cdot I_1(x_4) - I_1(x_3)\cdot K_1(x_4)$$

$$R = K_1(x_3)\cdot I_0(x_4) + I_1(x_3)\cdot K_0(x_4)$$

and $I_n(x)$, $K_n(x)$: modified Bessel function of order n

$$x_1 = m_1 a, \quad x_2 = m_1 b, \quad x_3 = m_2 b', \quad x_4 = m_2 c \tag{2.56}$$

$$m_1 = \sqrt{j\omega\mu_1/\rho_1},\ m_2 = \sqrt{j\omega\mu_2/\rho_2}$$

When there is no insulator, that is, $b = b'$ in Figure 2.8, the following result is obtained for $z_{12} = 0$.

$$Z_{11} = z_{10} + z_{2i} + z_{20} - 2z_{2m}, Z_{12} = z_{20} - z_{2m}, Z_{22} = z_{20} \tag{2.57}$$

2.4.2 *Impedance of Two-layered Conductor*

When conductors 1 and 2 are short-circuited at both ends of a cable with the receiving end grounded, and voltage V is applied to the sending end, then the following relation is obtained.

$$\begin{bmatrix} V \\ V \end{bmatrix} = \begin{bmatrix} Z_{11} & Z_{12} \\ Z_{12} & Z_{22} \end{bmatrix} \begin{bmatrix} I_1 \\ I_2 \end{bmatrix} \tag{2.58}$$

where I_1, I_2 are the current of conductors 1 and 2, respectively.

The total current I of the conductors is given by:

$$I = I_1 + I_2 \tag{2.59}$$

The total (outer surface) impedance of the two-layered conductor seen from the sending end is obtained from (2.58) and (2.59).

$$Z_{out} = V/I = (Z_{11}Z_{22} - Z_{12}^2)/(Z_{11} + Z_{22} - 2Z_{12}) \tag{2.60}$$

Substitution of (2.57) into the above equation yields the impedance of the two-layered conductor in the following form.

$$Z_{out} = z_{20} - z_{2m}^2/(z_{10} + z_{2i}) \tag{2.61}$$

Similarly to the above, the inner surface impedance and the mutual impedance between the inner and the outer surfaces of the two-layered conductor are given by:

$$Z_{in} = z_{1i} - z_{1m}^2/(z_{10} + z_{2i}) \tag{2.62a}$$

$$Z_m = z_{1m} \cdot z_{2m}/(z_{10} + z_{2i}) \tag{2.62b}$$

The above equation is a general formula of the impedance of a two-layered conductor, and is applicable to a semiconducting layer on the outer surface of a core, on the inner surface of a sheath, and on the outer surface of the sheath of a cable.

2.4.3 *Discussion of the Impedance Formula*

2.4.3.1 Comparison with the Accurate Formula in Appendix 2.D

In the solid conductor case, that is, $a = 0$ in conductor 1, z_{10} in (2.55) is simplified as:

$$z_{10} = (m_1\rho_1/2\pi b)I_0(x_2)/I_1(x_2) = (m_1\rho_1/2\pi b)I_{02}/I_{12} \tag{2.63}$$

Then, each term of (2.61) is rewritten by substituting (2.55).

$$z_{10} + z_{2i} = (m_1\rho_1 I_{02}/I_{12} + m_2\rho_2 E/F)/2\pi b$$
$$= (m_1\rho_1 F I_{02} + m_2\rho_2 E I_{12})/2\pi b F I_{12}$$
$$z_{2m}^2/(z_{10} + z_{2i}) = \rho_2^2 I_{12}/2\pi bc^2 F(m_1\rho_1 F I_{02} + m_2\rho_2 E I_{12})$$
$$= \rho_2^2 I_{12}/2\pi bc^2 FT$$

Thus, (2.61) is expressed in the following form.

$$Z_{out} = (\rho_2/2\pi cF)\cdot(m_2 R - \rho_2 I_{12}/bcT)$$
$$= (m_2\rho_2/2\pi cT)\{m_1\rho_1 R I_{02} + m_2\rho_2 I_{12}(ER - 1/m_2^2 bc)/F\}$$

Applying the characteristic of Bessel's functions,

$$1/x = I_0(x)\cdot K_1(x) + I_1(x)\cdot K_0(x)$$

The last term of Z_{out} is rewritten as:

$$ER - 1/m_2^2 bc = ER - 1/x_3 x_4$$
$$= (K_{13}I_{14} - I_{13}K_{14})\cdot(K_{03}I_{04} - I_{03}K_{04}) = F\cdot S$$

Substituting the above into Z_{out}, the following result is obtained.

$$Z_{out} = (m_2\rho_2/2\pi c)\cdot\{m_1\rho_1 R I_0(x_2) + m_2\rho_2 S I_1(x_2)\} / \{m_1\rho_1 F I_0(x_2) + m_2\rho_2 E I_1(x_2)\} \quad (2.64)$$

The above impedance is identical to Z_{out} of (2.D.11) with $a = 0$. Thus, it is proven that Z in (2.61) is the same as the accurate formula given in (2.D.11).

In the same manner as the above, (2.62) is proven to be the same as Z_{in} and Z_m of (2.D.11) for a two-layered conductor for $a = 0$ and $a \neq 0$.

2.4.3.2 Comparison with Schelkunoff's Formula

The second and the third terms of (2.64) are rewritten in the following form.

$$m_1\rho_1 R I_{02} + m_2\rho_2 S I_{12} = m_1\rho_1(K_{13}I_{04} + I_{13}K_{04})\mathrm{I}_{02} + m_2\rho_2(K_{03}I_{04} - I_{03}K_{04})I_{12}$$
$$= A' I_{14} - b' K_{04}$$
$$m_1\rho_1 F I_{02} + m_2\rho_2 E I_{12} = m_1\rho_1(K_{13}I_{14} - I_{13}K_{14})\mathrm{I}_{02} + m_2\rho_2(K_{03}I_{14} + I_{03}K_{14})I_{12}$$
$$= A' I_{14} - b' K_{14}$$

where and $.B' = m_2\rho_2 I_{12}K_{03} - m_1\rho_1 I_{02}K_{13}$.

Considering $\mu_3 = \mu_2 = \mu$, $\rho_3 = \rho_2 = \rho$ in a single-layer conductor and thus $m_1 = m_2 = m$, $A\prime$ and b' are rewritten as:

$$A' = m\rho(I_{02}K_{12} + I_{12}K_{02}) = m\rho/mb = \rho/b$$
$$B' = m\rho(I_{12}I_{02} - I_{02}I_{12}) = 0$$

Therefore, (2.64) is expressed in the single-layer conductor case as:

$$Z_{out} = (\rho/2\pi c) \cdot I_{04}/I_{14} = (\rho/2\pi c) I_0(x_4)/I_1(x_4) \tag{2.65}$$

The above impedance is identical to that of a solid conductor with radius c derived by Shelkunoff [5].

The above discussions in Sections 2.4.3.1 and 2.4.3.2 make it clear that (2.61) and (2.62) are a general expression of the impedance of a two-layered hollow conductor from which Schelkunoff's formula of a conductor impedance [5] is derived as a specific case. The expression is more advantageous than (2.D.11), because the two-layered conductor impedance can be evaluated from an existing impedance of the single-layer impedance and it is numerically more stable than a direct calculation of (2.D.11).

2.4.4 *Admittance of Semiconducting Layer*

The admittance of a semiconducting layer is given in the following form.

$$y_s = j\omega 2\pi\varepsilon_s / \ln(c/b) \tag{2.66}$$

where

$$\varepsilon_s = \varepsilon_s' + 1/j\omega\rho_2$$

Thus, the total admittance Y between the core outer and the sheath inner surfaces is given by:

$$1/Y = 1/y_s + 1/y_i \tag{2.67}$$

Where

$$y_i = j\omega 2\pi\varepsilon_i / \ln(r_0/c)$$

and r_0 is the radius of insulator outer surface (=sheath inner surface).

2.4.5 *Wave Propagation Characteristic of Cable with Core Outer Semiconducting Layer*

2.4.5.1 Impedance

Outer Surface Impedance of a Two-layered Conductor

Figure 2.9 and Figure 2.10 compare the outer surface impedance of a cable conductor having a semiconducting layer [Case 1, Z_{out} in (2.61)] with those of the cable conductor only [Case 2, z_{10} in (2.54a)] and of the semiconducting layer only [Case 3, z_{10} in (2.54a)] for a solid conductor and a hollow conductor, respectively. The figures clearly show that the impedance of Case 1 is the same as that of Case 2 in a low frequency region, and tends to approach that of Case 3 as frequency increases. The behavior physically satisfies the well-known skin effect. That is, a current penetrates deeply in a two-layered conductor in a low frequency region and

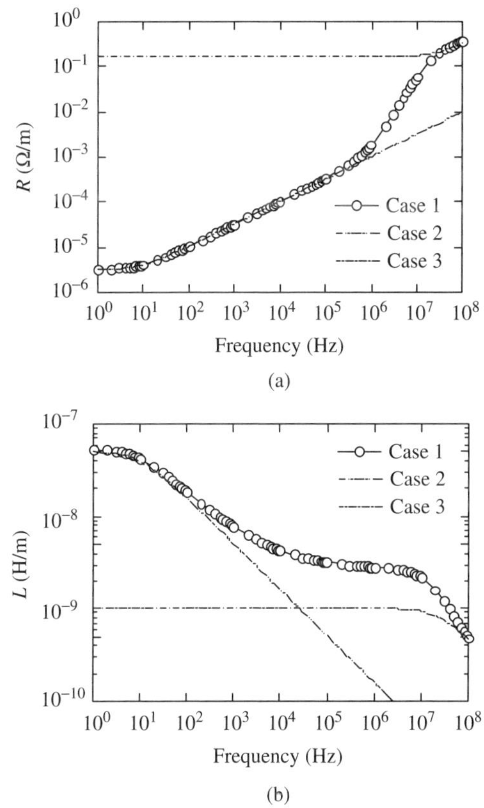

Figure 2.9 Outer surface impedance of a solid two-layered conductor ($\mu_1 = \mu_2 = \mu_0$): (a) resistance vs. frequency; and (b) inductance vs. frequency. Case 1: Two-layered conductor: $a = 0, b = 9.9$ cm, $c = 10$ cm, $\rho_1 = 10^{-7}\ \Omega$m, $\rho_2 = 10^{-4}\ \Omega$m. Case 2: Inner conductor only: $a = 0$, $b = c = 9.9$ cm, $\rho = 10^{-7}\Omega$m. Case 3: Outer conductor only: $a = b = 9.9$ cm, $c = 10$ cm, $\rho = 10^{-4}\ \Omega$m

its impedance is dominated by the low impedance conductor. Thus, the impedance of Case 1 becomes identical to that of Case 2. In a high frequency region, in contrast, most currents flow in the surface area, that is, the semiconducting layer, and its impedance tends to be the same as that of Case 3. The frequency, at which the penetration depth becomes the same as the thickness of the semiconducting layer in Case 1, is evaluated to be 12.7 MHz in the solid conductor case (Figure 2.9) and 203 kHz in the hollow case (Figure 2.10) from (2.53). The values agree with the frequency at which the resistance of Case 1 approaches that of Case 3.

At $f = 1$ Hz, the internal impedances of the core and the semiconducting layer are given independently by:

$$z_1 = 3.25 \times 10^{-6} + j3.38 \times 10^{-7}, z_2 = 1.59 + j1.09 \times 10^{-7}\ \Omega/\text{m}$$

The total impedance of the parallel circuit of z_1 and z_2 are evaluated as:

$$z = z_1//z_2 = 3.25 \times 10^{-6} + j3.38 \times 10^{-7}\ \Omega/\text{m},\ L = 5.38 \times 10^{-8}\ \text{H/m}$$

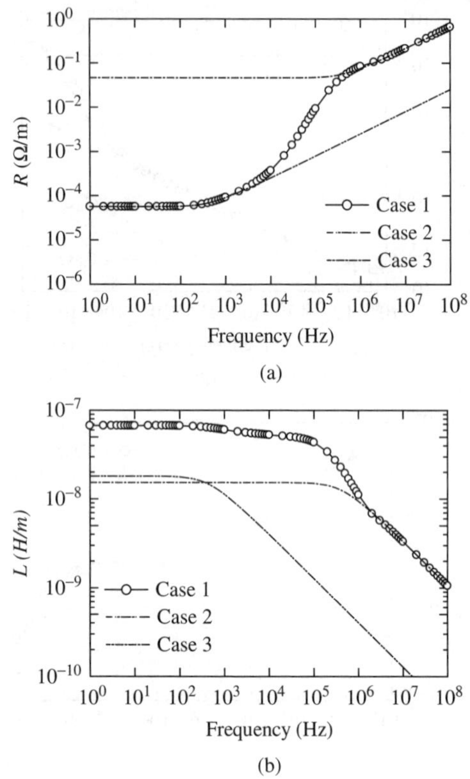

Figure 2.10 Outer surface impedance of a hollow two-layered conductor: (a) resistance vs. frequency; and (b) inductance vs. frequency. Case 1: Two-layered conductor: $a = 1$ cm, $b = 1.25$ cm, $c = 1.5$ cm, $\rho_1 = 10^{-8}\ \Omega$m, $\rho_2 = 10^{-5}\ \Omega$m. Case 2: Inner conductor only: $a = 1$ cm, $b = c = 1.25$ cm, $\rho = 10^{-8}\ \Omega$m. Case 3: Outer conductor only: $a = b = 1.25$ cm, $c = 1.5$ cm, $\rho = 10^{-5}\ \Omega$m

The above result agrees with the impedance of Case 1 at $f = 1$ Hz in Figure 2.9. In the same manner, the impedance of the parallel circuit composed of the hollow core and the semiconducting layer in the case of Figure 2.10 at 1 Hz is evaluated as:

$$R = 5.65 \times 10^{-5}\ \Omega/\text{m},\ L = 5.0 \times 10^{-8}\ \text{H/m}$$

The value agrees with the impedance of Case 1 at 1 Hz in Figure 2.10. The observation indicates the formula of (2.61) is correct.

Core Internal Impedance: Z_{11} in (2.54a)

Figure 2.11 illustrates the cross section of one phase of an underground 275 kV cable with a semiconducting layer on the core outer surface.

Figure 2.12a shows the core internal impedance of the cable with $d = 5$ mm as a function of the semiconducting layer resistivity ρ neglecting the earth. The resistance takes a peak (maximum) which becomes more distinctive as frequency increases, at a certain value

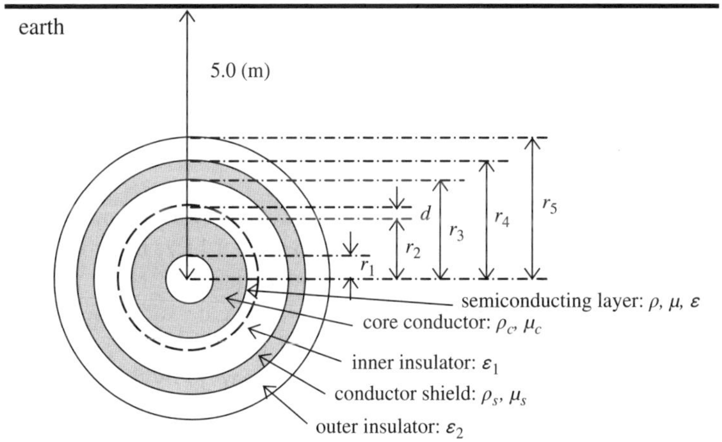

$r_1 = 0$, $r_2 = 30.45$ mm, $r_3 = 71.15$ mm, $r_4 = 74.80$ mm,
$r_5 = 81.61$ mm, $\varepsilon_1 = 3.1$, $\varepsilon_2 = 4.0$, $\rho_c = 1.82 \cdot 10^{-8}$,
$\rho_s = 2.83 \cdot 10^{-8}$, $\mu_c = \mu_s = \mu_0$, $\rho_e = 100$
d : thickness of semiconductinglayer

Figure 2.11 Cross section of an underground 275 kV XLPE cable

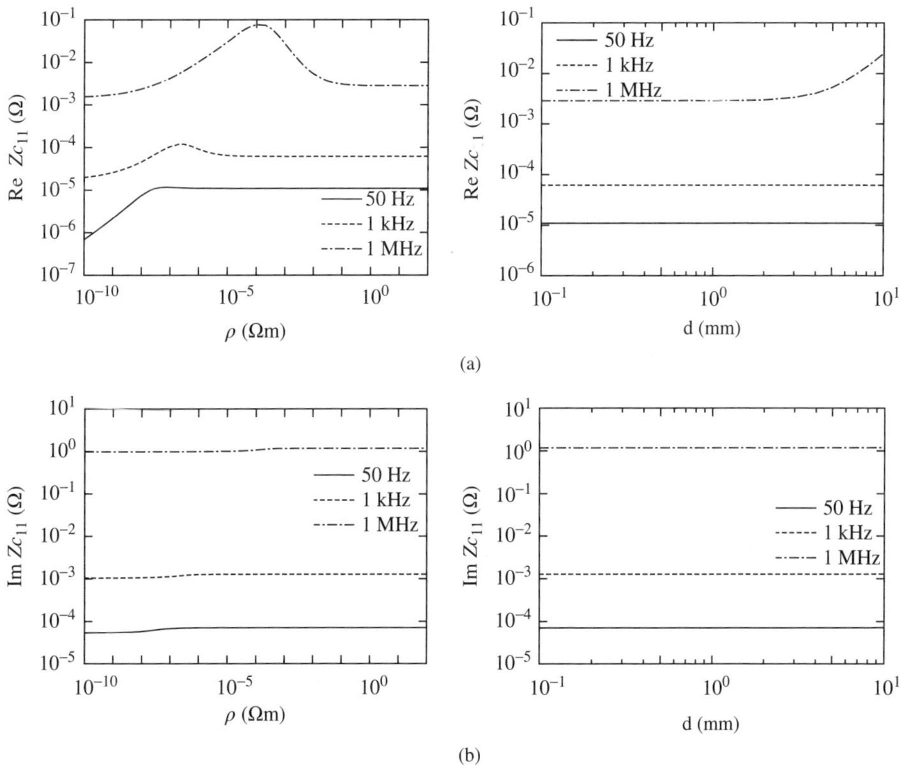

Figure 2.12 Conductor internal impedance. Real part (top) and imaginary part (bottom): (a) as a function of resistivity ρ ($d = 5$ mm); and (b) as a function of thickness d ($\rho = 0.01$ Ωm)

of ρ. The resistivity corresponds to the condition that the penetration depth becomes equal to the thickness of the semiconducting layer. For example, the resistivity is evaluated to be $1.97 \times 10^{-4}\,\Omega$m from (2.53) by $f = 1$ MHz and $d = 5$ mm, which agrees with the resistivity at which the resistance at $f = 1$ MHz becomes maximum in Figure 2.12. The inductance shows a similar trend, but the variation is very small. The resistivity causes no effect on the impedance of the insulation shield.

Figure 2.12b shows the internal impedance of the conductor as a function of the thickness d of the semiconducting layer. At frequencies 50 Hz and 1 kHz, the impedance stays constant in all the range of the thickness in the figure. At frequency 1 MHz, the resistance starts to increase at a thickness of about 3 mm. When $\rho = 1\,\Omega$m, no variation of the resistance is observed although the figure is not shown. As the resistivity increases, the penetration depth increases and thus the impedance is determined almost always by the conductor, that is, the thickness of the semiconducting layer causes no effect on the impedance. The thickness at which the resistance starts to increase relates to the penetration depth of the core. The inductance stays almost constant.

Total Impedance: Z_{11}, Z_{12}, Z_{22} in (2.57)

The internal impedance of the insulation shield Z_{22} and the mutual impedance Z_{12} between the conductor and the conductor shield are not affected by the semiconducting layer on the conductor's outer surface as is clear from (2.54a–c). The total impedance, that is, the elements of the Zmatrix in (2.58), is dominated by the earth-return impedance, which is far greater than the conductor's internal impedance. Thus, the total impedance is not affected by the semiconducting layer.

2.4.5.2 Propagation Constant

It should be now clear from the observations in the previous section that a semiconducting layer affects only the so-called coaxial mode of wave propagation of a cable, which is determined by the internal impedance of the conductor and the admittance between the conductor and the conductor shield as far as the semiconducting layer of the core outer surface is concerned. For the earth-return and the inter-phase modes are determined by the impedance and the admittance of the insulator shield, and thus the semiconducting layer has no effect on the propagation modes. In fact, calculated results of the propagation constants of the three-phase cable in Figure 2.11 show no effect of the semiconducting layer on the earth-return and the inter-phase modes, although the results are not given here. Thus, the following discussions concern only the coaxial mode.

Figure 2.13a shows the attenuation constant, the propagation velocity and the characteristic impedance of the coaxial mode on the three-phase 275 kV XLPE cable in Figure 2.11 as a function of the semiconducting layer resistivity ρ with the thickness $d = 5$ mm. The attenuation constant α shows quite a similar characteristic to the resistance of the conductor's internal impedance in Figure 2.12a. This is physically quite reasonable, because the attenuation is proportional to the resistance. The propagation velocity decreases and the characteristic impedance (real part) increases as the resistivity increases where the attenuation constant starts to increase and becomes constant again. This region relates to the penetration depth as explained in the previous section.

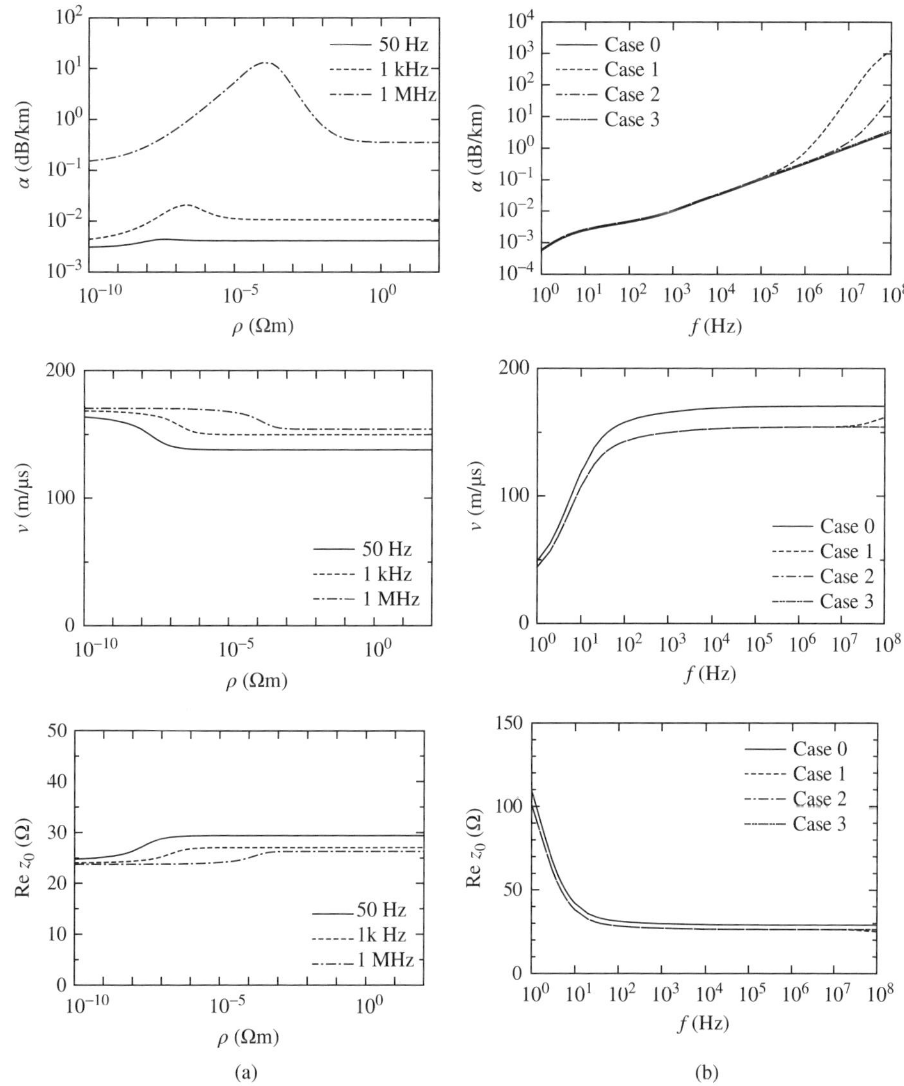

Figure 2.13 Propagation constants of the coaxial mode of an underground 500 kV cable ($d = 5$ mm). Attenuation constant (top), velocity (middle), and characteristic impedance (real part) (bottom): (a) as a function of resistivity ρ; and (b) as a function of frequency f. Case 0: no semiconducting layer. Case 1: $\rho = 0.01\ \Omega$m. Case 2: $\rho = 1\ \Omega$m. Case 3: $\rho = 100\ \Omega$m

Figure 2.13b is the frequency characteristic of the propagation constants, which is a typical characteristic of the propagation constants, and corresponds to the effect of the resistivity in Figure 2.13a. The attenuation constant shows no significant difference between the cases of no semiconducting layer and of the resistivity $\rho = 0.01$–$100\,\Omega$m in the region of frequency less than 1 MHz. The effect of the semiconducting layer on the characteristic impedance is not noticeable.

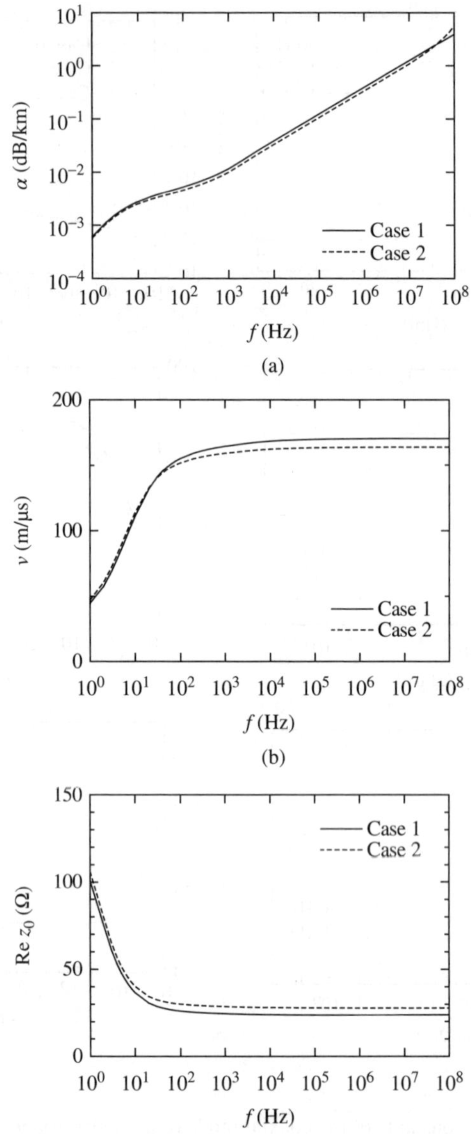

Figure 2.14 Frequency characteristic of coaxial mode ($\rho = 0.01\ \Omega$m, $d = 2$ mm): (a) attenuation constant; (b) velocity; and (c) characteristic impedance (real part) as a function of frequency. Case 1: neglecting the semiconducting layer impedance. Case 2: considering the semiconducting layer impedance

2.4.5.3 Effect of the Semiconducting Layer Admittance

Figure 2.14 shows the propagation constants of the coaxial mode neglecting the impedance of the semiconducting layer. No significant difference is observed between the cases of considering and neglecting the impedance in Figure 2.14 for $d = 2$ mm. On the contrary for

$d = 5$ mm and $\rho = 0.01$ Ωm, a similar difference to that between Cases 0 and 1 in Figure 2.13b is observed, though the results are not given here. The difference becomes smaller as the resistivity increases. The observation indicates that the effect of the semiconducting layer on the propagation constants is dominated by its admittance but not by the impedance when the layer thickness is small and the resistivity is high. Thus the semiconducting layer can be treated as an admittance in such cases. When the thickness is large and the resistivity is low, the impedance has to be considered.

2.4.6 Concluding Remarks

The impedance formula of the semiconducting layer in this section is expressed as a combination of existing impedance formulas of a cylindrical conductor, the impedance of a cable with a semiconducting layer is easily evaluated by an existing computer program for calculating cable parameters.

By applying the impedance formula together with a well-known admittance, the effect of the semiconducting layer on wave propagation and transient characteristics of a cable have been investigated and the following points were noted.

1. The resistance of a two-layered conductor composed of a metallic conductor and its outer semiconducting layer (real part of the core internal impedance) shows a peak as a function of the semiconducting layer resistivity. The resistivity, at which the peak appears, corresponds to the penetration depth of the core. The inductance also shows a similar trend, but the variation is not as noticeable as the resistance. The thickness of the semiconducting layer causes a minor effect on the impedance.
2. The conductor's internal impedance explained above contributes only to the conductor impedance, but not to the conductor shield impedance and the mutual impedance between the conductor and the conductor shield. Since the earth-return impedance of a cable is far greater than the conductor impedance, the effect of the conductor impedance cannot be observed in the total impedances of a cable, that is, the conductor-self, the conductor shield-self, and the conductor-to-shield mutual impedances.
3. The semiconducting layer affects only the coaxial propagation mode of a cable as far as the semiconducting layer of a core is concerned. The attenuation constant of the coaxial mode shows nearly the same trends as the resistance of the conductor impedance. The propagation velocity and the characteristic impedance of the coaxial mode are lower than those in the case of no semiconducting layer. The former decreases and the latter increases as the resistivity of the semiconducting layer increases.
4. The above explained effects of the semiconducting layer are dominated by the admittance of the layer, when the layer thickness is small and the resistivity is high.

2.5 Discussion of the Formulation

2.5.1 Discussion of the Formulas

The formulation of impedances and admittances of various cables given in the previous sections includes some approximations. It may be important to discuss these approximations so as to make the limit of applicability clear when the formulation is used.

First of all, the major assumptions made for the formulation of impedances and admittances should be discussed. One of the major assumptions is constant permeability. Quite often, a pipe and armor are ferromagnetic. It, however, seems to be rather unusual to have high currents to cause saturation of the pipe or the armor. Thus, the saturation can be neglected. When one needs to take the saturation into account, methods proposed in References [11] and [12] can be used. In regard to the first assumption, displacement currents are negligible as far as low frequencies (less than about 1 MHz) are concerned. In the analysis of transients and wave propagations on the cable system, the frequency of interest is, in most cases, less than 1 MHz. The dielectric losses are small in comparison with the losses in conducting media of cables and earth. Thus, the assumption is valid. The third assumption will be discussed later.

No approximation is made for the impedances and admittances of an SC cable as far as Carson's and Pollaczek's earth impedances and Scheikunoff's cylindrical conductor impedances are concerned. It has been said that Carson's and Pollaczek's formulas of the earth-return impedance are not applicable at frequencies higher than some megahertz because the effect of displacement currents is not included in the formulas as is well-known, although the displacement current is included in a modified Pollaczek's formula [22]. Thus, the formulation of the impedances of both SC and PT cables is correct only up to some megahertz, unless the modified formula is adopted.

One can easily find that the formulation of the impedances and admittances of an SC cable is identical to that given in Reference [3] for the case of a coaxial cable consisting of a core and a metallic sheath.

Two assumptions are included in the PT cable case. The first one is that the electric cable positions within the pipe do not affect the internal impedances and admittances of the inner conductors (SC cable) and outer surface of the pipe. Thus, the inner conductor impedance and admittance of a PT cable become the same as those of an SC cable. The same assumption has been made in References [11] and [12]. If one needs to take into account the effect of the eccentricity on the inner conductor impedance, the formula of the outer surface impedance of the inner conductor given in Reference [13] can be used.

The second assumption concerns the case of infinite pipe thickness. It is assumed that the pipe thickness will be greater than the penetration depth in the pipe wall. If the pipe thickness is smaller than the penetration depth, the formula of the pipe internal impedance given in (2.34) and (2.35) and potential coefficient given in (2.49) cannot be used. Accurate formulas of the impedance and potential coefficient were derived in Reference [14]. Since these formulas are too complicated for practical usage, the assumption of infinite pipe wall thickness may be used, but only to calculate the impedance and potential coefficients of the pipe. Note that earth-return currents are not neglected and complete shielding is not assumed. This assumption introduces negligible error for actual PT cables and for frequencies above 10 Hz. Figure 2.15 shows a comparison of the pipe impedances for the cases of the pipe thickness being finite and infinite. It is clear that the impedance for the finite pipe thickness case approaches that for the infinite thickness case at a frequency of 1 kHz. When the pipe thickness is 4 mm, which is nearly equivalent to the penetration depth at 10 Hz, its impedance is nearly the same as that for the infinite thickness case in the frequency range shown in the figure. The pipe thickness is, in most cases, greater than the penetration depth. Thus, the assumption is valid.

Calculated results of admittances of a single-phase SC cable are shown in Figure 2.16. From the results, it is clear that the admittance of an underground cable is much greater than that of an overhead cable. The impedance shows no significant difference between underground

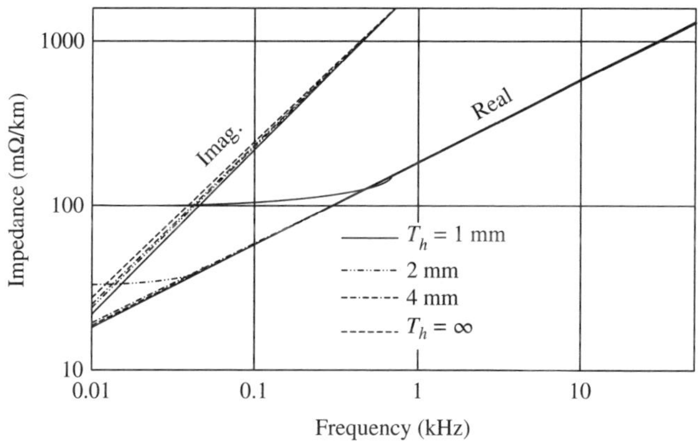

Figure 2.15 Effect of pipe thickness T_h on the pipe inner surface impedance

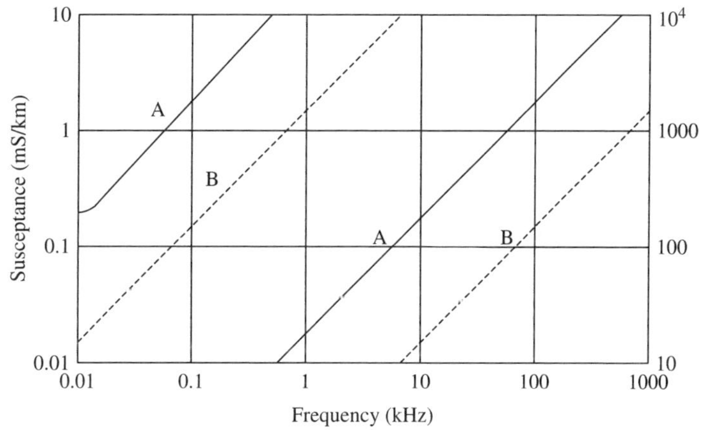

Figure 2.16 Susceptances (imaginary Y_{22}) of (A) underground and (B) overhead SC cables

and overhead cables. Thus, it should be expected that the attenuation of the underground cable is much higher than that of the overhead cable, and the propagation velocity is lower in the underground case.

The internal impedances of SC cables are shown in Figure 2.17. Significant differences are observed for the cases of SC cables consisting only of a core, of a core and a sheath, and of a core, a sheath and an armor.

2.5.2 *Parameters Influencing Cable Impedance and Admittance*

2.5.2.1 Stranded Conductor

A core conductor of a high-voltage cable is composed of a number of strands (thin wire) which are twisted altogether very tightly. The same is true for a phase conductor of a high-voltage overhead line such as an ACSR (Aluminum Conductor Steel Reinforced). Therefore the real

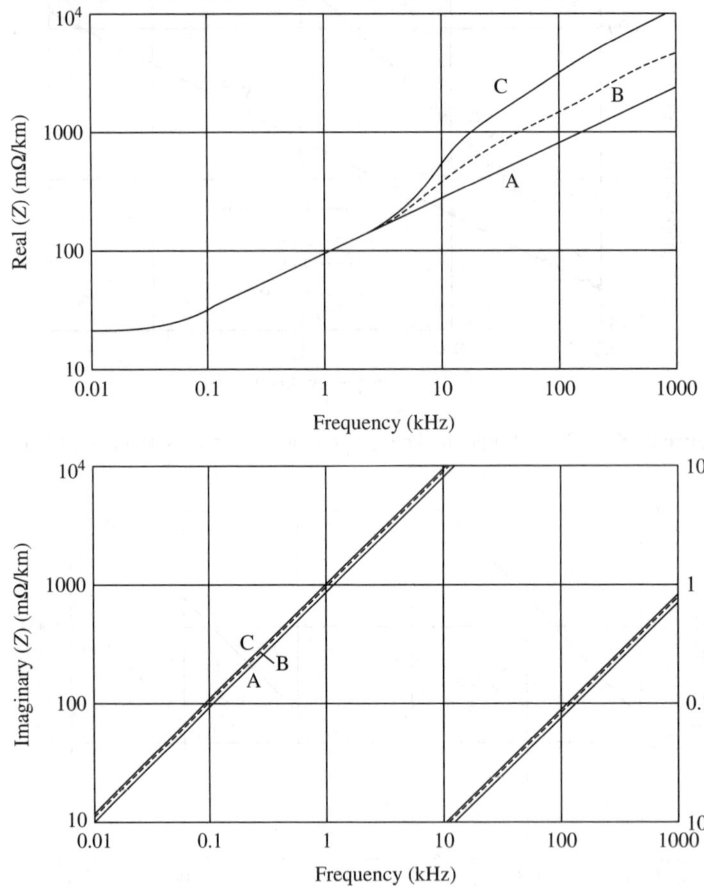

Figure 2.17 Effect of sheath and armor on the internal impedances Z_{cc} of SC cables. A, Core and its outer insulator; B, core, sheath and its outer insulator: and C, core, sheath, armor and its outer insulator

length of the strands is greater than the cable length. Thus, the real length of the strands has to be considered to evaluate the core impedance. A formula considering the strand length is given by the following [1, 23].

$$Z = R_{dc}\sqrt{1 + K(j\omega\mu_c/4\pi R_{dc})} \tag{2.68a}$$

$$K = 4.52r(m + 2)r_0 \tag{2.68b}$$

where r_0 is the strand radius, r is the stranded conductor radius and m is the number of strands at the outer surface.

2.5.2.2 Enamel Coated Strand

In a high-voltage cable, say above 200 kV, each strand is coated by enamel to insulate it from the other strands so as to reduce its impedance in a high frequency region, that is, to reduce an increase of the impedance due to the skin effect. Assume there are no "n" strands with a radius of r_0 as for a core conductor. The cross-sectional area S of the core with the radius r is given by:

$$S = n \cdot \pi r_0^2 = \pi_r^2 \quad \therefore r = \sqrt{n} \cdot r_0 \tag{2.69}$$

At a high frequency, a conductor internal impedance is given approximately in (2.C.4) as:

$$Z_i = \sqrt{j\omega\mu\rho/l^2} = \sqrt{j\omega\mu\rho}/2\pi r$$

Thus, the following internal impedance is obtained.

(a) Enamel coated

$$Z_a = \sqrt{j\omega\mu e}/n2\pi r_0 \tag{2.70}$$

(b) No enamel coating

$$Z_b = \sqrt{j\omega\mu\rho}/\sqrt{n}2\pi r_0 = \sqrt{n}Z_a \tag{2.71}$$

It should be clear that the high-frequency impedance due to the skin effect is reduced by $1/\sqrt{n}$ by using enamel coated strands. Thus, if a cable core is composed of "n" enamel coated strands, the core outer surface impedance, Z_{11} in Section 2.1.1, has to be modified by the factor $1/\sqrt{n}$.

2.5.2.3 Proximity Effect between Conductors

There are a few papers discussing the proximity effect between conductors and strands in the stranded conductor case [24–28]. For a PT cable, the proximity effect between an inner conductor and the pipe conductor has been taken into account as in (2.34) in Section 2.2.1. In the stranded conductor case, the proximity effect might be significant in an impedance calculation if each strand is enamel coated or insulated. This is the same for a segmented core conductor as studied in Section 2.3.2. However, in this case, the segmented conductor can be represented as a PT cable so that the proximity effect can be taken into account.

If a stranded conductor is not insulated between each strand, the proximity effect seems to be neglected, because the strands touch each other tightly and thus each strand becomes electrically connected.

However, the proximity effect in stranded and segmented conductors needs further investigation, and an improved approach is expected to be developed.

2.5.2.4 Snaking

Snaking is similar to transposition of a three-phase overhead line, and is often observed in a cable installed in a tunnel or trench. The cable snaking is dealt with as transposed impedance and admittance matrices in the same manner as those of an overhead line. The cable snaking option has been installed in the ATP Subroutine Cable Parameters [29].

2.5.2.5 Geometrical Configuration

Similarly to an overhead line with sag, the buried depth or the suspended height of a cable is not the same. The accurate data of the depth or the height "h" along a cable is not available

Table 2.3 Frequency- and temperature-dependent permittivity

	Water						Paper					
T (°C)	1.5	1.5	25	25	85	85	25	25	25	82	82	82
f (Hz)	1×10^5	1×10^8	1×10^5	1×10^8	1×10^5	1×10^8	100	1×10^5	1×10^8	100	1×10^5	1×10^8
ε_t	87	87	78.2	78	58	58	3.3	3.1	2.77	3.57	3.4	3.08
tan δ	1900	70	4000	50	12400	30	58	200	660	170	85	680

in most cases. The earth-return/space impedance and admittance of a cable are dependent on the height (depth) "h" being proportional to ln(h). This fact has to be remembered before discussing the accuracy of Carson's or Pollaczek's formulas. Snaking is a typical example. Also, the separation distance between phases of a three-phase cable is not the same along a cable buried horizontally.

This is engineering. An engineering practice is very different from a mathematically or physically ideal condition.

2.5.2.6 Physical Parameters of a Cable

In general, most of the physical parameters of a cable such as the resistivity, the permittivity and tan δ are frequency- and temperature-dependent, and such data are not available even from a cable manufacturer. The same is true for the earth resistivity and permittivity. A typical example is given in Table 2.3 [29, 30].

The frequency and temperature dependence of tan δ are far more significant than the real part of a complex permittivity, and the data are almost impossible to obtain from a cable manufacturer.

2.6 EMTP Subroutines "Cable Constants" and "Cable Parameters"

Cable Constants (CC) [17] in EMTP and Cable Parameters (CP) [29] in ATP are subroutines to calculate the following.

2.6.1 Overhead Line

1. Conductor internal impedance by Schelkunoff
2. Earth-return impedance by Carson
3. Formulation for computer calculation and bundled conductor by Galloway *et al.*

2.6.2 Underground/Overhead Cable

1. Conductor internal impedance by Schelkunoff
2. Underground/overhead earth-return impedance by Pollaczek with some approximations

3. Formulation for computer calculation of SC cable by Wedepohl *et al.*
4. PT cable impedance by Brown *et al.*
5. General formulation of underground/overhead SC and PT cables by Ametani.

The CC was developed by Ametani at Bonneville Power Administration, US Department of Energy (originally Department of Interior) from 1976 to 1981. The details of the CC is described elsewhere [17, 31].

The CP was developed by Ametani at Bonneville Power Administration in 1991 and 1992 to deal with cable snaking, distributed admittance and arbitrary cross-section conductors [29].

The basic structure of the CC and CP is summarized as:

1. Main routine
   (a) physical parameters preparation : $c_o, \mu_0, \varepsilon_0, \pi,$ and so on
   (b) data input
   (c) pre-processing of the input data
2. Calculations of geometrical parameters: SIMP
   (a) $d_{ij}^2 = (hi - hj)^2 + y_{ij}^2$, $D_{ij}^2 = (hi + hj)^2 + y_{ij}^2$
   (b) potential coefficient calculation
3. Conductor internal impedance calculation as a function of frequency: SKIN, SCZY2
4. Earth-return impedance: ZEGEN, ZEST
5. Formulation of impedance/admittance for overhead line: OLZY
6. Formulation of earth-return and conductor internal impedance for SC cable: SCZY1
7. Formulation of impedance/admittance for PT cable: PTZY
8. Formulation of Z and Y matrices: ZYMX
9. Eigen values/vectors calculations: EIGEN
10. Calculations of propagation constants: PRCON
11. Output of calculated results.

The basic subroutines in the CC and the CP are:

ZYMX: formulation of Z and Y
OLZY: formulation of space/earth-return and conductor internal impedances, and space admittance for overhead line
SCZY1: formulation of earth-return impedance and admittance for SC cable
SCZY2: formulation of SC cable internal impedance and admittance
SKIN: calculation of conductor internal impedance for overhead line
SIMP: calculation of potential coefficient
ZEGEN: calculation of earth-return impedance
ZEST: stratified earth impedance for overhead line
BSKIN: Bessel functions calculation
PRCON: calculation of propagation constants
EIGEN: eigenvalues/vectors calculations
PTZY: formulation of PT cable impedance/admittance

Appendix 2.A Impedance of an SC Cable Consisting of a Core, a Sheath and an Armor

In the case of an SC cable with a core, a sheath and an armor, an equivalent circuit for impedances is given in Figure 2.A.1.

Define currents flowing into the core, sheath, armor and outer medium (earth) by I_c, I_s, I_a and I_e at x. Also inner and outer surface currents of the sheath and the armor are I_2, I_3, I_4 and I_5 as shown in Figure 2.A.1. Voltages between the core, sheath, armor and outer medium are V_{12}, V_{23} and V_{34} at x, and are $V_{12} + \Delta V_{12}$, $V_{23} + \Delta V_3$, and $V_{34} + \Delta V_{34}$, at $x = x + \Delta x$.

Then, the following relation for currents are obtained.

$$I_2 = -I_c, I_3 = -I_4, I_5 = -I_e \tag{2.A.1}$$

$$I_s = I_2 + I_3 = -(I_c + I_4) \tag{2.A.2a}$$

$$I_a = I_4 + I_5 = I_4 - I_e \tag{2.A.2b}$$

From the above equations,

$$I_4 = -(I_c + I_s) \tag{2.A.3a}$$

$$I_e = -(I_c + I_s + I_a) \tag{2.A.3b}$$

For voltage V_{12} between the core and the sheath,

$$V_{12} = z_{11}\,\Delta x I_c - z_{12}\Delta x I_2 - z_{2i}\Delta x I_2 - z_{2m}\Delta x I_3 + V_{12} + \Delta V_{12}$$

$$\therefore -\Delta V_{12}/\Delta x = (z_{11} + z_{12} + z_{2i})I_c + z_{2m}I_4$$

Define z_{cs} by:

$$z_{cs} = z_{11} + z_{12} + z_{2i} \qquad (2.A.4) = (2.10)$$

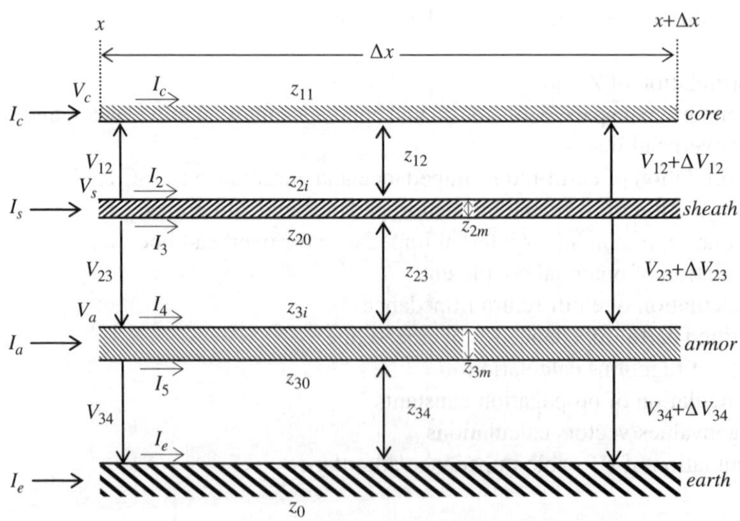

Figure 2.A.1 An equivalent circuit for impedances of an SC cable

Using the above equation,

$$-\Delta V_{12}/\Delta x = z_{cs}I_c + z_{2m}I_4 \tag{2.A.5}$$

For voltage V_{23},

$$-\Delta V_{23}/\Delta x = (z_{20} + z_{23} + z_{3i})I_4 + z_{2m}I_c - z_{3m}I_e$$

Define z_{sa} by:

$$z_{sa} = z_{20} + z_{23} + z_{3i} \qquad (2.A.6) = (2.10)$$

Then,

$$-\Delta V_{23}/\Delta x = z_{sa}I_4 + z_{2m}I_c + z_{3m}I_e \tag{2.A.7}$$

For voltage V_{34},

$$-\Delta V_{34}/\Delta x = (z_{a4} + z_0)I_e + z_{3m}I_4 \tag{2.A.8}$$

where

$$z_{a4} = z_{30} + z_{34} \qquad (2.A.9) = (2.10)$$

Take the earth voltage of zero potential as reference,

$$\begin{aligned} V_a &= -V_{34} \\ V_s &= -(V_{23} + V_{34}) = V_a - V_{23} \\ V_c &= V_{12} + V_s \end{aligned} \tag{2.A.10}$$

Substituting (2.A.3) and (2.A.10) into (2.A.8).

$$-\Delta V_a/\Delta x = (z_{a4} - z_{a3} + Z_0) \cdot (I_c + I_s) + (z_{a4} + Z_0)I_a \tag{2.A.11}$$

Substitute (2.A.3), (2.A.10) and (2.A.11) into (2.A.7).

$$\begin{aligned} -\Delta V_s/\Delta x = (z_{sa} + z_{a4} - z_{2m} - 2z_{3m} + Z_0)I_c + (z_{sa} + z_{a4} - 2z_{3m} + Z_0)I_s \\ +(z_{a4} - z_{3m} + Z_0)I_a \end{aligned} \tag{2.A.12}$$

In the same manner,

$$\begin{aligned} -\Delta V_c/\Delta x = (z_{cs} + z_{sa} + z_{a4} - 2z_{2m} - 2z_{3m} + Z_0)I_c \\ + (z_{sa} + z_{a4} - z_{2m} - 2z_{3m} + Z_0)I_s \\ + (z_{a4} - z_{3m} + Z_0)I_a \end{aligned} \tag{2.A.13}$$

Finally from (2.A.11), (2.A.12) and (2.A.13) with x converting to zero,

$$d(V)/dx = -[Z] \cdot (I) \tag{2.A.14}$$

where [Z] is given by:

$$[Z] = [Z_i] + [Z_0] \qquad (2.A.15) = (2.5)$$

and

$$[Z_i] = \begin{bmatrix} Z_{cc} & Z_{cs} & Z_{ca} \\ Z_{cs} & Z_{ss} & Z_{sa} \\ Z_{ca} & Z_{sa} & Z_{aa} \end{bmatrix} \qquad (2.A.16a) = (2.8)$$

$$[Z_0] = \begin{bmatrix} Z_0 & Z_0 & Z_0 \\ Z_0 & Z_0 & Z_0 \\ Z_0 & Z_0 & Z_0 \end{bmatrix} \qquad (2.A.16b) = (2.9)$$

where

$$Z_{cc} = z_{cs} + z_{sa} + z_{a4} - 2(z_{2m} + z_{3m}) \qquad (2.A.17a)$$

$$Z_{ss} = z_{sa} + z_{a4} - 2z_{3m} \qquad (2.A.17b)$$

$$Z_{aa} = z_{a4} \qquad (2.A.17c)$$

$$Z_{cs} = z_{sa} + z_{a4} - z_{2m} - 2z_{3m} \qquad (2.A.17d)$$

$$Z_{sa} = z_{a4} - z_{3m} \qquad (2.A.17e)$$

(2.A.17a–e = (2.9) and (2.14)

Appendix 2.B Potential Coefficient

An equivalent circuit for the admittance of an underground SC cable with a core, a sheath and an armor is illustrated in Figure 2.B.1. From the figure,

$$I_c = y_{cs}\Delta x(V_c - V_s) + I_c + \Delta I_c \qquad (2.B.1a)$$

$$I_s = y_{cs}\Delta x(V_s - V_c) + y_{sa}\Delta x(V_s - V_a) + I_s + \Delta I_s \qquad (2.B.1b)$$

$$I_a = y_{sa}\Delta x(V_a - V_s) + y_{a4}\Delta x V_a + I_a + \Delta I_a \qquad (2.B.1c)$$

Rewriting the above equations,

$$-\Delta I_c/\Delta x = y_{cs}V_c - y_{cs}V_s \qquad (2.B.2a)$$

$$-\Delta I_s/\Delta x = -y_{cs}V_c + (y_{cs} + y_{sa})V_s - y_{sa}V_a \qquad (2.B.2b)$$

$$-\Delta I_a/\Delta x = -y_{sa}V_a + (y_{sa} + y_{a4})V_a \qquad (2.B.2c)$$

Put x converting to 0 in the above equations,

$$\frac{d}{dx}\begin{bmatrix} I_c \\ I_s \\ I_a \end{bmatrix} = -\begin{bmatrix} y_{cs} & & 0 \\ -y_{cs} & (y_{cs} + y_{sa}) & -y_{sa} \\ 0 & -y_{sa} & (y_{sa} + y_{a4}) \end{bmatrix} \cdot \begin{bmatrix} V_c \\ V_s \\ V_a \end{bmatrix} = -[\mathrm{Y_i}](\mathrm{V}) \qquad (2.B.3)$$

where

$$y_{cs} = s2_{\pi}\varepsilon_0\varepsilon_{I_1}/\ln(r_3/r_2)$$

$$y_{sa} = s2_{\pi}\varepsilon_0\varepsilon_{i2}/\ln(r_5/r_4)$$

$$y_{a4} = s2_{\pi}\varepsilon_0\varepsilon_{i3}/\ln(r_7/r_6) \qquad (2.B.4)$$

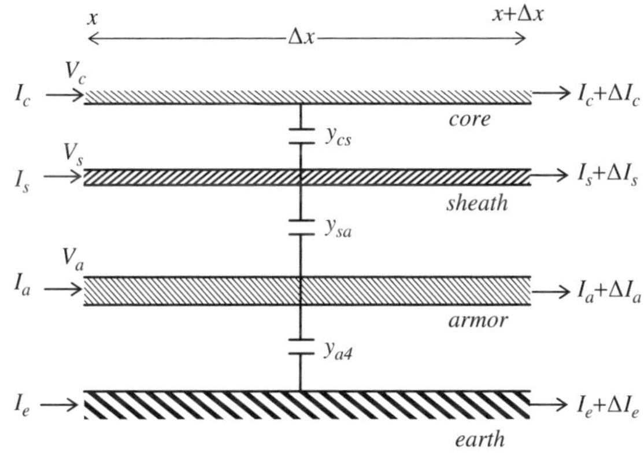

Figure 2.B.1 An equivalent circuit for admittance of an SC cable

Potential coefficients being inversely related to admittances,

$$[p_i] = \begin{bmatrix} p_c + p_s + p_a & p_s + p_a & p_a \\ p_s + p_a & p_s + p_a & p_a \\ p_a & p_a & p_a \end{bmatrix} \tag{2.B.5}$$

$$p_c = s/y_{cs}, p_s = s/y_{sa}, p_a = s/y_{a4} \qquad (2.B.6) = (2.22)$$

When a cable is overhead, considering a space admittance being connected in series of y_{a4}, in Figure 2.B.1, the potential coefficient matrix is derived in the same manner as the underground cable case.

Appendix 2.C Internal Impedances of Arbitrary Cross-section Conductor

The d.c. impedance (resistance) of a conductor shown in Figure 2.3 is given by the following equation irrespective of the cross-sectional shape:

$$Z_{DC} \fallingdotseq \rho_c/S = R_{DC} \quad (\Omega/\text{m}) \tag{2.C.1}$$

where S is the cross-sectional area () and ρ_c is the resistivity of the conductor (Ω/m).

In the high-frequency range, the current inside the conductor is concentrated around the surface due to the skin effect. The skin effect depth is defined by:

$$h_c = \sqrt{\rho_c/j\omega\mu_c} \ () \tag{2.C.2}$$

where $\omega = 2\pi f$ is the angular frequency, and $\mu_c = \mu_{cr}\mu_0$ is the permeability (H/m).

The skin depth is complex and it is assumed that the current density is uniform in the range covered by the skin depth. Under this assumption, the internal impedance can be derived similarly to the d.c. resistance and it is given approximately by:

$$Z_{hf} \fallingdotseq \rho_c/S_{hf} = \rho_c/(l \times h_c) \quad (\Omega/\text{m}) \tag{2.C.3}$$

where l is the circumferential length of the conductor (), and S_{hf} is the cross-sectional area of the current layer in the high-frequency range.

Equations (2.C.2) and (2.C.3) give the high-frequency impedance as:

$$Z_{hf} \fallingdotseq \sqrt{j\omega\mu_c\rho_c}/l \quad (\Omega/\text{m}) \tag{2.C.4}$$

Suppose that the function $f(x)$ defined over the range $[0, \infty]$ is not equal to zero and both $f(0)$ and $f(x \to \infty)$ are known. Then, the following Rolle's averaging theorem [32] is applied.

$$f(x) \fallingdotseq \sqrt{\{f(0)\}^2 + \{f(x \to \infty)\}^2}\ ;\quad 0 \leqq x \leqq \infty \tag{2.C.5}$$

The resistance in (2.C.1) corresponds to $\varepsilon_{f(0)}$ and the impedance in (2.C.4) corresponds to $f(x \to \infty)$. Therefore, the impedance of the conductor with arbitrary cross-section is given as a function of frequency:

$$Z_c \fallingdotseq \sqrt{{Z_{DC}}^2 + {Z_{hf}}^2} \fallingdotseq R_{DC}\sqrt{1 + j\omega\mu_c \cdot (S/R_{DC}l)^2} \quad (\Omega/\text{m}) \tag{2.C.6}$$

Appendix 2.D Derivation of Semiconducting Layer Impedance

To solve Maxwell's equation in a cylindrical coordinate illustrated in Figure 2.D.1, the following assumptions for the conductor are made.

$d/dt = j\omega$ for electric and magnetic fields with sinusoidal excitation $\exp(j\omega t)$.
$H_z = H_r = E_\theta = 0$ for the infinitely long conductor along the z axis.

and

$$\partial/\partial\theta = 0 \text{ for the circular symmetry.} \tag{2.D.1}$$

where $\omega = 2\pi f$ is the angular frequency, $j = \sqrt{-1}$, and E, H are the electric and magnetic field strength, respectively.

Under the above assumptions, the following differential equations with respect to x are obtained from Maxwell's equation [33].

$$\partial^2 H_\theta/\partial x^2 + (\partial H_\theta/\partial x)/x - (1 + 1/x^2) \cdot H_\theta = 0 \tag{2.D.2a}$$

$$\partial^2 E_r/\partial x^2 + (\partial E_r/\partial x)/x - (1 + 1/x^2) \cdot E_r = 0 \tag{2.D.2b}$$

$$\partial^2 E_z/\partial x^2 + (\partial E_z/\partial x)/x - E_z = 0 \tag{2.D.2c}$$

where $x = mr$, $m^2 = j\omega\mu\eta - \Gamma^2$,

$$\eta = \sigma + j\omega\varepsilon \text{ is the intrinsic admittance} \tag{2.D.3}$$

($\sigma = 1/\rho$ is the conductivity, ε is the permittivity) and

$$\Gamma \text{ is the propagation constant}$$

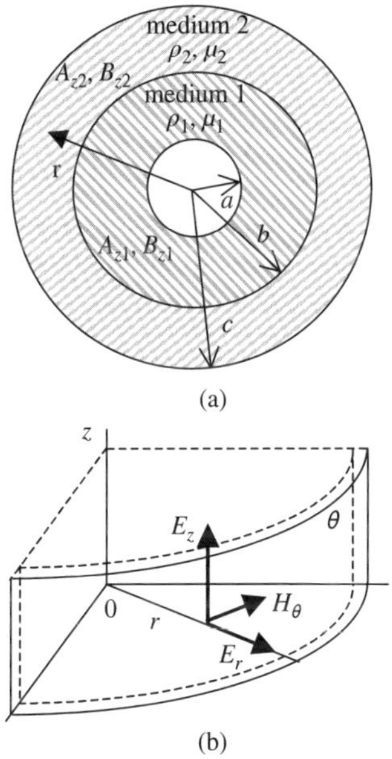

Figure 2.D.1 (a) Two-layered coaxial cylindrical conductor and (b) cylindrical coordinate

The solutions of the above Bessel's equations are:

$$H_\theta = A_\theta \cdot I_1(x) + B_\theta \cdot K_1(x) \tag{2.D.4a}$$

$$E_r = A_r \cdot I_1(x) + B_r \cdot K_1(x) \tag{2.D.4b}$$

$$E_z = A_z \cdot I_0(x) + B_z \cdot K_0(x) \tag{2.D.4c}$$

where $I_n(x)$, $K_n(x)$ is the modified Bessel function.

The constants A and B are related by substituting H_θ, E_r and E_z into (2.D.3).

$$A_\theta = (\eta/m) \cdot A_z \tag{2.D.5a}$$

$$B_\theta = -(\eta/m) \cdot B_z \tag{2.D.5b}$$

$$A_r = (\Gamma/m) \cdot A_z \tag{2.D.5c}$$

$$B_r = -(\Gamma/m) \cdot B_z \tag{2.D.5d}$$

Equations (2.D.4a–c) and (2.D.5a–d) are defined in media 1 and 2 in Figure 2.D.1 with subscript n=1 and 2, that is, $H_{\theta 1}, H_{\theta 2}, \ldots, Bz_1$, and B_{z_2}.

For the modified Bessel function $K_0(x)$ becomes infinite with $x = mr = 0$ ($m \neq 0$) and E_z should not be infinite, the following condition is obtained.

$$B_{z_1} = 0 \tag{2.D.6}$$

At the boundary of media 1 and 2, the following condition has to be satisfied.

$$E_{z_1}(x_2) = E_{z_2}(x_3) \tag{2.D.7a}$$

$$H_{\theta 1}(x_2) = H_{\theta 2}(x_3) \tag{2.D.7b}$$

Let us define a total current flowing through media 1 and 2 as I, which returns partially through the inner surface of medium 1 (I_{in}) and through the outer surface of medium 2 (I_{out}). The magnetic field $H_{\theta 1}(x_1)$ at $r = a$ is given with $x_1 = m_{1a}$ by:

$$H_{\theta 1}(x_1) = -I_{in}/2\pi a \tag{2.D.8}$$

At the outer surface of medium 2, the magnetic field $H_{\theta 2}(x_4)$ is determined by the current $I_{in} + I_{out} - I_{in} = I_{out}$ in the following form with $x_4 = m_2 c$.

$$H_{\theta 2}(x_4) = I_{out}/2\pi c \tag{2.D.9}$$

E_{z_1} at $r = a$ and Ez_2 at $r = c$ are related with component impedances and the currents I_{in} and I_{out} by:

$$E_{z_1}(x_1) = Z_{in} \cdot I_{in} + Z_m \cdot I_{out} \tag{2.D.10a}$$

$$E_{z_2}(x4) = Z_m \cdot I_{in} + Z_{out} \cdot I_{out} \tag{2.D.10b}$$

where Z_{in}, Z_{out} are the inner and outer surface impedances, respectively, and Z_m is the inner/outer surfaces mutual impedance.

Solving (2.D.4)–(2.D.10) simultaneously, the following impedances are obtained.

$$\begin{aligned} Z_{in} &= (m_1\rho_1/2\pi aD) \cdot (m_1\rho_1 FQ + m_2\rho_2 EP) \\ Z_{out} &= (m_2\rho_2/2\pi cD) \cdot (m_1\rho_1 GR + m_2\rho_2 HS) \\ Z_m &= \rho_1\rho_2/2\pi abcD \end{aligned} \tag{2.D.11}$$

Considering the following condition for a conductor,

$$j\omega\mu\sigma >> \omega^2\mu\varepsilon + \Gamma^2$$

$$\therefore\ m^2 \approx j\omega\mu\sigma = j\omega\mu/\rho$$

Parameters D–H and P–S in (2.D.11) are simplified as follows:

$$D = m_1\rho_1 FG + m_2\rho_2 EH \tag{2D.12}$$

$$E = I_0(x_3)\cdot K_1(x_4) + I_1(x_4)\cdot K_0(x_3), \quad F = I_1(x_4)\cdot K_1(x_3) - I_1(x_3)\cdot K_1(x_4)$$

$$G = I_0(x_2)\cdot K_1(x_1) + I_1(x_1)\cdot K_0(x_2), \quad H = I_1(x_2)\cdot K_1(x_1) - I_1(x_1)\cdot K_1(x_2)$$

$$P = I_0(x_1)\cdot K_1(x_2) + I_1(x_2)\cdot K_0(x_1), \quad Q = I_0(x_2)\cdot K_0(x_1) - I_0(x_1)\cdot K_0(x_2)$$

$$R = I_0(x_4)\cdot K_1(x_3) + I_1(x_3)\cdot K_0(x_4), \quad S = I_0(x_4)\cdot K_0(x_3) - I_0(x_3)\cdot K_0(x_4)$$

where $m_1 = \sqrt{j\omega\mu_1/\rho_1}$ and $m_2 = \sqrt{j\omega\mu_2/\rho_2}$.

References

[1] Ametani, A. (1990) *Distributed-Parameter Circuit Theory*, Corona Publishing Co. (in Japanese).

[2] Ametani, A. (1980) A general formulation of impedance and admittance of cables. *IEEE Transactions on Power Apparatus and Systems*, **99** (3), 902–10.

[3] Wedepohl, L.M. and Wilcox, D.J. (1973) Transient analysis of underground power-transmission systems. *Proceedings of the IEE*, **120**, 253–60.

[4] Wedepohl, L.M. and Wilcox, D.J. (1973) Estimations of transient sheath overvoltages in power cable transmission systems. *Proceedings of the IEE*, **120**, 877–82.

[5] Schelkunoff, S.A. (1934) The electromagnetic theory of coaxial transmission line and cylindrical shields. *Bell System Technical Journal*, **13**, 532–79.

[6] Carson, J.R. (1926) Wave propagation in overhead wires with ground return. *Bell System Technical Journal*, **5**, 539–54.

[7] Pollaczek, F. (1926) Uber das Feld einer unendlich langen wechsel stromdurchlossenen Einfachleitung. *Elektrishe Nachrichten Technik*, **3** (9), 339–60.

[8] Nakagawa, M., Ametani, A., and Iwamoto, K. (1973) Further studies on wave propagation in overhead lines with ground return. *Proceedings of the IEE*, **120**, 1521–8.

[9] Ametani, A. (1977) Generalized Program for Line and Cable Constants. Bonneville Power Administration, Purchase Order No. 70249, Report No. 2.

[10] Ametani, A. (1978) Extension of Generalized Program for Line and Cable Constants in EMTP. Bonneville Power Administration, Contract No. EW-780-C-80-1500, Report No. 1.

[11] Brown, G.W. and Rocamora, R.G. (1976) Surge propagation in three-phase pipe-type cables, Part I. Unsaturated pipe. *IEEE Transactions on Power Apparatus and Systems*, **PAS-95**, 89–95.

[12] Dugan, R.C., Rocamora, R.G., Brown, G.W. *et al.* (1977) Surge propagation in three phase pipe-type cables, Part II. Duplication of field tests including the effects of neutral wires and pipe saturation. *IEEE Transactions on Power Apparatus and Systems*, **PAS-96**, 826–33.

[13] Schinzinger, R. and Ametani, A. (1978) Surge propagation characteristics of pipe enclosed underground cables. *IEEE Transactions on Power Apparatus and Systems*, **PAS-97**, 1680–7.

[14] Amekawa, N., Nagaoka, N., and Ametani, A. (2004) Impedance derivation and wave propagation characteristics of pipe-enclosed and tunnel-installed cables. *IEEE Transactions on Power Delivery*, **19** (1), 380–6.

[15] Ametani, A. and Fuse, I. (1992) Approximate method for calculating impedance of multiconductor with arbitrary cross-section. *Electrical Engineering in Japan*, **112** (2), 117–23.

[16] Vandesompele, G. (1989) EMTP and low voltage cables. *EMTP News*, **1**, 19–49.

[17] Scott Meyer, W. (1984) *EMTP Rule Book*, Bonneville Power Administration.

[18] Iizuka, K. (ed.) (1989) *Handbook of Power Cable Technologies*, Denki-Shoin, Tokyo.

[19] Ametani, A., Miyamoto, Y., and Nagaoka, N. (2004) Semiconducting layer impedance and its effect on cable propagation and transient characteristics. *IEEE Transactions on Power Delivery*, **19** (4), 1523–31.

[20] Yasui, T. and Miyauchi, H. (1966) Investigation of a semiconducting layer for coating an EHV CV cable. *Journal of Sumitomo Electric*, **91**, 43–8.

[21] IEE of Japan (2001) High Temperature Capability of OF and CV Cables and Factors Controlling the Capability. Technical Report 858.

[22] Ametani, A., Ohe, M., Miyamoto, Y., and Tanabe, K. (2012) The effect of the earth-return admittance on wave propagation along an overhead conductor in a high frequency region. Proceedings of the EEUG Conference, Zwickau, Germany, September 2012.

[23] Galloway, R., Shorrocks, W., and Wedepohl, L.M. (1964) Calculation of electrical parameters for short and long polyphase transmission lines. *Proceedings of the IEE*, **111** (12), 2051–9.

[24] Carson, J.R. (1921) Wave propagation over parallel wires: The proximity effect. *Philosophical Magazine*, **607**.

[25] Gustavsen, B., Sletbak, J., and Henriksen, T. (1995) Simulation of transient sheath voltages in the presence of proximity effects. *IEEE Transactions on Power Delivery*, **10** (2), 1066–75.

[26] Brandão Faria, J.A. (2008) Application of harmonic expansion method approach to the computation of l and c matrices for open-boundary inhornogeous multiconductor transmission-line structures with long proximity effects present. *Electrical Engineering*, **90** (5), 313–21.

[27] Gustavsen, B., Bruaset, A., Bremnes, J., and Hassel, A. (2009) A finite element approach for calculating electrical parameters of umbilical cables. *IEEE Transactions on Power Delivery*, **24** (4), 2375–84.

[28] Pagnetti, A., Xemard, A., Paladian, F., and Nucci, C.A. (2012) An improved method for the calculation of the international impedances of solid and hollow conductors with the inclusion of proximity effect. *IEEE Transactions on Power Delivery*, **27** (4), 2063–72.
[29] Ametani, A. (1994) *Cable Parameter Rule Book*, Bonneville Power Administration.
[30] Ametani, A. (1988) Further improvement of cable constants and an investigation of cable parameters. *EMTP News*, **1** (4), 4–14.
[31] Dommel, H.W. (1986) *EMTP Theory Book*, Bonneville Power Administration.
[32] Okamoto, T. (ed.) *Handbook of Mathematics* (1960), Maruzen Co., pp. 45–56 (in Japanese).
[33] Amekawa, A., Nagaoka, N., Baba, Y., and Ametani, A. (2003) Derivation of a semiconducting layer impedance and its effect on wave propagation characteristics on a cable. *IEE Proceedings – Generation, Transmission and Distribution*, **150** (4), 434–40.

3

Theory of Wave Propagation in Cables

Akihiro Ametani
Graduate School of Electrical Engineering, Doshisha University, Kyoto, Japan

3.1 Modal Theory

A single-phase cable is, in general, composed of two conductors, a core and a metallic sheath (shield) as explained in Chapter 2. Thus, it is necessary to deal with a multi-conductor system in an analysis of voltages and currents of the cable. A basic theory of the multi-conductors involves matrix algebra for the impedance and the admittance which are given in a matrix form. A theory of eigenvalues and vectors or diagonalization of a matrix is learnt as one of basic mathematics in any university engineering department. In a distributed-parameter circuit, such as a cable, the theory of eigenvalues and vectors relates to the wave propagation along a conductor. This is called "Modal theory" or "Natural mode theory" [1–3], because an eigenvalue of the product of the impedance and the admittance gives wave propagation related parameters, that is, the attenuation constant and the propagation velocity on the conductor in a modal domain, but not in a phase domain.

The modal theory is used to transform voltages, currents, impedances and admittances in a phase domain to those in a modal domain, and is used to evaluate propagation constants, characteristic impedances and transformation matrices in the modal domain. By transforming those back to the phase domain, we can obtain voltages, currents, characteristic impedances, and so on in the actual phase domain.

3.1.1 Eigenvalues and Vectors

Let us consider as an example the following 2×2 matrix.

$$[P] = \begin{bmatrix} P_{11} & P_{12} \\ P_{21} & P_{22} \end{bmatrix} \tag{3.1}$$

where the element in the matrix are not zero, and P_{11} and $P_{22} > 0$.

Cable System Transients: Theory, Modeling and Simulation, First Edition.
Akihiro Ametani, Teruo Ohno and Naoto Nagaoka.

The eigenvalues and vectors of the above matrix P are obtained by solving the characteristic equation of P. Assume the eigenvector matrix of P is given by A, and the eigenvalue matrix is Q. Then, the matrix P is diagonalized by the following matrix algebra.

$$[A]^{-1}[P][A] = [Q] = [U](Q) \tag{3.2}$$

Or

$$[P] = [A][Q][A]^{-1} \tag{3.3}$$

where U is the unit diagonal matrix

[ ] for $n \times n$ matrix, ( ) for vector with n elements.

The above equation is rewritten as:

$$[P][A] = [A][Q] \quad \therefore [P][A] - [A][Q] = [0] \tag{3.4}$$

where [0] is the null matrix.

For Q is a diagonal matrix, the following relation is satisfied between the kth eigenvalue Q_k and the corresponding kth vector (A_k).

$$A_k \cdot Q_k = Q_k \cdot A_k \ ; \ k = 1, 2, \cdots, n \tag{3.5}$$

Substitution of the above equation into (3.4) gives the following result for the kth column.

$$(P - Q_k \cdot U)A_k = 0 \tag{3.6}$$

A solution $A_k = 0$ of the above equation is meaningless. Thus,

$$\det(P - Q_k \cdot U) = 0 = \text{characteristic equation} \tag{3.7}$$

Once the eigenvalue Q_k is determined by the above equation, then the eigenvector A_k is calculated by (3.4).

Example 3.1

Obtain the eigenvalues of the following matrices.

1. $[P] = \begin{bmatrix} P_{11} & P_{12} \\ P_{21} & P_{22} \end{bmatrix}$
2. $[P] = \begin{bmatrix} P_{11} & P_{12} & P_{13} \\ P_{21} & P_{22} & P_{21} \\ P_{13} & P_{12} & P_{11} \end{bmatrix}$

Solution

1. $\det(P - Q_k \cdot U) = \det \begin{vmatrix} (P_{11} - Q) & P_{12} \\ P_{21} & (P_{22} - Q) \end{vmatrix} = (P_{11} - Q)(P_{22} - Q) - P_{12}P_{21}$

$$= Q^2 - (P_{11} + P_{22})Q + (P_{11}P_{22} - P_{12}P_{21}) = 0$$

$$Q_1, Q_2 = \left\{ P_{11} + P_{22} \pm \sqrt{(P_{11} - P_{22})^2 + 4P_{12}P_{21}} \right\} /2$$

(i) $k = 1 \;:\; A_{11} = 1$

$$\begin{vmatrix} (P_{11} - Q_1) & P_{12} \\ P_{21} & (P_{22} - Q_1) \end{vmatrix} \begin{pmatrix} 1 \\ A_{21} \end{pmatrix} = \begin{pmatrix} 0 \\ 0 \end{pmatrix}$$

$$P_{11} - Q_1 + P_{12} \cdot A_{21} = 0, \quad P_{21} + (P_{22} - Q_1)A_{21} = 0$$

$$\therefore A_{21} = (Q_1 - P_{11})/P_{12} = P_{21}/(Q_1 - P_{22})$$

(ii) $k = 2 \;:\; A_{12} = 1$

$$A_{22} = P_{21}/(Q_2 - P_{22}) = (Q_2 - P_{11})/P_{12}$$

2. $$\det(P - Q_k U) = \det \begin{vmatrix} (P_{11} - Q) & P_{12} & P_{13} \\ P_{21} & (P_{22} - Q) & P_{21} \\ P_{13} & P_{12} & (P_{11} - Q) \end{vmatrix}$$

$$= \det \begin{vmatrix} (P_{11} + P_{13} - Q) & P_{12} \\ 2P_{21} & (P_{22} - Q) \end{vmatrix} \cdot (P_{11} - P_{13} - Q) = 0$$

$$Q_2 = P_{11} - P_{13}, Q_1, Q_3 = \{P_{11} + P_{22} + P_{13} \pm \sqrt{(P_{11} - P_{22} + P_{13})^2 + 8P_{12}P_{21}}\}/2$$

$$[A] = \begin{bmatrix} 1 & 1 & 1 \\ -2P_{21}/\left(P_{22} - Q_1\right) & 0 & -2P_{21}/(P_{22} - Q_3) \\ 1 & -1 & 1 \end{bmatrix}$$

3.1.2 Calculation of a Matrix Function by Eigenvalues/Vectors

It is well-known that a function of a matrix P is calculated by using series expansion. For example,

$$([U] + [P])^n \risingdotseq [U] + [P]/n \cdots \tag{3.8a}$$

$$\sinh([P]) \risingdotseq [P] + [P]^3/6 \cdots \tag{3.8b}$$

$$\cosh([P]) \risingdotseq [U] + [P]^2/2 \cdots \tag{3.8c}$$

$$\exp([P]) \risingdotseq [U] + [P] + [P]^2/2! \cdots \tag{3.8d}$$

where each element of $[P]$, that is, $|P_{ij}| \ll 1$.

Applying eigenvalues $[Q]$ and vectors $[A]$, any matrix function is accurately calculated by:

$$f([P]) = [A] \cdot f([Q]) \cdot [A]^{-1} \tag{3.9}$$

For example, $[P]^{1/2}$ is calculated by:

$$[P]^{1/2} = [A] \cdot [Q]^{1/2} \cdot [A]^{-1}$$

$$\text{where} \quad [Q]^{1/2} = \begin{bmatrix} \sqrt{Q_1} & 0 & \cdots & 0 \\ 0 & \sqrt{Q_2} & \cdots & 0 \\ \vdots & \vdots & \ddots & \vdots \\ 0 & 0 & \cdots & \sqrt{Q_n} \end{bmatrix}$$

Similarly,

$$\exp([P]) = [A] \cdot \exp([Q]) \cdot [A]^{-1}$$

$$\text{where} \quad \exp([Q]) = \begin{bmatrix} \exp(Q_1) & 0 & \cdots & 0 \\ 0 & \exp(Q_2) & \cdots & 0 \\ \vdots & \vdots & \ddots & \vdots \\ 0 & 0 & \cdots & \exp(Q_n) \end{bmatrix}$$

3.1.3 Direct Application of Eigenvalue Theory to a Multi-conductor System

In a multi-conductor system, define matrix $[P]$ as:

$$[P] = [Z] \cdot [Y] \tag{3.10}$$

where
$[Z]$ = series impedance matrix of a multi-conductor
$[Y]$ = shunt admittance matrix of a multi-conductor

Assume that eigenvalue (Q) and an eigenvector matrix $[A]$ are obtained. Then, the propagation constant is calculated by the following matrix equations.

$$(\gamma) = (Q)^{1/2} \tag{3.11}$$

$$[\Gamma] = [P]^{1/2} = [A] \cdot [Q]^{1/2} \cdot [A]^{-1} = [A] \cdot [\gamma] \cdot [A]^{-1} \tag{3.12}$$

where

$$[\gamma] = [U](\gamma), [Q] = [U](Q)$$

γ_k = modal propagation constant for mode k
Γ_{ij} = propagation constant in a phase domain for phases i and j

The modal propagation constant involves real and imaginary parts.

$$\gamma = \alpha + j\beta \tag{3.13}$$

where
α = modal attenuation constant
β = modal phase constant

In a semi-infinite line with a voltage source (E) at the sending end, voltage (V) at distance x from the sending end is given by:

$$(V) = \exp(-[\Gamma] \cdot x) \cdot (E) \tag{3.14}$$

The above equation is theoretically or mathematically correct, but it cannot be calculated in practice unless $\exp(-[\Gamma] \cdot x)$ is given as a numeric value. Since it is calculated by (3.12) and (3.9), the above equation is rewritten in the following form which can be calculated in practice.

$$(V) = [A] \cdot \exp(-[\gamma] \cdot x) \cdot \ [A]^{-1}(E) \tag{3.15}$$

3.1.4 Modal Theory

The eigenvalues/vectors theory explained in the previous section is just a mathematical method to diagonalize a given matrix containing off-diagonal elements. When we deal with an "n" conductor electric circuit, impedance Z and admittance Y are given as an "$n \times n$" matrix with non-zero off-diagonal elements due to mutual coupling between the conductors. Correspondingly, matrix $P = \Gamma^2 = Z \cdot Y$ becomes an "$n \times n$" matrix. By applying the eigenvalues/vectors theory, the matrix P is diagonalized, and the diagonal elements are the eigenvalues of P which is expressed as a diagonal matrix $Q = \gamma^2$, where γ_k is a propagation constant of mode "k" in the diagonalized domain called the "modal domain".

The above means physically that mutual coupling in the n-conductor system is eliminated by the diagonalization, and wave propagation defined by γ or Q is independent. Thus, the wave propagation characteristic in the modal domain is discussed as that on a single conductor which is explained in many textbooks. Voltages and currents in the original n-conductor system are obtained by transforming the voltage and currents in the modal domain (modal voltages and currents) back to the phase domain by using the transformation matrix given as the eigenvectors.

Well-known symmetrical component theories such as Fortesque [4] and Clark [5] transformation are, in fact, just a specific case of the modal theory. Such a transformation can diagonalize only a balanced (transposed) three-phase line, because the theories are based on a balanced three-phase a.c. source.

In summary, the modal theory is an application of an eigenvalues/vectors theory to an "n"-conductor system to make it be independent n conductors. Then, by applying an electric circuit theory, the wave propagation characteristics on the multi-conductors are analyzed as "n" single conductors. The transformation matrix between phase and modal domains explains the voltage and current distribution on the n-conductors.

3.1.4.1 Voltage

Multiplying transformation matrix $A^{-1}([A]^{-1})$ with (3.15), the following equation is obtained.

$$A^{-1} \cdot V = \exp(-\gamma \cdot x)A^{-1} \cdot E \tag{3.16}$$

Define modal voltages v and e as:

$$v = A^{-1} \cdot V, e = A^{-1} \cdot E \tag{3.17}$$

Substituting the above equations into (3.16),

$$v = \exp(-\gamma \cdot x) \cdot e \tag{3.18}$$

Or

$$v_k = \exp(-\gamma_k \cdot x) \cdot e_k \; ; \; k = 1, \; 2 \cdots, \; n \tag{3.19}$$

In the above equation, all the elements (voltages and propagation constant) are defined in the modal domain. It should be noted that the equation is in the same form as (3.14) in the actual phase domain. Similarly, any law such as Ohm's law can be applied to the modal quantities. Thus, we can solve voltages and currents for each mode, that is, for mode 1 to mode n in the modal domain by applying a theory of a single-conductor electric circuit, because each modal quantity is independent. Once the modal voltages are solved, then the actual phase voltages are given in the following form from (3.17).

$$V = A \cdot v \tag{3.20}$$

It is noteworthy that the solution of (3.19) is obtained by "n" calculations, while that of (3.14) in the actual domain requires "$n \times n$" calculations because Γ is an "$n \times n$" full matrix. Also, the wave propagation characteristic in the modal domain can be discussed exactly in the same manner as that in a single-conductor circuit.

3.1.4.2 Current

Let us consider currents in the modal domain. The propagation constant Γ_i for the current is defined similarly to that for the voltage in (3.10).

$$P_i = \Gamma_i^2 = Y \cdot Z \tag{3.21}$$

Because the matrices Z and Y are symmentrical, the transposed matrix of P_i is given as:

$$P_{it} = (Y \cdot Z)_t = Z_t \cdot Y_t = Z \cdot Y = P \text{ or } P_t = P_i \tag{3.22}$$

Assume that the eigenvector matrix of $P_i = P_t$ is given as B. Then,

$$P_t = B \cdot Q' \cdot \mathrm{B}^{-1} \tag{3.23a}$$

or

$$Q' = \mathrm{B}^{-1} \cdot P_t \cdot B \tag{3.23b}$$

where Q' is the eigenvalue matrix of P_t.

Taking the transposition at the characteristic equation of P_t,

$$\det(P_t - Q'_k \cdot U)_t = \det(P - Q'_k \cdot U) = 0 \tag{3.24}$$

It is clear from (3.7) that

$$Q'_k = Q_k \tag{3.25}$$

The above equation means that the eigenvalue of matrix $P = Z \cdot Y$ is the same as those of $P_i = Y \cdot Z = P_t$. Thus, the propagation characteristic of a current along a conductor is the same as that of a voltage along the conductor. This means that the wave propagation on a distributed-parameter line is of transverse electromagnetic (TEM) mode. However, the current transformation matrix B (= eigenvector matrix of P_t) is different from the voltage transformation matrix A, and thus current distribution on a multi-conductor differs from voltage distribution. Taking the transposition of (3.23a) considering (3.25),

$$P = B_t^{-1} \cdot Q \cdot B_t \tag{3.26}$$

A comparison of the above equation with (3.3) gives the following relation.

$$B_t^{-1} = A,\ B_t = \mathrm{A}^{-1} \tag{3.27}$$

For the eigenvector is not unique but proportional, the above relation should be defined exactly by introducing an arbitrary diagonal matrix D.

$$B_t^{-1} = A \cdot D^{-1},\ B_t = D \cdot A^{-1} \tag{3.28}$$

If we adopt a unit matrix U instead of D, then (3.27) is obtained.

By adopting the current transformation matrix B, actual phase current I and modal current i are related by:

$$i = B^{-1}I,\ I = B \cdot i \tag{3.29}$$

3.1.4.3 Power

The total instantaneous power in a multi-conductor line is given as $(V)_t \cdot (I)$. By applying (3.20) and (3.29),

$$(V)_t \cdot I = (A \cdot v)_t \cdot B \cdot i = v_t \cdot A_t \cdot B \cdot i$$

From (3.27),

$$A_t \cdot B = B^{-1} \cdot B = U$$

Thus,

$$V_t \cdot I = v_t \cdot i \tag{3.30}$$

The above equation means that power is not changed either in an actual phase domain or in a modal domain. This is called the "Power Invariant" condition.

3.1.5 Formulation of Multi-conductor Voltages and Currents

In an "n" conductor system as shown in Figure 3.1, voltages (V) and currents (I) along the conductor are related in the following form.

$$d(V)/dx = -[Z] \cdot (I) \tag{3.31}$$

$$d(I)/dx = -[Y] \cdot (V) \tag{3.32}$$

where

$[Z] = n \times n$ impedance matrix

$[Y] = n \times n$ admittance matrix

$(V),\ (I)$ = voltages and currents in phase domain

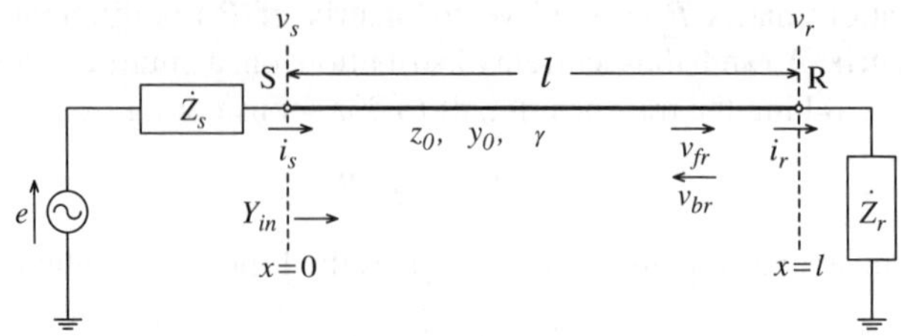

Figure 3.1 A distributed-parameter circuit

Multiplying the voltage transformation matrix $[A]^{-1}$ with (3.31) and substituting the relation $I = B \cdot i$, (3.29), into the right-hand side,

$$d\{[A]^{-1}(V)\}/dx = d(v)/dx = -[A]^{-1} \cdot [Z] \cdot [B] \cdot (i) = -[z] \cdot (i)$$

Thus,

$$d(v)/dx = -[z] \cdot (i) \tag{3.33}$$

Similarly,

$$d(i)/dx = -[y] \cdot (v) \tag{3.34}$$

where

$$[z] = [A]^{-1}[Z] \cdot [B] = \text{modal impedance matrix} \tag{3.35}$$

$$[y] = [B]^{-1}[Y] \cdot [A] = \text{modal admittance matrix} \tag{3.36}$$

$$(v),\ (i) = \text{modal voltages and currents}$$

For the matrices $[z]$ and $[y]$ are diagonal in the modal domain, (3.33) and (3.34) are rewritten in the following form for each mode.

$$dv_k/dx = -z_k \cdot i_k \tag{3.37a}$$

$$di_k/dx = -y_k \cdot v_k \tag{3.37b}$$

where $k = 1, 2, \cdots, n$.

The above equation is the same as that for a single conductor. This fact means that a well-known theory of a single-phase distributed-parameter line is directly applied to the above modal equation. Thus, the wave propagation characteristic of the multi-conductor system is discussed in the modal domain in the same manner as that of the single conductor.

We know the solution of (3.37a,b) is given as:

$$v = \exp(-\gamma \cdot x) \cdot v_f + \exp(\gamma \cdot x) \cdot v_b \tag{3.38a}$$

$$i = \exp(-\gamma \cdot x) \cdot i_f + \exp(\gamma \cdot x) \cdot i_b \tag{3.38b}$$

where

$$\gamma = \sqrt{z \cdot y} = \text{propagation constant in a modal domain} \tag{3.39}$$

Equation (3.37) is a simultaneous equation. Differentiating the voltage in (3.38) by x, and substituting into (3.37a), the following current is obtained.

$$-z \cdot i = -\gamma \cdot \exp(-\gamma \cdot x) \cdot v_f + \gamma \cdot \exp(\gamma x) \cdot v_b$$

$$\therefore i = z^{-1} \gamma \{\exp(-\gamma \cdot x) \cdot v_f - \exp(\gamma x) \cdot v_b\}$$

Considering the following relation,

$$z^{-1} \cdot \gamma = \sqrt{z \cdot y}/z = \sqrt{y/z} = y_0, \ y_0 \cdot v_f = i_f, \ y_0 v_b = i_b \tag{3.40}$$

the following solution of the current is given.

$$i = y_0\{\exp(-\gamma \cdot x) \cdot v_f - \exp(\gamma \cdot x) \cdot v_b\} \tag{3.41}$$

where

$$y_0 = \sqrt{y/z} = \text{characteristic admittance (S)} \tag{3.42a}$$

$$z_0 = 1/y_0 = \sqrt{z \cdot y} = \text{characteristic impedance } (\Omega) \tag{3.42b}$$

Equations (3.38) and (3.41) are a general solution of a voltage and a current in a modal domain which are exactly the same as those for a single conductor. Variables v_f and v_b are determined by given boundary conditions.

Once the voltage and current solutions in the modal domain are obtained in the same manner as that of the single conductor, the solution in a phase domain are easily given by transforming those into the phase domain by (3.17), (3.20) and (3.29).

3.1.6 Boundary Conditions and Two-port Theory

Let us consider a circuit illustrated in Figure 3.1 with a source impedance Z_s, a terminating impedance Z_r and a sinusoidal a.c. source e.

3.1.6.1 Reflection Coefficient Method

At $x = l$, the receiving end, (3.38) and (3.41) become,

$$v(x = l) = v_r = v_f(x = l) + v_b(x = l) = v_{fr} + v_{br} \tag{3.43a}$$

$$i(x = l) = i_r = y_0(v_{fr} - v_{br}) \tag{3.43b}$$

where

$$v_{fr} = \exp(-\gamma \cdot l)\, v_f \tag{3.44a}$$

$$v_{br} = \exp(\gamma \cdot l)\, v_b \tag{3.44b}$$

Also there exists the following relation at the receiving end.

$$v_r = Z_r \cdot i_r \tag{3.45}$$

Assume that v_{fr} and v_{br} are related by:

$$v_{br} = K_r \cdot v_{fr} \tag{3.46}$$

The above equation means that v_{br} is defined as the forward traveling wave v_{fr} mutiplied by the coefficient K_r. The coefficient is called the "reflection coefficient", and has been learnt as reflection and refraction of light at the boundary of air and water in a basic physics lesson at school.

Substituting the above equation into the (3.43a),

$$v_r = (U + K_r)\, v_{fr} \tag{3.47}$$

where U is the unit matrix, and $U = 1$ in a single conductor.

Substituting (3.43b) into (3.45) considering (3.46), the following relation is obtained.

$$v_r = Z_r \cdot i_r = Z_r \cdot y_0(U - K_r)\, v_{fr} \tag{3.48}$$

From (3.47) and (3.48),

$$(U + K_r)v_{fr} = Z_r y_0(U - K_r)\, v_{fr}$$

$$\therefore (Z_r \cdot y_0 + U)K_r = Z_r y_0 - U$$

Thus, K_r is defined as:

$$K_r = (Z_r \cdot y_0 + U)^{-1}(Z_r y_0 - U) = (Z_r y_0 - U)\,(Z_r y_0 + U)^{-1} \tag{3.49}$$

Rewriting the above equation by using $Z_r{}^{-1} = Y_r$ and $y_0{}^{-1} = z_0$,

$$K_r = (y_0 + Y_r)^{-1}(y_0 - Y_r) = (y_0 - Y_r)(y_0 + Y_r)^{-1} \tag{3.50}$$

The above derivation is rewritten in a matrix form where the sequence of product and division cannot be changed. In a single conductor,

$$K_r = (y_0 - Y_r)/(y_0 + Y_r) = (Z_r - z_0)/(Z_r + z_0) \tag{3.51}$$

At $x = 0$, the sending end, the following relation is obtained from (3.38) and (3.41).

$$v_s = v(x = 0) = v_f + v_b \tag{3.52a}$$

$$i_s = i(x = 0) = y_0(v_f + v_b) \tag{3.52b}$$

From (3.44b),

$$v_b = \exp(-\gamma \cdot l) \cdot v_{br}$$

Substituting (3.46) and (3.44a) into the above equation,

$$v_b = \exp(-\gamma \cdot l) \cdot K_r \cdot v_{fr} = \exp(-\gamma \cdot l) \cdot K_r \cdot \exp(-\gamma \cdot l) \cdot v_f = C \cdot v_f \tag{3.53}$$

where $C = \exp(-\gamma \cdot l) \cdot K_r \cdot \exp(-\gamma \cdot l)$.

Substituting (3.53) into (3.52a),

$$v_s = v_f + C \cdot v_f = (U + C)v_f$$

$$i_s = y_0(U - C)v_f = y_0(U - C)(U + C)^{-1}v_s = Y_{in} \cdot v_s \tag{3.54}$$

where

$$Y_{in} = y_0(U - C)(U + C)^{-1} = \text{input admittance} \tag{3.55}$$

In Figure 3.1, the following relation is obtained by using Y_{in},

$$e = (Z_s + Y_{in}^{-1})i_s = (Z_s + Y_{in}^{-1}) \cdot Y_{in} \cdot v_s = (Z_s Y_{in} + U) \cdot v_s$$

$$\therefore v_s = (Z_s Y_{in} + U)^{-1} \cdot e \tag{3.56}$$

Substituting (3.53) into (3.38),

$$v = [\exp(-\gamma x) + \exp\{-\gamma(l - x)\} \cdot K_r \cdot \exp(-\gamma l)](U + C)^{-1} v_s \tag{3.57}$$

Similarly,

$$i = y_0[\exp(-\gamma x) - \exp\{-\gamma(l - x)\} \cdot K_r \cdot \exp(-\gamma l)](U + C)^{-1} v_s \tag{3.58}$$

At the receiving end, the voltage v_r and the current i_r are given by taking $x = l$ in (3.57) and (3.58).

$$v_r = (U + K_r) \cdot \exp(-\gamma l) \cdot (U + C)^{-1} v_s \tag{3.59a}$$

$$i_r = y_0(U - K_r) \cdot \exp(-\gamma l) \cdot (U + C)^{-1} v_s \tag{3.59b}$$

The receiving-end current is also calculated by (3.45).

The advantage of the reflection coefficient method explained above is being able to calculate a voltage and a current at an arbitrary position of a line, that is, at the distance x from the sending end. A well-known approach based on a two-port theory, such as a four-terminal parameter (F-parameter) method, can calculate voltages and currents only at both ends of a given circuit. It should be noted that the reflection coefficient method adopts $\exp(-\gamma \cdot x)$ to evaluate the voltage and the current. In a long line (large x) and in a high frequency region (large γ), γx becomes very large, and $\exp(-\gamma x)$ converges to zero, and thus the method is numerically very stable. On the contrary, the F-parameter method involves $\sinh(\gamma x)$ and $\cosh(\gamma x)$ which diverges for large γx.

3.1.6.2 F-parameter Method

In a single conductor circuit, the following F-parameter is well-known.

$$\begin{pmatrix} v_s \\ i_s \end{pmatrix} = \begin{bmatrix} f_1 & f_2 \\ f_3 & f_4 \end{bmatrix} \begin{pmatrix} v_r \\ i_r \end{pmatrix} \tag{3.60}$$

where

$$f_1 = \cosh(\theta), \quad f_2 = \sinh(\theta) \cdot z_0, \quad f_3 = y_0 \cdot \sinh(\theta), \quad f_4 = y_0 \cdot \cosh(\theta) \cdot z_0 \tag{3.61}$$

and $\theta = \gamma \cdot l$

The above equation is also applicable to a multi-conductor circuit in a modal domain. By applying the relation in (3.17), (3.20) and (3.29), the above equation is rewritten as:

$$V_s = A \cdot v_s = A \cdot f_1 \cdot v_r + A \cdot f_2 \cdot i_r = A \cdot f_1 \cdot A^{-1} \cdot A \cdot v_r + A \cdot f_2 \cdot B^{-1} \cdot B \cdot i_r$$

$$I_s = B \cdot i_s = B \cdot f_3 \cdot A^{-1} \cdot A \cdot v_r + B \cdot f_4 \cdot B^{-1} \cdot B \cdot i_r$$

Thus, the following F-parameter for a multi-conductor circuit is obtained.

$$\begin{pmatrix} V_s \\ I_s \end{pmatrix} = \begin{bmatrix} F_1 & F_2 \\ F_3 & F_4 \end{bmatrix} \begin{pmatrix} V_r \\ I_r \end{pmatrix} \tag{3.62}$$

where

$$F_1 = A \cdot f_1 \cdot \mathrm{A}^{-1}, \ F_2 = A \cdot f_2 \cdot \mathrm{B}^{-1}, \ F_3 = B \cdot f_3 \cdot \mathrm{A}^{-1}, \ F_4 = B \cdot f_4 \cdot \mathrm{B}^{-1}$$

and

$$V_r = A \cdot v_r, \ I_r = B \cdot i_r \tag{3.63}$$

From (3.9) and (3.61), the above equation is rewritten as:

$$F_1 = \cosh[A \cdot (\gamma \cdot l) \cdot A^{-1}] = \cosh([\Gamma \cdot l]) \tag{3.64a}$$

$$F_2 = A \cdot \sinh(\gamma \cdot l)\, z_0 \cdot B^{-1} = A \cdot \sinh(\gamma \cdot l) \cdot A^{-1} \cdot A \cdot z_0 \cdot B^{-1} = \sinh([\Gamma \cdot l]) \cdot Z_0 \tag{3.64b}$$

$$F_3 = B \cdot y_0 \cdot \sinh(\gamma \cdot l) \cdot A^{-1} = B \cdot y_0 \cdot A^{-1} \cdot A \sinh(\gamma \cdot l) A^{-1} = Y_0 \cdot \sinh([\Gamma \cdot l]) \tag{3.64c}$$

$$\begin{aligned} F_4 &= B \cdot y_0 \cosh(\gamma \cdot l)\, z_0 \cdot B^{-1} = B \cdot y_0 \cdot A^{-1} \cdot \mathrm{A} \cosh(\gamma \cdot l) \cdot \mathrm{A}^{-1} \cdot A \cdot z_0 \cdot B^{-1} \\ &= Y_0 \cdot \cosh([\Gamma \cdot l]) \cdot Z_0 \end{aligned} \tag{3.64d}$$

where

$$\Gamma = A \cdot \gamma \cdot \mathrm{A}^{-1} = \text{propagation constant in an actual phase domain} \tag{3.65}$$

$$Z_0 = A \cdot z_0 \cdot B^{-1} = \text{characteristic impedance in an actual phase domain} \tag{3.66a}$$

$$Y_0 = Z_0{}^{-1} = B \cdot y_0 \cdot A^{-1} = \text{characteristic admittance in an actual phase domain} \tag{3.66b}$$

Rewriting the above equation, the modal component is given by:

$$z_0 = A^{-1} \cdot Z_0 \cdot B = \text{modal characteristic impedance} \tag{3.67a}$$

$$y_0 = B^{-1} \cdot Y_0 \cdot A = \text{modal characteristic admittance} \tag{3.67b}$$

Similarly, the modal impedance z and admittance y are obtained as already explained in (3.35) and (3.36).

$$z = A^{-1} \cdot Z \cdot B, y = B^{-1} \cdot Y \cdot A$$

Rewriting the equations, the impedance Z and admittance Y in an actual phase domain are related to the modal ones by:

$$Z = A \cdot z \cdot B^{-1}, Y = B \cdot y \cdot A^{-1} \tag{3.68}$$

It should be noted that the series impedance Z and the shunt admittance Y of a multi-conductor are calculated first by using the formulas explained in Chapter 2. Then, the modal impedance z and admittance y are evaluated based on (3.35) and (3.36). By using z and y, the modal propagation constant γ is calculated by (3.39), and the modal characteristic impedance z_0 and y_0 are calculated based on (3.42).

Finally, the propagation constant Γ, the characterisitic impedance Z_0 and admittance Y_0 in an actual domain are calculated by (3.65) and (3.66).

In a single conductor and also in a modal domain, the sequence of product and division can be changed. Thus, (3.61) is rewritten as follows:

$$f_2 = \sinh(\theta) \cdot z_0 = z_0 \cdot \sinh(\theta) \tag{3.69a}$$

$$f_3 = y_0 \cdot \sinh(\theta) = \sinh(\theta) \cdot y_0 \tag{3.69b}$$

$$f_4 = y_0 \cdot \cosh(\theta) \cdot z_0 = y_0 \cdot z_0 \cdot \cosh(\theta) = \cosh(\theta) = f_1 \tag{3.69c}$$

The sequence cannot be changed in an actual phase domain for a multi-conductor circuit. However, the following relation is given from (3.63) and (3.69).

$$F_2 = A \cdot z_0 \cdot \sinh(\theta) \cdot B^{-1} = A \cdot z_0 \cdot B^{-1} \cdot B \sinh(\theta) \cdot B^{-1} = Z_0 \cdot A_t^{-1} \cdot \sinh(\theta)_t \cdot A_t$$
$$= Z_{0t}\{A \cdot \sinh(\theta) \cdot A^{-1}\}_t = \{\sinh([\Gamma \cdot l]) \cdot Z_0\}_t = F_{2t} \tag{3.70a}$$

$$F_3 = B \cdot \sinh(\theta) \cdot B^{-1} \cdot B \cdot y_0 \cdot A^{-1} = A_t^{-1} \sinh(\theta) \cdot A_t \cdot Y_0 = \sinh([\Gamma \cdot l])_t \cdot Y_{0t}$$
$$= \{Y_0 \cdot \sinh([\Gamma \cdot l])\}_t = F_{3t} \tag{3.70b}$$

$$F_4 = B \cdot \cosh(\theta) \cdot B^{-1} = A_t^{-1} \cosh(\theta) \cdot A_t = \{A \cdot \cosh(\theta) \cdot A^{-1}\}_t = F_{1t} \tag{3.70c}$$

The above equations mean that:

- F_2 and F_3 are symmetrical matrices.
- F_4 is a transposed matrix of F_1, although $f_1 = f_4$ in a single conductor and in a modal domain for a multi-conductor.

By applying the F-parameter method, the voltage and current solution in the circuit of Figure 3.1 are easily obtained. In the figure, the following current relation is given.

$$i_s = Z_s^{-1}(e - v_s) = Y_s(e - v_s) \tag{3.71a}$$

$$i_r = Z_r^{-1} v_r = Y_r v_r \tag{3.71b}$$

By eliminating i_s and i_r in (3.60) considering the above equation, the voltage v_s and v_r are evaluated as follows:

$$v_s = \{U + Z_s(f_3 + f_4 Y_r)(f_1 + f_2 Y_r)^{-1}\}^{-1} \cdot e \tag{3.72a}$$

$$v_r = (f_1 + f_2 Y_r)^{-1} \cdot v_s \tag{3.72b}$$

Exactly the same form of the voltages and currents as (3.71a,b) and (3.72a,b) are obtained in a multi-conductor circuit by adopting (3.62) provided that the sequence of matrix product is not changed.

3.1.6.3 Z- and Y-parameter Methods

Rewriting (3.60) or (3.62) to express voltages as a function of currents,

$$i_r = -y_0 \sinh(\theta) \cdot v_s + y_0 \cosh(\theta) z_0 \cdot i_s$$

$$\therefore v_s = \{y_0 \sinh(\theta)\}^{-1} \cdot y_0 \cdot \cosh(\theta) z_0 \cdot i_s - \{y_0 \sinh(\theta)\}^{-1} \cdot i_r$$

$$= \{\sinh(\theta)\}^{-1} \cdot \cosh(\theta)\, z_0 \cdot i_s - \{\sinh(\theta)\}^{-1} \cdot y_0^{-1} \cdot i_r$$

$$= \coth(\theta) \cdot z_0 \cdot i_s - \mathrm{cosech}(\theta) \cdot z_0 \cdot i_r$$

Substituting the above results into the first part of (3.60),

$$v_r = \mathrm{cosech}(\theta) \cdot z_0 \cdot i_s - \coth(\theta) \cdot z_0 \cdot i_r$$

In Figure 3.1, currents are defined to be positive along the positive direction of the x-axis. Defining the currents as in Figure 3.2, the above equations are rewritten as:

$$\begin{pmatrix} v_s \\ v_r \end{pmatrix} = \begin{bmatrix} z_{11} & z_{12} \\ z_{12} & z_{11} \end{bmatrix} \begin{pmatrix} i_s \\ i_r \end{pmatrix} \tag{3.73}$$

where

$$z_{11} = \coth(\theta) \cdot z_0 \tag{3.74a}$$

$$z_{12} = \mathrm{cosech}(\theta) \cdot z_0 \tag{3.74b}$$

The above is called the "impedance parameter (Z-parameter)", because the coefficients z_{11} and z_{12} are in the dimension of the impedance. The Z-parameter is a symmetrical matrix.

In most circuit analysis, currents are to be obtained from given voltages. Taking the inverse of (3.73), the following "admittance parameter (Y-parameter)" is obtained.

$$\begin{pmatrix} i_s \\ i_r \end{pmatrix} = \begin{bmatrix} y_{11} & y_{12} \\ y_{12} & y_{11} \end{bmatrix} \begin{pmatrix} v_s \\ v_r \end{pmatrix} \tag{3.75}$$

where

$$y_{11} = y_0 \coth(\theta) \tag{3.76a}$$

$$y_{12} = y_0\, \mathrm{cosech}(\theta) \tag{3.76b}$$

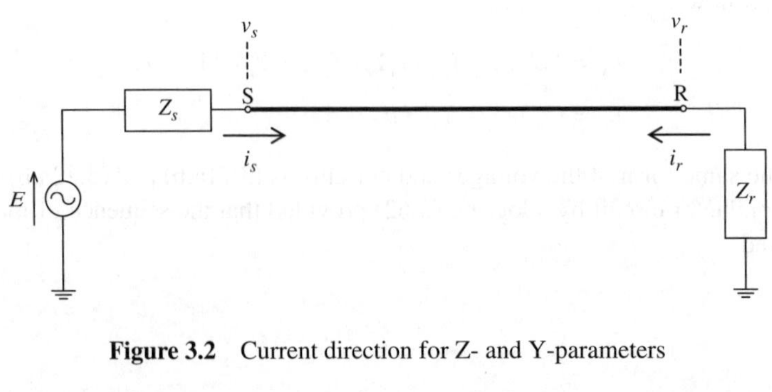

Figure 3.2 Current direction for Z- and Y-parameters

Let us obtain voltage and currents in Figure 3.2 by applying the Y-parameter method. Considering the sign of i_r in (3.71b) being changed, the following currents are given.

$$i_s = Y_s(e - v_s) \tag{3.77a}$$

$$i_r = -Y_r v_r \tag{3.77b}$$

By applying (3.75), the following results are obtained.

$$v_r = (y_{11} + Y_r)^{-1} y_{12} \cdot v_s \tag{3.78a}$$

$$v_s = \{Y_s + y_{11} - y_{12}(y_{11} + Y_r)^{-1} \cdot y_{12}\}^{-1} \cdot Y_s \cdot e \tag{3.78b}$$

The Z- and Y-parameter methods adopt converging functions coth(θ) and cosech(θ), and thus are numerically stable in a high frequency region. On the contrary, the F-parameter method is numerically stable in a low frequency region. Thus, the Z- and Y-parameter methods are effective for transient and high frequency phenomena, while the F-parameter method is advantageous for a low frequency steady-state analysis.

3.1.7 Problems

3.1 Obtain a condition of reciprocity in (3.62).
3.2 Prove a multi-conductor in Figure 3.1 is symmetrical based on (3.62).
3.3 Obtain the eigenvalues and vectors of the following matrices.

$$\text{(a)}\begin{bmatrix} P_{11} & P_{12} & P_{13} \\ P_{21} & P_{22} & P_{23} \\ P_{31} & P_{32} & P_{33} \end{bmatrix} \quad \text{(b)}\begin{bmatrix} 35 & 8 & 5 \\ 8 & 32 & 8 \\ 5 & 8 & 35 \end{bmatrix} \quad \text{(c)}\begin{bmatrix} 7 & \sqrt{3} & 1 \\ \sqrt{3} & 7 & \sqrt{3} \\ 1 & \sqrt{3} & 7 \end{bmatrix}$$

Solution

3.1 $F_2 \cdot F_3 - F_1 \cdot F_{4t} = U$

3.2 Change ports, that is, $\begin{pmatrix} v_r \\ i_r \end{pmatrix} = [F'] \begin{pmatrix} v_s \\ i_s \end{pmatrix}$

3.3 (a) $Q_1, Q_{13} = -(b_2/2) \pm \sqrt{(b_2/2)^2 - c_2}$

$$Q_2 = -(a_1/3) + \left(\sqrt{b_1^3 + c_1^2} - c_1\right)^{1/3} - \left(\sqrt{b_1^3 - c_1^2} + c_1\right)^{1/3}$$

$$A_{1k} = 1, A_{2k} = \{P_{23}(P_{11} - Q_k) - P_{13}P_{21}\}/d$$

$$A_{3k} = \{P_{12}P_{21} - (P_{11} - Q_k)(P_{22} - Q_k)\}/d, d = P_{13}(P_{22} - Q_k) - P_{12}P_{23}$$

$$a_1 = -(P_{11} + P_{22} + P_{33}),$$

$$a_2 = P_{11}P_{22} + P_{22}P_{33} + P_{33}P_{11} - (P_{12}P_{21} + P_{23}P_{32} + P_{13}P_{31})$$

$$a_3 = P_{33}P_{12}P_{21} + P_{22}P_{13}P_{31} + P_{11}P_{23}P_{32} - (P_{12}P_{23}P_{31} + P_{13}P_{32}P_{21} + P_{11} \cdot P_{22} \cdot P_{33})$$

$$b_1 = a_2/3 - a_1^3/9, \quad c_1 = \{a_3 - (a_1a_2/3) + (2a_1^3/27)\}/2$$

$$b_2 = a_1 + Q_2, c_1 = b_2Q_3 + a_2$$

(b) $Q_1 = 48,\ Q_2 = 30,\ Q_3 = 24$

$$A_{11} = A_{21} = A_{31} = 1,\ A_{12} = 1,\ A_{22} = 0,\ A_{32} = -1,\ A_{13} = A_{33} = 1,\ A_{23} = -2$$

(c) $Q_1 = 10,\ Q_2 = 6,\ Q_3 = 5$

$$A_{11} = A_{12} = A_{13} = 1,\ A_{21} = 2/\sqrt{3},\ A_{22} = 0,\ A_{23} = -\sqrt{3},$$

$$A_{31} = A_{33} = 1,\ A_{32} = -1$$

3.2 Basic Characteristics of Wave Propagation on Single-phase SC Cables

3.2.1 Basic Propagation Characteristics for a Transient

As explained in Chapter 2, a single-phase single-core coaxial (SC) cable is composed of two conductors, and thus the modal theory explained in the previous section is necessary to discuss its wave propagation characteristics.

When discussing a transient, the frequency involved is high, and the so-called "coaxial mode" of wave propagation is completed on the SC cable as illustrated in Figure 3.3 and Figure 3.4. That is, all the currents flowing through a core conductor return only through a sheath conductor as illustrated in Figure 3.4. Thus, no current appears outside of the cable, and this means physically that no outer medium (air or soil) affects the wave propagation along the cable. This is the reason why the SC cable has been widely used as a telecommunication cable since the 1860s. In a power cable, the sheath conductor works as a metallic shield which reduces electromagnetic induction from the power cable to an adjacent conducting material, such as a gas pipe and a telecommunication line.

The above explained situation is expressed by a current transformation matrix as follows:

$$(I) = [B](i) \quad \text{or} \quad \begin{pmatrix} I_c \\ I_s \end{pmatrix} = \begin{bmatrix} 0 & 1 \\ 1 & -1 \end{bmatrix} \begin{pmatrix} i_0 \\ i_1 \end{pmatrix} \tag{3.79}$$

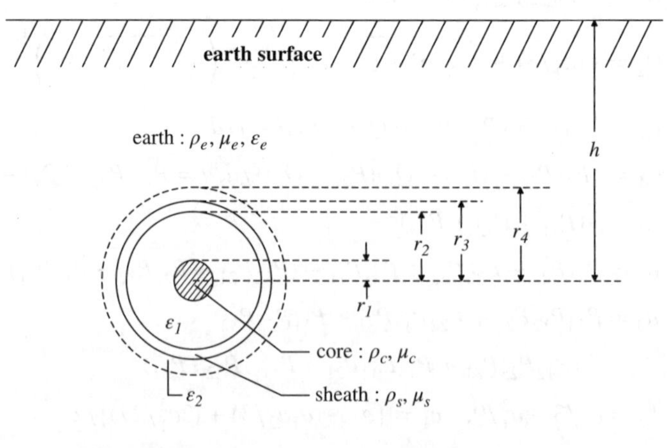

Figure 3.3 Cross-section of an underground SC cable

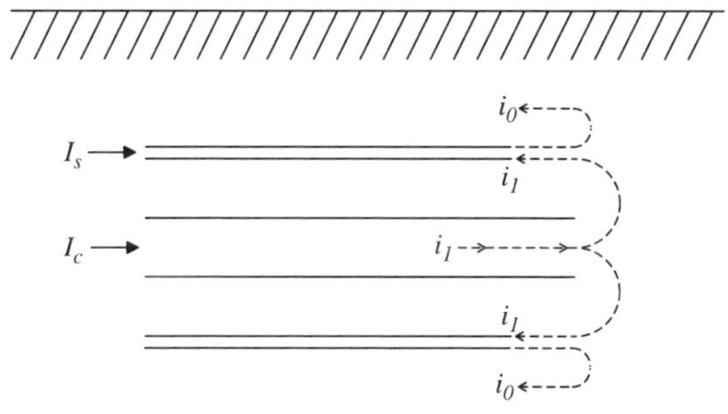

Figure 3.4 Modal current distribution along an SC cable

The above equation means physically that the coaxial mode current i_1 flows through the core and returns through the sheath. Thus, the core current I_c is composed only of the coaxial mode current, that is, $I_c = i_1$. The sheath current I_s is composed of the coaxial mode current i_1 and the earth-return mode current i_0, that is, $I_s = i_0 - i_1$, as illustrated in Figure 3.4.

Then, the voltage transformation matrix A^{-1} is given from (3.79) by:

$$A^{-1} = B_t = \begin{bmatrix} 0 & 1 \\ 1 & -1 \end{bmatrix} \tag{3.80}$$

Thus, the modal voltage v_0 and v_1 are given as:

$$\begin{pmatrix} v_0 \\ v_1 \end{pmatrix} = A^{-1}(V) = \begin{bmatrix} 0 & 1 \\ 1 & -1 \end{bmatrix} \begin{pmatrix} V_c \\ V_s \end{pmatrix} \tag{3.81}$$

The above equation means that the mode 0 (earth-return) voltage v_0 is given as the sheath voltage V_s, and the coaxial mode (mode 1) voltage v_1 is given as $v_1 = V_c - V_s$.

The frequency at which the coaxial mode becomes dominant is determined by using the penetration depth h_s defined as:

$$h_s = \sqrt{\rho_s/(j\omega\mu_s)} = H_s \cdot \exp(-j\pi/4) \tag{3.82}$$

where

$$H_s = |h_s| = \sqrt{\rho_s/(\omega\mu_s)}$$

When the above penetration depth H_s is smaller than the thickness of the metallic sheath, that is, $r_3 - r_2$ in Figure 3.3, the metallic sheath becomes a shield and the coaxial mode of wave propagation becomes dominant. Thus,

$$r_3 - r_2 > H_s = \sqrt{\rho_s/\omega\mu_s} \text{ or } f > f_c$$

where

$$f_c = \rho_s/\{2\pi\mu_s(r_3 - r_2)^2\} = \text{critical frequency} \tag{3.83}$$

In general, $\mu_s = \mu_0$. Then, f_c becomes approximately:

$$f_c \fallingdotseq 1.27 \times 10^5 \rho_s/(r_3 - r_2)^2$$

Having defined the voltage and current transformation matrices, the modal impedance and admittance are calculated from the transformation matrices.

In the frequency region $f > f_c$, the series impedance Z and the shunt admittance Y are given approximately in the following form [see (2.12) for $z_{2m} \fallingdotseq 0$, and (2.23) in Chapter 2].

$$[Z] = \begin{bmatrix} Z_c & Z_s \\ Z_s & Z_s \end{bmatrix} \tag{3.84}$$

$$[Y] = j\omega[P]^{-1} \tag{3.85a}$$

$$[P] = \begin{bmatrix} p_c + p_s & p_s \\ p_s & p_s \end{bmatrix} \tag{3.85b}$$

$$[Y] = \left(\frac{j\omega}{p_c p_s}\right) \begin{bmatrix} p_s & -p_s \\ -p_s & p_c + p_s \end{bmatrix} \tag{3.85c}$$

where
$p_c = (1/2\pi\varepsilon_0\varepsilon_1) \cdot \ln(r_2/r_1)$
$p_s = (1/2\pi\varepsilon_0\varepsilon_2) \cdot \ln(r_4/r_3)$

The modal impedance z and admittance y are obtained from the above equation as follows:

$$[z] = A^{-1} \cdot Z \cdot B = \begin{bmatrix} Z_s & 0 \\ 0 & Z_c - Z_s \end{bmatrix} \quad \text{or } z_0 = Z_s, z_1 = Z_c - Z_s \tag{3.86}$$

$$[y] = B^{-1} \cdot Y \cdot A = j\omega \begin{bmatrix} 1/p_s & 0 \\ 0 & 1/p_c \end{bmatrix} \tag{3.87}$$

From (3.86) and (3.87), the modal propagation constant $\gamma_k (k = 0 \text{ and } 1)$ and modal characteristic impedance z_{0k} are obtained.

$$\gamma_0 = \sqrt{j\omega Z_s/p_c},\ \gamma_1 = \sqrt{j\omega(Z_c - Z_s)/p_c} \tag{3.88a}$$

$$z_{00} = \sqrt{Z_s p_s/j\omega},\ z_{01} = \sqrt{(Z_c - Z_s)p_c/j\omega} \tag{3.88b}$$

An approximate formula of the impedances Z_c and Z_s in a high frequency region are given in the following form [3, 6].

$$Z_s \fallingdotseq j\omega(\mu_0/2\pi)[\ln\{2(h + h_e)/r_4\} - 4h/3h_e - 0.077 + \ln(r_4/r_3)]$$
$$\fallingdotseq j\omega(\mu_0/2\pi) \cdot \ln\{2(h + h_e)/r_3\} \tag{3.89a}$$

$$Z_c \fallingdotseq Z_s + j\omega(\mu_0/2\pi) \cdot \ln(r_2/r_1) \tag{3.89b}$$

where

h = buried depth of cable from the earth surface
r_1–r_4 = radii of cable as in Figure 3.3
$h_e = \sqrt{\rho_e/j\omega\mu_0}$ = complex penetration depth of earth
$H_e = |h_e| = \sqrt{\rho_e/\omega\mu_0}$ = penetration depth (3.90)

Substituting the potential coefficient p_c and p_s in (3.85), and Z_c and Z_s in (3.89) into (3.88), the following results are obtained considering $h \gg H_e$.

$$\gamma_0 \fallingdotseq j\omega\sqrt{\mu_0\varepsilon_0\varepsilon_2/\ln\{2(h+H_e)/r_3\}/\ln(r_4/r_3)} = j\beta_0$$

$$\gamma_1 \fallingdotseq j\omega\sqrt{\mu_0\varepsilon_0\varepsilon_1} = j\beta_1$$

Modal velocity:

$$v_0 = \omega/\beta_0 = \left(\frac{c_0}{\sqrt{\varepsilon_2}}\right) \cdot \sqrt{\ln(r_4/r_3)/\ln\{2(h+H_e)/r_3\}} \tag{3.91a}$$

$$v_1 = \omega/\beta_1 = c_0/\sqrt{\varepsilon_1} \tag{3.91b}$$

Modal characteristic impedance:

$$z_{00} = (60/\sqrt{\varepsilon_2})\sqrt{\ln\{2(h+H_e)/r_3\} \cdot \ln(r_4/r_3)} \tag{3.92a}$$

$$z_{01} = (60/\sqrt{\varepsilon_1}) \cdot \ln(r_2/r_1) = z_{0c} \tag{3.92b}$$

The above equations clearly show the following typical characteristics of wave propagation on an SC cable.

- Coaxial mode (mode 1). The propagation velocity ranges from 150 to 180 m/μs for ε_1 ranges from 2 to 4, and the characteristic impedance from 15 to 30 Ω for $\ln(r_2/r_1)$ being 0.5–1 in practice.
- Earth-return mode (mode 0). The propagation velocity is very low because $\ln(r_4/r_3)$ is less than 0.25 and $\ln\{2(h+h_e)/r_3\}$ is greater than 3.5. In a field test, the velocity was measured as less than 70 m/μs, and the characteristic impedance was less than 20 Ω [7].

3.2.2 *Frequency-dependent Characteristics*

It is well-known that conductor impedance is frequency-dependent due to the skin effect of the conductor and transformation matrix being frequency-dependent.

3.2.2.1 Impedance of a Single-phase SC Cable

Table 3.1 shows an example of a frequency-dependent impedance for an underground single-phase SC cable as illustrated in Figure 3.3. It is clear in the table that the resistance increases nearly proportional to the square root of frequency f. The reactance also increases as the frequency increases. However, it is noteworthy that the inductance decreases as the frequency increases. This is caused by decreasing flux linkage to currents because the currents concentrate at the conductor surface due to the skin effect as the frequency increases. In contrast, the resistance increases because the cross-sectional area of the flowing currents decreases.

It should be noted that Z_{cs} becomes nearly the same as Z_{ss} as the frequency increases as explained in (3.84).

Table 3.1 Frequency-dependent impedance of a single-phase SC cable (Ω/km)

Frequency (Hz)	Core-self Z_{cc}		Sheath-self Z_{ss}		Mutual Z_{cs}	
50	0.0855+j	0.625	0.487+j	0.569	0.0492+j	0.570
100	0.14	1.20	0.536	1.09	0.0982	1.10
1 k	1.08	10.4	1.41	9.51	0.970	9.53
5 k	5.10	46.9	5.24	42.6	4.82	42.7
10 k	10.1	89.4	9.98	80.9	9.62	81.1
50 k	49.4	395	47.2	356	47.3	356
100 k	95.9	747	92.9	671	93.0	671
500 k	480	3270	474	2900	474	2900
1 M	1010	6130	1000	5390	1000	5390

$r_1 = 2.54$ cm, $r_2 = 4.56$ cm, $r_3 = 5.08$ cm, $r_4 = 5.59$ cm, $h = 75$ cm
$\rho_c = 6.89 \times 10^{-8}$ Ωm, $\rho_s = 6.89 \times 10^{-7}$ (Pb), $\rho_e = 20$ Ωm, $\varepsilon_1 = 3.52$, $\varepsilon_2 = 3.30$

3.2.2.2 Transformation Matrix

In a coaxial cable, the voltage transformation matrix A^{-1} is defined, in general, in the following form.

$$\begin{pmatrix} v_0 \\ v_1 \end{pmatrix} = [A]^{-1}(V) = \begin{bmatrix} A_{11} & A_{12} \\ A_{12} & A_{22} \end{bmatrix} \begin{pmatrix} V_c \\ V_s \end{pmatrix} \qquad (3.93)$$

where
v_0 = mode 0 voltage
v_1 = mode 1 voltage
V_c = core to earth voltage
V_s = sheath to earth voltage

Table 3.2 shows the frequency dependence of the matrix elements and the critical frequency explained in Section 3.2.1. It is observed that A_{11} converges to zero, A_{12} to 1, and A_{22} to −1. The critical frequency given in Table 3.2 agrees with that defined in (3.83).

3.2.2.3 Attenuation and Velocity

Table 3.3 shows the frequency dependence of the modal attenuation constant α and the modal propagation velocity c. It is observed in the table that the attenuation of an underground cable is far greater and the velocity is far smaller than those of an overhead line. If the cable is overhead, the attenuation and the velocity become similar to those of the overhead line except those of the coaxial mode.

3.2.2.4 Characteristic Impedance

Table 3.4 shows the frequency dependence of characteristic impedances. The "actual" values are those for an actual phase domain, that is, Z_{0c} for the core self-impedance, Z_{0s} for the sheath

Table 3.2 Frequency dependence of transformation matrix A^{-1}

Pb sheath f (kHz)	$d = 1$ mm; $f_c \cong 50$ kHz A_{11}	A_{12}	A_{22}	$d = 5$ mm; $f_c \cong 5$ kHz A_{11}	A_{12}	A_{22}
0.05	−0.158	1.012	−0.198	−0.113	1.087	−0.837
0.1	−0.155	1.036	−0.375	−0.068	1.046	−0.960
0.5	−0.086	1.064	−0.943	−0.015	1.004	−0.995
1	−0.046	1.024	−0.980	−0.008	1.001	−0.998
5	−0.011	1.002	−0.996	−0.002	1	−1
10	−0.006	1.001	−0.998			
50	−0.001	1	−1			

$d = 10$ mm; $f_c \cong 2$ kHz
Al sheath
$d = 1$ mm; $f_c \cong 2$ kHz $d = 5$ mm; $f_c \cong 200$ Hz $d = 10$ mm; $f_c \cong 50$ Hz

$$[A]^{-1} = \begin{bmatrix} A_{11} & A_{12} \\ A_{12} & A_{22} \end{bmatrix} \qquad [A]^{-1} = \begin{bmatrix} 0 & 1 \\ 1 & -1 \end{bmatrix} \; ; f > f_c$$

Table 3.3 Frequency dependence of modal attenuation and propagation velocity

Frequency (Hz)	Attenuation (dB/km)		Velocity (m/μs)	
	α_0	α_1	c_0	c_1
50	6.93E−2	2.26E−2	16.04	52.86
100	7.75E−2	4.26E−2	17.00	70.87
1 k	0.218	0.111	18.51	138.20
10 k	1.67	0.191	20.08	151.73
100 k	17.1	0.646	22.04	156.75
1 M	204.6	2.079	24.55	158.82

E−2 = 10^{-2}.

Table 3.4 Frequency dependence of characteristic impedances (Ω)

Frequency (Hz)	Actual			Modal	
	Z_{0c}	Z_{0s}	Z_{0m}	z_{00}	z_{01}
50	73.03−j16.13	33.22−j11.65	27.29+j2.99	32.48−j13.37	66.07−j28.93
100	65.43−j15.05	31.12−j7.01	28.53+j1.75	31.19−j7.53	45.37−j25.09
1 k	49.62−j5.87	28.15−j2.08	28.10−j0.95	28.16−j2.08	21.66−j6.06
10 k	45.74−j2.43	25.95−j1.59	26.00−j1.49	25.95−j1.56	19.68−j1.04
100 k	42.71−j1.99	23.65−j1.63	23.65−j1.63	23.65−j1.63	19.06−j0.35
1 M	40.04−j2.07	21.23−j1.95	21.23−j1.95	21.23−j1.95	18.82−j0.11

self-impedance, and Z_{0m} for the mutual impedance. The "modal" values are the modal characteristic impedances z_{00} and z_{01}. The frequency dependence of the characteristic impedance is far greater than that for an overhead line.

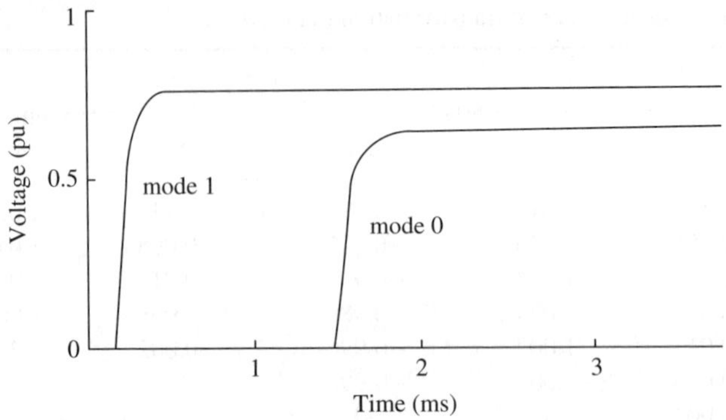

Figure 3.5 Modal step responses of an underground SC cable with length of $x = 30$ km

It is observed in the table that $Z_{0s} \fallingdotseq Z_{0m} \fallingdotseq z_{00}$, and z_{01} is nearly equal to $Z_{0c} - Z_{0s}$ in the frequency range above $f = 1$ kHz. This means physically that the wave propagation is reaching to the coaxial mode. The impedance above $f = 100$ kHz exactly corresponds to the coaxial mode impedance given in (3.92), $z_{01} = (60/\sqrt{\varepsilon_1}) \cdot \ln(r_2/r_1) = 18.71\ \Omega$.

3.2.3 *Time Response of Wave Deformation*

The frequency dependence of the propagation constant explained in Section 3.2.2.3 is expressed as a step response $s(t)$ of wave deforamation in a time domain. The step response is defined as a time response at the receiving end of a line with the length x, when a 1 pu step voltage is applied at the sending end.

$$s(t) = L^{-1}[\exp\{-\gamma(\omega) \cdot x\}/j\omega] \tag{3.94}$$

where $\omega = 2\pi f$ is the angular frequency, and L^{-1} is the Laplace (Fourier) inverse transform.

Figure 3.5 shows the modal step responses for $x = 30$ km. The wave deformation is far heavier than that of an overhead line. In the overhead line case, the mode 1 (aerial mode) response converges to 1 pu within 0.1 ms, and the mode 0 response reaches 1 pu in several tenths of a millisecond, while they take more than a few milliseconds in the cable case.

3.3 Three-phase Underground SC Cables

3.3.1 *Mutual Coupling between Phases*

A significant difference between an overhead line and an underground cable is that there is electrostatic coupling between phases of the overhead line, while there is no electrostatic coupling in the underground cable because the earth is conductive. Approximate formulas of the earth-return impedance Z_e and the space admittance Y_s are given in the following form [2, 6, 8].

(a) Overhead:

$$Z_e = j\omega(\mu_0/2\pi)\ln(S_{ij}/d_{ij}) \tag{3.95a}$$

$$Y_s = j\omega 2\pi\varepsilon_0[P]^{-1} \tag{3.95b}$$

$$P_{ij} = \ln(D_{ij}/d_{ij}) \tag{3.95c}$$

(b) Underground

$$Z_e = j\omega(\mu_0/2\pi)\{\ln(S_{ij}/d_{ij}) - 0.077 - 2(h_i + h_j)/3h_e\} \tag{3.96a}$$

$$Y_s = 0 \tag{3.96b}$$

where

$S_{ij}^2 = (h_i + h_j + 2h_e)^2 + y_{ij}^2$

$D_{ij}^2 = (h_i + h_j)^2 + y_{ij}^2, d_{ij}^2 = (h_i - h_j)^2 + y_{ij}^2$

$h_e^2 = \rho_e/j\omega\mu_e$ = complex penetration depth

h = height or depth of a conductor from the earth surface

y = horizontal separation between conductors

In general, h and y are much greater in the overhead line than those in an underground cable. In addition, since h_e is the same in the overhead and underground cases and $0.077 + 2(h_i + h_j)/3h_e$ is small, the earth-return impedance of the underground cable is greater than that of the overhead line as far as h_e is sufficiently greater than h, that is, frequency f is not sufficiently high. For example, when $\rho_e = 100\ \Omega$m, the absolute value $H_e = |h_e|$ of the complex penetration depth for different frequencies is calculated as:

$$H_e = 503 \text{ m at } f = 50 \text{ Hz}, \quad 112 \text{ m at } f = \text{kHz}, \quad 11.2 \text{ m at } f = 100 \text{ kHz}$$

Thus, the earth-return impedance Z_e in (3.95) and (3.96) is evaluated roughly approximating S_{ij} by its absolute value.

Overhead line: $h_i = h_j = 20$ m, $d_{ij} = 4$ m

$$\ln\{|S_{ij}|/d_{ij}\} = 5.566 \text{ at 50 Hz}, \ 4.19/1 \text{ kHz}, \ 2.75/100 \text{ kHz}$$

Underground cable: $h_i = h_j = 0.75$ m, $d_{ij} = 0.15$ m

$$\ln(|S_{ij}|/d_{ij}) - 0.077 - 4h_i/3H_e = 8.682 \text{ at 50 Hz}, \ 7.23/1 \text{ kHz}, \ 4.904/100 \text{ kHz}$$

It is clear from the above results that the earth-return impedance of an underground cable is greater than that of an overhead line even at $f = 100$ kHz. Also, it should be noted that the impedance is very much dependent on S_{ij}/d_{ij} as is clear from (3.95) and (3.96).

Except for the mutual coupling, that is, the mutual earth-return impedance between phases discussed above, the wave propagation characteristics on a multi-phase cable are nearly the same as those on a single phase cable as explained in the previous section.

It is noteworthy that the wave propagation characteristics on a cross-bonded cable (three-phase SC cable) differ significantly from those explained above for a solidly bonded cable. Also, the wave propagation characteristics on a pipe-enclosed type (PT) cable are quite different from those on a SC cable. These characteristics are explained later.

3.3.2 Transformation Matrix

The voltage transformation matrix A^{-1} of a three-phase SC cable is given, in general, in the following form.

$$(v) = \begin{pmatrix} (v_e) \\ (v_c) \end{pmatrix} = [A]^{-1}(V) = [A]^{-1}\begin{pmatrix} (V_c) \\ (V_s) \end{pmatrix} \tag{3.97}$$

where

$(v_e) = \begin{pmatrix} v_0 \\ v_1 \\ v_2 \end{pmatrix}$ is the modal voltages between phases involving earth-return

$(v_c) = \begin{pmatrix} v_3 \\ v_4 \\ v_5 \end{pmatrix}$ is the coaxial mode voltages between a core and a sheath in one phase

$(V_c) = \begin{pmatrix} V_{ca} \\ V_{cb} \\ V_{cc} \end{pmatrix}$ is the core voltage on each phase

$(V_s) = \begin{pmatrix} V_{sa} \\ V_{sb} \\ V_{sc} \end{pmatrix}$ is the sheath voltages on each phase

$$[A]^{-1} \fallingdotseq \begin{bmatrix} [\Delta] & [a] \\ [a] & [\Delta] \end{bmatrix} : \quad \text{type A for } f < f_s \tag{3.98a}$$

$$\fallingdotseq \begin{bmatrix} [0] & [a] \\ [a] & -[a] \end{bmatrix} : \quad \text{type B for } f_s < f < f_c \tag{3.98b}$$

$$\fallingdotseq \begin{bmatrix} [0] & [a] \\ [U'] & -[U'] \end{bmatrix} : \quad \text{type C for } f_c < f \tag{3.98c}$$

Where

$$[a] = (1/3)\begin{bmatrix} 1 & 1 & 1 \\ 3/2 & 0 & -3/2 \\ -1 & 2 & -1 \end{bmatrix}, \quad [\Delta] \fallingdotseq \begin{bmatrix} 0 & 0 & 0 \\ 0 & 0 & 0 \\ 0 & 0 & 0 \end{bmatrix}, \quad [U'] = \begin{bmatrix} 0 & 0 & 1 \\ 1 & 0 & 0 \\ 0 & 1 & 0 \end{bmatrix}$$

and f_s is the critical (highest) frequency for type A matrix and f_c is the critical (lowest) frequency for type C matrix.

Table 3.5 gives an example of the type A matrix at $f = 50$ Hz and the critical frequencies f_s and f_c for Pb and Al sheath cables with various sheath thickness d. It should be clear from the table and (3.98) that the frequency dependence of a transformation matrix on an underground three-phase SC cable is very strong. In (3.98), the first column submatrices, [0] and [a] in type C for example, are related to the earth-return and inter-sheath modes (modes 0–2) which are significantly affected by the earth-return impedance. The second column submatrices correspond to a propagation mode (modes 3–5) between a core and a sheath on one phase.

In the frequency range lower than f_s, a metalic sheath being not able to shield a core current completely, modes 3–5 interact with the cores on the other phases and the type A matrix appears. In the frequency range higher than f_c, the metalic sheath becoming a complete shield, the core and the sheath on each phase are independent of those on the other phases, and thus

Table 3.5 Frequency-dependent transformation matrix A^{-1} for a three-phase SC cable

Pb sheath					
$d = 1$ mm; $f_s \cong 5$ kHz, $f_c \cong 500$ kHz					
(i) $f = 50$ Hz					
−0.051	−0.052	−0.051	0.357	0.357	0.357
−0.080	0	0.080	0.5	0	−0.5
0.054	−0.104	0.054	−0.336	0.653	−0.336
0.357	0.377	0.357	−0.174	−0.179	−0.174
0.5	0	−0.5	−0.019	0	0.019
−0.336	0.653	−0.336	0.012	−0.012	0.006
(ii) 5 kHz $\leqq f <$ 500 kHz; type B matrix					
(iii) $f \geqq$ 500 kHz; type C matrix					
$d = 5$ mm: $f_s \cong 1$ kHz, $f_c \cong 100$ kHz					
$d = 10$ mm: $f_s \cong 500$ kHz, $f_c \cong 20$ kHz					
Al sheath					
$d = 1$ mm: $f_s \cong 500$ kHz, $f_c \cong 20$ kHz					
$d = 5$ mm: $f_s \cong 100$ kHz, $f_c \cong 1$ kHz					
$d = 10$ mm: $f_s \cong 50$ kHz, $f_c \cong 2$ kHz					

a coaxial mode is completed on each phase. Thus, the type C matrix appears. The critical frequency f_c is given in (3.83). The other critical frequency f_s is given empirically in the following form.

$$f_s \approx {\rho_s}^{3/4}/10\pi\mu_s dy \text{ (Hz)} \tag{3.99}$$

where ρ_s, μ_s are the resistivity and permeability of the sheath, respectively, d is the sheath thickness, and y is the separation between phases.

3.3.3 Attenuation and Velocity

Table 3.6 shows the modal attenuation constants and propagation velocities on a three-phase cable as illustrated in Figure 3.6. It is clear from the table that modes 3–5 show the nature of the coaxial mode propagation which is independent of the other phases above a frequency of 1 kHz. In fact, the attenuation and the velocity are nearly the same as those for a single-phase cable as shown in Table 3.3.

In a low frequency range, less than several hundred hertz in Table 3.6, the earth-return mode (mode 0) is not distinguished from the inter-sheath mode (modes 1 and 2) as far as the attenuation constants (α_0, α_1, and α_2) are concerned. However, there is a clear difference between the propagation velocities (v_0, v_1, and v_2). The mode 0 velocity is about 10 m/μs which is less than half that of the inter-sheath mode. This very small velocity is caused by the fact that most of the mode 0 current returns through the earth, while the inter-sheath mode current returns partially through the other phase sheaths. This is similar to the characteristics of the earth-return mode and the inter-circuit modes of a double-circuit overhead line [3, 9].

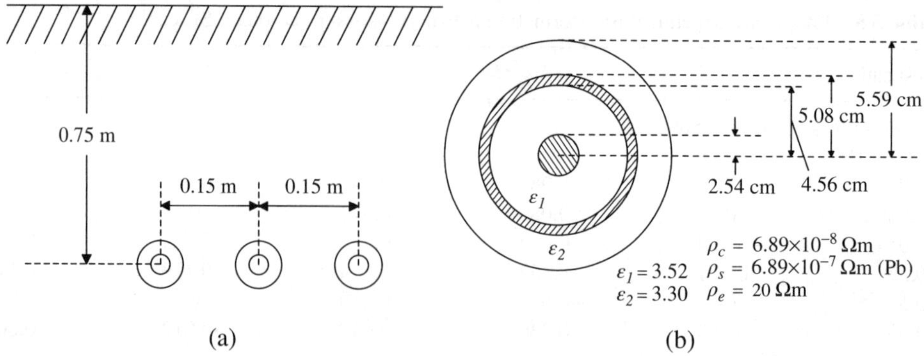

Figure 3.6 A three-phase underground SC cable: (a) arrangement (flat); and (b) cross-section

3.3.4 Characteristic Impedance

The characteristic impedance of a three-phase SC cable is given in the following form.

$$[Z_0] = \begin{bmatrix} [Z]_c & [Z]_m \\ [Z]_m & [Z]_s \end{bmatrix} \tag{3.100}$$

where subscripts c, s and m are for the core, sheath and mutual between the core and sheath, respectively.

Each submatrix is given in the following form.

$$[Z] = \begin{bmatrix} Z_{aa} & Z_{ab} & Z_{ac} \\ Z_{ab} & Z_{bb} & Z_{bc} \\ Z_{ac} & Z_{bc} & Z_{cc} \end{bmatrix} \tag{3.101}$$

Table 3.6 Modal attenuation constants and propagation velocities on a three-phase SC cable as shown in Figure 3.6

Attenuation (dB/km)						
f (Hz)	α_0	α_1	α_2	α_3	α_4	α_5
50	5.031–E2	9.812–E2	0.1302	3.443–E2	5.391–E3	4.819–E3
100	6.417–E2	0.1250	0.1391	5.092–E2	1.141–E2	6.959–E3
1 k	0.3181	0.1386	0.2020	9.379–E2	0.1012	0.1113
10 k	3.006	0.1647	0.2397	0.1935	0.1918	0.1905
100 k	31.86	0.6177	0.8440	0.6465	0.6465	0.6465
1 M	391.3	9.273	4.086	2.079	2.079	2.079
Velocity (m/μs)						
f (Hz)	v_0	v_1	v_2	v_3	v_4	v_5
50	10.20	22.99	24.08	54.34	79.55	97.42
100	10.49	29.75	32.55	75.17	81.51	98.91
1 k	11.39	37.89	54.11	129.0	133.0	139.0
10 k	12.49	37.96	54.49	151.7	151.7	151.7
100 k	13.91	38.17	55.16	156.8	156.8	156.8
1 M	15.85	38.30	55.52	158.8	158.8	158.8

The form in (3.100) is the same as the series impedance matrix and the characteristic impedance matrix of a single-phase SC cable explained in the previous sections except for the fact that a submatrix consists of a single element in the single-phase case.

In (3.101), there exists the following relation approximately.

$$Z_{aa} \fallingdotseq Z_{bb} \fallingdotseq Z_{cc} \tag{3.102a}$$

$$Z_{ab} \fallingdotseq Z_{ac} \text{ for a flat or trefoil arranged cable} \tag{3.102b}$$

At a frequency higher than f_c in (3.83), the coaxial mode is completed on each phase, and also the following relation is satisfied.

$$[Z]_s \fallingdotseq [Z]_m, [Z_{ij}]_c \fallingdotseq [Z_{ij}]_s \fallingdotseq [Z_{ij}]_m \;:\; i, j = a,\ b \text{ and } c \tag{3.103}$$

Table 3.7 shows the frequency dependence of the actual characteristic impedances on a three-phase SC cable as shown in Figure 3.6. It is observed in the table that Z_{aa} of the core ($\fallingdotseq Z_{bb} \fallingdotseq Z_{cc}$) decreases by about 40% from 60.0 Ω at $f = 50$ Hz to 37.8 Ω at $f = 1$ MHz. The sheath impedance decreases by about 40%, and the mutual impedance decreases by about 15–40% from 50 Hz to 1 MHz.

Figure 3.7 shows the frequency dependence of modal characteristic impedances. It is clearly observed in the figure that the impedances of the earth-return mode (mode 0) and the inter-sheath mode (modes 1 and 2) are monotonously decreasing as frequency increases. However, the coaxial mode (modes 3–5) impedances decrease until about $f = 1$ kHz, and then stay constant up to 20 kHz. After around 20 kHz, they increase and reach the complete coaxial mode impedance z_{0c} which is given in (3.92) as $z_{0c} \fallingdotseq 18.71$ Ω.

Now, we investigate the oscillating nature of the coaxial mode characteristic impedances. In a frequency region of $f > f_c$, the actual characteristic impedance matrix in (3.100) is given approximately in the following form as in (3.103).

$$[Z_0] \fallingdotseq \begin{bmatrix} [Z]_c & [Z]_s \\ [Z]_s & [Z]_s \end{bmatrix} \text{ for } f > f_c \tag{3.104}$$

where

$$[Z]_c = \begin{bmatrix} Z_c & Z_m & Z_n \\ Z_m & Z_c & Z_m \\ Z_n & Z_m & Z_c \end{bmatrix}, \quad [Z]_s = \begin{bmatrix} Z_s & Z_m & Z_n \\ Z_m & Z_s & Z_m \\ Z_n & Z_m & Z_s \end{bmatrix}$$

It should be noted that the above submatrices are in the same form as the impedance matrix of an untransposed horizontal three-phase overhead line [3, 10]. By applying the type C

Table 3.7 Actual characteristic impedances of a three-phase SC cable [real part only (Ω)]

f (Hz)	$(Zaa)_c$	$(Zab)_c$	$(Zac)_c$	$(Zaa)_s$	$(Zab)_s$	$(Zaa)_m$	$(Zab)_m$
50	60.04	20.89	16.40	30.91	10.73	21.96	14.09
100	55.53	17.24	13.09	27.55	11.71	22.67	13.93
1 k	44.99	12.15	9.60	23.51	12.12	23.35	12.24
10 k	41.90	10.81	8.29	22.05	10.87	22.13	10.84
100 k	39.61	9.50	6.89	20.54	9.50	20.54	9.50
1 M	37.76	8.01	5.34	18.94	8.01	18.94	8.01

transformation matrix in (3.98), the modal characteristic impedance $z_0 = A^{-1} \cdot Z_0 \cdot B$ is obtained as:

$$z_{00} = (3Z_s + 4Z_m + 2Z_n)/6 \tag{3.105a}$$

$$z_{01} = (Z_s - Z_m)/2 \tag{3.105b}$$

$$z_{02} = (Z_s - 4Z_m + Z_n)/3 \tag{3.105c}$$

$$z_{03} = (Z_c - Z_s)/3 \tag{3.105d}$$

$$z_{04} = (Z_c - Z_s)/2 \tag{3.105e}$$

$$z_{05} = 2(Z_c - Z_s)/3 \tag{3.105f}$$

At a much higher frequency, that is, $f \gg f_c$, the following coaxial mode impedance is completed as in (3.92).

$$z_{03} = z_{04} = z_{05} = z_{0c} = Z_c - Z_s \tag{3.106}$$

For example, from Table 3.7, the following result is obtained.

$$z_{03} \fallingdotseq 6.6\ \Omega, z_{04} \fallingdotseq 9.9\ \Omega, z_{05} \fallingdotseq 13.2\ \Omega \text{ at } f = 10 \text{ kHz}$$

$$z_{03} = z_{04} = z_{05} \fallingdotseq 19.8\ \Omega \text{ at } f = 1 \text{ MHz}$$

The above result explains the frequency response of the modal characteristic impedance in Figure 3.7. The oscillating nature of the coaxial mode impedance is a result of the frequency dependence of the transformation matrix. In general, the following relation is satisfied.

$$Z_c - Z_s > z_{0c} = (60/\sqrt{\varepsilon_1}) \cdot \ln(r_2/r_1) > 2(Z_c - Z_s)/3 \tag{3.107}$$

3.4 Effect of Various Parameters of an SC Cable

The explanation in the previous sections concerns only specific physical and geometrical parameters. The parameters certainly affect the wave propagation characteristics on a cable.

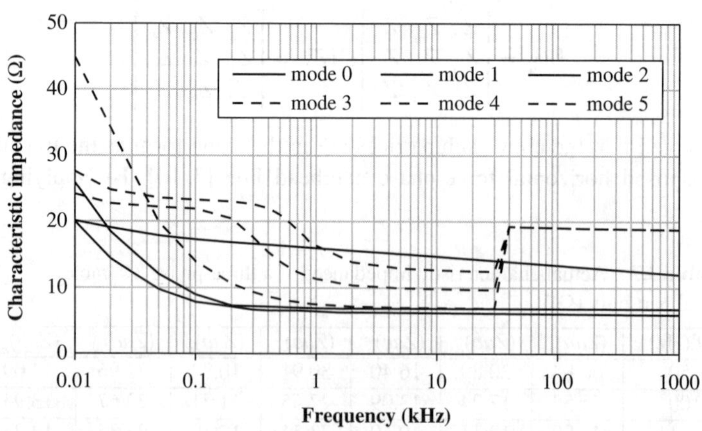

Figure 3.7 Frequency dependence of modal characteristic impedance

In this section, the effect of various physical and geometrical parameters is investigated. For simplicity, the investigation concerns only a single-phase SC cable except the case of the geometrical arrangement of a three-phase cable.

Table 3.8 shows the effect of various parameters on wave propagation characteristics of a single-phase cable. The parameters of the base case are:

$$h = 75\text{ cm}, r_1 = 2.54\text{ cm}, r_2 = 4.56\text{ cm}, r_3 = 5.08\text{ cm}, r_4 = 5.59\text{ cm},$$
$$\rho_c = 6.89 \times 10^{-8}\ \Omega\text{m}, \rho_s = 6.89 \times 10^{-7}\ \Omega\text{m (Pb)}, \rho_e = 20\ \Omega\text{m}, \varepsilon_1 = 3.52,$$
$$\varepsilon_2 = 3.30,\ f_c \fallingdotseq 3.23\text{ kHz},\ f_s \fallingdotseq 116.5/y(\text{Hz})$$

The critical frequency for various cases in Table 3.8 is: Cases 1–4, $f_c = 3.23$ kHz; Case 5, $f_c = 87.3$ kHz; Case 6, $f_c = 8.73$ kHz; Case 7, $f_c = 323$ Hz.

3.4.1 Buried Depth *h*

It is observed in Table 3.8 that there is no noticeable effect of the buried depth on the attenuation, the propagation velocity and the characteristic impedance of both the earth-return and the coaxial modes in the frequency range from 50 Hz to 100 kHz.

3.4.2 Earth Resistivity ρ_e

When the earth resistivity becomes greater, the attenuation and the velocity of the earth-return mode becomes smaller, and the characteristic impedance becomes greater. It causes no effect on the coaxial mode in a high frequency region.

3.4.3 Sheath Thickness *d*

When the sheath thickness becomes smaller, the attenuation of the earth-return and coaxial modes becomes greater. The earth-return mode velocity decreases in a low frequency region. The coaxial mode velocity decreases. The earth-return and coaxial modes characteristic impedances become greater, especially at 50 Hz.

The sheath thickness affects the critical frequencies f_c as already explained.

3.4.4 Sheath Resistivity ρ_s

When the sheath resistivity decreases, the attenuation decreases significantly, and the velocity increases. The characteristic impedance also decreases significantly.

The effect of the sheath resistivity, that is, the metal used as the shield, has been already explained in Table 3.2 and Table 3.5. As is clear from (3.82), the penetration depth h_s is propotional to the resistivity and thus the critical frequency f_c becomes lower as the resistivity decreases.

Table 3.8 Effect of various parameters on a single-phase SC cable

Case	h (m)	ρ_e (Ωm)	d (mm)	$\rho_s \times 10^{-8}$ (Ωm)	Attenuation (dB/km)		Velocity (m/μs)		Charact. imp. (Ω)	
					α_0	α_1	v_0	v_1	z_{00}	z_{01}
$f = 50$ Hz										
Base	0.75	20	5.2	68.9	6.92×10^{-2}	2.26×10^{-2}	16.04	52.86	32.835	66.070
1	1.5	20	5.2	68.9	6.92×10^{-2}	2.26×10^{-2}	16.04	52.85	32.839	66.078
2	3.0	20	5.2	68.9	6.92×10^{-2}	2.26×10^{-2}	16.04	52.84	32.847	66.094
3	0.75	100	5.2	68.9	6.66×10^{-2}	2.40×10^{-2}	15.52	52.50	34.116	65.900
4	0.75	1000	5.2	68.9	6.31×10^{-2}	2.57×10^{-2}	14.84	52.22	35.844	65.321
5	0.75	20	1.0	68.9	2.25×10^{-1}	9.13×10^{-2}	10.25	42.73	47.296	83.561
6	0.75	20	1.0	6.89	3.91×10^{-2}	2.31×10^{-2}	17.32	67.44	33.327	48.228
7	0.75	20	5.2	6.89	1.31×10^{-2}	1.27×10^{-2}	16.92	109.68	30.826	27.361
$f = 1$ kHz										
Case	h (m)	ρ_e (Ωm)	d (mm)	$\rho_s \times 10^{-8}$ (Ωm)	Attenuation (dB/km)		Velocity (m/μs)		Charact. imp. (Ω)	
					α_0	α_1	v_0	v_1	z_{00}	z_{01}
Base	0.75	20	5.2	68.9	2.18×10^{-1}	1.11×10^{-1}	18.51	138.20	28.156	21.662
1	1.5	20	5.2	68.9	2.15×10^{-1}	1.11×10^{-1}	18.50	138.20	28.179	21.661
2	3.0	20	5.2	68.9	2.12×10^{-1}	1.11×10^{-1}	18.47	138.21	28.223	21.659
3	0.75	100	5.2	68.9	2.08×10^{-1}	1.11×10^{-1}	17.62	138.32	29.588	21.634
4	0.75	1000	5.2	68.9	1.96×10^{-1}	1.11×10^{-1}	16.53	138.46	31.536	21.604
5	0.75	20	1.0	68.9	4.95×10^{-1}	3.22×10^{-1}	19.08	95.56	29.930	32.494
6	0.75	20	1.0	6.89	1.78×10^{-1}	7.39×10^{-2}	19.23	145.44	29.425	20.569
7	0.75	20	5.2	6.89	1.58×10^{-1}	3.49×10^{-2}	18.54	145.17	28.111	20.571
f=5 kHz										
Case	h (m)	ρ_e (Ωm)	d (mm)	$\rho_s \times 10^{-8}$ (Ωm)	Attenuation (dB/km)		Velocity (m/μs)		Charact. imp. (Ω)	
					α_0	α_1	v_0	v_1	z_{00}	z_{01}
Base	0.75	20	5.2	68.9	8.54×10^{-1}	1.53×10^{-1}	19.58	149.93	26.620	19.921
1	1.5	20	5.2	68.9	8.36×10^{-1}	1.53×10^{-1}	19.54	149.93	26.671	19.921
2	3.0	20	5.2	68.9	8.16×10^{-1}	1.53×10^{-1}	19.47	149.93	26.762	19.921
3	0.75	100	5.2	68.9	8.21×10^{-1}	1.53×10^{-1}	18.53	149.94	28.117	19.920
4	0.75	1000	5.2	68.9	7.73×10^{-1}	1.53×10^{-1}	17.28	149.95	30.154	19.920
5	0.75	20	1.0	68.9	1.114	0.548	20.28	145.02	27.909	20.682
6	0.75	20	1.0	6.89	0.782	0.107	20.31	153.79	27.852	19.431
7	0.75	20	5.2	6.89	0.800	8.10×10^{-2}	19.60	152.33	26.584	19.614
$f = 100$ kHz										
Case	h (m)	ρ_e (Ωm)	d (mm)	$\rho_s \times 10^{-8}$ (Ωm)	Attenuation (dB/km)		Velocity (m/μs)		Charact. imp. (Ω)	
					α_0	α_1	v_0	v_1	z_{00}	z_{01}
Base	0.75	20	5.2	68.9	17.057	0.646	22.04	156.75	23.648	19.063
1	1.5	20	5.2	68.9	17.283	0.646	21.89	156.75	23.803	19.063
2	3.0	20	5.2	68.9	19.429	0.646	21.80	156.75	23.903	19.063
3	0.75	100	5.2	68.9	16.375	0.646	20.65	156.75	25.236	19.063
4	0.75	1000	5.2	68.9	15.549	0.646	19.00	156.75	27.434	19.063
5	0.75	20	1.0	68.9	16.388	0.800	22.79	157.67	24.823	18.951
6	0.75	20	1.0	6.89	16.063	0.369	22.80	158.05	24.815	18.905
7	0.75	20	5.2	6.89	16.862	0.371	22.06	158.06	23.627	18.905

Charact. imp., characteristic impedance.

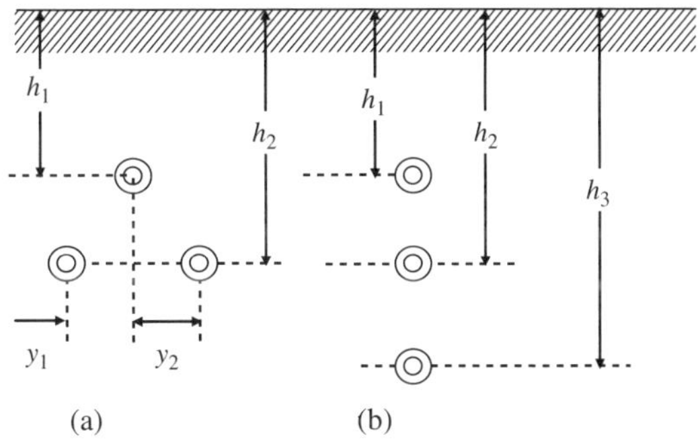

Figure 3.8 Three-phase cable arrangement: (a) trefoil; and (b) vertical

Table 3.9 Effect of three-phase cable arrangement on the earth-return and inter-sheath modes

Case 1 (Flat arrangement) 1.03×10^{-1}									
f (Hz)	Modal attenuation constant (dB/km)			Modal velocity (m/μs)			Modal characteristic impedance (Ω)		
	α_0	α_1	α_2	v_0	v_1	v_2	z_{00}	z_{01}	z_{02}
50	5.03×10^{-2}	9.81×10^{-2}	1.03×10^{-1}	10.20	22.99	24.08	17.875	9.758	12.036
1 k	3.18×10^{-1}	1.39×10^{-1}	2.02×10^{-1}	11.39	37.89	54.11	15.982	6.925	6.445
5 k	1.495	0.145	0.209	12.13	37.94	54.43	15.102	6.869	6.217
100 k	31.865	0.618	0.844	13.91	38.17	55.16	13.390	6.827	6.071
Case 2 (Vertical arrangement)									
f (Hz)	Modal attenuation constant (dB/km)			Modal velocity (m/μs)			Modal characteristic impedance (Ω)		
	α_0	α_1	α_2	v_0	v_1	v_2	z_{00}	z_{01}	z_{02}
50	5.03×10^{-2}	9.81×10^{-2}	1.03×10^{-1}	10.20	22.99	24.08	17.875	9.758	12.036
1 k	3.18×10^{-1}	1.39×10^{-1}	2.02×10^{-1}	11.39	37.89	54.11	15.982	6.927	6.445
5 k	1.495	0.145	0.209	12.13	37.94	54.43	15.102	6.872	6.217
100 k	31.865	0.714	0.844	13.91	38.18	55.16	13.391	6.839	6.071
Case 3 (Trefoil arrangement)									
f (Hz)	Modal attenuation constant (dB/km)			Modal velocity (m/μs)			Modal characteristic impedance (Ω)		
	α_0	α_1	α_2	v_0	v_1	v_2	z_{00}	z_{01}	z_{02}
50	4.99×10^{-2}	1.02×10^{-1}	1.02×10^{-1}	10.11	23.80	23.80	17.352	9.336	12.449
1 k	3.15×10^{-1}	1.78×10^{-1}	1.78×10^{-1}	11.26	48.13	48.15	15.424	5.514	7.355
5 k	1.479	0.184	0.184	11.98	48.30	48.29	14.505	5.398	7.202
100 k	31.431	0.703	0.735	13.69	48.76	48.75	12.702	5.343	7.134

3.4.5 Arrangement of a Three-phase SC Cable

Figure 3.6 and Figure 3.8 illustrate a typical arrangement of a three-phase cable. Figure 3.6 is the flat (horizontal) arrangement. Figure 3.8a is the triangular (trefoil) arrangement, and Figure 3.8b is the vertical arrangement. Table 3.9 shows the effect of cable arrangement on the attenuation constants, the propagation velocity and the characteristic impedance related to the earth-return and the inter-sheath modes. The physical and geometrical parameters of each cable is the same as the base case (Case 1) in Table 3.8. The depth h and separation y in Figure 3.7 are:

(a) $h_1 = 0.62$ m, $h_2 = 0.75$ m, $y_1 = y_2 = 0.075$ m
(b) $h_1 = 0.60$ m, $h_2 = 0.75$ m, $h_3 = 0.9$ m

The critical frequency f_s is the same for all the arrangements with the above parameters.

$$f_s = 776.7 \text{ Hz}$$

No significant difference is observed for the earth-return and coaxial modes (not shown) between the three arrangements. The intersheath mode attenuation of 100 kHz shows a minor difference between the flat and the vertical arrangement in Table 3.9. The two intersheath modes in the triangular arrangement show nearly the same characteristic except the characteristic impedance where the average becomes similar to those in the flat and vertical arrangements.

3.5 Cross-bonded Cable

3.5.1 Introduction of Cross-bonded Cable

Figure 3.9 illustrates a cross-bonded cable. One major section, of which the three-phase sheaths are short-circuited and grounded at both ends, is composed of three minor sections of which the length x is several hundred meters to 1 or 2 km. At the boundary of the minor sections, each phase sheath in a section is connected to the other phase sheath in the next section. For example, the phase a sheath of the first minor section is connected to the phase b sheath of the second minor section as in Figure 3.9. This is called "cross-bonding". Thus, within one major section, the three-phase sheath impedances are balanced similarly to the three-phase impedances of a transposed overhead line. As already explained, the core-to-sheath impedance being the same in each phase, the total impedance matrix of a cross-bonded cable is balanced. Therefore, induced currents from the core to the sheath are cancelled out for a symmetrical (balanced) three phase a.c. source in a steady state. This leads to no heat increase due to the sheath currents. As a result, a transmission capacity becomes larger in comparison with that of a solidly bonded (not cross-bonded) cable. However, the cross-bonded points become a transition (boundary)

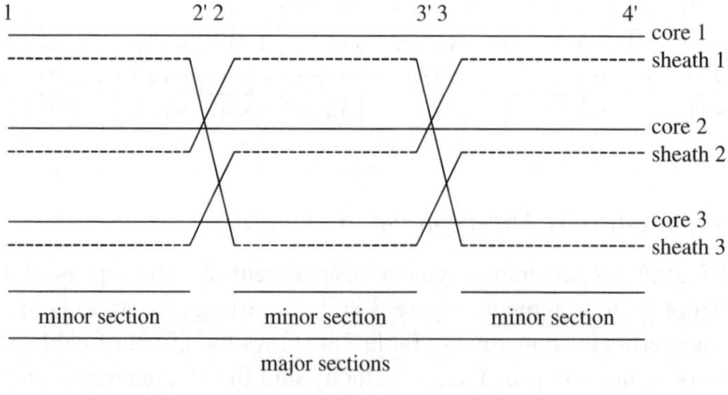

Figure 3.9 Cross-bonded cable with one major section

point which produces multiple reflection and refraction of travelling waves during a transient. Thus, a transient overvoltage becomes greater than that on the solidly bonded cable, and a transient analysis on the cross-bonded cable becomes a significant problem [11–17]. Also, fault location on the cross-bonded cable becomes far more complicated than that on the solidly bonded cable [17–19].

3.5.2 Theoretical Formulation of a Cross-bonded Cable

3.5.2.1 Rotation Matrix

In Figure 3.10, relations between nodes k' and k are given in the following form.

$$(V'_k) = [R] \cdot (V_k) \tag{3.108a}$$

$$(I'_k) = [R] \cdot (I_k) \tag{3.108b}$$

where

$$(V'_k) = \begin{pmatrix} (V'_{kc}) \\ (V'_{ks}) \end{pmatrix}, \ (V_k) = \begin{pmatrix} (V_{kc}) \\ (V_{ks}) \end{pmatrix}, \ [R] = \begin{bmatrix} [U] & [0] \\ [0] & [U'] \end{bmatrix} : \ \text{rotation matirx}$$

$$(V_{kc}) = \begin{pmatrix} V_{kca} \\ V_{kcb} \\ V_{kcc} \end{pmatrix}, \ (V_{ks}) = \begin{pmatrix} V_{ksa} \\ V_{ksb} \\ V_{ksc} \end{pmatrix}, \ [U] = \begin{bmatrix} 1 & 0 & 0 \\ 0 & 1 & 0 \\ 0 & 0 & 1 \end{bmatrix}, \ [U'] = \begin{bmatrix} 0 & 0 & 1 \\ 1 & 0 & 0 \\ 0 & 1 & 0 \end{bmatrix}$$

and first subscript $k = 1$–4 for nodes in Figure 3.10, second subscript c is for core, second subscript s is for sheath, and third subscripts a, b and c are for phase.

(I'_k) and (I_k) are defined in the same manner. The rotation matrix $[R]$ has the following characteristics.

$$[R]^3 = [U] \tag{3.109a}$$

$$[R]^2 = [R]^{-1} \tag{3.109b}$$

$$[R]_t = [R]^{-1} \tag{3.109c}$$

where $[U] = 6 \times 6$ unit matrix and subscript t is for a transposed matrix

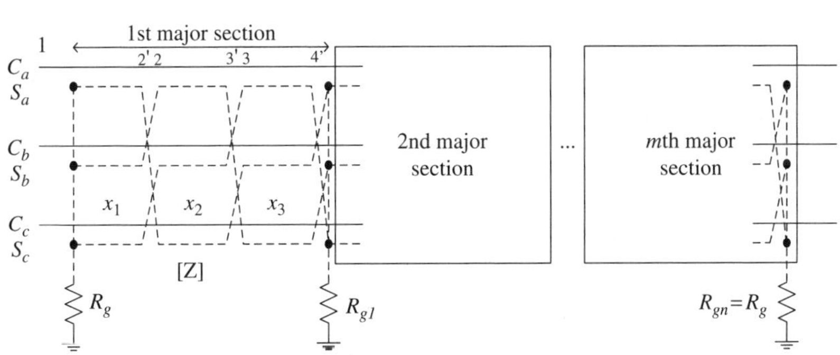

Figure 3.10 Cross-bonded cable system with "m" major sections

3.5.2.2 Accurate Formulation by Adopting a Travelling-wave Theory

The reflection coefficient at node 4' in Figure 3.10 is given by:

$$[K_{4'}] = ([Z_T] - [Z_0]) \cdot ([Z_T] + [Z_0])^{-1} \tag{3.110}$$

where Z_T is the impedance looking from node 4′ to the right-hand side, and Z_0 is the characteristic impedance of the cable.

Define $[S_{fk}]$ as:

$$[S_{fk}] = \exp(-[\Gamma]x_k) \tag{3.111}$$

where x_k is the length of one minor section.

The admittance $[Y_{in3}]$ seen from node 3 to the right-hand side is given in the following equation [12, 15].

$$[Y_{in3}] = [Y_0] \cdot ([U] - [S_{f3}] \cdot [K_{4'}] \cdot [S_{f3}]) \cdot ([U] + [S_{f3}] \cdot [K_{4'}] \cdot [S_{f3}])^{-1} \tag{3.112}$$

The admittance $[Y_{in3'}]$ seen from node 3′ to the right-hand side is given in the following form [12].

$$[Y_{in3'}] = [R] \cdot [Y_{in3'}] \cdot [\mathrm{R}]^{-1} \tag{3.113}$$

The reflection coefficient at node 3′ is:

$$[K_{3'}] = ([Y_0] + [Y_{in3'}])^{-1} \cdot ([Y_0] - [Y_{in3'}]) \tag{3.114}$$

In the same manner, the following equations are obtained.

$$[Y_{ink}] = [Y_0] \cdot ([U] - [S_{fk}] \cdot [K_{k+1'}] \cdot [S_{fk}]) \cdot ([U] + [S_{fk}] \cdot [K_{k+1'}] \cdot [S_{fk}])^{-1} \tag{3.115a}$$

$$[K_{k+1'}] = ([Y_0] + [Y_{ink+1'}])^{-1} \cdot ([Y_0] - [Y_{ink+1'}]) \tag{3.115b}$$

$$[Y_{ink+1'}] = [R] \cdot [Y_{ink+1'}] \cdot [R]^{-1} \tag{3.115c}$$

The sending-end voltage (V_1) is evaluated using source voltage (V_s), source impedance $[Z_s]$, and $[Y_{in1}]$.

$$(V_1) = ([Z_s] \cdot [Y_{in1}] + [U])^{-1} \cdot (V_s) \tag{3.116}$$

The voltage at each node is:

$$(V'_k) = ([S_{fk-1}] + [K_{k'}] \cdot [S_{fk-1}]) \cdot ([U] + [S_{fk-1}] \cdot [K_{k'}] \cdot [S_{fk-1}])^{-1} \cdot (V_{k-1}) \tag{3.117a}$$

$$(V_k) = [R]^{-1} \cdot (V'_k) \tag{3.117b}$$

The above method is useful when the lengths of minor sections are different from each other ($x_1 \neq x_2 \neq x_3$). But it is necessary to evaluate every node voltage, and the computation time increases rapidly as the number of major sections increases. In such a case, a method based on a four-terminal Y-parameter theory is more efficient.

3.5.2.3 Accurate Formulation by Y-parameter

In Figure 3.10, if the length of minor sections are the same ($x = x_1 = x_2 = x_3$), a relation of voltages and currents at each node can be evaluated using Y-parameters [3].

$$\begin{pmatrix} (I_k) \\ (I'_{k+1}) \end{pmatrix} = \begin{bmatrix} [A] & -[B] \\ -[B] & [A] \end{bmatrix} \cdot \begin{pmatrix} (V_k) \\ (V'_{k+1}) \end{pmatrix} \tag{3.118}$$

where

$$[A] = [Y_0] \cdot \coth([\Gamma] \cdot x)$$

and

$$[B] = [Y_0] \cdot \operatorname{cosech}([\Gamma] \cdot x)$$

Y-parameters being numerically quite stable, (3.118) is more stable than those obtained using other parameter representations. From the equation, relations for (I_2) and (I_3) are:

$$-(I'_2) = -[B] \cdot (V_1) + [A] \cdot (V'_2) \tag{3.119a}$$

$$(I_2) = [A] \cdot (V_2) - [B] \cdot (V'_3) \tag{3.119b}$$

$$-(I'_3) = -[B] \cdot (V_2) + [A] \cdot (V'_3) \tag{3.119c}$$

$$(I_3) = [A] \cdot (V_3) - [B] \cdot (V'_4) \tag{3.119d}$$

Eliminating (I_2) and (I_3) from the above equation using (3.108) and solving the above equation for (V_2) and (V_3), the following relations are obtained.

$$(V_2) = [E] \cdot (V_1) + [\mathrm{F}] \cdot (V'_4) \tag{3.120a}$$

$$(V_3) = [G] \cdot (V_1) + [\mathrm{H}] \cdot (V'_4) \tag{3.120b}$$

where

$[E] = ([C] - [D_1] \cdot [D_2])^{-1} \cdot [D_2]$
$[F] = ([C] - [D_1] \cdot [D_2])^{-1} \cdot [D_1] \cdot [B]$
$[G] = ([C] - [D_2] \cdot [D_3])^{-1} \cdot [D_2] \cdot [C]^{-1} \cdot [D_2]$
$[H] == ([C] - [D_2] \cdot [D_3])^{-1} \cdot [B]$

and

$[C] = [A] + [R]^{-1} \cdot [A] \cdot [R]$
$[D_1] = [B] \cdot [R] \cdot [C]^{-1}$
$[D_2] = [R]^{-1} \cdot [B]$
$[D_3] = [C]^{-1} \cdot [B] \cdot [R]$

The Y-parameters of one major section are evaluated by substituting (3.120) into (3.118).

$$\begin{pmatrix} (I_1) \\ -(I'_4) \end{pmatrix} = \begin{bmatrix} [Y_{11}] & -[Y_{12}] \\ -[Y_{21}] & [Y_{22}] \end{bmatrix} \cdot \begin{pmatrix} (V_1) \\ (V'_4) \end{pmatrix} \tag{3.121}$$

where

$[Y_{11}] = [A] - [B] \cdot [R] \cdot [E]$
$[Y_{12}] = [B] \cdot [R] \cdot [F]$
$[Y_{21}] = [B] \cdot [G]$
$[Y_{22}] = [A] - [B] \cdot [H]$

On a homogeneous line, $[Y_{11}]$ equals $[Y_{22}]$ and $[Y_{12}]$ equals $[Y_{21}]$, and each matrix is symmentrical. On a cross-bonded cable, $[Y_{21}] = [Y_{12}]_t$ and are symmentrical.

At both ends of a major section with the three-phase sheath connected together, the number of independent voltage and current elements is four, that is:

$$V_{1s} = V_{1s1} = V_{1s2} = V_{1s3} \tag{3.122a}$$

$$V_{4s} = V_{4s1} = V_{4s2} = V_{4s3} \tag{3.122b}$$

$$I_{1s} = I_{1s1} + I_{1s2} + I_{1s3} \tag{3.122c}$$

$$I_{4s} = I_{4s1} + I_{4s2} + I_{4s3} \tag{3.122d}$$

Rewriting the above equation,

$$(V_k) = [T]_t \cdot (V''_k) \tag{3.123a}$$

$$(I''_k) = [T] \cdot (I_k) \tag{3.123b}$$

where $k = 1$ and 4, and V''_k and I''_k are the voltage and current in an equivalent four-conductor system, respectively, where

$$[T] = \begin{bmatrix} 1 & 0 & 0 & 0 & 0 & 0 \\ 0 & 1 & 0 & 0 & 0 & 0 \\ 0 & 0 & 1 & 0 & 0 & 0 \\ 0 & 0 & 0 & 1 & 1 & 1 \end{bmatrix}$$

From the above equations, each Y-parameter matrix is reduced from a 6×6 to 4×4 matrix.

$$[Y'_{ij}] = [T] \cdot [Y_{ij}] \cdot [T]_t \tag{3.124}$$

When two major sections are connected in cascade as illustrated in Figure 3.11, the Y-parameters of the system are given in the following equation.

$$\begin{pmatrix} (I_A) \\ -(I_B) \end{pmatrix} = \begin{bmatrix} [Y_{11T}] & -[Y_{12T}] \\ -[Y_{21T}] & [Y_{22T}] \end{bmatrix} \cdot \begin{pmatrix} (V_A) \\ (V_B) \end{pmatrix} \tag{3.125}$$

where

$$[Y_{11T}] = [Y''_{11A}] - [Y''_{12A}] \cdot ([Y''_{11B}] + [Y''_{22A}])^{-1} \cdot [Y''_{21A}]$$
$$[Y_{12T}] = [Y''_{12A}] \cdot ([Y''_{11B}] + [Y''_{22A}])^{-1} \cdot [Y''_{12B}]$$
$$[Y_{21T}] = [Y''_{21B}] \cdot ([Y''_{11B}] + [Y''_{22A}])^{-1} \cdot [Y''_{21A}]$$
$$[Y_{22T}] = [Y''_{22B}] - [Y''_{21B}] \cdot ([Y''_{11B}] + [Y'_{22A}])^{-1} \cdot [Y''_{12B}]$$
$$[Y''_{kkA}] = [Y'_{kkA}] + [G_g] \quad (k = 1,\ 2)$$
$$[Y''_{kkB}] = [Y'_{kkB}] + [G_g] \quad (k = 1,\ 2)$$
$$[G_g] = \begin{bmatrix} 0 & 0 & 0 & 0 \\ 0 & 0 & 0 & 0 \\ 0 & 0 & 0 & 0 \\ 0 & 0 & 0 & 1/2R_g \end{bmatrix}$$

When "n" major sections are connected in cascade, it is possible to reduce the computation time by converting "n" to a binary number and adopting (3.125), that is, when "n" is $2^{m-1} < n \leq 2^m$, the Y-parameters are evaluated with only "m" multiplications of intermediate products.

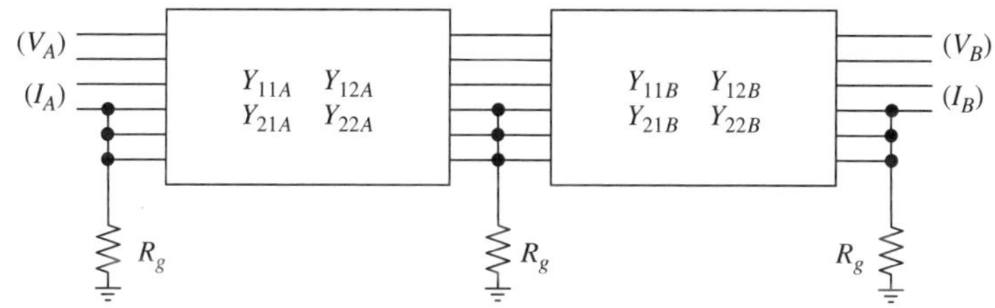

Figure 3.11 Two major sections

3.5.2.4 Approximate Formulation

Figure 3.12 shows another cross-bonding point added to the right-hand side of node 4' of Figure 3.10. Sheaths are, in general, connected together and grounded at both ends of one major section. Therefore, the circuit in Figure 3.12 is equivalent to that in Figure 3.10.

The length of one minor section being several hundred meters, it is possible to replace one minor section by one lumped-parameter circuit. Figure 3.13 shows an equivalent circuit of one major section using three lumped-parameter circuits. Derivation of voltage (ΔV_k) and current (ΔI_k) between two nodes is given in the following form.

$$(\Delta V_k) = -[Z] \cdot l \cdot (I_k) \tag{3.126a}$$

$$(\Delta I_k) \simeq [Y] \cdot l \cdot (V_k) \tag{3.126b}$$

where

$$[Z] = \begin{array}{c} \begin{array}{cccccc} \text{core a} & \text{b} & \text{c} & \text{sheath a} & \text{b} & \text{c} \end{array} \\ \left[\begin{array}{cccccc} Z_{11} & Z_{12} & Z_{13} & Z_{14} & Z_{15} & Z_{16} \\ Z_{12} & Z_{22} & Z_{23} & Z_{24} & Z_{25} & Z_{26} \\ Z_{13} & Z_{23} & Z_{33} & Z_{34} & Z_{35} & Z_{36} \\ Z_{14} & Z_{24} & Z_{34} & Z_{44} & Z_{45} & Z_{46} \\ Z_{15} & Z_{25} & Z_{35} & Z_{45} & Z_{55} & Z_{56} \\ Z_{16} & Z_{26} & Z_{36} & Z_{46} & Z_{56} & Z_{66} \end{array}\right] \end{array} \begin{array}{l} \text{core a} \\ \text{b} \\ \text{c} \\ \text{sheath a} \\ \text{b} \\ \text{c} \end{array}$$

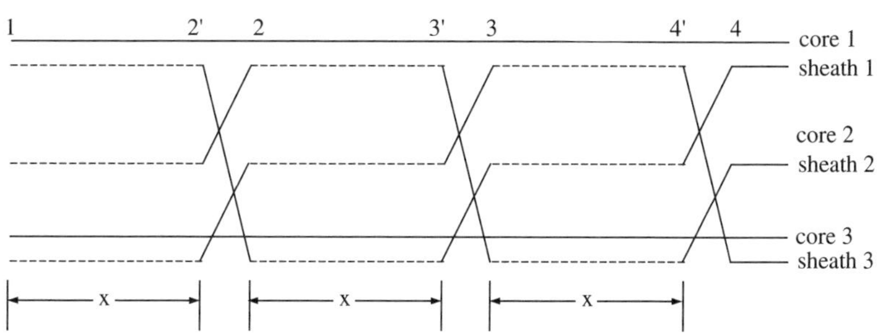

Figure 3.12 One major section with one more cross-bonding joint

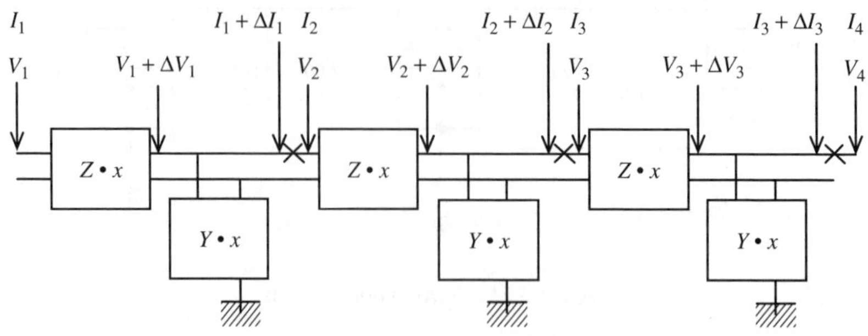

Figure 3.13 Equivalent lumped-parameter circuit

From Figure 3.13, the following relation is obtained.

$$(V'_k) = (V_k) + (\Delta V_k) \tag{3.127}$$

The voltage drop between the sending and receiving ends per unit length is given in the following form.

$$(\Delta V)/3x \simeq -[Z'] \cdot (I_1) \tag{3.128}$$

From (3.108) and (3.127),

$$\begin{aligned}\Delta V/3x &= \{(V_4) - (V_1)\}/3x = \{[R]^{-1} \cdot (V'_4) - (V_1)\}/3x \\ &= \{[R]^{-1} \cdot ((V_3) + (\Delta V_3)) - (V_1)\}/3x \\ &= \{[R]^{-1} \cdot (V_3) + [R]^{-1}(\Delta V_3) - (V_1)\}/3x\end{aligned}$$

In the same manner,

$$\Delta V/3x = \{(\Delta V_1) + [R]^{-2} \cdot (\Delta V_2) + [R]^{-1} \cdot (\Delta V_3)\}/3x \tag{3.129}$$

Substituting (3.128) and (3.108) into the above equation and neglecting the displacement of the current, the following relation is obtained.

$$\begin{aligned}\Delta V/3x &= \{[Z] \cdot (I_1) + [R]^{-2} \cdot [Z] \cdot (I_2) + [R]^{-1} \cdot [Z] \cdot (I_3)\}/3x \\ &\simeq \{([Z] + [R]^{-2} \cdot [Z] \cdot [R]^{-1} + [R]^{-1} \cdot [Z] \cdot [R]^{-2}) \cdot (I_1)\}/3x\end{aligned}$$

Substituting (3.109) into the above equation, the voltage drop between both ends is given in the following equation.

$$\Delta V/3x = \{([Z] + [R] \cdot [Z] \cdot [R]^{-1} + [R]^{-1} \cdot [Z] \cdot [R]) \cdot (I_1)\}/3 \tag{3.130}$$

Thus,

$$[Z'] = ([Z] + [R] \cdot [Z] \cdot [R]^{-1} + [R]^{-1}[Z] \cdot [R])/3 \tag{3.131}$$

The above impedance matrix is given as:

$$[Z'] = \begin{bmatrix} [Z_1] & [Z_2] \\ [Z_2]_t & [Z_3] \end{bmatrix}, \quad [Z_1] = \begin{bmatrix} Z_{11} & Z_{12} & Z_{13} \\ Z_{12} & Z_{22} & Z_{23} \\ Z_{13} & Z_{23} & Z_{33} \end{bmatrix}$$

$$[Z_2] = \begin{bmatrix} Z_a & Z_a & Z_a \\ Z_b & Z_b & Z_b \\ Z_c & Z_c & Z_c \end{bmatrix}, \quad [Z_3] = \begin{bmatrix} Z_d & Z_e & Z_e \\ Z_e & Z_d & Z_e \\ Z_e & Z_e & Z_d \end{bmatrix}$$

and

$$Z_a = (Z_{14} + Z_{15} + Z_{16})/3, \quad Z_b = (Z_{24} + Z_{25} + Z_{26})/3,$$
$$Z_c = (Z_{34} + Z_{35} + Z_{36})/3, \quad Z_d = (Z_{44} + Z_{55} + Z_{66})/3,$$
$$Z_e = (Z_{45} + Z_{46} + Z_{56})/3$$

In the same manner, the displacement current per unit length is given by:

$$(\Delta I)/3x \simeq -[Y'] \cdot (V_1) \tag{3.132}$$

where

$$[Y'] = ([Y] + [R] \cdot [Y] \cdot [R]^{-1} + [R]^{-1} \cdot [Y] \cdot [R])/3 \tag{3.133}$$

Thus, a nonhomogeneous major section can be represented by an equivalent homogeneous line with the series impedance $[Z']$ and shunt admittance $[Y']$. The series impedance matrix $[Z]$ is symmentrical. It is clear that the self and mutual impedances of the cores are not changed, but mutual impedances between the cores and sheaths are given as the mean values. This is similar to that of a transposed overhead line.

From (3.123), the following relation is obtained.

$$(\Delta V) = [T]_t \cdot (\Delta V'') \tag{3.134a}$$

$$(\Delta I'') = [T] \cdot (\Delta I) \tag{3.134b}$$

Substituting (3.123 a,b) and (3.134 a,b) into (3.128) and (3.132),

$$(\Delta V'')/3x = -[Z''] \cdot (I''_1), \quad (\Delta I'')/3x = -[Y''] \cdot (V''_1) \tag{3.135}$$

where

$$[Z''] = ([T] \cdot [Z']^{-1} \cdot [T]_t)^{-1} \tag{3.136a}$$

$$[Y''] = [T] \cdot [Y'] \cdot [T]_t \tag{3.136b}$$

The equivalent series impedance matrix $[Z'']$ and shunt admittance matrix $[Y'']$ explained above are 4×4 matrices.

From the above, one major section of a nonhomogeneous cross-bonded cable can be represented by a homogeneous line with the series impedance $[Z'']$ and shunt admittance $[Y'']$ which are 4×4 matrices.

The theoretical formulation explained in this section has been adopted in a transient analysis by a frequency-domain method [3, 12, 13, 15, 20, 21].

3.5.3 *Homogeneous Model of a Cross-bonded Cable*

It is not easy to discuss the wave propagation characteristic on a cross-bonded cable based on a theoretical formulation explained in the previous section. The wave propagation characteristic within one minor section is the same as those in a three-phase SC cable explained in Sections 3.3 and 3.4. If a cross-bonded cable with more than two major sections is concerned, the impedance Z and admitance Y in (3.136) of the homogeneous 4×4 model are very useful to discuss the wave propagation characteristic. By adopting the homogeneous model, the wave propagation characteristic on a 275 kV tunnel-installed three-phase SC cable (1400 mm^2) in Figure 3.14 is discussed in this section. Table 3.10 shows wave propagation related parameters of a solidly bonded cable and a cross-bonded cable.

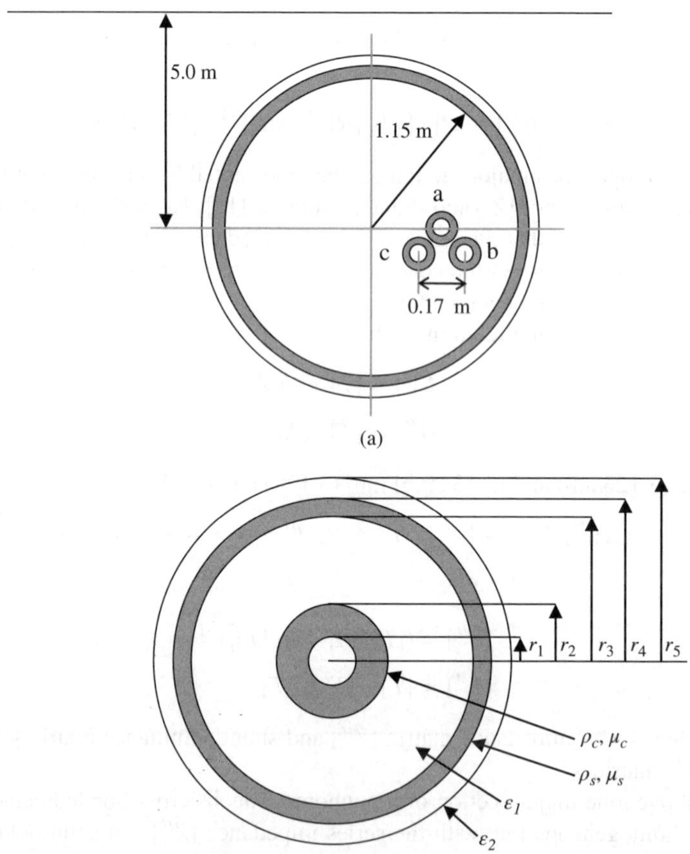

Figure 3.14 Tunnel-installed three-phase cable: (a) cable configuration; and (b) cross-section of a phase cable: $r_1 = 0$, $r_2 = 30.45$ mm, $r_3 = 71.15$ mm, $r_4 = 74.80$ mm, $r_5 = 81.61$ mm, $\varepsilon_1 = 3.1$, $\varepsilon_2 = 4.0$, $\rho_c = 1.82 \times 10^{-8}$ Ωm, $\rho_s = 2.83 \times 10^{-8}$ Ωm, $\mu_c = \mu_s = 1.0$, $\rho_e = 100$ Ωm

Table 3.10 Parameters of a tunnel-installed cable at 100 kHz

Solidly bonded cable

R (Ω/km) and L (0.1 mH/km)

i	1	2	3	4	5	6
		core-to-core			core-to-sheath	
R	2.867	2.060	1.850	2.187	2.060	1.850
L	6.951	3.399	3.519	5.242	3.399	3.519
		3.260	2.010	2.060	2.525	2.010
		6.767	3.454	3.399	5.058	3.454
			2.757	1.850	2.010	2.076
			7.015	3.519	3.454	5.306
				2.187	2.060	1.850
				5.242	3.399	3.519
					2.525	2.010
					5.058	3.454
						2.076
						5.306

C (0.1 μF/km)

i	1	2	3	4	5	6
		core-to-core			core-to-sheath	
	2.025	0	0	−2.025	0	0
		2.025	0	0	−2.025	0
		0	2.025	0	0	−2.025
				2.511	−0.194	−0.200
					2.529	−0.204
						2.513

Voltage transformation matrix A^{-1}

	core 1	2	3	sheath 1	2	3
core 1	0	0	0	−0.312	0.738	−0.347
2	0	0	0	0.559	−0.013	−0.506
3	0	0	0	0.369	0.307	0.359
sheath 1	0	1	0	0	−1	0
2	1	0	0	−1	0	0
3	0	0	1	0	0	−1

Modal propagation constants

Mode	1	2	3	4–6
α (0.1 dB/km)	0.2050	0.0698	0.8450	1.017
v (m/μs)	288.0	288.4	296.8	170.0

(*continued overleaf*)

Table 3.10 (*continued*)

Cross-bonded cable using a homogeneous model

R (Ω/km) and L (0.1 mH/km)

i	1	2	3	4
R	2.867	2.060	1.850	2.032
L	6.951	3.399	3.519	4.053
		3.260	2.010	2.199
		6.767	3.454	3.970
			2.757	1.979
			7.015	4.093
				2.070
				4.039

C (0.1 μF/km)

i	1	2	3	4
	2.025	0	0	−2.025
	0	2.025	0	−2.025
	0	0	2.025	−2.025
	−2.025	−2.025	−2.025	6.358

A^{-1}

core 1 2 3 sheath

$$\begin{bmatrix} 0.667 & -0.293 & -0.374 & 0 \\ -0.040 & 0.516 & -0.476 & -0.003 \\ -0.005 & 0.024 & -0.019 & 1.001 \\ 1/3 & 1/3 & 1/3 & -1 \end{bmatrix}$$

Modal propagation constants

Mode	1	2	3	4–6
α (0.1 dB/km)	1.015	1.024	0.754	1.017
v (m/μs)	119.3	119.9	296.8	170.0

3.5.3.1 Impedance

It is clear in Table 3.10 that the three-phase impedances for cores are the same in the solidly bonded and cross-bonded cables. The core-to-sheath mutual impedances in the cross-bonded cable are given as the average of the three-phase impedances. Also, the sheath self-impedance is the average of the three-phase sheath impedances on the solidly bonded cable, i. e.

$$Z(4, j)_{cross} = (1/3)\sum_{i=4}^{6} Z(i, j)_{solid}\ ; j = 1\text{–}4 \tag{3.137}$$

3.5.3.2 Capacitance

The upper 3×3 elements of the capacitance matrix are the same in the solidly and cross-bonded cable as observed in Table 3.10. The elements of the fourth row and colum in the cross-bonded cable are given as the sum of the upper right elements of the capacitance matrix in the solidly bonded cable, that is,

$$C(4, j)_{cross} = \sum_{i=4}^{6} C(i, j)_{solid}\ ; j = 1\text{–}4 \tag{3.138}$$

3.5.3.3 Voltage Transformation Matrix $[A]^{-1}$

The characteristic of the voltage transformation matrix has been explained in Section 3.3.2, (3.98), in a solidly bonded (homogeneous) cable. Table 3.10 follows the type C matrix for $f = 100$ kHz $> f_c = 270$ Hz. The upper right 3×3 matrix corresponds to the transformation matrix of a transposed three-phase overhead line defined in the following form [3].

$$[A_T]^{-1} = (1/3)\begin{bmatrix} -1 & 2 & -1 \\ 1.5 & 0 & -1.5 \\ 1 & 1 & 1 \end{bmatrix} \begin{matrix} :\ \text{first aerial mode} \\ :\ \text{second aerial mode} \\ :\ \text{earth-return mode} \end{matrix} \tag{3.139}$$

In a cross-bonded cable in Table 3.10, 1/3 for every core and −1 for sheath in the fourth column represent the coaxial mode. The third column which is roughly given by (0, 0, 0, 1) corresponds to the earth (pipe)-return mode. The first column $(2/3, -1/3, -1/3, 0)$ represents the first intersheath mode (corresponding to the first aerial mode in the above equation), and the second column $(0, 1/2, -1/2, 0)$ is the second intersheath mode corresponding to the second aerial mode.

3.5.3.4 Attenuation and Velocity

Because the cable system in Figure 3.14 is installed in a tunnel, that is, air insulated, the propagation characteristic becomes similar to that of an overhead line except the coaxial mode, that is, modes 4–6 in the solidly bonded cable and mode 4 in the cross-bonded cable which are shown to be identical.

Mode 3 in the solidly and cross-bonded cable is the earth-return mode. The propagation velocity in both cables is the same, but the attenuation is greater in the solidly bonded cable. Instead, the two intersheath modes (modes 1 and 2) in the cross-bonded cable show far greater attenuation than those in the solidly bonded cable, and far smaller propagation velocity. The difference in the intersheath mode propagation is the basic characteristic of a cross-bonded cable, and is caused by the difference of the sheath capacitance as is clear from Table 3.10, that is, the capacitance is about three times greater in the cross-bonded cable than that in the solidly bonded cable.

3.5.4 Difference between Tunnel-installed and Buried Cables

An example investigated in the previous section is a tunnel-installed cable, that is, an air-insulated cable. Thus, it shows a similar characteristic to that of an overhead line. A directly buried underground cable shows a quite different characteristic of wave propagation as explained in Sections 3.3 and 3.4. A difference between the tunnel-installed and directly buried cross-bonded cable is investigated in this section.

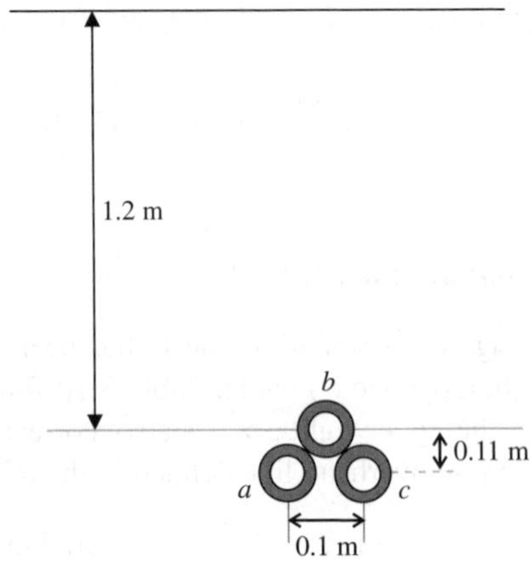

Figure 3.15 Configuration of a buried cable

Figure 3.15 illustrates a buried cable where the cable cross-section is the same as in Figure 3.14b with the following parameters.

$$r_1 = 0,\ r_2 = 2.075\ \text{cm},\ r_3 = 3.925\ \text{cm},\ r_4 = 4.276\ \text{cm},\ r_5 = 4.794\ \text{cm},$$

$$\varepsilon_1 = 2.68,\ \varepsilon_2 = 2.30,\ \rho_c = 3.19 \times 10^{-8}\ \Omega\text{m},\ \rho_s = 1.87 \times 10^{-7}\ \Omega\text{m},\ \rho_e = 100\ \Omega\text{m}$$

Data 3.1 shows the output list of the Cable Parameters of the EMTP-ATP (electromagnetic transients program-alternative transients program) at frequency 50 kHz in the case of solidly bonded cable. Data 3.2 is the output list of a homogeneous model of a cross-bonded cable.

Figure 3.16 shows modal attenuation and velocity for a solidly bonded cable. The attenuation and the velocity in Figure 3.16 are quite different from those in Table 3.10, that is, far greater attenuation and far smaller velocity except for the coaxial mode (modes 4–6) in the buried cable as expected. No difference is observed for the coaxial mode in any cable condition.

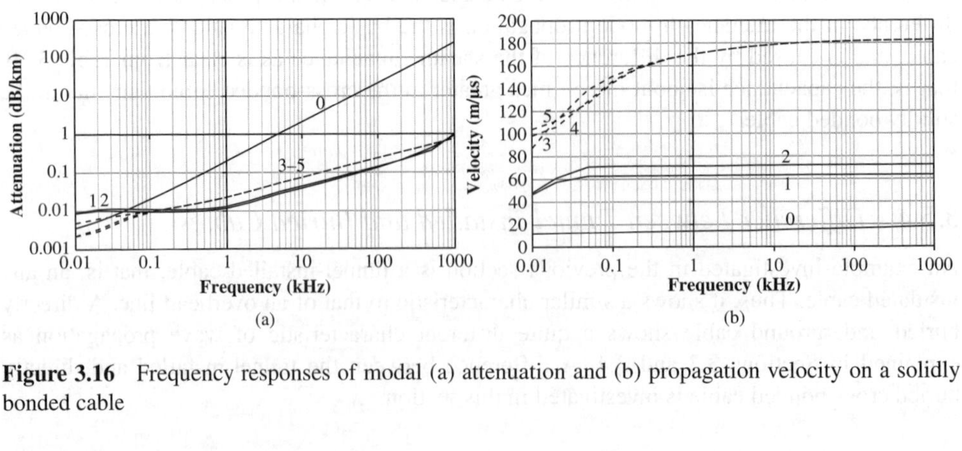

Figure 3.16 Frequency responses of modal (a) attenuation and (b) propagation velocity on a solidly bonded cable

Data 3.1 Cable Parameters output of an underground three-phase SC cable consisting of a core and a sheath

```
 ---   34    cards of disk file read into card cache cells   1    onward.
 1st card (Lists   1-10).      |   8000   15000   15000     340    60000    120
 2nd card (Lists 11-20).       |    360     900   36400     255   129600     38
 3rd card (Lists 21-29).       |   3000   15000  160000      24    30000   1000
 Supplemental offsets.          | 240000      742
Alternative Transients Program (ATP),  Watcom translation.   All rights reserved by Can/Am user group of Portland, Oregon, USA.
 Date (dd-mth-yy) and time of day (hh.mm.ss) = 25-Feb-14  12.18.05    Name of disk plot file is   C:¥ATPWNT¥UNTITLED.pl4
Consult the 860-page ATP Rule Book of the Can/Am EMTP User Group in Portland,  Oregon, USA.   Program is no older than Sept., 1998.
Total size of LABCOM tables =  3639004   INTEGER words.   VARDIM  List Sizes follow :   8002 15000 15000   340 60000  1200
 13140 120000 2250 1900  360  900 36400   255  129K   384    45   254 240000 100000  3000 15000  160K   24 30000 10000    600 252000
-----------------------------------------------+------------------------------------------------------------------------------
Descriptive interpretation of input data cards.   |  Input data card images are shown below, all 80 columns, character by character
                                                     0         1         2         3         4         5         6         7         8
                                                     012345678901234567890123456789012345678901234567890123456789012345678901234567890
-----------------------------------------------+------------------------------------------------------------------------------
Comment card.   NUMDCD = 1.                          |C data:C:¥ATPWNT¥UNTITLED.DAT
Comment card.   NUMDCD = 2.                          |C Cable Constants for underground RTE cable
Comment card.   NUMDCD = 3.                          |C ===Cable Constants===
Marker card preceding new EMTP data case.         |BEGIN NEW DATA CASE
Find parameters of cables.  Type = 0.             |CABLE CONSTANTS
Prof. Akihiro Ametani`s 1994 code.   Order = 38   |CABLE PARAMETERS
Comment card.   NUMDCD = 7.                          |C ==Miscellaneous Data Card==
Comment card.   NUMDCD = 8.                          |C TY][SYS][NPC][EAR][KMO][ZFL]
Misc.   2  -1   3   0   1   1   0   0   0   0   0|   2  -1    3          1   1
Comment card.   NUMDCD = 10.                         |C ==The number of conductors==
Comment card.   NUMDCD = 11.                         |C N1][N2 ][N3 ]
Cable conductor numbers.  2 2 2                   |   2    2    2
Comment card.   NUMDCD = 13.                         |C ==Physical parameters==
Comment card.   NUMDCD = 14.                         |C    r1    ][    r2    ][    r3    ][    r4    ][    r5    ]
Radii.    .00    .02    .04    .04    .05         |   0.0000  20.75E-3  39.25E-3  42.76E-3  47.94E-3
Comment card.   NUMDCD = 16.                         |C  rohc  ][  muc   ][  muI1  ][ epsI1 ][  rhos  ][   mus  ][  muI2  ][ epsI2 ]
Physical constants for conductors & insulators.  |  3.19E-8      1.0       1.0      2.68    2.83E-8     1.000     1.000      2.3
Comment card.   NUMDCD = 18.                         |C    r1    ][    r2    ][    r3    ][    r4    ][    r5    ]
Radii.    .00    .02    .04    .04    .05         |   0.0000  20.75E-3  39.25E-3  42.76E-3  47.94E-3
Comment card.   NUMDCD = 20.                         |C  rohc  ][  muc   ][  muI1  ][ epsI1 ][  rhos  ][   mus  ][  muI2  ][ epsI2 ]
Physical constants for conductors & insulators.  |  3.19E-8      1.0       1.0      2.68    2.83E-8     1.000     1.000      2.3
Comment card.   NUMDCD = 22.                         |C    r1    ][    r2    ][    r3    ][    r4    ][    r5    ]
Radii.    .00    .02    .04    .04    .05         |   0.0000  20.75E-3  39.25E-3  42.76E-3  47.94E-3
Comment card.   NUMDCD = 24.                         |C  rohc  ][  muc   ][  muI1  ][ epsI1 ][  rhos  ][   mus  ][  muI2  ][ epsI2 ]
Physical constants for conductors & insulators.  |  3.19E-8      1.0       1.0      2.68    2.83E-8     1.000     1.000      2.3
Comment card.   NUMDCD = 26.                         |C ==Location==
Comment card.   NUMDCD = 27.                         |C   Vert1 ][ Horiz1 ][   Vert2 ][ Horiz2 ][   Vert3 ][ Horiz3 ][   Vert4 ][ Horiz4 ]
(Y,X)  1.310E+00 -5.000E-02  1.200E+00  0.000E+00 |     1.31     -5.E-2      1.20       0.00       1.31      5.E-2
Comment card.   NUMDCD = 29.                         |C ==Frequency Card==
Comment card.   NUMDCD = 30.                         |C       rhoe      ][       Freq.    ][DEC][PNT][ DIST ][   IPUN   ]
Frequency card 1.000E+02 5.000E+04   0   1 2.00E+03 |        100.0             50.E3                  2.00E3          2
          3 -phase SC cable with 6   conductors.   0   grounded.   Total # of conductors = 6
            Under-ground cable
            Cable constants
            Phase 1   Boundary radii ;
             1  0.00000E+00   2   2.07500E-02   3   3.92500E-02   4   4.27600E-02   5   4.79400E-02
                   Resistivity(ohm-m): Core 3.19000000E-08, Sheath 2.83000000E-08, Armor 9.99999994E-09
                   Relative permeability;   1.00000000E+00, 1.00000000E+00, 1.00000000E+00
                   Insulator:   1; (Relative) Permeability   1.0    Permittivity   2.68    Velocity(m/s) 1.83127E+08
                   Insulator:   2; (Relative) Permeability   1.0    Permittivity   2.30    Velocity(m/s) 1.97677E+08
                   DC resistance(ohm/m),          2.35833E-05              3.12941E-05
            Phase 2   Boundary radii ;
             1  0.00000E+00   2   2.07500E-02   3   3.92500E-02   4   4.27600E-02   5   4.79400E-02
                   Resistivity(ohm-m): Core 3.19000000E-08, Sheath 2.83000000E-08, Armor 9.99999994E-09
                   Relative permeability;   1.00000000E+00, 1.00000000E+00, 1.00000000E+00
                   Insulator:   1; (Relative) Permeability   1.0    Permittivity   2.68    Velocity(m/s) 1.83127E+08
                   Insulator:   2; (Relative) Permeability   1.0    Permittivity   2.30    Velocity(m/s) 1.97677E+08
                   DC resistance(ohm/m),          2.35833E-05              3.12941E-05
            Phase 3   Boundary radii ;
             1  0.00000E+00   2   2.07500E-02   3   3.92500E-02   4   4.27600E-02   5   4.79400E-02
                   Resistivity(ohm-m): Core 3.19000000E-08, Sheath 2.83000000E-08, Armor 9.99999994E-09
                   Relative permeability;   1.00000000E+00, 1.00000000E+00, 1.00000000E+00
                   Insulator:   1; (Relative) Permeability   1.0    Permittivity   2.68    Velocity(m/s) 1.83127E+08
                   Insulator:   2; (Relative) Permeability   1.0    Permittivity   2.30    Velocity(m/s) 1.97677E+08
                   DC resistance(ohm/m),          2.35833E-05              3.12941E-05
```

Distance from Earth surface ; 1 1.3100 2 1.2000 3 1.3100
Distance between phases : -.0500 .0000 .0500
Earth ; resistivity 1.00000000E+02
Punched cards: 1) pi-circuit (IPUNCH=1) or distributed line (IPUNCH=2);
IPUNCH = 2 Total length = 2.00000000E+03
Frequency 5.00000000E+04 Hz

Earth impedance ZE

4.63980E-02	4.64649E-02	4.63961E-02	4.63980E-02	4.64649E-02	4.63961E-02
4.07829E-01	3.49581E-01	3.61634E-01	4.07829E-01	3.49581E-01	3.61634E-01
4.64649E-02	4.65438E-02	4.64649E-02	4.64649E-02	4.65438E-02	4.64649E-02
3.49581E-01	4.07498E-01	3.49581E-01	3.49581E-01	4.07498E-01	3.49581E-01
4.63961E-02	4.64649E-02	4.63980E-02	4.63961E-02	4.64649E-02	4.63980E-02
3.61634E-01	3.49581E-01	4.07829E-01	3.61634E-01	3.49581E-01	4.07829E-01
4.63980E-02	4.64649E-02	4.63961E-02	4.63980E-02	4.64649E-02	4.63961E-02
4.07829E-01	3.49581E-01	3.61634E-01	4.07829E-01	3.49581E-01	3.61634E-01
4.64649E-02	4.65438E-02	4.64649E-02	4.64649E-02	4.65438E-02	4.64649E-02
3.49581E-01	4.07498E-01	3.49581E-01	3.49581E-01	4.07498E-01	3.49581E-01
4.63961E-02	4.64649E-02	4.63980E-02	4.63961E-02	4.64649E-02	4.63980E-02
3.61634E-01	3.49581E-01	4.07829E-01	3.61634E-01	3.49581E-01	4.07829E-01

Conductor internal impedance ZC

1.19571E-03	0.00000E+00	0.00000E+00	2.79470E-04	0.00000E+00	0.00000E+00
4.84240E-02	0.00000E+00	0.00000E+00	7.46289E-03	0.00000E+00	0.00000E+00
0.00000E+00	1.19571E-03	0.00000E+00	0.00000E+00	2.79470E-04	0.00000E+00
0.00000E+00	4.84240E-02	0.00000E+00	0.00000E+00	7.46289E-03	0.00000E+00
0.00000E+00	0.00000E+00	1.19571E-03	0.00000E+00	0.00000E+00	2.79470E-04
0.00000E+00	0.00000E+00	4.84240E-02	0.00000E+00	0.00000E+00	7.46289E-03
2.79470E-04	0.00000E+00	0.00000E+00	2.79425E-04	0.00000E+00	0.00000E+00
7.46289E-03	0.00000E+00	0.00000E+00	7.46283E-03	0.00000E+00	0.00000E+00
0.00000E+00	2.79470E-04	0.00000E+00	0.00000E+00	2.79425E-04	0.00000E+00
0.00000E+00	7.46289E-03	0.00000E+00	0.00000E+00	7.46283E-03	0.00000E+00
0.00000E+00	0.00000E+00	2.79470E-04	0.00000E+00	0.00000E+00	2.79425E-04
0.00000E+00	0.00000E+00	7.46289E-03	0.00000E+00	0.00000E+00	7.46283E-03

Eigenvector rotation has been made.

Total impedance [Zc]

4.75937E-02	4.64649E-02	4.63961E-02	4.66775E-02	4.64649E-02	4.63961E-02
4.56253E-01	3.49581E-01	3.61634E-01	4.15292E-01	3.49581E-01	3.61634E-01
4.64649E-02	4.77395E-02	4.64649E-02	4.64649E-02	4.68232E-02	4.64649E-02
3.49581E-01	4.55922E-01	3.49581E-01	3.49581E-01	4.14961E-01	3.49581E-01
4.63961E-02	4.64649E-02	4.75937E-02	4.63961E-02	4.64649E-02	4.66775E-02
3.61634E-01	3.49581E-01	4.56253E-01	3.61634E-01	3.49581E-01	4.15292E-01
4.66775E-02	4.64649E-02	4.63961E-02	4.66775E-02	4.64649E-02	4.63961E-02
4.15292E-01	3.49581E-01	3.61634E-01	4.15292E-01	3.49581E-01	3.61634E-01
4.64649E-02	4.68232E-02	4.64649E-02	4.64649E-02	4.68232E-02	4.64649E-02
3.49581E-01	4.14961E-01	3.49581E-01	3.49581E-01	4.14961E-01	3.49581E-01
4.63961E-02	4.64649E-02	4.66775E-02	4.63961E-02	4.64649E-02	4.66775E-02
3.61634E-01	3.49581E-01	4.15292E-01	3.61634E-01	3.49581E-01	4.15292E-01

Total admittance [Yc]

0.00000E+00	0.00000E+00	0.00000E+00	0.00000E+00	0.00000E+00	0.00000E+00
7.34848E-05	0.00000E+00	0.00000E+00	-7.34848E-05	0.00000E+00	0.00000E+00
0.00000E+00	0.00000E+00	0.00000E+00	0.00000E+00	0.00000E+00	0.00000E+00
0.00000E+00	7.34848E-05	0.00000E+00	0.00000E+00	-7.34848E-05	0.00000E+00
0.00000E+00	0.00000E+00	0.00000E+00	0.00000E+00	0.00000E+00	0.00000E+00
0.00000E+00	0.00000E+00	7.34848E-05	0.00000E+00	0.00000E+00	-7.34848E-05
0.00000E+00	0.00000E+00	0.00000E+00	0.00000E+00	0.00000E+00	0.00000E+00
-7.34848E-05	0.00000E+00	0.00000E+00	4.25030E-04	0.00000E+00	0.00000E+00
0.00000E+00	0.00000E+00	0.00000E+00	0.00000E+00	0.00000E+00	0.00000E+00
0.00000E+00	-7.34848E-05	0.00000E+00	0.00000E+00	4.25030E-04	0.00000E+00
0.00000E+00	0.00000E+00	0.00000E+00	0.00000E+00	0.00000E+00	0.00000E+00
0.00000E+00	0.00000E+00	-7.34848E-05	0.00000E+00	0.00000E+00	4.25030E-04

Characteristic impedance matrix [Zc] in phase variables, in [ohm]:

51.11234	14.12907	15.29044	27.50136	14.12907	15.14746
-1.44680	-1.16314	-1.15932	-1.18277	-1.16314	-1.15046
14.12907	51.63391	14.12907	14.12907	28.02294	14.12907
-1.16314	-1.46762	-1.16314	-1.16314	-1.20359	-1.16314
15.29044	14.12907	50.82967	15.29044	14.12907	27.35838
-1.15932	-1.16314	-1.42949	-1.15932	-1.16314	-1.17391
27.50136	14.12907	15.29044	27.50136	14.12908	15.14746
-1.18277	-1.16314	-1.15932	-1.18277	-1.16315	-1.15046
14.12907	28.02294	14.12907	14.12908	28.02293	14.12908
-1.16314	-1.20359	-1.16314	-1.16315	-1.20359	-1.16315
15.14746	14.12907	27.35838	15.14746	14.12908	27.50136
-1.15046	-1.16314	-1.17391	-1.15046	-1.16315	-1.18277

Modal transformation matrices follow. These are complex, with the real part displayed above the imaginary part.
The current transformation matrix [Ti] follows. The transpose of this is the inverse of the voltage transformation matrix.
By definition, [Ti] gives the mapping from modal to phase variables: i-phase = [Ti] * i-mode

.00000	.00000	.00000	.00000	.00000	1.00000
.00000	.00000	.00000	.00000	.00000	.00000
.00000	.00000	.00000	.00000	1.00000	.00000
.00000	.00000	.00000	.00000	.00000	.00000
.00000	.00000	.00000	1.00000	.00000	.00000
.00000	.00000	.00000	.00000	.00000	.00000
.33589	-.33203	-.50000	-.01158	.00000	-1.00000
.00019	.00039	.00000	.00069	.00000	.00000
.33203	.67178	.00000	.00000	-1.00000	.00000
-.00039	.00038	.00000	.00000	.00000	.00000
.33589	-.33203	.50000	-.98843	.00000	.00000
.00019	.00039	.00000	-.00069	.00000	.00000

Voltage transformation matrix [Tv]. This gives the mapping from modal to phase variables: v-phase = [Tv] * v-mode

1.00000	-.49426	-1.00000	.00000	.00000	1.00000
.00057	.00058	.00000	.00000	.00000	.00000
.98851	1.00000	.00000	.00000	1.00000	.00000
-.00115	.00057	.00000	.00000	.00000	.00000
1.00000	-.49426	.97685	1.00000	.00000	.00000
.00057	.00058	.00138	.00000	.00000	.00000
1.00000	-.49426	-1.00000	.00000	.00000	.00000
.00057	.00058	.00000	.00000	.00000	.00000
.98851	1.00000	.00000	.00000	.00000	.00000
-.00115	.00057	.00000	.00000	.00000	.00000
1.00000	-.49426	1.00000	.00000	.00000	.00000
.00057	.00058	.00000	.00000	.00000	.00000

Modal components

Mode	Attenuation	Velocity	Impedance (Ohm/m)		Admittance (S/m)		Charact. imp. (Ohm)		Charact. Admit. (S)	
No.	(db/km)	(m/mic.s)	Real	Imag.	Real	Imag.	Real	Imag.	Real	Imag.
1	1.07095E+01	15.79	4.68921E-02	3.77008E-01	0.00000E+00	1.04660E-03	19.016	-1.178	5.23862E-02	3.24538E-03
2	9.02352E-02	63.57	1.96221E-04	4.66732E-02	0.00000E+00	5.23301E-04	9.444	-.020	1.05886E-01	2.22579E-04
3	9.88946E-02	72.33	1.40656E-04	2.68275E-02	0.00000E+00	7.03129E-04	6.177	-.016	1.61891E-01	4.24392E-04
4	1.68524E-01	181.07	9.16191E-04	4.09610E-02	0.00000E+00	7.34848E-05	23.611	-.264	4.23479E-02	4.73547E-04
5	1.68524E-01	181.07	9.16190E-04	4.09610E-02	0.00000E+00	7.34848E-05	23.611	-.264	4.23479E-02	4.73547E-04
6	1.68524E-01	181.07	9.16190E-04	4.09610E-02	0.00000E+00	7.34848E-05	23.611	-.264	4.23479E-02	4.73547E-04

Blank card terminating frequency cards. |BLANK ENDING CABLE CONSTANTS CASE
Request for flushing of punch buffer. |$PUNCH

```
A listing of 80-column card images now being flushed from punch buffer follows.
================================================================================
12345678901234567890123456789012345678901234567890123456789012345678901234567890
================================================================================
C  <++++++>  Cards punched by support routine on  25-Feb-14  12.18.05  <++++++>
C CABLE CONSTANTS
C CABLE PARAMETERS
C C ==Miscellaneous Data Card==
C C TY][SYS][NPC][EAR][KMO][ZFL]
C    2   -1    3         1    1
C C ==The number of conductors==
C C N1][N2 ][N3 ]
C    2    2    2
C C ==Physical parameters==
C C   r1    ][   r2    ][   r3    ][   r4    ][   r5    ]
C     0.0000  20.75E-3  39.25E-3  42.76E-3  47.94E-3
C C  rohc  ][  muc   ][  mul1  ][ epsI1 ][  rhos  ][   mus  ][  mul2  ][ epsI2
C    3.19E-8      1.0       1.0      2.68   2.83E-8     1.000     1.000       2
C C   r1    ][   r2    ][   r3    ][   r4    ][   r5    ]
C     0.0000  20.75E-3  39.25E-3  42.76E-3  47.94E-3
C C  rohc  ][  muc   ][  mul1  ][ epsI1 ][  rhos  ][   mus  ][  mul2  ][ epsI2
C    3.19E-8      1.0       1.0      2.68   2.83E-8     1.000     1.000       2
C C   r1    ][   r2    ][   r3    ][   r4    ][   r5    ]
C     0.0000  20.75E-3  39.25E-3  42.76E-3  47.94E-3
C C  rohc  ][  muc   ][  mul1  ][ epsI1 ][  rhos  ][   mus  ][  mul2  ][ epsI2
C    3.19E-8      1.0       1.0      2.68   2.83E-8     1.000     1.000       2
C C ==Location==
C C  Vert1 ][ Horiz1 ][  Vert2 ][ Horiz2 ][  Vert3 ][ Horiz3 ][  Vert4 ][ Horiz4
C       1.31     -5.E-2      1.20      0.00      1.31     5.E-2
C C ==Frequency Card==
C C       rhoe     ][      Freq.    ][DEC][PNT][ DIST ][  IPUN  ]
C             100.0           50.E3              2.00E3           2
$UNITS, 0.0, 0.0
$VINTAGE, 1
C    Branch cards for constant-parameter distributed:
-1                              4.68921E-02 1.90160E+01 1.57851E+07-2.00000E+03 1 16
-2                              1.96221E-04 9.44406E+00 6.35681E+07-2.00000E+03 1 16
-3                              1.40656E-04 6.17695E+00 7.23337E+07-2.00000E+03 1 16
-4                              9.16191E-04 2.36110E+01 1.81067E+08-2.00000E+03 1 16
-5                              9.16190E-04 2.36110E+01 1.81067E+08-2.00000E+03 1 16
-6                              9.16190E-04 2.36110E+01 1.81067E+08-2.00000E+03 1 16
    .00000000  -.00000007  -.00000013   .00000009   .00000007   .99999983
    .00000000   .00000005   .00000010  -.00000006  -.00000005   .00000000
    .00000000   .00000014   .00000001   .00000006   .99999985   .00000000
    .00000000  -.00000010  -.00000001  -.00000004   .00000000   .00000000
    .00000000  -.00000007   .00000013   .99999984   .00000000   .00000000
    .00000000   .00000005  -.00000009   .00000000   .00000000   .00000000
    .33589158  -.33203238  -.49999987  -.01157526   .00000026 -1.00000070
    .00019145   .00038729   .00000156   .00068656  -.00000019   .00000063
    .33203245   .67178303  -.00000001   .00000026 -1.00000056   .00000032
   -.00038734   .00038299   .00000001  -.00000019   .00000051  -.00000024
    .33589158  -.33203238   .49999988  -.98842504   .00000032   .00000050
    .00019145   .00038729  -.00000157  -.00068622  -.00000023  -.00000037
$VINTAGE, 0
$UNITS, -1., -1.,
==========<  End of LUNIT7 punched cards as flushed by  $PUNCH  request  >=======

Blank card ending  "CABLE CONSTANTS"  cases.        |BLANK ENDING SEMLYEN SETUP
Timing figures characterizing central processor (CP) solution speed.  ---------------------------  CP sec   Wait sec   Real sec
                                                                               Totals     0.060      0.000      0.060
```

Data 3.2 Cable Constants output of a homogeneous model of a three-phase cross-bonded cable

```
--- 37 cards of disk file read into card cache cells 1 onward.
 1st card (Lists 1-10).      | 8000 15000 15000 340 60000 120
 2nd card (Lists 11-20).     | 360 900 36400 255 129600 38
 3rd card (Lists 21-29).     | 3000 15000 160000 24 30000 1000
 Supplemental offsets.       | 240000 742
Alternative Transients Program (ATP), Watcom translation. All rights reserved by Can/Am user group of Portland, Oregon, USA.
 Date (dd-mth-yy) and time of day (hh.mm.ss) = 13-Feb-14 14.58.27 Name of disk plot file is C:¥ATPWNT¥UNTITLED.pl4
Consult the 860-page ATP Rule Book of the Can/Am EMTP User Group in Portland, Oregon, USA. Program is no older than Sept., 1998.
Total size of LABCOM tables = 3639004 INTEGER words. VARDIM List Sizes follow : 8002 15000 15000 340 60000 1200
 13140 120000 2250 1900 360 900 36400 255 129K 384 45 254 240000 100000 3000 15000 160K 24 30000 10000 600 252000
-------------------------------------------------+--------------------------------------------------------------------------------
Descriptive interpretation of input data cards. | Input data card images are shown below, all 80 columns, character by character
                                                  0         1         2         3         4         5         6         7         8
                                                  012345678901234567890123456789012345678901234567890123456789012345678901234567890
-------------------------------------------------+--------------------------------------------------------------------------------
Comment card. NUMDCD = 1.                        |C data:C:¥ATPWNT¥UNTITLED.DAT
Comment card. NUMDCD = 2.                        |C Cable Constants for underground RTE cable
Comment card. NUMDCD = 3.                        |C ===Cable Constants===
Marker card preceding new EMTP data case.        |BEGIN NEW DATA CASE
Erase all of 0 cards in the punch buffer.        |$ERASE
Find parameters of cables. Type = 0.             |CABLE CONSTANTS
Special request for crossbonded cable.           |PUNCH
Comment card. NUMDCD = 8.                        |C ==Miscellaneous Data Card==
Comment card. NUMDCD = 9.                        |C TY][SYS][NPC][EAR][KMO][ZFL]
Comment card. NUMDCD = 11.                       |C NP][NCR][IRS][ XMAJOR ][ RSG ]!
Crossbonded. 6 1 0 6.000E+03 5.000E+00           | 6 1 0 6.E3 5.A
 *** Warning. Out of order. Preceding crossbonded cable card actually follows the upcoming misc. data card.
Misc. data for cables. 2 -1 3 0 1 1 0 0          | 2 -1 3 1 1
Comment card. NUMDCD = 13.                       |C ==The number of conductors==
Comment card. NUMDCD = 14.                       |C N1][N2 ][N3 ]
Cable conductor numbers. 2 2 2                   | 2 2 2
Comment card. NUMDCD = 16.                       |C ==Physical parameters==
Comment card. NUMDCD = 17.                       |C r1 ][ r2 ][ r3 ][ r4 ][ r5 ]
Radii. .00 .02 .04 .04 .05 .00 .00               | 0.0000 20.75E-3 39.25E-3 42.76E-3 47.94E-3
Comment card. NUMDCD = 19.                       |C rohc ][ muc ][ mul1 ][ epsl1 ][ rhos ][ mus ][ mul2 ][ epsl2 ]
Physical constants for conductors & insulators.  | 3.19E-8 1.0 1.0 2.68 2.83E-8 1.000 1.000 2.3
Comment card. NUMDCD = 21.                       |C r1 ][ r2 ][ r3 ][ r4 ][ r5 ]
Radii. .00 .02 .04 .04 .05 .00 .00               | 0.0000 20.75E-3 39.25E-3 42.76E-3 47.94E-3
Comment card. NUMDCD = 23.                       |C rohc ][ muc ][ mul1 ][ epsl1 ][ rhos ][ mus ][ mul2 ][ epsl2 ]
Physical constants for conductors & insulators.  | 3.19E-8 1.0 1.0 2.68 2.83E-8 1.000 1.000 2.3
Comment card. NUMDCD = 25.                       |C r1 ][ r2 ][ r3 ][ r4 ][ r5 ]
Radii. .00 .02 .04 .04 .05 .00 .00               | 0.0000 20.75E-3 39.25E-3 42.76E-3 47.94E-3
Comment card. NUMDCD = 27.                       |C rohc ][ muc ][ mul1 ][ epsl1 ][ rhos ][ mus ][ mul2 ][ epsl2 ]
Physical constants for conductors & insulators.  | 3.19E-8 1.0 1.0 2.68 2.83E-8 1.000 1.000 2.3
Comment card. NUMDCD = 29.                       |C ==Location==
Comment card. NUMDCD = 30.                       |C Vert1 ][ Horiz1 ][ Vert2 ][ Horiz2 ][ Vert3 ][ Horiz3 ][ Vert4 ][ Horiz4 ]
Y,X locations. 1.310E+00 -5.000E-02 0.000E+00    | 1.31 -5.E-2 1.20 0.00 1.31 5.E-2
Comment card. NUMDCD = 32.                       |C ==Frequency Card==
Comment card. NUMDCD = 33.                       |C rhoe ][ Freq. ][DEC][PNT][ DIST ][ IPUN ]
Frequency card 1.000E+02 5.000E+04 0 1 0.00E+00  | 100.0 5.E4

Table of underground cable parameters follows ....
   Phase number 1 boundary radii in [meters] follow, from the center outward ....
      1 0.00000E+00 2 2.07500E-02 3 3.92500E-02 4 4.27600E-02 5 4.79400E-02 6 4.79400E-02 7 4.79400E-02
     Resistivities in [ohm-m] of core, sheath, and armor : 3.190000E-08 2.830000E-08 2.830000E-08
     Relative permeabilities of the same three components : 1.000000E+00 1.000000E+00 1.000000E+00
     Insulator 1. Relative permittivity (imaginary, real), and velocity in [m/sec] : 1.000000E+00 2.680000E+00 1.831274E+08
     Insulator 2. Relative permittivity (imaginary, real), and velocity in [m/sec] : 1.000000E+00 2.300000E+00 1.976773E+08
   Phase number 2 boundary radii in [meters] follow, from the center outward ....
      1 0.00000E+00 2 2.07500E-02 3 3.92500E-02 4 4.27600E-02 5 4.79400E-02 6 4.79400E-02 7 4.79400E-02
     Resistivities in [ohm-m] of core, sheath, and armor : 3.190000E-08 2.830000E-08 2.830000E-08
     Relative permeabilities of the same three components : 1.000000E+00 1.000000E+00 1.000000E+00
     Insulator 1. Relative permittivity (imaginary, real), and velocity in [m/sec] : 1.000000E+00 2.680000E+00 1.831274E+08
     Insulator 2. Relative permittivity (imaginary, real), and velocity in [m/sec] : 1.000000E+00 2.300000E+00 1.976773E+08
   Phase number 3 boundary radii in [meters] follow, from the center outward ....
      1 0.00000E+00 2 2.07500E-02 3 3.92500E-02 4 4.27600E-02 5 4.79400E-02 6 4.79400E-02 7 4.79400E-02
     Resistivities in [ohm-m] of core, sheath, and armor : 3.190000E-08 2.830000E-08 2.830000E-08
     Relative permeabilities of the same three components : 1.000000E+00 1.000000E+00 1.000000E+00
     Insulator 1. Relative permittivity (imaginary, real), and velocity in [m/sec] : 1.000000E+00 2.680000E+00 1.831274E+08
     Insulator 2. Relative permittivity (imaginary, real), and velocity in [m/sec] : 1.000000E+00 2.300000E+00 1.976773E+08
```

Distance in [m] of each SC cable from earth`s surface: 1.3100 1.2000 1.3100
Distance in [m] between phases, for conductors: -0.05000 .00000 .05000
Beginning frequency in [Hz] = 5.00000000E+04 Number of decades IK = 0. Number of points/decade = 1.
Uniform (non-stratified) earth model has resistivity in [ohm-m] = 1.00000000E+02
Punched LUNIT7 cards for Pi-circuit modeling have the following specifications. Node name = "A".
Total length of cable, and length of each major section, in [m] : 36000.000 6000.000
Number of major sections of cable = 6. Sheath grounding resistance in [ohm] at major sections = 5.00000000E+00
The cable is crossbonded.
Homogeneous Pi-circuit modeling is used for the cable.

-------------------- Cable constants matrices for frequency 5.00000000E+04 Hertz. -------------------

Impedance matrix [Z] in [ohm/m] follows :

4.7593748E-02	4.6464915E-02	4.6396140E-02	4.6512856E-02
4.5625408E-01	3.4958170E-01	3.6163499E-01	3.7550324E-01
4.6464915E-02	4.7739478E-02	4.6464915E-02	4.6584358E-02
3.4958170E-01	4.5592286E-01	3.4958170E-01	3.7137507E-01
4.6396140E-02	4.6464915E-02	4.7593748E-02	4.6512856E-02
3.6163499E-01	3.4958170E-01	4.5625408E-01	3.7550324E-01
4.6512856E-02	4.6584358E-02	4.6512856E-02	4.6536675E-02
3.7550324E-01	3.7137507E-01	3.7550324E-01	3.7412716E-01

Conductance [G] in [mho/m] and capacitance [C] in [farad/m] follows :

0.0000000E+00	0.0000000E+00	0.0000000E+00	0.0000000E+00
2.3390940E-10	0.0000000E+00	0.0000000E+00	-2.3390940E-10
0.0000000E+00	0.0000000E+00	0.0000000E+00	0.0000000E+00
0.0000000E+00	2.3390940E-10	0.0000000E+00	-2.3390940E-10
0.0000000E+00	0.0000000E+00	0.0000000E+00	0.0000000E+00
0.0000000E+00	0.0000000E+00	2.3390940E-10	-2.3390940E-10
0.0000000E+00	0.0000000E+00	0.0000000E+00	0.0000000E+00
-2.3390940E-10	-2.3390940E-10	-2.3390940E-10	4.0587352E-09

Modal transformation matrices follow. These are complex, with the magnitude displayed above the angle in degrees. Transpose of the current transformation matrix [Ti] follows. This is the inverse of the voltage transformation matrix. By definition, [Ti] gives the mapping from modal to phase variables: i-phase = [Ti] * i-mode

.0002596	.0005192	.0002596	.9999944
8.2163892	-171.7826496	8.2163892	-.0000937
.3335884	.6671769	.3335884	.0074577
-179.9937209	.0062788	-179.9937209	8.2233420
.5000000	.0000000	.5000000	.0000000
-180.0000000	-19.3670124	.0000000	123.1915141
.3333333	.3333333	.3333333	.9999999
.0000001	.0000005	.0000001	179.9999966

Voltage transformation matrix [Tv]. This gives the mapping from modal to phase variables: v-phase = [Tv] * v-mode

1.0036880	.4988446	1.0000000	1.0000000
.0303984	179.9808515	-180.0000000	.0000000
.9926251	1.0000000	.0000000	1.0000000
-.0614801	.0000000	2.1716485	.0000004
1.0036880	.4988446	1.0000000	1.0000000
.0303984	179.9808515	.0000000	.0000000
1.0000000	.0007783	.0000000	.0000000
.0000000	8.2106493	167.6244470	151.0188924

```
Characteristic impedance matrix  [Zc]   in phase variables,   in [ohm]:
         51.2651409      13.7580127      15.3812685      18.9311519
         -1.3972881      -1.1909448      -1.1702033      -1.1648032

         13.7580127      52.3461929      13.7580127      18.7504173
         -1.1909448      -1.4157321      -1.1909448      -1.1778650

         15.3812685      13.7580127      51.2651409      18.9311519
         -1.1702033      -1.1909448      -1.3972881      -1.1648032

         18.9311519      18.7504173      18.9311519      18.8709061
         -1.1648032      -1.1778650      -1.1648032      -1.1691565

          Modal     Propagation       Modal impedance              Modal        Characteristic   impedance      Characteristic admittance
Mode   attenuation   velocity        real         imaginary      susceptance        real          imaginary          real          imaginary
        [ db/km ]    [ m/sec ]      [ ohms / meter ]            [mho/meter]   SQRT ( z / y )   in   [ohm]     SQRT ( y / z )   in   [mho]
  1   1.07098E+01  1.57853E+07   4.653669E-02  3.741271E-01  1.054641E-03   1.887089E+01 -1.169161E+00   5.278904E-02  3.270586E-0
  2   1.35325E-01  1.10277E+08   7.902781E-04  7.373929E-02  1.100581E-04   2.588481E+01 -1.358443E-01   3.863163E-02  2.027400E-0
  3   1.44944E-01  1.19139E+08   5.988044E-04  4.730954E-02  1.469696E-04   1.794194E+01 -1.135424E-01   5.573311E-02  3.526973E-0
  4   1.68522E-01  1.81067E+08   3.053968E-04  1.365367E-02  2.204544E-04   7.870321E+00 -8.800818E-02   1.270437E-01  1.420639E-0

Blank card terminating frequency cards.           |BLANK ENDING CABLE CONSTANTS CASE
Request for flushing of punch buffer.             |$PUNCH

A listing of 80-column card images now being flushed from punch buffer follows.
================================================================================
12345678901234567890123456789012345678901234567890123456789012345678901234567890123456789
================================================================================
C   <++++++>   Cards punched by support routine on   13-Feb-14   14.58.27   <++++++>
C CABLE CONSTANTS
C PUNCH
C C ==Miscellaneous Data Card==
C C TY][SYS][NPC][EAR][KMO][ZFL]
C       2   -1     3              1      1
C C NP][NCR][IRS][ XMAJOR ][     RSG    ]!
C       6     1     0        6.E3          5.A
C C ==The number of conductors==
C C N1][N2 ][N3 ]
C       2     ?     ?
C C ==Physical parameters==
C C     r1     ][    r2     ][    r3     ][    r4     ][    r5     ]
C        0.0000   20.75E-3   39.25E-3   42.76E-3   47.94E-3
C C   rohc    ][   muc     ][   muI1    ][   epsI1 ][   rhos    ][     mus    ][   muI2    ][   epsI2
C      3.19E-8           1.0           1.0          2.68     2.83E-8         1.000         1.000            2
C C     r1     ][    r2     ][    r3     ][    r4     ][    r5     ]
C        0.0000   20.75E-3   39.25E-3   42.76E-3   47.94E-3
C C   rohc    ][   muc     ][   muI1    ][   epsI1 ][   rhos    ][     mus    ][   muI2    ][   epsI2
C      3.19E-8           1.0           1.0          2.68     2.83E-8         1.000         1.000            2
C C     r1     ][    r2     ][    r3     ][    r4     ][    r5     ]
C        0.0000   20.75E-3   39.25E-3   42.76E-3   47.94E-3
C C   rohc    ][   muc     ][   muI1    ][   epsI1 ][   rhos    ][     mus    ][   muI2    ][   epsI2
C      3.19E-8           1.0           1.0          2.68     2.83E-8         1.000         1.000            2
C C ==Location==
C C   Vert1 ][ Horiz1 ][   Vert2 ][ Horiz2 ][   Vert3 ][ Horiz3 ][   Vert4 ][ Horiz4
C           1.31       -5.E-2          1.20          0.00            1.31          5.E-2
C C ==Frequency Card==
C C         rhoe        ][          Freq.     ][DEC][PNT][ DIST ][    IPUN    ]
C                     100.0                    5.E4
$VINTAGE, 1
  AIN    4                        5.0000000E+00
 1AIN    1A 11 1                  2.8556249E+02   8.7138111E+00   1.4034564E+00
 2AIN    2A 11 2                  2.7878949E+02   6.6765187E+00   0.0000000E+00
                                  2.8643687E+02   8.7074852E+00   1.4034564E+00
 3AIN    3A 11 3                  2.7837684E+02   6.9067196E+00   0.0000000E+00
                                  2.7878949E+02   6.6765187E+00   0.0000000E+00
                                  2.8556249E+02   8.7138111E+00   1.4034564E+00
 4AIN    4A 11 4                  2.7907713E+02   7.1715836E+00  -1.4034564E+00
                                  2.7950615E+02   7.0927413E+00  -1.4034564E+00
                                  2.7907713E+02   7.1715836E+00  -1.4034564E+00
                                  2.7922005E+02   7.1453024E+00   2.4352411E+01
```

```
  A11 4       AIN  4
1A11 1A21 1AIN   1A11 1
2A11 2A21 2
3A11 3A21 3
4A11 4A21 4
  A21 4       AIN  4
1A21 1A31 1AIN   1A11 1
2A21 2A31 2
3A21 3A31 3
4A21 4A31 4
  A31 4       AIN  4
1A31 1A41 1AIN   1A11 1
2A31 2A41 2
3A31 3A41 3
4A31 4A41 4
  A41 4       AIN  4
1A41 1A51 1AIN   1A11 1
2A41 2A51 2
3A41 3A51 3
4A41 4A51 4
  A51 4       AIN  4
1A51 1AOUT 1AIN   1A11 1
2A51 2AOUT 2
3A51 3AOUT 3
4A51 4AOUT 4
  AOUT 4       AIN  4
$VINTAGE, 0
$VINTAGE, 1
-1                  4.65361E-02 1.88346E+01 1.57853E+07 0.00000E+00 1  4
-2                  8.06561E-04 2.58844E+01 1.10277E+08 0.00000E+00 1  4
-3                  5.98804E-04 1.79416E+01 1.19139E+08 0.00000E+00 1  4
-4                  3.05397E-04 7.86983E+00 1.81067E+08 0.00000E+00 1  4
$VINTAGE, 0
  .00025696  -.33358843  -.50000000   .33333333
  .00003710   .00000027   .00000000   .00000000
 -.00051391   .66717687   .00000000   .33333333
 -.00007421  -.00000055   .00000000   .00000000
  .00025696  -.33358843   .50000000   .33333333
  .00003710   .00000027   .00000000   .00000000
  .99999442   .00738115   .00000000  -.99999994
 -.00000081   .00106588   .00000000   .00000006
==========<  End of LUNIT7 punched cards as flushed by  $PUNCH  request  >=======

Blank card ending  "CABLE CONSTANTS"  cases.        |BLANK ENDING SEMLYEN SETUP
Timing figures characterizing central processor (CP) solution speed.  ---------------------------  CP sec  Wait sec  Real sec
                                                              Totals     0.070     0.000     0.070
```

3.6 PT Cable

3.6.1 *Introduction of PT Cable*

There are various cables specified as a PT cable. For example, a tunnel-installed cable explained in Section 3.5.3, Figure 3.14, is a PT cable. The three-phase SC cables are enclosed within a conducting pipe composed of steel frames and cement. The pipe is buried underground, in fact not buried but constructed. As a typical PT cable, an oil-filled (OF) cable and a pressured oil-filled (POF) cable are well-known. Also, a gas insulated bus/line, in which three-phase conductors are installed, is an overhead PT cable.

In general, a PT cable is composed of multi-phase SC cables and a conducting pipe enclosure. A typical PT cable consists of six inner conductors (three cores/sheath) and a pipe conductor, and thus becomes a seven conductor system.

A significant difference of the PT cable from an SC cable comes from the conducting (metallic) pipe which affects the wave propagation characteristic of the cable. Because the resistivity (10^{-7}–10^{-8} Ωm) of the metallic pipe is far smaller than the earth resistivity (10–5000 Ωm), most zero-sequence currents of the inner conductors flow back through the pipe conductor. Thus, the influence of the earth becomes far smaller in the PT cable than in the SC cable, especially during a transient. This fact suggests that the earth can be neglected if the pipe thickness is electrically great enough, that is, if the pipe thickness d is greater than the penetration depth h_p.

$$d > H_p = \sqrt{\rho_p/\omega\mu_p} \tag{3.140}$$

where

$$h_p = \sqrt{\rho_p/j\omega\mu_p} = H_p \cdot \exp(-j\pi/4) \text{ : penetration depth} \tag{3.141}$$

and ρ_p is the pipe resistivity and μ_p is the pipe permeability.

For example, assume a steel pipe with the following parameters.

$$\rho_p = 1.5 \times 10^{-7}\ \Omega\text{m}, \mu_p = \mu_{pr} \cdot \mu_0, \mu_{pr} = 300$$

Then,

$$H_p = 7.96 \times 10^{-3}/\sqrt{f}$$

$$H_p = 1.13 \text{ mm at } f = 50 \text{ Hz}, H_p = 0.251 \text{ mm at 1 kHz}$$

The above result indicates that the earth can be neglected at $f = 50$ Hz, if the pipe thickness is more than 2 mm. Even in the case of an overhead gas-insulated bus where the pipe conductor is aluminum, that is, the resistivity is about 2×10^{-8} Ωm,

$$H_p = 50.3 \times 10^{-3}/\sqrt{f}$$

$$H_p = 7.12 \text{ mm at } f = 50\text{Hz}/H_p, H_p = 1.59 \text{ mm at 1 kHz}$$

Therefore, if the aluminum pipe thickness is about 1 cm, again the earth can be neglected. When the earth is neglected, the pipe thickness is assumed to be infinite. In other words, we can assume the pipe to be the earth with the resistivity much smaller than soil. In fact, the tunnel-installed cable in Figure 3.14 shows a similar characteristic to an overhead line as explained in Section 3.5.1, because the three-phase SC cable is laid in the air.

Another significant difference from the SC cable is an inner conductor being eccentric to the pipe centre as in Figure 3.14. Quite often, the inner SC cable is placed at the bottom of the pipe enclosure, and the eccentricity affects significantly the impedance between the SC cable and the pipe as observed from (2.34). When the three-phase SC cable is arranged symmetrically to the pipe center, the wave propagation characteristic becomes quite similar to that of a transposed three-phase overhead line, and the transformation matrix is nearly the same as that of the transposed line which is not frequency-dependent. A similar situation is often observed in a long PT cable where inner cables show snaking which is similar to transposition of a three-phase overhead line. In a gas-insulated bus with three inner conductors, the three conductors are symmetrically positioned to the pipe center.

3.6.2 PT Cable with Finite-pipe Thickness

3.6.2.1 Symmetrical Configuration of Inner Conductors

An underground three-phase PT cable with symmetrical configuration of inner conductors is illustrated in Figure 3.17. Each inner conductor consists only of a core. Thus, the PT cable

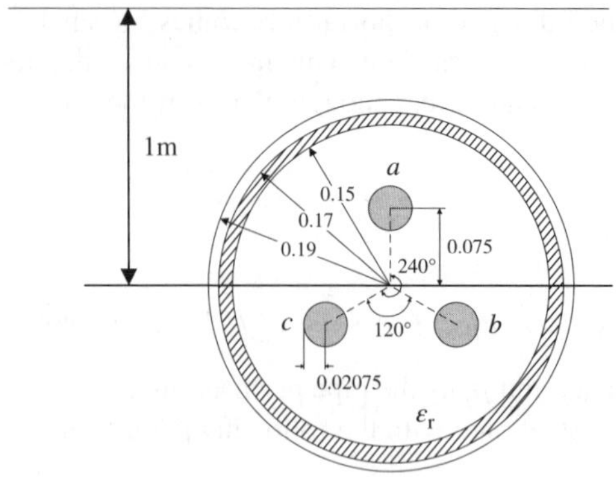

Figure 3.17 A three-phase underground PT cable

composes a four conductor system. Calculated results of a voltage transformation matrix and propagation constants are given in Table 3.11 and Figure 3.18, respectively, where the relative permittivity of a dielectric medium inside the pipe is unity. The output list of the EMTP-ATP Cable Parameters is given in Data 3.3.

Table 3.11 Voltage transformation matrix A^{-1} for a PT cable with symmetrically arranged inner conductors as shown in Figure 3.17 for 10 Hz $\leq f \leq$ 1 MHz

Conductor / Mode	Phase			Pipe
	a	b	c	
0	0	0	0	1
1	1/3	1/3	1/3	−1
2	−1/3	−1/3	2/3	0
3	1/2	−1/2	0	0

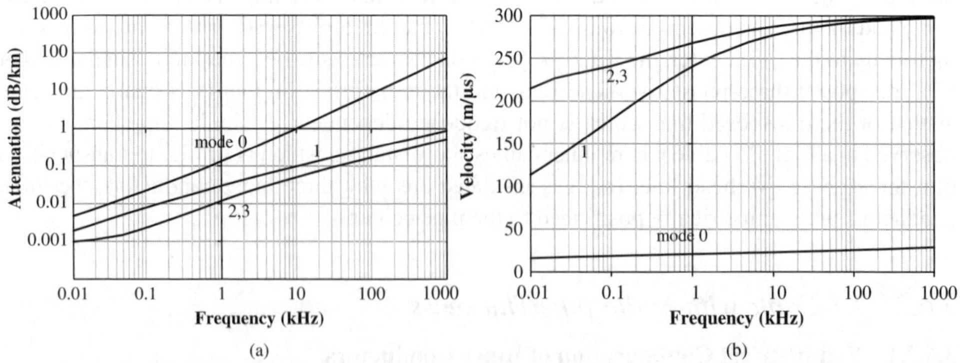

Figure 3.18 Propagation constants on a PT cable with symmentrically arranged three-phase cores ($\varepsilon_r = 1$): (a) attenuation vs. frequency; and (b) propagation velocity vs. frequency

Data 3.3 Cable Parameters output of a PT cable with symmetrically arranged three-phase inner conductors

```
 ···  39    cards of disk file read into card cache cells   1      onward.
 1st card (Lists  1-10).          |    8000    15000    15000      340    60000      120
 2nd card (Lists 11-20).          |     360      900    36400      255   129600       38
 3rd card (Lists 21-29).          |    3000    15000   160000       24    30000     1000
 Supplemental offsets.            |  240000      742
Alternative Transients Program (ATP),  Watcom translation.   All rights reserved by Can/Am user group of Portland, Oregon, USA.
 Date (dd-mth-yy) and time of day (hh.mm.ss) = 24-Feb-14   14.54.56     Name of disk plot file is   C:¥ATPWNT¥UNTITLED.pl4
Consult the 860-page ATP Rule Book of the Can/Am EMTP User Group in Portland,   Oregon, USA.    Program is no older than Sept., 1998.
Total size of LABCOM tables =   3639004    INTEGER words.    VARDIM   List Sizes follow :    8002 15000 15000     340 60000   1200
 13140 120000 2250 1900   360   900 36400    255   129K    384     45    254 240000 100000   3000 15000   160K     24 30000 10000     600 252000
--------------------------------------------------+--------------------------------------------------------------------------------
Descriptive interpretation of input data cards.    |  Input data card images are shown below, all 80 columns, character by character
                                                     0         1         2         3         4         5         6         7         8
                                                     012345678901234567890123456789012345678901234567890123456789012345678901234567890
--------------------------------------------------+--------------------------------------------------------------------------------
Comment card.    NUMDCD = 1.                       |C data:C:¥ATPWNT¥UNTITLED.DAT
Comment card.    NUMDCD = 2.                       |C Cable Constants for underground PT cable 2014.2.8 asym.
Comment card.    NUMDCD = 3.                       |C ===Cable Constants===
Marker card preceding new EMTP data case.         |BEGIN NEW DATA CASE
Find parameters of cables.  Type = 0.             |CABLE CONSTANTS
Prof. Akihiro Ametani`s 1994 code.    Order = 38  |CABLE PARAMETERS
Comment card.    NUMDCD = 7.                       |C ==Miscellaneous Data Card==
Comment card.    NUMDCD = 8.                       |C TY][SYS][NPC][EAR][KMO][ZFL][YFL][NPP][NGD]
Misc.   3  -1   3   0   0   0   0   1   0   0   0 |  3   -1    3          0     0            1
Comment card.    NUMDCD = 10.                      |C ==pipe physical data==
Comment card.    NUMDCD = 11.                      |C    Rp1   ][    Rp2   ][    Rp3  ][  rohp   ][   mur    ][  eps1   ][  eps2  ]
Pipe radii.    1.5000E-01  1.7000E-01  1.9000E-01  |    0.150       0.170       0.190    0.15E-6        300.          1.0          3.0
Comment card.    NUMDCD = 13.                      |C ==Location==
Comment card.    NUMDCD = 14.                      |C   Dist1 ][ Theta1 ][  Dist2 ][ Theta2 ][   Dist3 ][ Theta3 ][   Dist4 ][ Theta4 ]
Cable to pipe.   7.500E-02  0.000E+00   7.500E-02  |    0.075         0.0       0.075        120.       0.075       -120.
Comment card.    NUMDCD = 16.                      |C ==The number of conductors==
Comment card.    NUMDCD = 17.                      |C N1][N2 ][N3 ]
Cable conductor numbers.   1 1 1                   |  1    1    1
Comment card.    NUMDCD = 19.                      |C ==Physical parameters==
Comment card.    NUMDCD = 20.                      |C    r1    ][    r2    ][    r3    ][    r4    ][    r5    ]
Radii.      .00     .02     .04                    | 0.0000   20.75E-3   39.25E-3
Comment card.    NUMDCD = 22.                      |C   rohc   ][   muc    ][  muI1   ][  epsI1 ][  rhos   ][    mus   ][  muI2   ][  epsI2 ]
Physical constants for conductors & insulators.    |  3.19E-8          1.0          1.0        1.00
Comment card.    NUMDCD = 24.                      |C    r1    ][    r2    ][    r3    ][    r4    ][    r5    ]
Radii.      .00     .02     .04                    | 0.0000   20.75E-3   39.25E-3
Comment card.    NUMDCD = 26.                      |C   rohc   ][   muc    ][  muI1   ][  epsI1 ][  rhos   ][    mus   ][  muI2   ][  epsI2 ]
Physical constants for conductors & insulators.    |  3.19E-8          1.0          1.0        1.00
Comment card.    NUMDCD = 28.                      |C    r1    ][    r2    ][    r3    ][    r4    ][    r5    ]
Radii.      .00     .02     .04                    | 0.0000   20.75E-3   39.25E-3
Comment card.    NUMDCD = 30.                      |C   rohc   ][   muc    ][  muI1   ][  epsI1 ][  rhos   ][    mus   ][  muI2   ][  epsI2 ]
Physical constants for conductors & insulators.    |  3.19E-8          1.0          1.0        1.00
Comment card.    NUMDCD = 32.                      |C   Vert1 ][ Horiz1 ][  Vert2 ][ Horiz2 ][   Vert3 ][ Horiz3 ][   Vert4 ][ Horiz4 ]
(Y,X)   1.000E+00   0.000E+00                      |     1.00         0.0
Comment card.    NUMDCD = 34.                      |C ==Frequency Card==
Comment card.    NUMDCD = 35.                      |C         rhoe         ][         Freq.      ][DEC][PNT][ DIST ][   IPUN   ]
Frequency card 1.000E+02 5.000E+04   0   1 0.00E+00 |           100.0                5.E4                                        2
          3 -phase PT cable with 4    conductors.   0   grounded.     Total # of conductors = 4
             Under-ground cable
             Cable constants
          Pipe and its insulator
          Pipe : inner radius 1.50000000E-01, outer radius 1.70000000E-01.   Outer radius of pipe insulator 1.90000000E-01
                   Resistivity (ohm-m): 1.50000000E-07, Relative permeability:   3.00000000E+02
           Insulator ; Relative permittivity, inner: 1.00000000E+00, outer: 3.00000000E+00
           Pipe 1 ; Dist. from Earth surface: 1.00000000E+00    m; Dist. from pipe 1: 0.00000000E+00    m
          Inner conductors (SC cables)
           Dist. from pipe center :        1    0.0750      2    0.0750      3    0.0750
           Angle to first cable :          0.0000          120.0000          -120.0000
           Phase 1    Boundary radii ;
            1   0.00000E+00   2   2.07500E-02   3   3.92500E-02
                    Resistivity(ohm-m): Core 3.19000000E-08, Sheath 9.99999994E-09, Armor 9.99999994E-09
                    Relative permeability;   1.00000000E+00, 1.00000000E+00, 1.00000000E+00
                    Insulator:   1; (Relative) Permeability   1.0      Permittivity   1.00     Velocity(m/s) 2.99793E+08
                    DC resistance(ohm/m),            2.35833E-05
           Phase 2    Boundary radii ;
            1   0.00000E+00   2   2.07500E-02   3   3.92500E-02
                    Resistivity(ohm-m): Core 3.19000000E-08, Sheath 9.99999994E-09, Armor 9.99999994E-09
                    Relative permeability;   1.00000000E+00, 1.00000000E+00, 1.00000000E+00
                    Insulator:   1; (Relative) Permeability   1.0      Permittivity   1.00     Velocity(m/s) 2.99793E+08
                    DC resistance(ohm/m),            2.35833E-05
           Phase 3    Boundary radii ;
            1   0.00000E+00   2   2.07500E-02   3   3.92500E-02
                    Resistivity(ohm-m): Core 3.19000000E-08, Sheath 9.99999994E-09, Armor 9.99999994E-09
                    Relative permeability;   1.00000000E+00, 1.00000000E+00, 1.00000000E+00
                    Insulator:   1; (Relative) Permeability   1.0      Permittivity   1.00     Velocity(m/s) 2.99793E+08
                    DC resistance(ohm/m),            2.35833E-05
```

```
        Earth ; resistivity 1.00000000E+02
     Punched cards:  1) pi-circuit (IPUNCH=1) or distributed line (IPUNCH=2);
       IPUNCH = 2          Total length = 0.00000000E+00
         Frequency 5.00000000E+04   Hz
        Eigenvector rotation has been made.

         Total impedance [Zc]
 5.52509E-02  5.19751E-02  5.19751E-02  4.96262E-02
 4.42206E-01  3.49979E-01  3.49979E-01  3.30135E-01
 5.19751E-02  5.52509E-02  5.19751E-02  4.96262E-02
 3.49979E-01  4.42206E-01  3.49979E-01  3.30135E-01
 5.19751E-02  5.19751E-02  5.52509E-02  4.96262E-02
 3.49979E-01  3.49979E-01  4.42206E-01  3.30135E-01
 4.96262E-02  4.96262E-02  4.96262E-02  4.96262E-02
 3.30135E-01  3.30135E-01  3.30135E-01  3.30135E-01

         Total admittance [Yc]
 0.00000E+00  0.00000E+00  0.00000E+00  0.00000E+00
 1.08493E-05 -1.54081E-06 -1.54081E-06 -7.76768E-06
 0.00000E+00  0.00000E+00  0.00000E+00  0.00000E+00
-1.54081E-06  1.08493E-05 -1.54081E-06 -7.76768E-06
 0.00000E+00  0.00000E+00  0.00000E+00  0.00000E+00
-1.54081E-06 -1.54081E-06  1.08493E-05 -7.76768E-06
 0.00000E+00  0.00000E+00  0.00000E+00  0.00000E+00
-7.76768E-06 -7.76768E-06 -7.76768E-06  4.94709E-04

Characteristic impedance matrix  [Zc]  in phase variables,   in [ohm]:
 130.68333   44.39325   44.39325   26.53780
  -4.58841   -3.05643   -3.05643   -1.98345
  44.39325  130.68333   44.39325   26.53780
  -3.05643   -4.58841   -3.05643   -1.98345
  44.39325   44.39325  130.68333   26.53780
  -3.05643   -3.05643   -4.58841   -1.98345
  26.53780   26.53780   26.53780   26.53780
  -1.98345   -1.98345   -1.98345   -1.98345

Modal transformation matrices follow.   These are complex, with the real part displayed above the imaginary part.
The current transformation matrix [Ti] follows.   The transpose of this is the inverse of the voltage transformation matrix.
By definition, [Ti] gives the mapping from modal to phase variables: i-phase = [Ti] * i-mode
       .00000     .33333    -.33333     .50000
       .00000     .00000     .00000     .00000
       .00000     .33333    -.33333    -.50000
       .00000     .00000     .00000     .00000
       .00000     .33333     .66667     .00000
       .00000     .00000     .00000     .00000
      1.00000   -1.00000     .00000     .00000
       .00000     .00000     .00000     .00000

 Voltage transformation matrix [Tv].   This gives the mapping from modal to phase variables: v-phase = [Tv] * v-mode
      1.00000    1.00000    -.50000    1.00000
       .00000     .00000     .00000     .00000
      1.00000    1.00000    -.50000   -1.00000
       .00000     .00000     .00000     .00000
      1.00000    1.00000    1.00000     .00000
       .00000     .00000     .00000     .00000
      1.00000     .00000     .00000     .00000
       .00000     .00000     .00000     .00000

              Modal components
 Mode  Attenuation  Velocity      Impedance (Ohm/m)          Admittance (S/m)       Charact. imp. (Ohm)       Charact. Admit. (S)
 No.       (db/km) (m/mic.s)      Real         Imag.         Real         Imag.       Real       Imag.         Real         Imag.
    1  8.12150E+00     25.11  4.96262E-02  3.30135E-01  0.00000E+00  4.71406E-04     26.538     -1.983  3.74728E-02  2.80074E-03
    2  3.20545E-01    289.19  3.44080E-03  5.05864E-02  0.00000E+00  2.33030E-05     46.619     -1.584  2.14258E-02  7.27834E-04
    3  1.64872E-01    293.84  2.18387E-03  6.14849E-02  0.00000E+00  1.85851E-05     57.527     -1.021  1.73777E-02  3.08522E-04
    4  1.64872E-01    293.84  1.63791E-03  4.61137E-02  0.00000E+00  2.47802E-05     43.145      -.766  2.31703E-02  4.11363E-04
Blank card terminating frequency cards.              |BLANK ENDING CABLE CONSTANTS CASE
Request for flushing of punch buffer.               |$PUNCH

A listing of 80-column card images now being flushed from punch buffer follows.
===============================================================================
1234567890123456789012345678901234567890123456789012345678901234567890123456789
===============================================================================
C  <++++++>  Cards punched by support routine on   24-Feb-14   14.54.56   <++++++>
C CABLE CONSTANTS
C CABLE PARAMETERS
C C ==Miscellaneous Data Card==
C C TY][SYS][NPC][EAR][KMO][ZFL][YFL][NPP][NGD]
C      3   -1     3           0     0           1
```

```
C C ==pipe physical data==
C C   Rp1  ][   Rp2  ][   Rp3  ][  rohp  ][  mur    ][  eps1  ][  eps2  ]
C       0.150      0.170      0.190    0.15E-6        300.          1.0          3.0
C C ==Location==
C C  Dist1 ][ Theta1 ][  Dist2 ][ Theta2 ][  Dist3 ][ Theta3 ][  Dist4 ][ Theta4
C       0.075          0.0       0.075          120.        0.075        -120.
C C ==The number of conductors==
C C N1][N2 ][N3 ]
C       1    1     1
C C ==Physical parameters==
C C    r1    ][    r2    ][    r3    ][    r4    ][    r5    ]
C       0.0000   20.75E-3   39.25E-3
C C  rohc  ][  muc    ][  mul1  ][  epsl1 ][  rhos  ][    mus  ][  mul2  ][  epsl2
C      3.19E-8           1.0            1.0          1.00
C C    r1    ][    r2    ][    r3    ][    r4    ][    r5    ]
C       0.0000   20.75E-3   39.25E-3
C C  rohc  ][  muc    ][  mul1  ][  epsl1 ][  rhos  ][    mus  ][  mul2  ][  epsl2
C      3.19E-8           1.0            1.0          1.00
C C    r1    ][    r2    ][    r3    ][    r4    ][    r5    ]
C       0.0000   20.75E-3   39.25E-3
C C  rohc  ][  muc    ][  mul1  ][  epsl1 ][  rhos  ][    mus  ][  mul2  ][  epsl2
C      3.19E-8           1.0            1.0          1.00
C C  Vert1 ][ Horiz1 ][  Vert2 ][ Horiz2 ][  Vert3 ][ Horiz3 ][  Vert4 ][ Horiz4
C        1.00             0.0
C C ==Frequency Card==
C C        rhoe       ][        Freq.    ][DEC][PNT][ DIST ][  IPUN  ]
C                  100.0                  5.E4                                    2
$UNITS, 0.0, 0.0
$VINTAGE, 1
C     Branch cards for constant-parameter distributed:
-1                                   4.96262E-02 2.65378E+01 2.51125E+07 0.00000E+00 1 14
-2                                   3.44080E-03 4.66188E+01 2.89185E+08 0.00000E+00 1 14
-3                                   2.18387E-03 5.75267E+01 2.93842E+08 0.00000E+00 1 14
-4                                   1.63791E-03 4.31450E+01 2.93842E+08 0.00000E+00 1 14
     .00000000   .33333333  -.33333333   .50000000
     .00000000   .00000000   .00000000   .00000000
     .00000000   .33333333  -.33333333  -.50000000
     .00000000   .00000000   .00000000   .00000000
     .00000000   .33333333   .66666667   .00000000
     .00000000   .00000000   .00000000   .00000000
    1.00000000 -1.00000000   .00000000   .00000000
     .00000000   .00000000   .00000000   .00000000
$VINTAGE, 0
$UNITS, -1., -1.,
=========< End of LUNIT7 punched cards as flushed by  $PUNCH  request  >=======

Blank card ending  "CABLE CONSTANTS"  cases.          | BLANK ENDING SEMLYEN SETUP
Timing figures characterizing central processor (CP) solution speed.  ---------------------------  CP sec  Wait sec  Real sec
                                                                              Totals      0.060      0.000      0.060
```

Mode 0 is observed from Table 3.11 to be the earth-return mode where currents flow through the pipe and return through the earth. The mode has characteristics quite similar to those of the earth-return mode in the underground SC cable case, that is, characteristics of very high attenuation and very low velocity.

Mode 1 is the pipe-return mode, that is, the propagation mode of currents flowing through all the inner conductors and returning through the pipe, and is seen to have characteristics similar to those of the earth-return mode in an overhead line. The attenuation is not as high as that of the earth-return mode in the underground cable case, and the propagation velocity reaches the velocity of light in free space as frequency increases. This characteristic is due to the relative permittivity of the dielectric medium within the pipe being unity, in other words, SF_6 gas being assumed to be an insulator inside the pipe.

Mode 2 is a propagation mode of currents flowing through the phase-c conductor and returning through the phase-a and -b conductors, and mode 3 is that of currents flowing through the phase-a conductor and returning through the phase-b conductor, that is, the interconductor modes. Theses modes correspond to the aerial modes of a transposed overhead line, and show identical attenuation and propagation velocity to each other. The velocities approach the velocity of light in free space as frequency increases.

3.6.2.2 Asymmetrical Configuration of Inner Conductors

A PT cable with an asymmetrical configuration of inner conductors is illustrated in Figure 3.19. A calculated result of a voltage transformation matrix is given in Table 3.12. Propagation constants are shown in Figure 3.20 and compared with those for the symmetrical configuration case in Figure 3.17. The output list is given in Data 3.4.

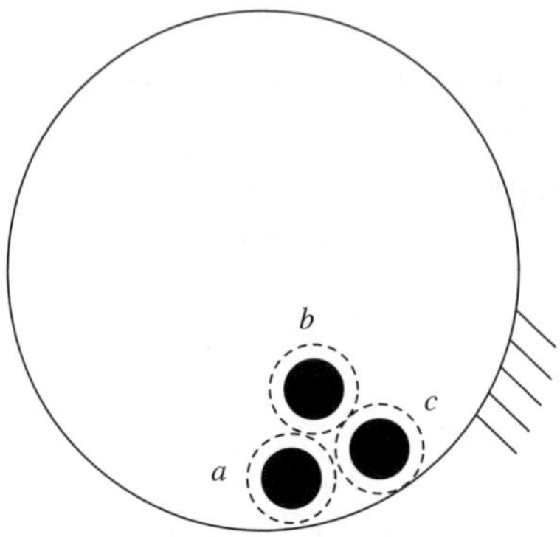

Figure 3.19 A PT cable with an asymmetrical configuration of inner conductors

Table 3.12 Voltage transformation matrix A^{-1} for a PT cable as shown in Figure 3.19 at $f = 50$ kHz

Mode \ Conductor	Phase a	Phase b	Phase c	Pipe
0	0	0	0	1
1	0.775	−0.833	0.775	−0.717
2	−0.512	3/2	−0.512	−0.526
3	−1/2	0	−1/2	0

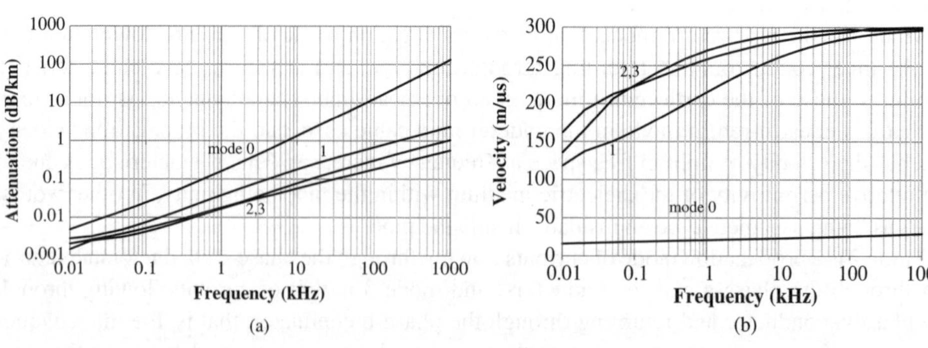

Figure 3.20 Propagation constants on a PT cable with asymmetrically arranged three-phase cores ($\varepsilon_r = 1$): (a) attenuation vs. frequency; and (b) propagation velocity vs. frequency

Data 3.4 Output of a PT cable with asymmetrically arranged three-phase inner conductors

```
 ---  39   cards of disk file read into card cache cells  1   onward.
 1st card (Lists  1-10).        |   8000   15000   15000    340   60000    120
 2nd card (Lists 11-20).        |    360     900   36400    255  129600     38
 3rd card (Lists 21-29).        |   3000   15000  160000     24   30000   1000
 Supplemental offsets.          | 240000     742
Alternative Transients Program (ATP),  Watcom translation.  All rights reserved by Can/Am user group of Portland, Oregon, USA.
 Date (dd-mth-yy) and time of day (hh.mm.ss) = 24-Feb-14  13.54.52   Name of disk plot file is  C:¥ATPWNT¥UNTITLED.pl4
Consult the 860-page ATP Rule Book of the Can/Am EMTP User Group in Portland,  Oregon, USA.   Program is no older than Sept., 1998.
Total size of LABCOM tables =  3639004   INTEGER words.   VARDIM  List Sizes follow :   8002 15000 15000   340 60000  1200
 13140 120000 2250 1900  360  900 36400   255  129K   384   45   254 240000 100000  3000 15000  160K   24 30000 10000   600 252000
--------------------------------------------------+--------------------------------------------------------------------------------
Descriptive interpretation of input data cards.  |  Input data card images are shown below, all 80 columns, character by character
                                                     0         1         2         3         4         5         6         7         8
                                                     012345678901234567890123456789012345678901234567890123456789012345678901234567890
--------------------------------------------------+--------------------------------------------------------------------------------
Comment card.   NUMDCD = 1.                        |C data:C:¥ATPWNT¥UNTITLED.DAT
Comment card.   NUMDCD = 2.                        |C Cable Constants for underground PT cable asymmetrically
Comment card.   NUMDCD = 3.                        |C ===Cable Constants===
Marker card preceding new EMTP data case.         |BEGIN NEW DATA CASE
Find parameters of cables.  Type = 0.              |CABLE CONSTANTS
Prof. Akihiro Ametani`s 1994 code.   Order = 38    |CABLE PARAMETERS
Comment card.   NUMDCD = 7.                        |C ==Miscellaneous Data Card==
Comment card.   NUMDCD = 8.                        |C TY][SYS][NPC][EAR][KMO][ZFL][YFL][NPP][NGD]
Misc.  3  -1   3   0   0   0   0   1   0   0   0 |  3   -1    3          0     0           1
Comment card.   NUMDCD = 10.                       |C ==pipe physical data==
Comment card.   NUMDCD = 11.                       |C   Rp1  ][   Rp2  ][   Rp3  ][  rohp  ][  mur   ][  eps1  ][  eps2  ]
Pipe radii.   1.5000E-01  1.7000E-01  1.9000E-01  |  0.150      0.170      0.190    0.15E-6        300.         1.0         3.0
Comment card.   NUMDCD = 13.                       |C ==Location==
Comment card.   NUMDCD = 14.                       |C  Dist1 ][ Theta1 ][  Dist2 ][ Theta2 ][  Dist3 ][ Theta3 ][  Dist4 ][ Theta4 ]
Cable to pipe.  1.000E-01 -9.000E+01  7.500E-02   |  0.100      -90.0      0.075       -80.     0.100       -70.
Comment card.   NUMDCD = 16.                       |C ==The number of conductors==
Comment card.   NUMDCD = 17.                       |C N1][N2 ][N3 ]
Cable conductor numbers.  1 1 1                    |  1    1    1
Comment card.   NUMDCD = 19.                       |C ==Physical parameters==
Comment card.   NUMDCD = 20.                       |C   r1    ][   r2    ][   r3    ][   r4    ][   r5    ]
Radii.    .00   .02   .04                          |  0.0000   20.75E-3   39.25E-3
Comment card.   NUMDCD = 22.                       |C  rohc  ][  muc   ][  mu11  ][  eps11 ][  rhos  ][   mus  ][  mu12  ][  eps12 ]
Physical constants for conductors & insulators.  |  3.19E-8         1.0          1.0          1.00
Comment card.   NUMDCD = 24.                       |C   r1    ][   r2    ][   r3    ][   r4    ][   r5    ]
Radii.    .00   .02   .04                          |  0.0000   20.75E-3   39.25E-3
Comment card.   NUMDCD = 26.                       |C  rohc  ][  muc   ][  mu11  ][  eps11 ][  rhos  ][   mus  ][  mu12  ][  eps12 ]
Physical constants for conductors & insulators.  |  3.19E-8         1.0          1.0          1.00
Comment card.   NUMDCD = 28.                       |C   r1    ][   r2    ][   r3    ][   r4    ][   r5    ]
Radii.    .00   .02   .04                          |  0.0000   20.75E-3   39.25E-3
Comment card.   NUMDCD = 30.                       |C  rohc  ][  muc   ][  mu11  ][  eps11 ][  rhos  ][   mus  ][  mu12  ][  eps12 ]
Physical constants for conductors & insulators.  |  3.19E-8         1.0          1.0          1.00
Comment card.   NUMDCD = 32.                       |C  Vert1 ][ Horiz1 ][  Vert2 ][ Horiz2 ][  Vert3 ][ Horiz3 ][  Vert4 ][ Horiz4 ]
(Y,X)  1.000E+00   0.000E+00                       |     1.00        0.0
Comment card.   NUMDCD = 34.                       |C ==Frequency Card==
Comment card.   NUMDCD = 35.                       |C      rhoe      ][       Freq.     ][DEC][PNT][ DIST ][  IPUN  ]
Frequency card 1.000E+02 5.000E+04   0   1 0.00E+00 |        100.0            5.E4                                 2
      3 -phase PT cable with 4   conductors.   0   grounded.   Total # of conductors = 4
        Under-ground cable
        Cable constants
       Pipe and its insulator
       Pipe : inner radius 1.50000000E-01, outer radius 1.70000000E-01.  Outer radius of pipe insulator 1.90000000E-01
              Resistivity (ohm-m): 1.50000000E-07, Relative permeability:   3.00000000E+02
        Insulator ; Relative permittivity, inner: 1.00000000E+00, outer: 3.00000000E+00
        Pipe 1 ; Dist. from Earth surface: 1.00000000E+00   m; Dist. from pipe 1: 0.00000000E+00   m
       Inner conductors (SC cables)
        Dist. from pipe center :      1    0.1000     2    0.0750     3    0.1000
        Angle to first cable :            -90.0000           -80.0000          -70.0000
        Phase 1   Boundary radii ;
         1   0.00000E+00   2   2.07500E-02   3   3.92500E-02
                Resistivity(ohm-m): Core 3.19000000E-08, Sheath 9.99999994E-09, Armor 9.99999994E-09
                Relative permeability;   1.00000000E+00, 1.00000000E+00, 1.00000000E+00
                Insulator:   1; (Relative) Permeability   1.0     Permittivity   1.00     Velocity(m/s) 2.99793E+08
                DC resistance(ohm/m),           2.35833E-05
        Phase 2   Boundary radii ;
         1   0.00000E+00   2   2.07500E-02   3   3.92500E-02
                Resistivity(ohm-m): Core 3.19000000E-08, Sheath 9.99999994E-09, Armor 9.99999994E-09
                Relative permeability;   1.00000000E+00, 1.00000000E+00, 1.00000000E+00
                Insulator:   1; (Relative) Permeability   1.0     Permittivity   1.00     Velocity(m/s) 2.99793E+08
                DC resistance(ohm/m),           2.35833E-05
```

```
Phase 3   Boundary radii ;
 1   0.00000E+00   2   2.07500E-02   3   3.92500E-02
         Resistivity(ohm-m): Core 3.19000000E-08, Sheath 9.99999994E-09, Armor 9.99999994E-09
         Relative permeability;   1.00000000E+00, 1.00000000E+00, 1.00000000E+00
         Insulator:   1; (Relative) Permeability   1.0    Permittivity   1.00    Velocity(m/s) 2.99793E+08
         DC resistance(ohm/m),          2.35833E-05
Earth ; resistivity 1.00000000E+02
Punched cards:   1) pi-circuit (IPUNCH=1) or distributed line (IPUNCH=2);
         IPUNCH = 2          Total length = 0.00000000E+00
           Frequency 5.00000000E+04   Hz
          Eigenvector rotation has been made.

           Total impedance [Zc]
 5.76591E-02  5.54102E-02  5.61541E-02  4.96262E-02
 4.26236E-01  4.14331E-01  3.97134E-01  3.30135E-01
 5.54102E-02  5.52509E-02  5.54102E-02  4.96262E-02
 4.14331E-01  4.42206E-01  4.14331E-01  3.30135E-01
 5.61541E-02  5.54102E-02  5.76591E-02  4.96262E-02
 3.97134E-01  4.14331E-01  4.26236E-01  3.30135E-01
 4.96262E-02  4.96262E-02  4.96262E-02  4.96262E-02
 3.30135E-01  3.30135E-01  3.30135E-01  3.30135E-01

           Total admittance [Yc]
 0.00000E+00  0.00000E+00  0.00000E+00  0.00000E+00
 3.68896E-05 -2.46885E-05 -3.29101E-06 -8.91006E-06
 0.00000E+00  0.00000E+00  0.00000E+00  0.00000E+00
-2.46885E-05  4.66218E-05 -2.46885E-05  2.75517E-06
 0.00000E+00  0.00000E+00  0.00000E+00  0.00000E+00
-3.29101E-06 -2.46885E-05  3.68896E-05 -8.91006E-06
 0.00000E+00  0.00000E+00  0.00000E+00  0.00000E+00
-8.91006E-06  2.75517E-06 -8.91006E-06  4.86470E-04

Characteristic impedance matrix   [Zc]   in phase variables,   in [ohm]:
 114.03860 103.92891  87.11746  26.53780
  -5.59401  -4.59633  -4.89835  -1.98345
 103.92891 130.67491 103.92891  26.53780
  -4.59633  -4.55623  -4.59633  -1.98345
  87.11746 103.92891 114.03860  26.53780
  -4.89835  -4.59633  -5.59401  -1.98345
  26.53780  26.53780  26.53780  26.53780
  -1.98345  -1.98345  -1.98345  -1.98345
```

Modal transformation matrices follow. These are complex, with the real part displayed above the imaginary part.
The current transformation matrix [Ti] follows. The transpose of this is the inverse of the voltage transformation matrix.
By definition, [Ti] gives the mapping from modal to phase variables: i-phase = [Ti] * i-mode

```
  .00000    .77508   -.51185   -.50000
  .00000    .00080    .00209    .00000
  .00000   -.83306   1.55015    .00000
  .00000   -.00243   -.00225    .00000
  .00000    .77508   -.51185    .50000
  .00000    .00080    .00209    .00000
 1.00000   -.71709   -.52644    .00000
  .00000    .00083   -.00194    .00000
```

Voltage transformation matrix [Tv]. This gives the mapping from modal to phase variables: v-phase = [Tv] * v-mode

```
 1.00000   1.00000    .53741  -1.00000
  .00000   -.00145    .00157    .00000
 1.00000    .66039   1.00000    .00000
  .00000   -.00270    .00104    .00000
 1.00000   1.00000    .53741   1.00000
  .00000   -.00145    .00157    .00000
 1.00000    .00000    .00000    .00000
  .00000    .00000    .00000    .00000
```

Modal components

Mode	Attenuation	Velocity	Impedance (Ohm/m)		Admittance (S/m)		Charact. imp. (Ohm)		Charact. Admit. (S)	
No.	(db/km)	(m/mic.s)	Real	Imag.	Real	Imag.	Real	Imag.	Real	Imag.
1	8.12150E+00	25.11	4.96262E-02	3.30135E-01	0.00000E+00	4.71406E-04	26.538	-1.983	3.74728E-02	2.80074E-03
2	5.57659E-01	279.88	6.45939E-03	5.62814E-02	0.00000E+00	2.23132E-05	50.305	-2.877	1.98139E-02	1.13329E-03
3	1.47306E-01	294.93	2.78827E-03	8.75439E-02	0.00000E+00	1.29576E-05	82.206	-1.309	1.21614E-02	1.93622E-04
4	2.42791E-01	290.43	7.52500E-04	1.45507E-02	0.00000E+00	8.03612E-05	13.461	-.348	7.42415E-02	1.91845E-03

```
Blank card terminating frequency cards.        |BLANK ENDING CABLE CONSTANTS CASE
Request for flushing of punch buffer.          |$PUNCH
```

```
A listing of 80-column card images now being flushed from punch buffer follows.
================================================================================
12345678901234567890123456789012345678901234567890123456789012345678901234567890123456789
================================================================================
C  <++++++>  Cards punched by support routine on  24-Feb-14  13.54.52  <++++++>
C CABLE CONSTANTS
C CABLE PARAMETERS
C C ==Miscellaneous Data Card==
C C TY][SYS][NPC][EAR][KMO][ZFL][YFL][NPP][NGD]
C     3   -1    3           0    0              1
C C ==pipe physical data==
C C    Rp1   ][    Rp2   ][     Rp3   ][   rohp   ][    mur    ][   eps1   ][   eps2   ]
C          0.150        0.170        0.190     0.15E-6          300.          1.0          3.0
C C ==Location==
C C   Dist1 ][ Theta1 ][   Dist2 ][ Theta2 ][   Dist3 ][ Theta3 ][   Dist4 ][ Theta4
C          0.100        -90.0       0.075          -80.        0.100          -70.
C C ==The number of conductors==
C C N1][N2 ][N3 ]
C     1    1    1
C C ==Physical parameters==
C C    r1     ][    r2     ][    r3     ][    r4     ][    r5     ]
C       0.0000   20.75E-3   39.25E-3
C C   rohc   ][   muc     ][   muI1   ][   epsI1 ][   rhos   ][    mus   ][   muI2   ][   epsI2
C      3.19E-8           1.0           1.0          1.00
C C    r1    ][    r2    ][    r3    ][    r4    ][    r5    ]
C       0.0000   20.75E-3   39.25E-3
C C   rohc   ][   muc     ][   muI1   ][   epsI1 ][   rhos   ][    mus   ][   muI2   ][   epsI2
C      3.19E-8           1.0           1.0          1.00
C C    r1    ][    r2    ][    r3    ][    r4    ][    r5    ]
C       0.0000   20.75E-3   39.25E-3
C C   rohc   ][   muc     ][   muI1   ][   epsI1 ][   rhos   ][    mus   ][   muI2   ][   epsI2
C      3.19E-8           1.0           1.0          1.00
C C   Vert1 ][ Horiz1 ][   Vert2 ][ Horiz2 ][   Vert3 ][ Horiz3 ][   Vert4 ][ Horiz4
C          1.00          0.0
C C ==Frequency Card==
C C        rhoe       ][        Freq.     ][DEC][PNT][ DIST ][   IPUN    ]
C                 100.0                5.E4                                      2
$UNITS, 0.0, 0.0
$VINTAGE, 1
C     Branch cards for constant-parameter distributed:
-1                                  1.96262E-02 2.65378E+01 2.31123E+07 0.00000E+00 1 14
-2                                  6.45939E-03 5.03052E+01 2.79882E+08 0.00000E+00 1 14
-3                                  2.78827E-03 8.22063E+01 2.94930E+08 0.00000E+00 1 14
-4                                  7.52500E-04 1.34606E+01 2.90429E+08 0.00000E+00 1 14
    .00000000    .77507612   -.51185337   -.50000000
    .00000000    .00080264    .00209295    .00000000
    .00000000   -.83306190   1.55015144    .00000000
    .00000000   -.00243082   -.00224953    .00000000
    .00000000    .77507612   -.51185337    .50000000
    .00000000    .00080264    .00209295    .00000000
   1.00000000   -.71709034   -.52644469    .00000000
    .00000000    .00082553   -.00193637    .00000000
$VINTAGE, 0
$UNITS, -1., -1.,
=========<   End of LUNIT7 punched cards as flushed by   $PUNCH   request   >=======

Blank card ending   "CABLE CONSTANTS"   cases.        |BLANK ENDING SEMLYEN SETUP
Timing figures characterizing central processor (CP) solution speed.  ---------------------------  CP sec   Wait sec   Real sec
                                                                       Totals      0.080       0.000       0.080
```

Different natures of the voltage transformation matrix in the asymmetrical case from the symmetrical configuration case are observed in Table 3.12. Only the eigenvectors of mode 3 are observed to be frequency-independent and the same as those in the symmetrical configuration case given in Table 3.11. The eigenvectors of all the other modes are frequency-dependent. Modes 1 and 2 show characteristics of currents flowing through phase conductors and returning through not only the remaining conductors but also the pipe. This differs from the symmetrical configuration case in which the whole currents return through the phase conductors.

The attenuations and velocities of modes 0 and 1 in the asymmetrical case are seen from Figure 3.20 to be similar to those in the symmetrical configuration case. However, those of mode 3 are different from those of mode 2, while modes 2 and 3 show the same in the symmetrical configuration case in Figure 3.18.

3.6.2.3 Inner Conductors with Core and Sheath

In the previous sections, cases of inner conductors consisting only of a core were discussed. In this section, a case of inner conductors having a core and a sheath will be investigated. Calculated results of a voltage transformation matrix and propagation constants for a PT cable as shown in Figure 3.21 are given in Table 3.13 and Figure 3.22, respectively. The output list is given in Data 3.5.

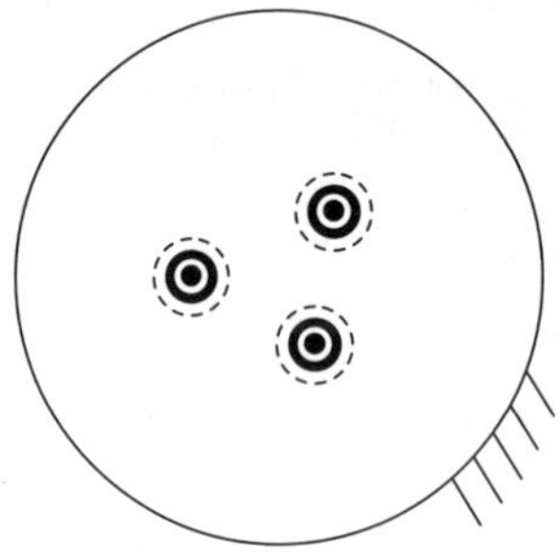

Figure 3.21 A PT cable with inner conductors consisting of a core and sheath

Table 3.13 Voltage transformation matrix A^{-1} a PT cable as shown in Figure 3.21 at $f = 50$ kHz

Mode	Core			Sheath			Pipe
	a	b	c	a	b	c	
0	0	0	0	0	0	0	1
1	1	0	0	−1	0	0	0
2	0	0	1	0	0	−1	0
3	0	1	0	0	−1	0	0
4	0	0	0	0	0	1	−1
5	0	0	0	0	1	0	−1
6	0	0	0	1	0	0	−1

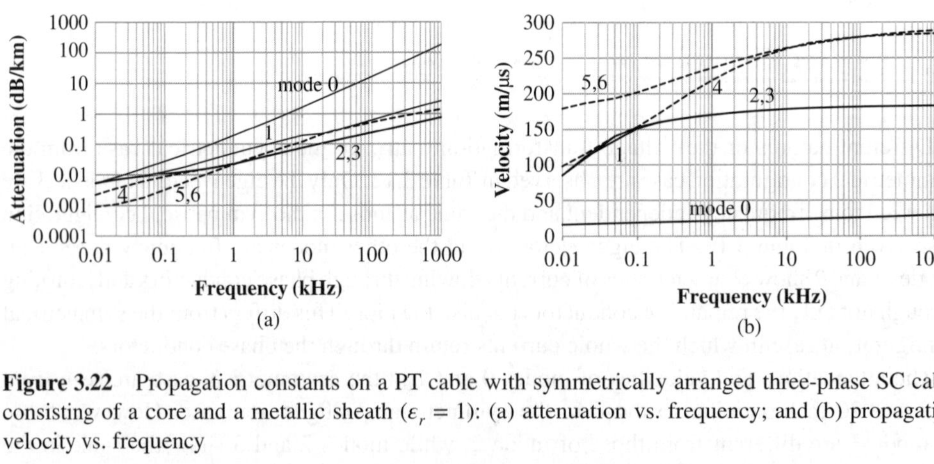

Figure 3.22 Propagation constants on a PT cable with symmetrically arranged three-phase SC cable consisting of a core and a metallic sheath ($\varepsilon_r = 1$): (a) attenuation vs. frequency; and (b) propagation velocity vs. frequency

Data 3.5 Output of a PT cable with symmetrically arranged three-phase inner SC cable consisting of a core and a metallic sheath

```
 ---   39    cards of disk file read into card cache cells   1     onward.
 1st card (Lists   1-10).          |     8000     15000     15000       340     60000       120
 2nd card (Lists 11-20).           |      360       900     36400       255    129600        38
 3rd card (Lists 21-29).           |     3000     15000    160000        24     30000      1000
 Supplemental offsets.             |   240000       742
Alternative Transients Program (ATP),   Watcom translation.    All rights reserved by Can/Am user group of Portland, Oregon, USA.
 Date (dd-mth-yy) and time of day (hh.mm.ss) = 24-Feb-14   15.03.30    Name of disk plot file is   C:¥ATPWNT¥UNTITLED.pl4
Consult the 860 page ATP Rule Book of the Can/Am EMTP User Group in Portland, Oregon, USA.    Program is no older than Sept., 1998.
Total size of LABCOM tables =   3639004    INTEGER words.    VARDIM   List Sizes follow :    8002 15000 15000     340 60000   1200
 13140 120000 2250 1900   360   900 36400     255   129K    384     45    254 240000 100000   3000 15000   160K     24 30000 10000     600 252000
--------------------------------------------------+--------------------------------------------------------------------------------
Descriptive interpretation of input data cards.   |  Input data card images are shown below, all 80 columns, character by character
                                                      0         1         2         3         4         5         6         7         8
                                                      012345678901234567890123456789012345678901234567890123456789012345678901234567890
--------------------------------------------------+--------------------------------------------------------------------------------
Comment card.   NUMDCD = 1.                       |C data:C:¥ATPWNT¥UNTITLED.DAT
Comment card.   NUMDCD = 2.                       |C Cable Constants for underground PT cable symmetrical arrangement core and shea
Comment card.   NUMDCD = 3.                       |C ===Cable Constants===
Marker card preceding new EMTP data case.         |BEGIN NEW DATA CASE
Find parameters of cables.   Type = 0.            |CABLE CONSTANTS
Prof. Akihiro Ametani`s 1994 code.    Order = 38  |CABLE PARAMETERS
Comment card.   NUMDCD = 7.                       |C ==Miscellaneous Data Card==
Comment card.   NUMDCD = 8.                       |C TY][SYS][NPC][EAR][KMO][ZFL][YFL][NPP][NGD]
Misc.   3  -1   3   0   0   0   0   1   0   0   0 |   3   -1     3            0     0              1
Comment card.   NUMDCD = 10.                      |C ==pipe physical data==
Comment card.   NUMDCD = 11.                      |C    Rp1   ][    Rp2   ][    Rp3   ][  rohp   ][  mur    ][  eps1  ][  eps2  ]
Pipe radii.    1.5000E-01   1.7000E-01   1.9000E-01  |  0.150      0.170      0.190     0.15E-6       300.          1.0         3.0
Comment card.   NUMDCD = 13.                      |C ==Location==
Comment card.   NUMDCD = 14.                      |C   Dist1 ][ Theta1 ][   Dist2 ][ Theta2 ][   Dist3 ][ Theta3 ][   Dist4 ][ Theta4 ]
Cable to pipe.   7.500E-02   0.000E+00   7.500E-02   |  0.075          0.0        0.075          120.        0.075        -120.
Comment card.   NUMDCD = 16.                      |C ==The number of conductors==
Comment card.   NUMDCD = 17.                      |C N1][N2 ][N3 ]
Cable conductor numbers.   2 2 2                  |  2     2     2
Comment card.   NUMDCD = 19.                      |C ==Physical parameters==
Comment card.   NUMDCD = 20.                      |C    r1    ][    r2    ][    r3    ][    r4    ][    r5    ]
Radii.    .00    .02    .04    .04    .05         |  0.0000    20.75E-3    39.25E-3    42.76E-3    47.94E-3
Comment card.   NUMDCD = 22.                      |C  rohc   ][  muc    ][  mul1   ][  epsl1  ][  rhos   ][   mus   ][  mul2   ][  epsl2 ]
Physical constants for conductors & insulators.   |  3.19E-8        1.0        1.0        2.68     2.83E-8       1.000       1.000         2.3
Comment card.   NUMDCD = 24.                      |C    r1    ][    r2    ][    r3    ][    r4    ][    r5    ]
Radii.    .00    .02    .04    .04    .05         |  0.0000    20.75E-3    39.25E-3    42.76E-3    47.94E-3
Comment card.   NUMDCD = 26.                      |C  rohc   ][  muc    ][  mul1   ][  epsl1  ][  rhos   ][   mus   ][  mul2   ][  epsl2 ]
Physical constants for conductors & insulators.   |  3.19E-8        1.0        1.0        2.68     2.83E-8       1.000       1.000         2.3
Comment card.   NUMDCD = 28.                      |C    r1    ][    r2    ][    r3    ][    r4    ][    r5    ]
Radii.    .00    .02    .04    .04    .05         |  0.0000    20.75E-3    39.25E-3    42.76E-3    47.94E-3
Comment card.   NUMDCD = 30.                      |C  rohc   ][  muc    ][  mul1   ][  epsl1  ][  rhos   ][   mus   ][  mul2   ][  epsl2 ]
Physical constants for conductors & insulators.   |  3.19E-8        1.0        1.0        2.68     2.83E-8       1.000       1.000         2.3
Comment card.   NUMDCD = 32.                      |C   Vert1 ][ Horiz1 ][   Vert2 ][ Horiz2 ][   Vert3 ][ Horiz3 ][   Vert4 ][ Horiz4 ]
(Y,X)   1.000E+00   0.000E+00                     |     1.00          0.0
Comment card.   NUMDCD = 34.                      |C ==Frequency Card==
Comment card.   NUMDCD = 35.                      |C        rhoe      ][        Freq.     ][DEC][PNT][ DIST ][  IPUN  ]
Frequency card 1.000E+02 5.000E+04   0    1 0.00E+00 |      100.0                5.E4                                   2
      3 -phase PT cable with 7    conductors.   0   grounded.    Total # of conductors = 7
         Under-ground cable
         Cable constants
        Pipe and its insulator
        Pipe : inner radius 1.50000000E-01, outer radius 1.70000000E-01.   Outer radius of pipe insulator 1.90000000E-01
                 Resistivity (ohm-m): 1.50000000E-07, Relative permeability:   3.00000000E+02
         Insulator ; Relative permittivity, inner: 1.00000000E+00, outer: 3.00000000E+00
         Pipe 1 ; Dist. from Earth surface: 1.00000000E+00   m; Dist. from pipe 1: 0.00000000E+00   m
        Inner conductors (SC cables)
         Dist. from pipe center :      1    0.0750      2    0.0750      3    0.0750
         Angle to first cable :            0.0000          120.0000         -120.0000
         Phase 1   Boundary radii ;
          1   0.00000E+00   2   2.07500E-02   3   3.92500E-02   4   4.27600E-02   5   4.79400E-02
                 Resistivity(ohm-m): Core 3.19000000E-08, Sheath 2.83000000E-08, Armor 9.99999994E-09
                 Relative permeability;   1.00000000E+00, 1.00000000E+00, 1.00000000E+00
                 Insulator:   1; (Relative) Permeability   1.0     Permittivity   2.68     Velocity(m/s) 1.83127E+08
                 Insulator:   2; (Relative) Permeability   1.0     Permittivity   2.30     Velocity(m/s) 1.97677E+08
                 DC resistance(ohm/m),          2.35833E-05              3.12941E-05
         Phase 2   Boundary radii ;
          1   0.00000E+00   2   2.07500E-02   3   3.92500E-02   4   4.27600E-02   5   4.79400E-02
                 Resistivity(ohm-m): Core 3.19000000E-08, Sheath 2.83000000E-08, Armor 9.99999994E-09
                 Relative permeability;   1.00000000E+00, 1.00000000E+00, 1.00000000E+00
                 Insulator:   1; (Relative) Permeability   1.0     Permittivity   2.68     Velocity(m/s) 1.83127E+08
                 Insulator:   2; (Relative) Permeability   1.0     Permittivity   2.30     Velocity(m/s) 1.97677E+08
                 DC resistance(ohm/m),          2.35833E-05              3.12941E-05
         Phase 3   Boundary radii ;
          1   0.00000E+00   2   2.07500E-02   3   3.92500E-02   4   4.27600E-02   5   4.79400E-02
                 Resistivity(ohm-m): Core 3.19000000E-08, Sheath 2.83000000E-08, Armor 9.99999994E-09
                 Relative permeability;   1.00000000E+00, 1.00000000E+00, 1.00000000E+00
```

Insulator: 1; (Relative) Permeability 1.0 Permittivity 2.68 Velocity(m/s) 1.83127E+08
Insulator: 2; (Relative) Permeability 1.0 Permittivity 2.30 Velocity(m/s) 1.97677E+08
DC resistance(ohm/m), 2.35833E-05 3.12941E-05
Earth ; resistivity 1.00000000E+02
Punched cards: 1) pi-circuit (IPUNCH=1) or distributed line (IPUNCH=2);
IPUNCH = 2 Total length = 0.00000000E+00
Frequency 5.00000000E+04 Hz
Eigenvector rotation has been made.

Total impedance [Zc]

5.58320E-02	5.19751E-02	5.19751E-02	5.49158E-02	5.19751E-02	5.19751E-02	4.96262E-02
4.37406E-01	3.49979E-01	3.49979E-01	3.96445E-01	3.49979E-01	3.49979E-01	3.30135E-01
5.19751E-02	5.58320E-02	5.19751E-02	5.19751E-02	5.49158E-02	5.19751E-02	4.96262E-02
3.49979E-01	4.37406E-01	3.49979E-01	3.49979E-01	3.96445E-01	3.49979E-01	3.30135E-01
5.19751E-02	5.19751E-02	5.58320E-02	5.19751E-02	5.19751E-02	5.49158E-02	4.96262E-02
3.49979E-01	3.49979E-01	4.37406E-01	3.49979E-01	3.49979E-01	3.96445E-01	3.30135E-01
5.49158E-02	5.19751E-02	5.19751E-02	5.49157E-02	5.19751E-02	5.19751E-02	4.96262E-02
3.96445E-01	3.49979E-01	3.49979E-01	3.96445E-01	3.49979E-01	3.49979E-01	3.30135E-01
5.19751E-02	5.49158E-02	5.19751E-02	5.19751E-02	5.49157E-02	5.19751E-02	4.96262E-02
3.49979E-01	3.96445E-01	3.49979E-01	3.49979E-01	3.96445E-01	3.49979E-01	3.30135E-01
5.19751E-02	5.19751E-02	5.49158E-02	5.19751E-02	5.19751E-02	5.49157E-02	4.96262E-02
3.49979E-01	3.49979E-01	3.96445E-01	3.49979E-01	3.49979E-01	3.96445E-01	3.30135E-01
4.96262E-02	4.96262E-02	4.96262E-02	4.96262E-02	4.96262E-02	4.96262E-02	4.96262E-02
3.30135E-01	3.30135E-01	3.30135E-01	3.30135E-01	3.30135E-01	3.30135E-01	3.30135E-01

Total admittance [Yc]

0.00000E+00	0.00000E+00	0.00000E+00	0.00000E+00	0.00000E+00	0.00000E+00	0.00000E+00
7.34848E-05	0.00000E+00	0.00000E+00	-7.34848E-05	-1.41761E-21	-8.42690E-24	0.00000E+00
0.00000E+00	0.00000E+00	0.00000E+00	0.00000E+00	0.00000E+00	0.00000E+00	0.00000E+00
0.00000E+00	7.34848E-05	0.00000E+00	-1.28611E-22	-7.34848E-05	-8.33204E-22	0.00000E+00
0.00000E+00	0.00000E+00	0.00000E+00	0.00000E+00	0.00000E+00	0.00000E+00	0.00000E+00
0.00000E+00	0.00000E+00	7.34848E-05	-1.05313E-21	-1.05313E-21	-7.34848E-05	0.00000E+00
0.00000E+00	0.00000E+00	0.00000E+00	0.00000E+00	0.00000E+00	0.00000E+00	0.00000E+00
-7.34848E-05	-1.28611E-22	-1.05313E-21	9.61739E-05	-5.36865E-06	-5.36865E-06	-1.19517E-05
0.00000E+00	0.00000E+00	0.00000E+00	0.00000E+00	0.00000E+00	0.00000E+00	0.00000E+00
-1.41761E-21	-7.34848E-05	-1.05313E-21	-5.36865E-06	9.61739E-05	-5.36865E-06	-1.19517E-05
0.00000E+00	0.00000E+00	0.00000E+00	0.00000E+00	0.00000E+00	0.00000E+00	0.00000E+00
-8.42690E-24	-8.33204E-22	-7.34848E-05	-5.36865E-06	-5.36865E-06	9.61739E-05	-1.19517E-05
0.00000E+00	0.00000E+00	0.00000E+00	0.00000E+00	0.00000E+00	0.00000E+00	0.00000E+00
0.00000E+00	0.00000E+00	0.00000E+00	-1.19517E-05	-1.19517E-05	-1.19517E-05	5.07261E-04

Characteristic impedance matrix [Zc] in phase variables, in [ohm]:

99.12777	24.27187	23.86221	75.51679	24.27185	23.86220	26.53780
1.29526	-5.84417	-4.83962	1.55931	-5.84416	-4.83961	-1.98345
24.27187	101.53549	24.51115	24.27185	77.92450	24.51114	26.53780
-5.84417	-6.96595	-4.89547	-5.84416	-6.70192	-4.89546	-1.98345
23.86221	24.51115	96.88904	23.86220	24.51114	73.27804	26.53780
-4.83962	-4.89547	-11.21465	-4.83961	-4.89546	-10.95061	-1.98345
75.51679	24.27185	23.86220	75.51677	24.27184	23.86219	26.53780
1.55931	-5.84416	-4.83961	1.55933	-5.84415	-4.83960	-1.98345
24.27185	77.92450	24.51114	24.27184	77.92448	24.51113	26.53780
-5.84416	-6.70192	-4.89546	-5.84415	-6.70190	-4.89546	-1.98345
23.86220	24.51114	73.27804	23.86219	24.51113	73.27802	26.53780
-4.83961	-4.89546	-10.95061	-4.83960	-4.89546	-10.95060	-1.98345
26.53780	26.53780	26.53780	26.53780	26.53780	26.53780	26.53780
-1.98345	-1.98345	-1.98345	-1.98345	-1.98345	-1.98345	-1.98345

Modal transformation matrices follow. These are complex, with the real part displayed above the imaginary part.
The current transformation matrix [Ti] follows. The transpose of this is the inverse of the voltage transformation matrix.
By definition, [Ti] gives the mapping from modal to phase variables: i-phase = [Ti] * i-mode

.00000	1.00000	.00000	.00000	.00000	.00000	.00000
.00000	.00000	.00000	.00000	.00000	.00000	.00000
.00000	.00000	.00000	1.00000	.00000	.00000	.00000
.00000	.00000	.00000	.00000	.00000	.00000	.00000
.00000	.00000	1.00000	.00000	.00000	.00000	.00000
.00000	.00000	.00000	.00000	.00000	.00000	.00000
.00000	-1.00000	.00000	.00000	-.02358	.00124	1.02732
.00000	.00000	.00000	.00000	-.08036	-.10176	-.05161
.00000	.00000	.00000	-1.00000	-.02254	1.02559	.03626
.00000	.00000	.00000	.00000	-.08010	.02388	.17761
.00000	.00000	-1.00000	.00000	1.01799	.07246	.04409
.00000	.00000	.00000	.00000	.06498	.14923	.16681
1.00000	.00000	.00000	.00000	-.97187	-1.09929	-1.10766
.00000	.00000	.00000	.00000	.09547	-.07135	-.29280

Voltage transformation matrix [Tv]. This gives the mapping from modal to phase variables: v-phase = [Tv] * v-mode

1.00000	1.00000	.00000	.00000	-.07120	-.02306	.99923
.00000	.00000	.00000	.00000	-.14451	-.18263	.03925
1.00000	.00000	.00000	1.00000	-.07120	.99933	.00754
.00000	.00000	.00000	.00000	-.14451	-.03651	.09053
1.00000	.00000	1.00000	.00000	.99668	.04328	.01828
.00000	.00000	.00000	.00000	-.08139	.06901	.08122
1.00000	.00000	.00000	.00000	-.07120	-.02306	.99923
.00000	.00000	.00000	.00000	-.14451	-.18263	.03925
1.00000	.00000	.00000	.00000	-.07120	.99933	.00754
.00000	.00000	.00000	.00000	-.14451	-.03651	.09053
1.00000	.00000	.00000	.00000	.99668	.04328	.01828
[illegible]	[illegible]	.00000	.00000	-.08140	.06901	.08122
1.00000	.00000	.00000	.00000	.00000	.00000	.00000
.00000	.00000	.00000	.00000	.00000	.00000	.00000

Modal components

Mode	Attenuation	Velocity	Impedance (Ohm/m)		Admittance (S/m)		Charact. imp. (Ohm)		Charact. Admit. (S)	
No.	(db/km)	(m/mic.s)	Real	Imag.	Real	Imag.	Real	Imag.	Real	Imag.
1	8.12150E+00	25.11	4.96262E-02	3.30135E-01	0.00000E+00	4.71406E-04	26.538	-1.983	3.74728E-02	2.80074E-03
2	1.68524E-01	181.07	9.16196E-04	4.09612E-02	0.00000E+00	7.34844E-05	23.611	-.264	4.23477E-02	4.73544E-04
3	1.68524E-01	181.07	9.16196E-04	4.09612E-02	0.00000E+00	7.34844E-05	23.611	-.264	4.23477E-02	4.73544E-04
4	1.68524E-01	181.07	9.16196E-04	4.09612E-02	0.00000E+00	7.34844E-05	23.611	-.264	4.23477E-02	4.73544E-04
5	3.42727E-01	276.52	3.79623E-03	5.45877E-02	0.00000E+00	2.36172E-05	48.106	-1.671	2.07626E-02	7.21084E-04
6	3.68748E-01	276.40	4.48420E-03	5.99448E-02	0.00000E+00	2.15213E-05	52.813	-1.973	1.89082E-02	7.06231E-04
7	3.89415E-01	275.63	4.61073E-03	5.85188E-02	0.00000E+00	2.21653E-05	51.422	-2.023	1.94170E-02	7.63754E-04

Blank card terminating frequency cards. |BLANK ENDING CABLE CONSTANTS CASE
Request for flushing of punch buffer. |$PUNCH

A listing of 80-column card images now being flushed from punch buffer follows.

```
================================================================================
12345678901234567890123456789012345678901234567890123456789012345678901234567890123456789
================================================================================
C  <++++++>  Cards punched by support routine on  24-Feb-14  15.03.30  <++++++>
C CABLE CONSTANTS
C CABLE PARAMETERS
C C ==Miscellaneous Data Card==
C C TY][SYS][NPC][EAR][KMO][ZFL][YFL][NPP][NGD]
C      3   -1    3          0    0          1
C C ==pipe physical data==
C C   Rp1  ][   Rp2  ][   Rp3  ][  rohp  ][   mur   ][  eps1  ][  eps2  ]
C       0.150     0.170     0.190   0.15E-6       300.        1.0       3.0
C C ==Location==
C C   Dist1 ][ Theta1 ][  Dist2 ][ Theta2 ][  Dist3 ][ Theta3 ][  Dist4 ][ Theta4
C       0.075        0.0     0.075       120.      0.075      -120.
C C ==The number of conductors==
C C N1][N2 ][N3 ]
C     2    2    2
C C ==Physical parameters==
C C   r1   ][   r2   ][   r3   ][   r4   ][   r5   ]
C      0.0000  20.75E-3  39.25E-3  42.76E-3  47.94E-3
C C  rohc  ][  muc   ][  mul1  ][ epsl1 ][  rhos  ][   mus  ][  mul2  ][  epsl2
C      3.19E-8       1.0        1.0       2.68    2.83E-8     1.000      1.000        2
C C   r1   ][   r2   ][   r3   ][   r4   ][   r5   ]
C      0.0000  20.75E-3  39.25E-3  42.76E-3  47.94E-3
C C  rohc  ][  muc   ][  mul1  ][ epsl1 ][  rhos  ][   mus  ][  mul2  ][  epsl2
C      3.19E-8       1.0        1.0       2.68    2.83E-8     1.000      1.000        2
C C   r1   ][   r2   ][   r3   ][   r4   ][   r5   ]
C      0.0000  20.75E-3  39.25E-3  42.76E-3  47.94E-3
C C  rohc  ][  muc   ][  mul1  ][ epsl1 ][  rhos  ][   mus  ][  mul2  ][  epsl2
C      3.19E-8       1.0        1.0       2.68    2.83E-8     1.000      1.000        2
C C  Vert1 ][ Horiz1 ][  Vert2 ][ Horiz2 ][  Vert3 ][ Horiz3 ][  Vert4 ][ Horiz4
C         1.00        0.0
C C ==Frequency Card==
C C        rhoe     ][        Freq.    ][DEC][PNT][ DIST ][  IPUN  ]
C              100.0              5.E4                              2
$UNITS, 0.0, 0.0
$VINTAGE, 1
C    Branch cards for constant-parameter distributed:
-1                                4.96262E-02 2.65378E+01 2.51125E+07 0.00000E+00 1 17
-2                                9.16196E-04 2.36111E+01 1.81067E+08 0.00000E+00 1 17
-3                                9.16196E-04 2.36111E+01 1.81067E+08 0.00000E+00 1 17
-4                                9.16196E-04 2.36111E+01 1.81067E+08 0.00000E+00 1 17
-5                                3.79623E-03 4.81055E+01 2.76520E+08 0.00000E+00 1 17
-6                                4.48420E-03 5.28134E+01 2.76399E+08 0.00000E+00 1 17
-7                                4.61073E-03 5.14218E+01 2.75632E+08 0.00000E+00 1 17
```

```
 .00000000  1.00000264   .00000000   .00000000   .00000026   .00000023
-.00000261
 .00000000   .00000000   .00000000   .00000000   .00000017   .00000028
 .00000219
 .00000000  -.00000003  -.00000003  1.00000264   .00000025  -.00000275
-.00000041
 .00000000   .00000000   .00000000   .00000000   .00000017   .00000198
-.00000039
 .00000000  -.00000003  1.00000264   .00000000  -.00000282  -.00000045
-.00000041
 .00000000   .00000000   .00000000   .00000000   .00000185  -.00000025
-.00000034
 .00000000 -1.00000182  -.00000020  -.00000020  -.02358376   .00123664
1.02731570
 .00000000  -.00000061   .00000014   .00000014  -.08036310  -.10176390
-.05161207
 .00000000  -.00000017  -.00000017 -1.00000182  -.02253611  1.02558948
 .03626029
 .00000000   .00000015   .00000015  -.00000061  -.08009756   .02387956
 .17760633
 .00000000  -.00000017 -1.00000182  -.00000020  1.01799468   .07246186
 .04409217
 .00000000   .00000015  -.00000061   .00000014   .06498363   .14922979
 .16680632
1.00000000  -.00000042  -.00000042  -.00000042  -.97187251 -1.09928500
-1.10766473
 .00000000   .00000033   .00000033   .00000033   .09547484  -.07134745
-.29280205
$VINTAGE, 0
$UNITS, -1., -1.,
=========<  End of LUNIT7 punched cards as flushed by  $PUNCH  request  >=======

Blank card ending  "CABLE CONSTANTS"  cases.      |BLANK ENDING SEMLYEN SETUP
Timing figures characterizing central processor (CP) solution speed.  --------------------------  CP sec  Wait sec  Real sec
                                                                                          Totals   0.070    0.000     0.070
```

In the case of inner conductors consisting of a core and a sheath, the whole system becomes a seven conductor system, that is, two conductors for each three-phase and the pipe conductor. Except for mode 0, the earth-return mode, the other six modes show similar eigenvectors compared with the four conductor system. Modes 1–3 are the coaxial modes between the core and sheath, and their velocities are lower than those of modes 4–6 in the frequency range from 10 Hz to 1 MHz. This characteristic is different from those of the underground SC cable case, and is due to the fact that the relative permittivity of the dielectric medium inside the pipe is smaller than that between the core and sheath.

3.6.3 *Effect of Eccentricity of Inner Conductor*

A PT cable to discuss the effect of the eccentricity is illustrated in Figure 3.23. The parameters are given by:

Pipe: $h = 7.3$ cm, $r_{pi} = 5$ cm, $r_{p0} = 5.3$ cm, $\rho_p = 2.0 \times 10^{-8}$ Ωm
Inner conductor: $r_1 = 3.1$ mm, $r_0 = 4.1$ mm, $\varepsilon_1 = 1.0$, $\rho_c = 2.0 \times 10^{-8}$ Ωm
Earth=Al plate: 10×5 m, $\rho_e = 2.8 \times 10^{-8}$ Ωm
Cable length: $x = 2.1$ m
Case 1 $d = 4.5$ cm,Case 2: $d = 4.0$ cm,Case 3: $d = 2.5$ cm,
Case 4: $d = 1.25$ cm,Case 5: $d = 0$ cm

3.6.3.1 Frequency Responses of the Impedance

Figure 3.24 shows the frequency response of the coaxial mode impedance z_c with various separation distance d, that is, the degree of the eccectricity of the inner conductor against the

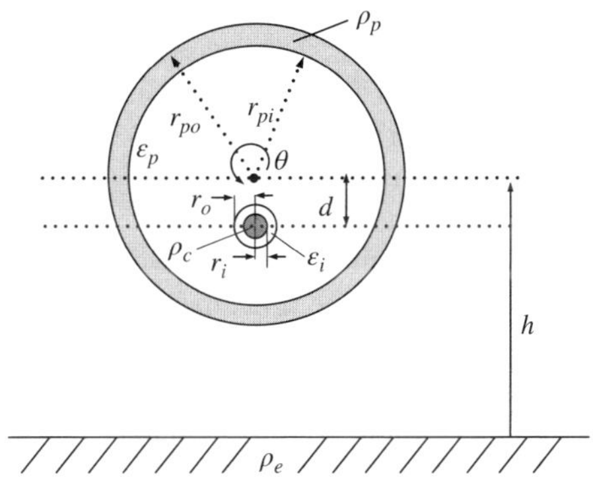

Figure 3.23 An overhead pipe-type cable

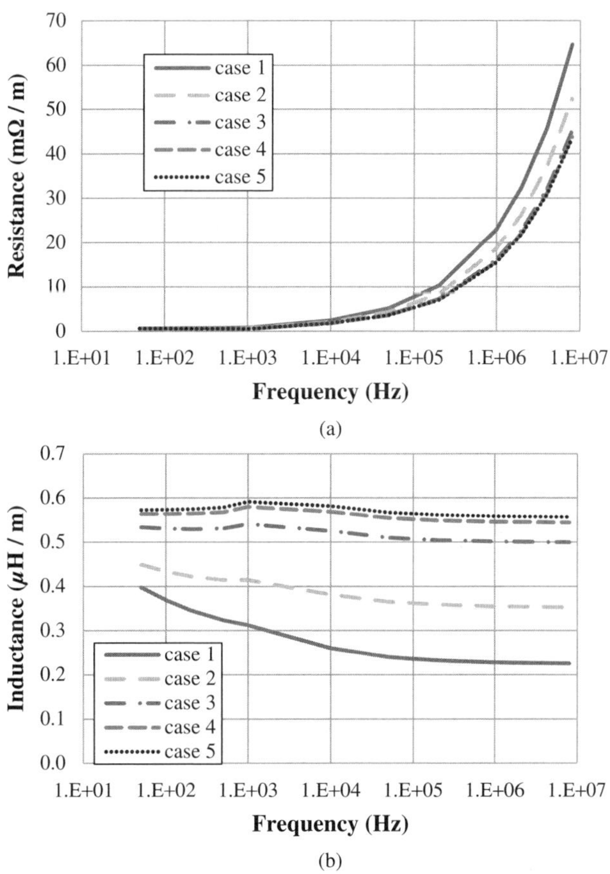

Figure 3.24 Frequency characteristic of coaxial mode impedance z_c: (a) resistance vs. frequency; and (b) inductance vs. frequency

center of the pipe. It is observed that the resistance increases as frequency increases due to the well-known skin effect. The resistance also increases as the eccentricity d/r_{pi} of the inner conductor increase, especially in a high frequency region. The increase of the resistance is somehow proportional to the distance from the pipe center.

In contrast, the inductance decrease as the eccentricity increases. This phenomenon is caused by the eccentricity of the inner conductor, and involves the so-called *proximity effect*. Because of the proximity, more charges are crowded together in the area where two conductors are facing each other. At the same time, there are less charges on the other side of the conductor surface. This results in the capacitance between the conductors being different from that based on the assumption of uniform charges across the conductor cross section as observed from (2.35). Similarly, current distribution changes in each part of the cross section and thus results in the impedance being different from that in the case of the uniform current distribution as observed in Figure 3.24. It is clear in the figure that the effect of the inner conductor eccentricity on the impedance becomes more pronounced as frequency increases. Also, it is obvious that the eccentricity d/r_{pi} less than 1/2 shows no significant effect on the impedance. In the resistance case, even the eccentricity being 1/2 shows no difference from eccentricity being 0, that is, the cable being coaxial. It should be noted that Z_p in (2.34) is rewritten for $d = 0$ as:

$$Z_p = j\omega(\mu_0/2\pi)[\mu_p K_0(m_2)/\{m_2 K_1(m_2)\} + \ln(r_{pi}/r_0)] \tag{3.142}$$

The above equation is the inner surface impedance of a metallic sheath for a coaxial cable explained in Section 2.1.1 in Chapter 2, which is a function of r_{pi} and r_0. Because r_{pi} and r_0 are constant in any pipe-type cable, the effect of r_{pi} and r_0 is not discussed in this section. However, there are a number of papers investigating the effect [22–38]. Remember that case 5 ($d = 0$) in the figure shows the contribution of r_{pi} to the propagation constant.

3.6.3.2 Frequency Responses of Characteristic Impedance

Figure 3.25 shows the frequency responses of the characteristic impedance of Z_{011} in the phase domain and z_{oc} of the coaxial mode. It is clear that the frequency response of the characteristic impedances are similar to that of the inductance L in Figure 3.24b. This is physically reasonable because of the following approximate relation between L and Z_0 [3].

$$Z_0 \approx cL = 60P \tag{3.143}$$

where c is the propagation velocity and P is the potential coefficient.

In the overhead conductor case, the above approximation is expressed as [3]:

$$Z_0 \approx 60 \ln\{2(h + h_e)/r\} = 60P \tag{3.144}$$

In the case of the pipe-type cable in Figure 3.23, the characteristic impedance Z_{011} for the eccentricity $d/r_{pi} = 0.9$ varies from about 210 Ω at 60 Hz to 130 Ω for a frequency higher than 100 kHz. When the eccentricity is less than 0.5, the characteristic impedance is always greater by about 100 Ω than that for $d/r_{pi} = 0.9$.

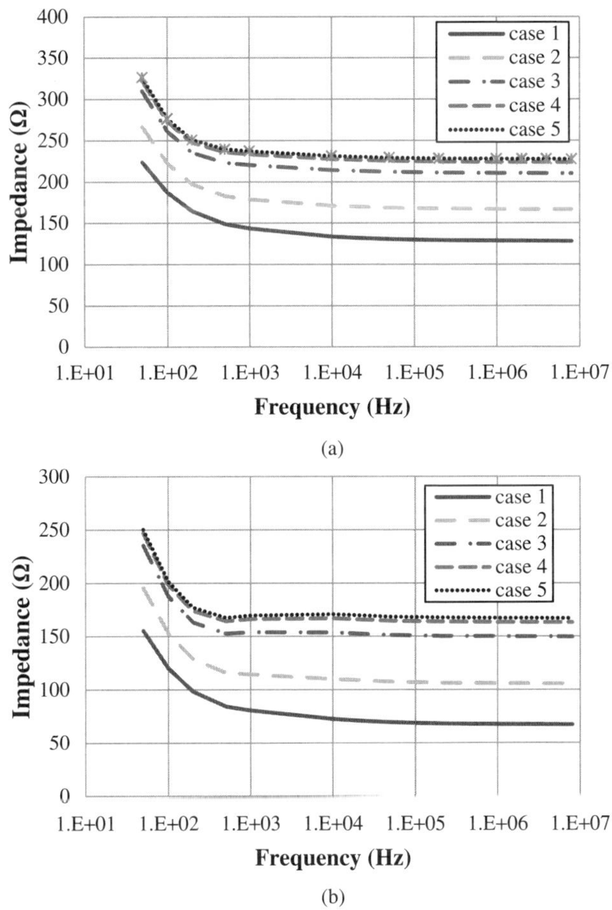

Figure 3.25 Frequency response of characteristic impedance z_c: (a) Z_{011}; and (b) z_{oc}

3.6.3.3 Propagation Constant

Figure 3.26 shows the frequency characteristic of the modal propagation constant γ. The attenuation constant α and the propagation velocity c are defined as:

$$\gamma = \alpha + j\beta \tag{3.145a}$$

$$c = \omega/\beta \tag{3.145b}$$

It is observed in Figure 3.26a that the frequency characteristic of the attenuation constant is quite similar to that of the resistance in Figure 3.24a. This is physically reasonable because of the following relation [3].

$$\alpha = R/2Z_0 \cong R/2\sqrt{L/C} \tag{3.146}$$

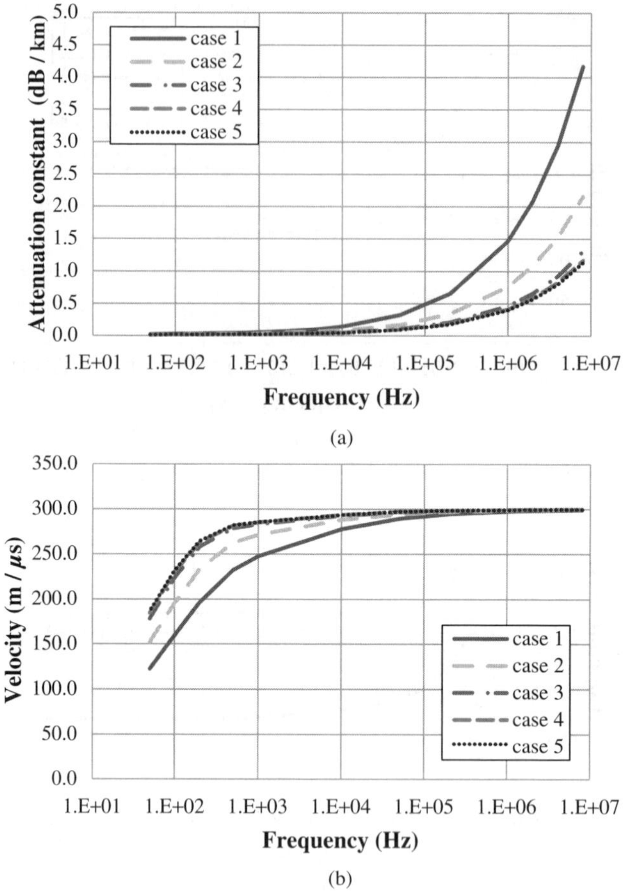

Figure 3.26 Frequency characteristic of coaxial mode propagation constant: (a) attenuation constant vs. frequency; and (b) velocity vs. frequency

The propagation velocity in Figure 3.26 shows a significant difference between the eccentricity $d/r_{pi} = 0.9$ and $d/r_{pi} \leq 0.5$ in a low frequency region. At 60 Hz, the velocity is about 120 m/μs for $d/r_{pi} = 0.9$, while it is 180 m/μs for $d/r_{pi} \leq 0.5$. For a frequency above 100 kHz, the difference becomes negligible, and thus velocity is assumed to be $c = c_0/\sqrt{\varepsilon_p}$ independently of the eccentricity for a high frequency transient.

3.6.3.4 Concluding Remarks

This section has investigated the proximity effect on the wave propagation on an inner conductor of a pipe-type cable as a function of the eccentricity of the conductor against the pipe inner surface. Based on the investigations, the following remarks are noted.

- The conductor impedance is significantly affected by the eccentricity when it is large. The resistance increases and the inductance decreases as the eccentricity increases. The effect of the eccentricity is more pronounced in a high frequency region.
- The conductor characteristic impedance and the propagation velocity decrease as the eccentricity increases in proportion to the decrease in the inductance. The attenuation constant increases in proportion to the increase in the resistance as the eccentricity increases. The difference of the characteristic impedance and the attenuation constant are more noticeable in a high frequency region, while that of the propagation velocity becomes more noticeable when the frequency becomes lower.

3.6.4 Effect of the Permittivity of the Pipe Inner Insulator

For an OF and POF cable, oil is used as an insulator within a pipe. Thus, the relative permittivity ε_r of the pipe inner insulator is around 3 but not unity as in a tunnel installed cable discussed in Section 3.6.1.

Figure 3.27 shows the attenuation and the propagation velocity on a PT cable illustrated in Figure 3.17 with $\varepsilon_r = 3$, corresponding to Figure 3.18 for $\varepsilon_r = 1$. It is observed that the mode 1 velocity starts from about 110 m/μs at $f = 10$ Hz and converges to 300 m/μs (light velocity c_0 in free space) for $\varepsilon_r = 1$ in Figure 3.18, while it starts from about 65 m/μs at 10 Hz and converges to 175 m/μs $\fallingdotseq c_0\sqrt{\varepsilon_r}$ for $\varepsilon_r = 3$ in Figure 3.27. Similarly the velocities of modes 2 and 3 range from 220 to 300 m/μs in Figure 3.18, and range from 130 to 175 m/μs in Figure 3.27. The attenuation in Figure 3.27 is slightly higher than that in Figure 3.18.

3.6.5 Overhead PT Cable

In the case of an overhead PT cable, only the earth-return mode shows a significant difference from the underground PT cable case. The attenuation is much smaller and the velocity is much greater than in the underground case. These characteristics are the same as those observed in the SC cable case. All the other propagation modes, that is, propagation modes inside the pipe, show identical characteristics to those of the underground PT cable case.

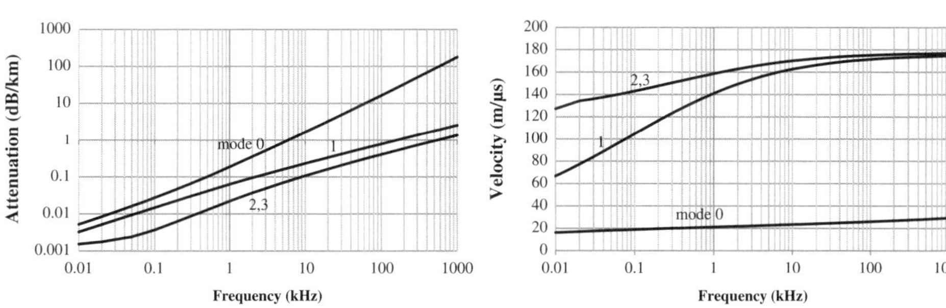

Figure 3.27 Effect of the permittivity of the pipe inner insulator ($\varepsilon_r = 3$) corresponding to Figure 3.18: (a) attenuation vs. frequency; and (b) propagation velocity vs. frequency

3.7 Propagation Characteristics of Intersheath Modes

There are many papers investigating wave propagation and transient characteristics on cables [6, 7, 11–20]. Most of the papers are related to coaxial mode wave propagation and transients. Only a few papers have investigated wave propagation and transient characteristics due to the earth-return and intersheath modes, that is, propagation between metallic sheaths including earth [20, 39–41].

References [40, 41] discuss the significance of a transient due to an intersheath mode based on a field measurement. Both ends of the metallic sheaths were open-circuited in the field test. In practice, all sheaths being grounded at every major section, it is not clear if the measured result is applicable to transient voltages and currents in a real cable.

This section investigates wave propagation characteristics of intersheath modes and resultant transient voltages and currents on underground cables with normal bonding and cross bonding. A theoretical analysis on a field measurement to observe the intersheath modes is made first. Then, wave propagation characteristics of the intersheath modes are investigated, and finally the transient voltages and currents due to the intersheath modes are discussed based on EMTP simulations.

3.7.1 Theoretical Analysis of Intersheath Modes

3.7.1.1 Intersheath Modes

In the case of a three-phase SC cable composed of a core conductor and a metallic sheath, the transformation matrices are given approximately in the following from in a high frequency region, that is, for a transient analysis as explained in Section 3.3.2.

$$(v) = [A]^{-1}(V) \tag{3.147a}$$

$$(V) = [A](v) \tag{3.147b}$$

$$[A]^{-1} \fallingdotseq \begin{bmatrix} [0] & [a] \\ [U'] & -[U'] \end{bmatrix}, [0] : \text{null matrix}$$

$$[a] = \frac{1}{3}\begin{bmatrix} 1 & 1 & 1 \\ -1 & 2 & -1 \\ -3/2 & 0 & 3/2 \end{bmatrix} \tag{3.148a}$$

$$[U'] = \begin{bmatrix} 0 & 0 & 1 \\ 1 & 0 & 0 \\ 0 & 1 & 0 \end{bmatrix} \tag{3.148b}$$

$$(v) = \begin{pmatrix} (v_e) \\ (v_c) \end{pmatrix} \tag{3.149a}$$

$$(v_e) = \begin{pmatrix} v_0 \\ v_1 \\ v_2 \end{pmatrix} \tag{3.149b}$$

$$(v_c) = \begin{pmatrix} v_3 \\ v_4 \\ v_5 \end{pmatrix} \tag{3.149c}$$

$$(V) = \begin{pmatrix} (V_c) \\ (V_s) \end{pmatrix}, (V_c) = \begin{pmatrix} V_{ca} \\ V_{cb} \\ V_{cc} \end{pmatrix}, (V_s) = \begin{pmatrix} V_{sa} \\ V_{sb} \\ V_{sc} \end{pmatrix}$$

where υ is the modal voltage, υ_e is the earth/intersheath mode voltage, υ_c is the coaxial mode voltage, V is the phase voltage, V_c is the core-to-ground voltage, and V_s is the sheath-to-ground voltage.

Considering only the sheath voltages, the above relation is simplified as:

$$(\upsilon_e) = [A]^{-1}(V_s) \tag{3.150a}$$

$$(V_s) = [A](\upsilon_e) \tag{3.150b}$$

$$[A]^{-1} = [a] \tag{3.150c}$$

For currents, the following relation is given.

$$(i_e) = [B]^{-1}(I_s), (I_s) = [B](i_e)$$

$$(i_e) = \begin{pmatrix} i_0 \\ i_1 \\ i_2 \end{pmatrix} \tag{3.151a}$$

$$(I_s) = \begin{pmatrix} I_{sa} \\ I_{sb} \\ I_{sc} \end{pmatrix} \tag{3.151b}$$

$$[B]^{-1} = [A]_t = \begin{bmatrix} 1 & 1 & 1 \\ -1/2 & 1 & -1/2 \\ -1 & 0 & 1 \end{bmatrix} \tag{3.151c}$$

3.7.1.2 Circuit for Intersheath Modes

Figure 3.28 illustrates intersheath mode circuits corresponding to (3.150). If node G is to be grounded in an experiment, that is, a PG (pulse generator) or IG (impulse generator) is to be grounded, current I_g flows from the node G to the ground. Thus, the intersheath mode becomes not complete, and the earth-return mode is involved in the circuits.

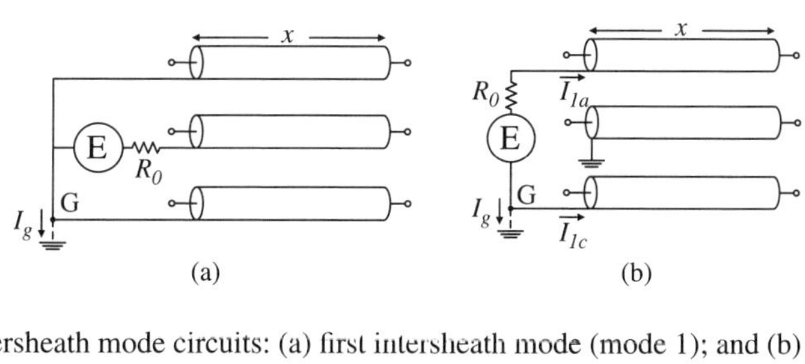

Figure 3.28 Intersheath mode circuits: (a) first intersheath mode (mode 1); and (b) second intersheath mode (mode 2)

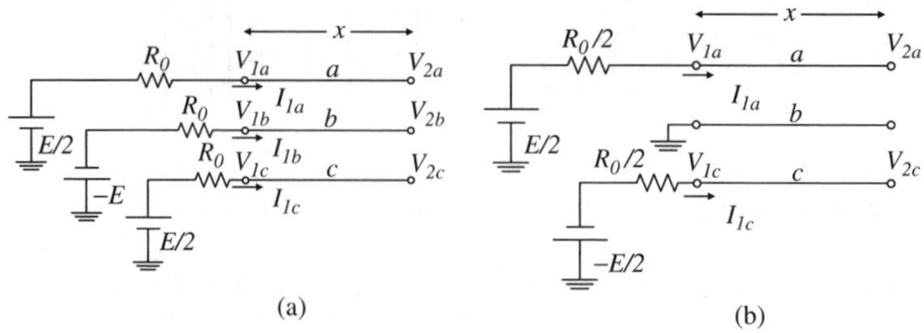

Figure 3.29 Equivalent intersheath mode circuits to those in Figure 3.28: (a) first intersheath mode; and (b) second intersheath mode

By applying Thevenin's theorem to Figure 3.28, equivalent circuits are obtained as in Figure 3.29.

First Intersheath Mode (Mode 1)

In Figure 3.29a, branch currents are given as:

$$I_{1a} = (E/2)/R_0 = (E/R_0)/2 = I/2 \qquad (3.152a)$$

$$I_{1b} = -I \qquad (3.152b)$$

$$I_{1c} = I/2 \qquad (3152c)$$

Then, modal currents are:

$$i_0 = 0 \qquad (3.153a)$$

$$i_1 = -3I/2 \qquad (3.153b)$$

$$i_2 = 0 \qquad (3.153c)$$

The equation shows only mode 1 current exists in the circuit of Figure 3.29a.

Second Intersheath Mode (Mode 2)

In Figure 3.29b, currents at the sending end are given by:

$$I_{1a} = I \qquad (3.154a)$$

$$I_{1b} = 0 \qquad (3.154b)$$

$$I_{1c} = -I \qquad (3.154c)$$

Modal currents of the above are evaluated by (3.151a).

$$i_0 = 0 \qquad (3.155a)$$

$$i_1 = 0 \qquad (31.55b)$$

$$i_2 = 2I \qquad (3.155c)$$

The above equation clearly shows that there exists only the second intersheath mode (mode 2) current in the circuit of Figure 3.29b. Similarly, the following node voltages are given in the circuit.

$$V_{1a} = E/2 - R_0 I \tag{3.156a}$$

$$V_{1b} = 0 \tag{3.156b}$$

$$V_{1c} = -E/2 - R_0 I_{1c} = -E/2 + R_0 I \tag{3.156c}$$

From (3.150a),

$$v_0 = 0 \tag{3.157a}$$

$$v_1 = 0 \tag{3.157b}$$

$$v_2 = (E - 2R_0 I)/2 \tag{3.157c}$$

There exists only mode 2 voltage in the circuit.

Grounded Voltage Source

When the voltage source in Figure 3.28 is grounded, the total current of the three-phase sheaths becomes not zero, and a current I_g flows to the ground. For example, in Figure 3.28b,

$$I_{1a} + I_{1c} + I_g = 0,\ I_g \neq 0 \tag{3.158}$$

Then, each modal current is given as:

$$\begin{pmatrix} i_0 \\ i_1 \\ i_2 \end{pmatrix} = \begin{bmatrix} 1 & 1 & 1 \\ -1/2 & 1 & -1/2 \\ -1 & 0 & 1 \end{bmatrix} \begin{pmatrix} I_{1a} \\ 0 \\ I_{1c} \end{pmatrix} = \begin{pmatrix} I_{1a} + I_{1c} \\ -\left(I_{1a} + I_{1c}\right)/2 \\ -I_{1a} + I_{1c} \end{pmatrix} \tag{3.159}$$

It is clear from the above equation that all the modal currents exist. That is, if the source is to be grounded in an experiment, it becomes very hard to distinguish the intersheath mode currents.

3.7.1.3 Modal Attenuation and Velocity

Figure 3.30 illustrates the cross section and the arrangement of a three-phase XLPE cable. Figure 3.16 shows the frequency responses of modal attenuation and velocity of a solidly bonded cable. No significant difference is observed between the earth-return mode and two intersheath modes in a low frequency region. However, in a high frequency region, the propagation velocities of the three modes show a noticeable difference. The coaxial mode (modes 3–5) attenuation and velocity are quite different from those of modes 0–2. Therefore, it might be possible to observe the effect of the intersheath modes on transient voltages and currents from a difference of traveling time of the modal components [3].

3.7.1.4 Theoretical Voltage and Current Waveforms

Table 3.14 gives conditions for theoretical calculations of transient currents and voltages by adopting a traveling wave theory [3, 42] in circuits illustrated in Figure 3.28 and Figure 3.29

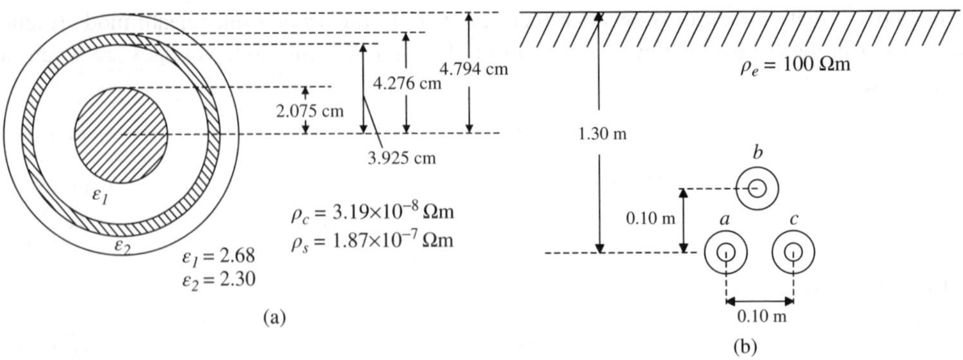

Figure 3.30 Underground three-phase XLPE cable: (a) cable cross section; and (b) conductor arrangement

Table 3.14 Conditions for theoretical calculations (cable length $x = 2$ km)

Case	Test circuit	Mode	R_0	Receiving end	Results
1-0	Figure. 3.29a	1	0	Open	Figure. 3.31
2-0	Figure. 3.29b	2	0	Open	Figure. 3.32
3-0	Figure. 3.28a	0,1,2	0	Open	Figure. 3.33
2-S	Figure. 3.29b	2	1000	Short	Figure. 3.34

with the cable length $x = 2$ km. Parameters required for the calculations are given in Table 3.15. A step function voltage E with the amplitude 1000 V is applied at the sending end.

First Intersheath Mode: Case 1-0

Figure 3.31 shows the theoretical waveforms for Case 1-0. The oscillating period T is about 126 μs. Since the cable is open-circuited, the oscillating period T is defined theoretically as four times the traveling time τ along the length x. Therefore,

$$\tau = T/4 = 31.5\,\mu\text{s} = x/c, \quad \therefore c = x/\tau = 2000/31.5 = 63\,\text{m}/\mu\text{s}$$

The above velocity agrees with that of the first intersheath mode in Table 3.15.

In Figure 3.31, current and voltage components other than those by the first intersheath mode are not observed, and thus Figure 3.29a is confirmed to be the circuit where only the first intersheath mode is generated.

Second Intersheath Mode: Case 2-0

Figure 3.32 shows theoretical waveforms of currents at the sending end and voltages on phases a and c. The currents and voltages are oscillating with the time period $T \fallingdotseq 110$ μs.

$$T = 4\tau = 110\mu\text{s}, \quad \tau = 27.5\mu\text{s} = x/c, \quad \therefore c = x/\tau = 2000/27.5 = 72.7\text{m}/\mu\text{s}$$

The above propagation velocity agrees with that of the second intersheath mode given in Table 3.15. Thus, it should be clear that the transient currents and voltages are generated only by the second intersheath mode, and Figure 3.29b is the circuit of the second intersheath mode.

Table 3.15 Cable constants at $f = 50$ kHz

(a) Characteristic impedance [Ω]

$$\left[\begin{array}{ccc|ccc} 51.3 & 14.1 & 15.1 & 27.5 & 14.1 & 15.1 \\ 14.1 & 54.6 & 14.1 & 14.1 & 28 & 14.1 \\ 15.1 & 14.1 & 51.1 & 15.1 & 14.1 & 27.5 \\ \hline 27.5 & 24.1 & 15.1 & 27.5 & 14.1 & 15.1 \\ 14.1 & 28 & 14.1 & 14.1 & 28 & 14.1 \\ 15.1 & 14.1 & 27.5 & 15.1 & 14.1 & 27.5 \end{array}\right] \begin{array}{r} \text{core-}a \\ b \\ c \\ \text{sheath-}a \\ b \\ c \end{array}$$

(b) Modal components

Model No.	Attenuation (dB/km)	Velocity (m/μs)	Impedance (Ω/km)		Admittance (S/km)	Characteristic impedance (Ω)		Characteristic admittance (S)	
			Real	Imag	Imag.	Real	Imag.	Real	Imag.
0	10.71	15.78	46.9	377	1.05	19.028	– 1.18	5.24×10^{-2}	3.25×10^{-3}
1	0.11	63.54	0.25	46.7	0.52	9.45	– 0.025	0.106	2.78×10^{-4}
2	0.17	72.19	0.25	27.2	0.69	6.245	–0.029	0.16	7.36×10^{-4}
3	0.25	180.04	1.37	41.4	7.35×10^{-2}	23.743	–0.393	4.21×10^{-2}	6.97×10^{-4}
4,5	0.17	181.07	0.92	40.9	7.35×10^{-2}	23.611	–0.264	4.23×10^{-2}	4.74×10^{-4}

The theoretical results in Figure 3.31 and Figure 3.32 are almost identical to those calculated by the EMTP.

Grounded Voltage Source: Case 3-0

Figure 3.33 shows EMTP simulation results, when the voltage source in Figure 3.28b for the second intersheath mode is grounded. In comparison with Figure 3.32, it is not easy to distinguish the second intersheath mode, because various modes are superimposed in the voltage and current waveforms. For example, let us investigate V_{2a}, the phase-a voltage at the open-circuited receiving end. At $t = 28$ μs, the fastest wave arrives at the receiving end and produces a voltage of about 1000 V. Then, the second fastest voltage appears at $t = 32$ μs and the overall voltage V_{2a} reaches 1600 V. Then, the earth-return mode wave arrives at $t = 127$ μs with a small positive voltage. The phase-c voltage also shows the same modal voltage components although the voltages are different polarity. The above are easily explained theoretically in the following manner.

Sending End: At the sending end, the modal voltages are calculated from (3.147) and (3.150) as:

$$(v_e) = [A]^{-1} \cdot (V_s) = \frac{1}{3}\begin{bmatrix} 1 & 1 & 1 \\ -1 & 2 & -1 \\ -3/2 & 0 & 3/2 \end{bmatrix}\begin{pmatrix} E \\ 0 \\ 0 \end{pmatrix} = \frac{1}{3}\begin{pmatrix} E \\ -E \\ -3E/2 \end{pmatrix} = (e_{12}) \tag{3.160}$$

From Table 3.15, the traveling time from the sending end to the receiving is:

$$\begin{aligned} &\text{mode } 0: \ c_0 \fallingdotseq 15.7\,\text{m}/\mu\text{s}, \quad \tau_0 = x/c_0 = 127\,\mu\text{s} \\ &\text{mode } 1: \ c_1 = 63.5\,\text{m}/\mu\text{s}, \tau_1 = 31.5\,\mu\text{s} \\ &\text{mode } 2: \ c_2 = 72\,\text{m}/\mu\text{s}, \tau_2 = 28\,\mu\text{s} \end{aligned}$$

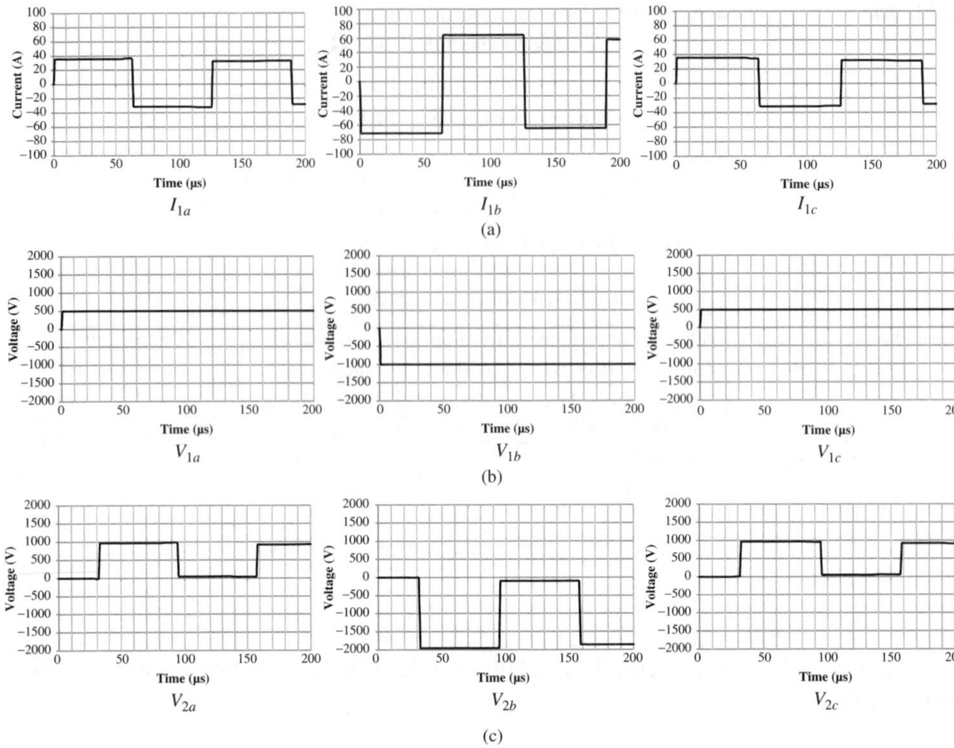

Figure 3.31 Theoretical waveforms for the first intersheath mode: Case 1-0: (a) sending-end currents; (b) sending-end voltages; and (c) receiving-end voltages

Therefore, the forward modal traveling wave (e_{12}) arrives as (e_{2f}) at the receiving end in the sequence of: (i) mode 2 at $t = 28$ μs: (ii) mode 1 at $t = 32$ μs: and (iii) mode 0 at $t = 127$ μs.

Receiving End: The modal traveling wave (e_{2f}) is transformed into phase traveling wave (E_{2f}) by transformation matrix A.

$$(E_{2f}) = [A] \cdot (e_{2f}) = \begin{bmatrix} 1 & -1/2 & -1 \\ 1 & 1 & 0 \\ 1 & -1/2 & 1 \end{bmatrix} \begin{pmatrix} e_{12}^{0}\left(t-\tau_0\right) \\ e_{12}^{1}(t-\tau_1) \\ e_{12}^{2}(t-\tau_2) \end{pmatrix} = \begin{pmatrix} e_{12}^{0}\left(t-\tau_0\right) - e_{12}^{1}(t-\tau_1)/2 - e_{12}^{2}(t-\tau_2) \\ e_{12}^{0}(t-\tau_0) + e_{12}^{1}(t-\tau_1) \\ e_{12}^{0}(t-\tau_0) - e_{12}^{1}(t-\tau_1)/2 + e_{12}^{2}(t-\tau_2) \end{pmatrix} \tag{3.161}$$

Substituting the result in (3.160) into the above equation,

$$(E_{2f}) = \frac{1}{3} \begin{pmatrix} E\left(t-\tau_0\right) + E(t-\tau_1)/2 + 3E(t-\tau_2)/2 \\ E(t-\tau_0) - E(t-\tau_1) \\ E(t-\tau_0) + E(t-\tau_1)/2 - 3E(t-\tau_2)/2 \end{pmatrix}$$

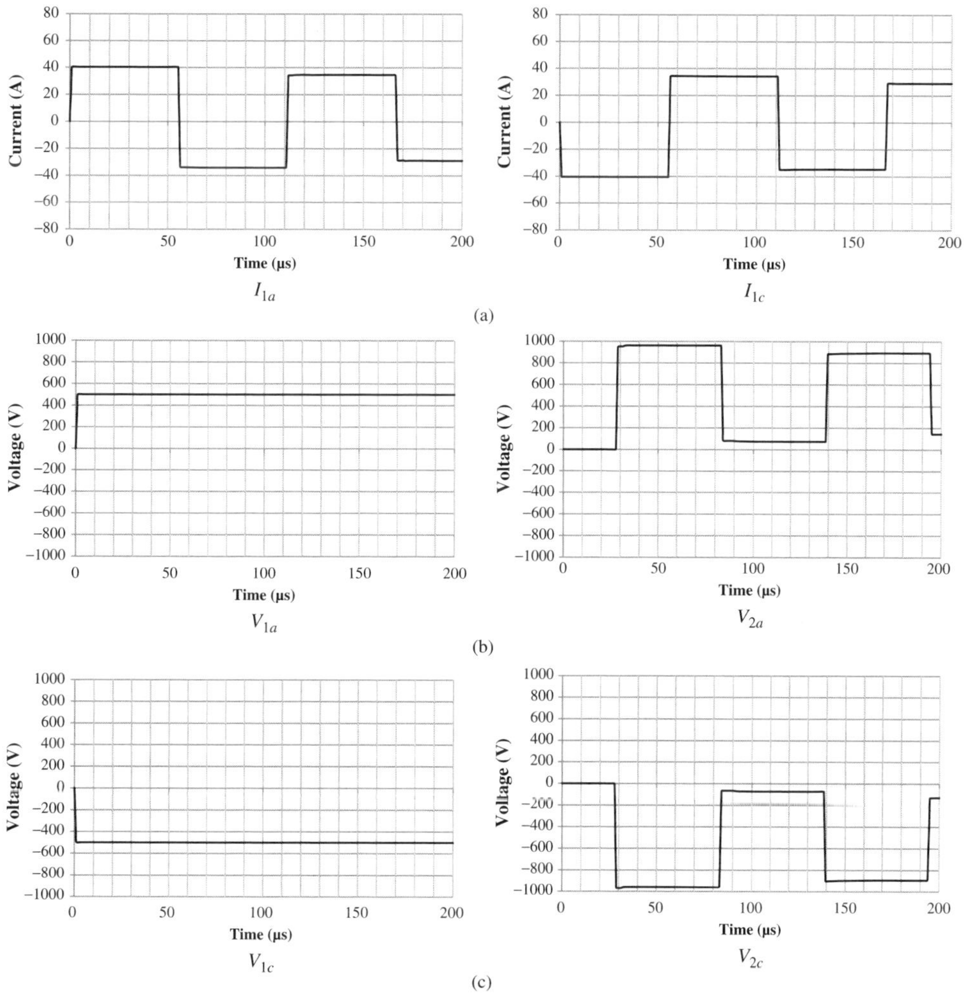

Figure 3.32 Theoretical current and voltage waveforms for the second intersheath mode: Case 2-0: (a) sending-end currents; (b) phase-a voltages; and (c) phase-c voltages

The receiving end being open-circuited, the refraction coefficient matrix becomes $2[U]$. Thus, the following phase voltages are obtained.

$$V_{2a} = E(t - \tau_2) + E(t - \tau_1)/3 + 2E(t - \tau_0)/3 \tag{3.162a}$$

$$V_{2b} = -2E(t - \tau_1)/3 + 2E(t - \tau_0)/3 \tag{3.162b}$$

$$V_{c2} = -E(t - \tau_2) + E(t - \tau_1)/3 + 2E(t - \tau_0)/3 \tag{3.162c}$$

The above result explains the observation made for the wavefront of V_{2a}.

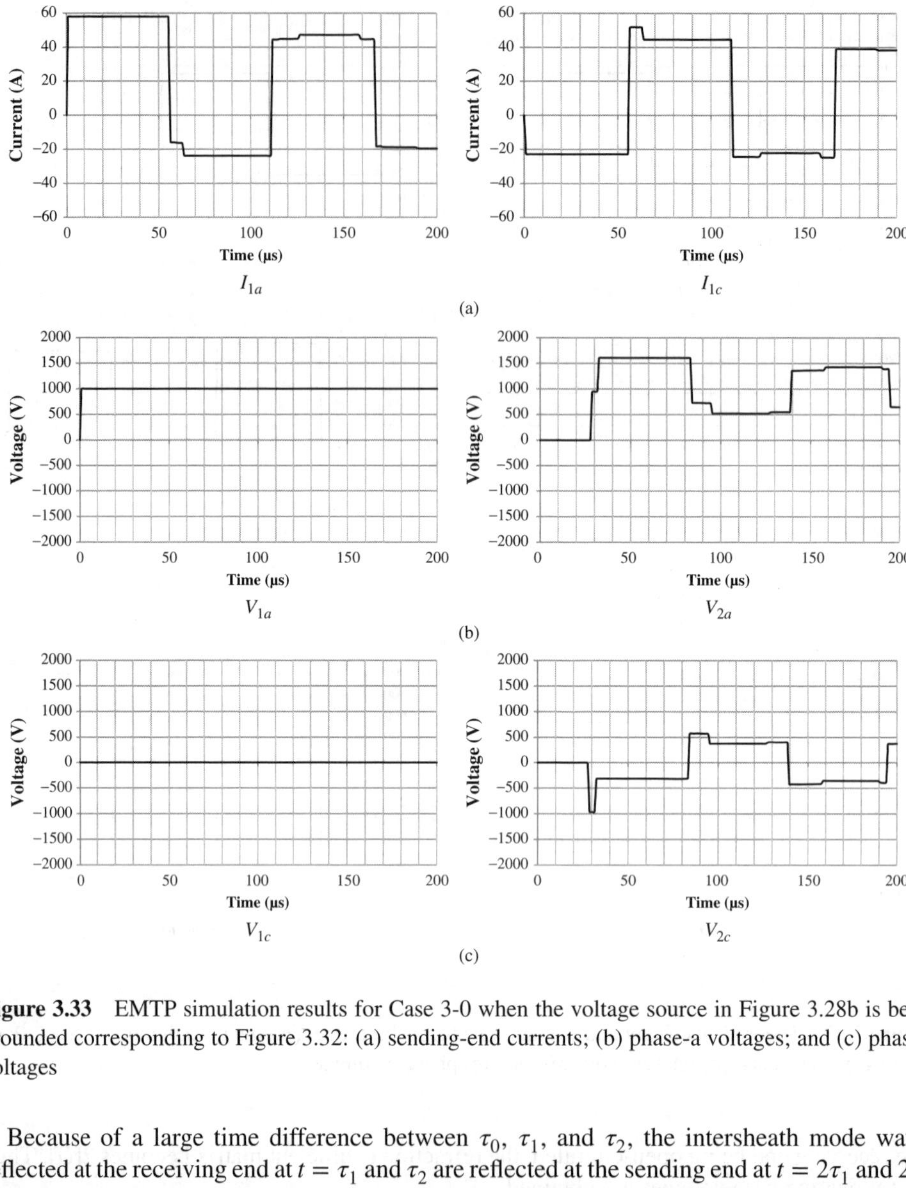

Figure 3.33 EMTP simulation results for Case 3-0 when the voltage source in Figure 3.28b is being grounded corresponding to Figure 3.32: (a) sending-end currents; (b) phase-a voltages; and (c) phase-c voltages

Because of a large time difference between τ_0, τ_1, and τ_2, the intersheath mode waves reflected at the receiving end at $t = \tau_1$ and τ_2 are reflected at the sending end at $t = 2\tau_1$ and $2\tau_2$, and come back to the receiving end before the first mode 0 wave arrives. The voltage decreases observed at around 84 μs ($\fallingdotseq 3\tau_2$) and 95 μs ($\fallingdotseq 3\tau_1$) are due to the multiple reflection of the intersheath modes.

Current Source: Case 2-S

It is quite common in an experiment to use a current source which is realized by a high resistance R_0 connected in series to a voltage source. Figure 3.34 shows EMTP simulation

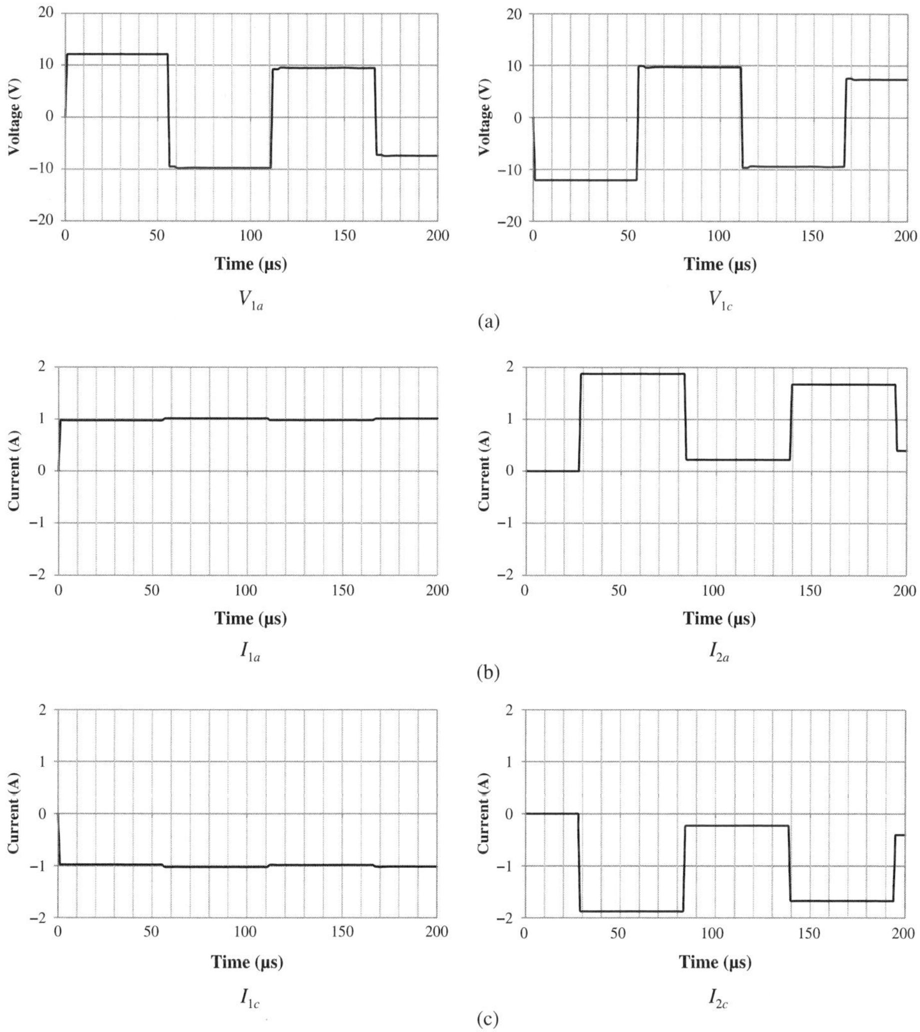

Figure 3.34 EMTP simulation results for Case 2-S equivalent to a current source corresponding to Figure 3.32 (Case 2-0): (a) phase-a voltages; (b) phase-a currents; and (c) phase-c currents

results in Figure 3.28b with $R_0 = 1$ kΩ and the remote end being short-circuited for the first intersheath mode, because an open-circuited line in the voltage source case is equivalent to a short-circuited line. It is observed that the waveforms in Figure 3.34 agree with those in Figure 3.32, reminding us that the voltages (current) in Figure 3.34 correspond to the current (voltage) in Figure 3.32.

3.7.2 Transients on a Cross-bonded Cable

Figure 3.35 illustrates a cross-bonded cable with one major section where the length of one minor section is 2 km. At both ends of the major section, three-phase sheaths are short-circuited and grounded through resistance $R_g = 5\ \Omega$ as in Figure 3.35. The lead wire for cross bonding the sheaths is represented by an inductance of 1 μH. An impulse voltage with a wavefront duration $T_f = 2$ μs and the tail $T_t = 70$ μs with amplitude of 1000 V which is expressed by double exponential functions is applied to a core or to sheaths as described in Table 3.16. An example of the EMTP output for a transient simulation on one major section cable is shown in Data 3.6 for a cross-bonded cable, and in Data 3.7 for its homogeneous model when an impulse voltage is applied to the phase-a core. To investigate the effect of intersheath modes, EMTP simulations of transients on a solidly bonded cable are also carried out.

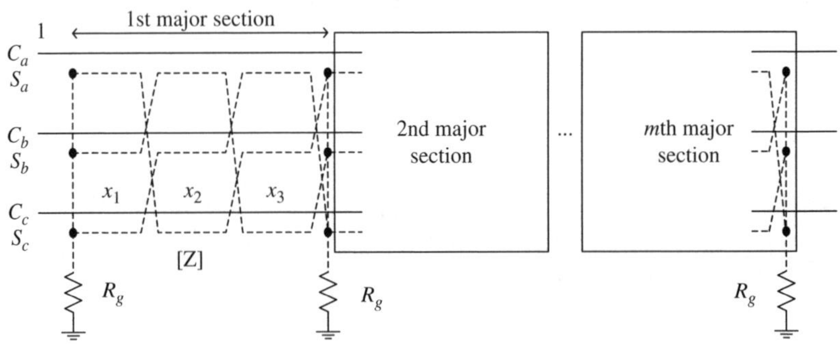

Figure 3.35 Cross-bonded cable system with "m" major sections

Table 3.16 EMTP simulation conditions

Case	Bonding	x(km)	Voltage applied to	Modes	Figure.
1N-1	Normal	6	Core-*a*	All	
1N-21	Normal	6	Sheaths *b-a/c*	1	3.36
1N-22	Normal	6	Sheaths *a/c*	2	
1N-3	Normal	6	Sheaths *a/b/c*	0	3.43a
1X-1	Cross	6	Core-*a*	All	3.40
1X-21	Cross	6	Sheaths *b-a/c*	1	3.37
1X-22	Cross	6	Sheaths *a/c*	2	3.38
1X-3	Cross	6	Sheaths *a/b/c*	0	3.43b
2X-1	Cross	18	Core-*a*	All	3.42
2X-21	Cross	18	Sheaths *b-a/c*	1	3.41
2X-22	Cross	18	Sheaths *a/c*	2	
2X-3	Cross	18	Sheaths *a/b/c*	0	3.43c

Data 3.6 EMTP output of a transient simulation when an impulse voltage ($T_f = 1\mu s$, $T_t = \mu s$) is applied to phase-a core on a cross-bonded cable with one major section, length 6 km

```
 ---   106   cards of disk file read into card cache cells  1    onward.
 1st card (Lists  1-10).        |   8000   15000   15000     340   60000     120
 2nd card (Lists 11-20).        |    360     900   36400     255  129600      38
 3rd card (Lists 21-29).        |   3000   15000  160000      24   30000    1000
 Supplemental offsets.          |  240000     742
Alternative Transients Program (ATP), Watcom translation.   All rights reserved by Can/Am user group of Portland, Oregon, USA.
 Date (dd-mth-yy) and time of day (hh.mm.ss) = 25-Feb-14  06.49.15    Name of disk plot file is   C:¥ATPWNT¥UNTITLED.pl4
Consult the 860-page ATP Rule Book of the Can/Am EMTP User Group in Portland,   Oregon, USA.    Program is no older than Sept., 1998.
Total size of LABCOM tables =   3639004   INTEGER words.   VARDIM   List Sizes follow :   8002 15000 15000   340 60000   1200
 13140 120000 2250 1900   360   900 36400    255   129K    384    45    254 240000 100000   3000 15000   160K    24 30000 10000    600 252000
--------------------------------------------------+--------------------------------------------------------------------------------
Descriptive interpretation of input data cards.   |  Input data card images are shown below, all 80 columns, character by character
                                                       0         1         2         3         4         5         6         7         8
                                                       012345678901234567890123456789012345678901234567890123456789012345678901234567890
--------------------------------------------------+--------------------------------------------------------------------------------
Comment card.    NUMDCD = 1.                      |C data:C:¥ATPWNT¥UNTITLED.DAT
Comment card.    NUMDCD = 2.                      |C 2013.12.18   A2X-1
Marker card preceding new EMTP data case.         |BEGIN NEW DATA CASE
Comment card.    NUMDCD = 4.                      |C    DT][ TMAX ][ XOPT ][ COPT ][EPSILN][TOLMAT]
Misc. data.     2.000E-08    1.000E-03    0.000E+00 |  2.0E-8     1.E-3
Comment card.    NUMDCD = 6.                      |C   IOUT][ IPLOT][IDOUBL][KSSOUT][MAXOUT][ IPUN ][MEMSAV][ ICAT ][NENERG][IPRSUP]
Misc. data.       10      1   1   0   1   0   0   0   0   0   |       10          1           1           0           1
Electric network, too.   But TACS data first ...  |TACS HYBRID
Blank card terminating all TACS data cards.       |BLANK ending TACS
Comment card.    NUMDCD = 10.                     |C "Core"
Comment card.    NUMDCD = 11.                     |C [BUS1][BUS2][BUS3][BUS4][   R ][   L ][   C ]                                  !
Comment card.    NUMDCD = 12.                     |C   BUS1CA                    1.E9
Series R-L-C.    1.000E+09   0.000E+00   0.000E+00   |   BUS4CA                   1.E9
Series R-L-C.    1.000E+09   0.000E+00   0.000E+00   |   BUS1CB                   1.E9
Series R-L-C.    1.000E+09   0.000E+00   0.000E+00   |   BUS4CB                   1.E9
Series R-L-C.    1.000E+09   0.000E+00   0.000E+00   |   BUS1CC                   1.E9
Series R-L-C.    1.000E+09   0.000E+00   0.000E+00   |   BUS4CC                   1.E9
Comment card.    NUMDCD = 18.                     |C "Sheath"
Comment card.    NUMDCD = 19.                     |C [BUS1][BUS2][BUS3][BUS4][   R ][   L ][   C ]                                  !
Series R-L-C.    0.000E+00   1.000E-06   0.000E+00   |   BUS1SABUS1SN                   1.E-3
Series R-L-C.    0.000E+00   1.000E-06   0.000E+00   |   BUS4SABUS4SN                   1.E-3
Series R-L-C.    0.000E+00   1.000E-06   0.000E+00   |   BUS1SBBUS1SN                   1.E-3
Series R-L-C.    0.000E+00   1.000E-06   0.000E+00   |   BUS4SBBUS4SN                   1.E-3
Series R-L-C.    0.000E+00   1.000E-06   0.000E+00   |   BUS1SCBUS1SN                   1.E-3
Series R-L-C.    0.000E+00   1.000E-06   0.000E+00   |   BUS4SCBUS4SN                   1.E-3
Comment card.    NUMDCD = 26.                     |C [BUS1][BUS2][BUS3][BUS4][   R ][   L ][   C ]                                  !
Series R-L-C.    5.000E+00   0.000E+00   0.000E+00   |   BUS1SN                   5.00
Series R-L-C.    5.000E+00   0.000E+00   0.000E+00   |   BUS4SN                   5.00
New MOLDAT = 1      (data vintage).               |$VINTAGE, 1
Comment card.    NUMDCD = 30.                     |C     Branch cards for constant-parameter distributed:
Comment card.    NUMDCD = 31.                     |C [BUS1][BUS2][BUS3][BUS4][      R     ][     Zs     ][     vel    ][   Leng    ] 1[]!!
 4.697E-02 1.903E+01 1.578E+07 1.903E+01 1.267E-04|-1BUS1CABUS2CA            4.69683E-02 1.90285E+01 1.57841E+07-2.00000E+03 1 16
 2.452E-04 9.450E+00 6.354E+07 9.450E+00 3.148E-05|-2BUS1CBBUS2CB            2.45167E-04 9.45021E+00 6.35364E+07-2.00000E+03 1 16
 2.497E-04 6.245E+00 7.219E+07 6.245E+00 2.770E-05|-3BUS1CCBUS2CC            2.49721E-04 6.24494E+00 7.21911E+07-2.00000E+03 1 16
 1.371E-03 2.374E+01 1.800E+08 2.374E+01 1.111E-05|-4BUS1SABUS2SA            1.37090E-03 2.37433E+01 1.80043E+08-2.00000E+03 1 16
 9.162E-04 2.361E+01 1.811E+08 2.361E+01 1.105E-05|-5BUS1SBBUS2SB            9.16190E-04 2.36110E+01 1.81067E+08-2.00000E+03 1 16
 9.162E-04 2.361E+01 1.811E+08 2.361E+01 1.105E-05|-6BUS1SCBUS2SC            9.16190E-04 2.36110E+01 1.81067E+08-2.00000E+03 1 16
TR(1,1 )...   0.000 0.000 0.000 1.000 0.000 0.000  |   .00000121  -.00001428  -.00003122   .99996074   .00000000   .00000000
TI(1,1 )...   0.000 0.000 0.000 0.000 0.000 0.000  |  -.00000305   .00005888   .00011720   .00000004   .00000000   .00000000
TR(2,1 )...   0.000 0.000 0.000 0.000 1.000 0.000  |   .00000000   .00000014   .00000000   .00000007   .99999985   .00000000
TI(2,1 )...   0.000 0.000 0.000 0.000 0.000 0.000  |   .00000000  -.00000010   .00000000  -.00000005   .00000000   .00000000
TR(3,1 )...   0.000 0.000 0.000 0.000 0.000 1.000  |   .00000000  -.00000007   .00000015   .00000010   .00000007   .99999982
TI(3,1 )...   0.000 0.000 0.000 0.000 0.000 0.000  |   .00000000   .00000005  -.00000011  -.00000007  -.00000005   .00000000
TR(4,1 )...   0.336-0.336-0.500-1.000 0.000 0.000  |   .33608786  -.33635803  -.49995761 -1.00014930   .00000032   .00000050
TI(4,1 )...   0.000 0.005-0.002 0.001 0.000 0.000  |   .00011709   .00479260  -.00246289   .00071106  -.00000024  -.00000037
TR(5,1 )...   0.332 0.672-0.004 0.000-1.000 0.000  |   .33208003   .67188477  -.00432966   .00006645 -1.00000056   .00000032
TI(5,1 )...   0.000 0.000 0.005 0.000 0.000 0.000  |  -.00034835   .00038263   .00450404  -.00026103   .00000051  -.00000024
TR(6,1 )...   0.336-0.328 0.504 0.000 0.000-1.000  |   .33594086  -.32764408   .50448213   .00010586   .00000025 -1.00000070
TI(6,1 )...   0.000-0.004-0.002 0.000 0.000 0.000  |   .00023026  -.00419636  -.00228600  -.00040971  -.00000019   .00000064
New MOLDAT = 0      (data vintage).               |$VINTAGE, 0
New MOLDAT = 1      (data vintage).               |$VINTAGE, 1
Comment card.    NUMDCD = 52.                     |C     Branch cards for constant-parameter distributed:
Comment card.    NUMDCD = 53.                     |C [BUS1][BUS2][BUS3][BUS4][      R     ][     Zs     ][     vel    ][   Leng    ] 1[]!!
 4.697E-02 1.903E+01 1.578E+07 1.903E+01 1.267E-04|-1BUS2CABUS3CA            4.69683E-02 1.90285E+01 1.57841E+07-2.00000E+03 1 16
 2.452E-04 9.450E+00 6.354E+07 9.450E+00 3.148E-05|-2BUS2CBBUS3CB            2.45167E-04 9.45021E+00 6.35364E+07-2.00000E+03 1 16
 2.497E-04 6.245E+00 7.219E+07 6.245E+00 2.770E-05|-3BUS2CCBUS3CC            2.49721E-04 6.24494E+00 7.21911E+07-2.00000E+03 1 16
 1.371E-03 2.374E+01 1.800E+08 2.374E+01 1.111E-05|-4BUS2SBBUS3SA            1.37090E-03 2.37433E+01 1.80043E+08-2.00000E+03 1 16
 9.162E-04 2.361E+01 1.811E+08 2.361E+01 1.105E-05|-5BUS2SCBUS3SB            9.16190E-04 2.36110E+01 1.81067E+08-2.00000E+03 1 16
 9.162E-04 2.361E+01 1.811E+08 2.361E+01 1.105E-05|-6BUS2SABUS3SC            9.16190E-04 2.36110E+01 1.81067E+08-2.00000E+03 1 16
TR(1,1 )...   0.000 0.000 0.000 1.000 0.000 0.000  |   .00000121  -.00001428  -.00003122   .99996074   .00000000   .00000000
TI(1,1 )...   0.000 0.000 0.000 0.000 0.000 0.000  |  -.00000305   .00005888   .00011720   .00000004   .00000000   .00000000
TR(2,1 )...   0.000 0.000 0.000 0.000 1.000 0.000  |   .00000000   .00000014   .00000000   .00000007   .99999985   .00000000
TI(2,1 )...   0.000 0.000 0.000 0.000 0.000 0.000  |   .00000000  -.00000010   .00000000  -.00000005   .00000000   .00000000
TR(3,1 )...   0.000 0.000 0.000 0.000 0.000 1.000  |   .00000000  -.00000007   .00000015   .00000010   .00000007   .99999982
TI(3,1 )...   0.000 0.000 0.000 0.000 0.000 0.000  |   .00000000   .00000005  -.00000011  -.00000007  -.00000005   .00000000
```

```
TR(4,1 )...  0.336-0.336-0.500-1.000 0.000 0.000  |  .33608786  -.33635803  -.49995761 -1.00014930  .00000032  .00000050
TI(4,1 )...  0.000 0.005-0.002 0.001 0.000 0.000  |  .00011709  .00479260  -.00246289  .00071106  -.00000024  -.00000037
TR(5,1 )...  0.332 0.672-0.004 0.000-1.000 0.000  |  .33208003  .67188477  -.00432966  .00006645 -1.00000056  .00000032
TI(5,1 )...  0.000 0.000 0.005 0.000 0.000 0.000  | -.00034835  .00038263  .00450404  -.00026103  .00000051  -.00000024
TR(6,1 )...  0.336-0.328 0.504 0.000 0.000-1.000  |  .33594086  -.32764408  .50448213  .00010586  .00000025 -1.00000070
TI(6,1 )...  0.000-0.004-0.002 0.000 0.000 0.000  |  .00023026  -.00419636  -.00228600  -.00040971  -.00000019  .00000064
New MOLDAT = 0      (data vintage).                |$VINTAGE, 0
New MOLDAT = 1      (data vintage).                |$VINTAGE, 1
Comment card.   NUMDCD = 74.                       |C    Branch cards for constant-parameter distributed:
Comment card.   NUMDCD = 75.                       |C [BUS1][BUS2][BUS3][BUS4][     R    ][     Zs     ][     vel    ][    Leng    ] 1[]!!
 4.697E-02 1.903E+01 1.578E+07 1.903E+01 1.267E-04|-1BUS3CABUS4CA             4.69683E-02 1.90285E+01 1.57841E+07-2.00000E+03 1 16
 2.452E-04 9.450E+00 6.354E+07 9.450E+00 3.148E-05|-2BUS3CBBUS4CB             2.45167E-04 9.45021E+00 6.35364E+07-2.00000E+03 1 16
 2.497E-04 6.245E+00 7.219E+07 6.245E+00 2.770E-05|-3BUS3CCBUS4CC             2.49721E-04 6.24494E+00 7.21911E+07-2.00000E+03 1 16
 1.371E-03 2.374E+01 1.800E+08 2.374E+01 1.111E-05|-4BUS3SBBUS4SA             1.37090E-03 2.37433E+01 1.80043E+08-2.00000E+03 1 16
 9.162E-04 2.361E+01 1.811E+08 2.361E+01 1.105E-05|-5BUS3SCBUS4SB             9.16190E-04 2.36110E+01 1.81067E+08-2.00000E+03 1 16
 9.162E-04 2.361E+01 1.811E+08 2.361E+01 1.105E-05|-6BUS3SABUS4SC             9.16190E-04 2.36110E+01 1.81067E+08-2.00000E+03 1 16
TR(1,1 )...  0.000 0.000 0.000 1.000 0.000 0.000  |  .00000121  -.00001428  -.00003122  .99996074  .00000000  .00000000
TI(1,1 )...  0.000 0.000 0.000 0.000 0.000 0.000  | -.00000305  .00005888  .00011720  .00000004  .00000000  .00000000
TR(2,1 )...  0.000 0.000 0.000 0.000 1.000 0.000  |  .00000000  .00000014  .00000000  .00000007  .99999985  .00000000
TI(2,1 )...  0.000 0.000 0.000 0.000 0.000 0.000  |  .00000000  -.00000010  .00000000  -.00000005  .00000000  .00000000
TR(3,1 )...  0.000 0.000 0.000 0.000 0.000 1.000  |  .00000000  -.00000007  .00000015  .00000010  .00000007  .99999982
TI(3,1 )...  0.000 0.000 0.000 0.000 0.000 0.000  |  .00000000  .00000005  -.00000011  -.00000007  -.00000005  .00000000
TR(4,1 )...  0.336-0.336-0.500-1.000 0.000 0.000  |  .33608786  -.33635803  -.49995761 -1.00014930  .00000032  .00000050
TI(4,1 )...  0.000 0.005-0.002 0.001 0.000 0.000  |  .00011709  .00479260  -.00246289  .00071106  -.00000024  -.00000037
TR(5,1 )...  0.332 0.672-0.004 0.000-1.000 0.000  |  .33208003  .67188477  -.00432966  .00006645 -1.00000056  .00000032
TI(5,1 )...  0.000 0.000 0.005 0.000 0.000 0.000  | -.00034835  .00038263  .00450404  -.00026103  .00000051  -.00000024
TR(6,1 )...  0.336-0.328 0.504 0.000 0.000-1.000  |  .33594086  -.32764408  .50448213  .00010586  .00000025 -1.00000070
TI(6,1 )...  0.000-0.004-0.002 0.000 0.000 0.000  |  .00023026  -.00419636  -.00228600  -.00040971  -.00000019  .00000064
New MOLDAT = 0      (data vintage).                |$VINTAGE, 0
Comment card.   NUMDCD = 95.                       |C [BUS1][BUS2][BUS3][BUS4][  R ][  L ][  C ]                                   !
Series R-L-C.   2.000E+00  0.000E+00  0.000E+00    |  SOURCEBUS1CA            2.00
Blank card ending branches.  IBR, NTOT = 32   28   |BLANK ending BRANCH
Blank card ending switches.   KSWTCH = 0.          |BLANK ending SWITCH
Comment card.   NUMDCD = 99.                       |C [BUS1][][  AMP.  ][  alpha ][  beta  ]                  [ TSTART ][  TSTOP ]
Source.   1.00E+03 -1.03E+04 -1.53E+06 -1.00E+00   |15SOURCE        1.E3    -1.03E4    -1.53E6                    -1          1
Blank card ends electric network sources.          |BLANK ending SOURCE

List of input elements that are connected to each node.   Only the physical connections of multi-phase lines are shown (capacitive
and inductive coupling are ignored).   Repeated entries indicate parallel connections.   Switches are included,   although sources
(including rotating machinery) are omitted -- except that U.M. usage produces extra,   internally-defined nodes   "UMXXXX".
--------------+------------------------------
From bus name |  Names of all adjacent busses.
--------------+------------------------------
        BUS4CA |TERRA *BUS3CA*
        BUS1CB |TERRA *BUS2CB*
        BUS4CB |TERRA *BUS3CB*
        BUS1CC |TERRA *BUS2CC*
        BUS4CC |TERRA *BUS3CC*
        BUS1SA |BUS1SN*BUS2SA*
        BUS1SN |TERRA *BUS1SA*BUS1SB*BUS1SC*
        BUS4SA |BUS4SN*BUS3SB*
        BUS4SN |TERRA *BUS4SA*BUS4SB*BUS4SC*
        BUS1SB |BUS1SN*BUS2SB*
        BUS4SB |BUS4SN*BUS3SC*
        BUS1SC |BUS1SN*BUS2SC*
        BUS4SC |BUS4SN*BUS3SA*
        BUS1CA |BUS2CA*SOURCE*
        BUS2CA |BUS1CA*BUS3CA*
        BUS2CB |BUS1CB*BUS3CB*
        BUS2CC |BUS1CC*BUS3CC*
        BUS2SA |BUS1SA*BUS3SC*
        BUS2SB |BUS1SB*BUS3SA*
        BUS2SC |BUS1SC*BUS3SB*
        BUS3CA |BUS4CA*BUS2CA*
        BUS3CB |BUS4CB*BUS2CB*
        BUS3CC |BUS4CC*BUS2CC*
        BUS3SA |BUS4SC*BUS2SB*
        BUS3SB |BUS4SA*BUS2SC*
        BUS3SC |BUS4SB*BUS2SA*
        SOURCE |BUS1CA*
        TERRA  |BUS4CA*BUS1CB*BUS4CB*BUS1CC*BUS4CC*BUS1SN*BUS4SN*
--------------+------------------------------
Comment card.   NUMDCD = 102.                      |C [BUS1][BUS2][BUS3][BUS4][BUS5][BUS6][BUS7][BUS8][BUS9][BUSA][BUSB][BUSC][BUSD]
Comment card.   NUMDCD = 103.                      |C  BUS1CABUS4CA
Card of names for time-step loop output.           |  BUS1SABUS4SB
Blank card ending requests for output variables.  |BLANK ending OUTPUT Specification
```

```
Column headings for the  2  EMTP output variables follow.  These are divided among the 5 possible classes as follows ....
  First  2    output variables are electric-network voltage differences (upper voltage minus lower voltage);
   Step       Time        BUS1SA       BUS4SB

      0        0.0          0.0          0.0
     10      .2E-6   62.4596979          0.0
     20      .4E-6   84.0791261          0.0
     30      .6E-6   97.0020038          0.0
     40      .8E-6   106.030081          0.0
     50      .1E-5   112.547563          0.0
     60     .12E-5   117.267501          0.0
     70     .14E-5   120.671211          0.0
     80     .16E-5   123.106863          0.0
     90     .18E-5   124.829945          0.0
    100      .2E-5   126.028466          0.0
  49950    .999E-3   1.58818738 -.18026241
  49960   .9992E-3   1.54876331 -.18304421
  49970   .9994E-3   1.62456986 -.15383808
  49980   .9996E-3    1.6674682 -.06600489
  49990   .9998E-3   1.72340621 -.05148944
  50000      1.E-3   1.78442759 -.02786865

Extrema of output variables follow.   Order and column positioning are the same as for the preceding time-step loop output.
Variable maxima :   127.905554 9.46782861
Times of maxima :        .302E-5  .13708E-3
Variable minima :   -68.645304 -15.842062
Times of minima :     .23406E-3    .7054E-4

Blank card terminating all plot cards.                     |BLANK ending PLOT

Memory storage figures for the preceding, now-completed data case.  ------------------------------  Present    Program
A value of  "-9999"  indicates that no figure is available.         25-Feb-14  06.49.15             figure      limit   (name)
      Size List 1.   Number of electric network nodes.                                                      28       8002  (LBUS)
      Size List 2.   Number of electric network branches.                                                   32      15000  (LBRNCH)
      Size List 3.   Number of data values in R, L, C tables.                                               14      15000  (LDATA)
      Size List 4.   Number of electric network sources.                                                     1        340  (LEXCT)
      Size List 5.   Storage for [Y] and triangularized [Y].   No. times =    1     Factors =   159    318      60000  (LYMAT)
      Size List 6.   Number of entries in switch table.                        No. flops = 0                 0       1200  (LSWTCH)
      Size List 7.   Number of distinct ALPHANUMERIC data names plus program SPY variables.                  45      13140  (LSIZE7)
      Size List 8.   History points of distributed lines.                                                32892     120000  (LPAST)
      Size List 9.   Number of nonlinear elements.                                                           0       2250  (LNONL)
      Size List 10.  Points of nonlinear characteristics.                                                    0       1900  (LCHAR)
      Size List 11.  Number of Type-59 S.M. outputs.                                                         0        360  (LSMOUT)
      Size List 12.  Total number of EMTP output variables.                                                  2        900  (LSIZ12)
      Size List 13.  Working space for batch/SPY plotting.                                                9999      36400  (LSIZ13)
      Size List 14.  S.M./U.M. connections to TACS.                                                      -9999        255  (LBSTAC)
      Size List 15.  Character storage in bytes for MODELS.                                              -9999     129600  (LCTACS)
      Size List 16.  Total number of Type-59 S.M. masses.                                                    0        384  (LIMASS)
      Size List 17.  Number of Type-59 Synchronous machines.                                                 0         45  (LSYN)
      Size List 18.  Branch and switch power/energy outputs.                                                 0        254  (MAXPE)
      Size List 19.  Total floating-point TACS table space.                                             235552     240000  (LTACST)
        Tacs table number        1          2          3          4          5          6          7          8
        Present figure           1          1          1         11          0          0          0       1345
        Program limit         2962      13333      14814       1344       4444      37037      51851       8888
      Size List 20.  Non-copied recursive convolution data.                                                  0     100000  (LFSEM)
      Size List 21.  Total modal/phase [T] matrix storage.                                                 108       3000  (LFD)
      Size List 22.  Total recursive convolution history.                                                    0      15000  (LHIST)
      Size List 23.  Giant vectors for renumbering, phasors.                                                67     160000  (LSIZ23)
      Size List 24.  Peak phases of compensation for data.                                                   0         24  (LCOMP)
      Size List 25.  Total table space for all U.M usage.                                                -9999      30000  (LSPCUM)
      Size List 26.  Square of max number of coupled phases.                                                36      10000  (LSIZ26)
      Size List 28.  MODELS.   Total work space is divided into INTEGER and REAL.   1st, REAL:           -9999     252000  (LRTACS)
                                  Second   and last,   statistics for INTEGER work space.                    0     252000  (LITACS)
      Size List 30.  Taku Noda frequency-dependent circuits.                                                 0         60  (LSIZ30)
Timing figures characterizing central processor (CP) solution speed.  ----------------------------  CP sec   Wait sec   Real sec
      Data input time (through blank card ending branches) ....                                      0.120      0.000      0.120
      Node renumbering and phasor solution   ....                                                    0.000      0.000      0.000
      After phasor solution, but before time-step loop   ....                                        0.000      0.000      0.000
      Integration of equations (time-step loop)   ....                                               0.210      0.000      0.210
      Plotting or STATISTICS termination overlays ....                                               0.000      0.000      0.000
                                                                                                    ----------------------------
                                                                                        Totals       0.330      0.000      0.330
```

Data 3.7 EMTP output of a transient simulation on a homogeneous model of a cross-bonded cable corresponding to Data 3.6

```
 ---   122    cards of disk file read into card cache cells   1     onward.
 1st card (Lists  1-10).        |    8000    15000    15000       340     60000       120
 2nd card (Lists 11-20).        |     360       900    36400       255   129600        38
 3rd card (Lists 21-29).        |    3000    15000   160000        24     30000     1000
 Supplemental offsets.          | 240000       742
Alternative Transients Program (ATP),  Watcom translation.   All rights reserved by Can/Am user group of Portland, Oregon, USA.
 Date (dd-mth-yy) and time of day (hh.mm.ss) = 25-Feb-14   06.59.27    Name of disk plot file is   C:¥ATPWNT¥UNTITLED.pl4
Consult the 860-page ATP Rule Book of the Can/Am EMTP User Group in Portland,   Oregon, USA.    Program is no older than Sept., 1998.
Total size of LABCOM tables =  3639004    INTEGER words.    VARDIM   List Sizes follow :    8002 15000 15000    340 60000   1200
 13140 120000 2250 1900   360   900 36400    255   129K    384     45    254 240000 100000   3000 15000   160K    24 30000 10000      600 252000
--------------------------------------------------+--------------------------------------------------------------------------------
Descriptive interpretation of input data cards.   |  Input data card images are shown below, all 80 columns, character by character
                                                   0         1         2         3         4         5         6         7         8
                                                   012345678901234567890123456789012345678901234567890123456789012345678901234567890
--------------------------------------------------+--------------------------------------------------------------------------------
Comment card.    NUMDCD = 1.                       |C data:C:¥ATPWNT¥UNTITLED.DAT
Comment card.    NUMDCD = 2.                       |C 2013.12.18   A2X-1
Marker card preceding new EMTP data case.          |BEGIN NEW DATA CASE
Comment card.    NUMDCD = 4.                       |C        DT][ TMAX ][ XOPT ][ COPT ][EPSILN][TOLMAT]
Misc. data.      2.000E-08    1.000E-03    0.000E+00 |  2.0E-8    1.E-3
Comment card.    NUMDCD = 6.                       |C    IOUT][ IPLOT][IDOUBL][KSSOUT][MAXOUT][ IPUN ][MEMSAV][ ICAT ][NENERG][IPRSUP]
Misc. data.      10     1   1   0   1   0   0   0   0   0   |        10            1            1            0            1
Electric network, too.   But TACS data first ...   |TACS HYBRID
Blank card terminating all TACS data cards.        |BLANK ending TACS
Comment card.    NUMDCD = 10.                      |C "Core"
Comment card.    NUMDCD = 11.                      |C [BUS1][BUS2][BUS3][BUS4][   R ][   L ][   C ]                               !
Comment card.    NUMDCD = 12.                      |C   BUS1CA                          1.E9
Series R-L-C.    1.000E+09   0.000E+00   0.000E+00   |  BUS7CA                            1.E9
Series R-L-C.    1.000E+09   0.000E+00   0.000E+00   |  BUS1CB                            1.E9
Series R-L-C.    1.000E+09   0.000E+00   0.000E+00   |  BUS7CB                            1.E9
Series R-L-C.    1.000E+09   0.000E+00   0.000E+00   |  BUS1CC                            1.E9
Series R-L-C.    1.000E+09   0.000E+00   0.000E+00   |  BUS7CC                            1.E9
Comment card.    NUMDCD = 18.                      |C [BUS1][BUS2][BUS3][BUS4][   R ][   L ][   C ]                               !
Series R-L-C.    5.000E+00   0.000E+00   0.000E+00   |  BUS1SN                            5.00
Series R-L-C.    5.000E+00   0.000E+00   0.000E+00   |  BUS7SN                            5.00
New MOLDAT = 1        (data vintage).              |$VINTAGE, 1
Comment card.    NUMDCD = 22.                      |C [BUS1][BUS2][BUS3][BUS4][        R      ][        Zs      ][       vel      ][      Leng     ] 1[]!!
 4.654E-02 1.883E+01 1.579E+07 1.883E+01 6.335E-05|-1BUS1CABUS2CA              4.65361E-02 1.88346E+01 1.57853E+07 1.00000E+03 1    4
 8.066E-04 2.588E+01 1.103E+08 2.588E+01 9.068E-06|-2BUS1CBBUS2CB              8.06561E-04 2.58844E+01 1.10277E+08 1.00000E+03 1    4
 5.988E-04 1.794E+01 1.191E+08 1.794E+01 8.394E-06|-3BUS1CCBUS2CC              5.98804E-04 1.79416E+01 1.19139E+08 1.00000E+03 1    4
 3.054E-04 7.870E+00 1.811E+08 7.870E+00 5.523E-06|-4BUS1SNBUS2SN              3.05397E-04 7.86983E+00 1.81067E+08 1.00000E+03 1    4
 TR(1,1 )...   0.000-0.334-0.500 0.333             |    .00025696   -.33358843   -.50000000    .33333333
 TI(1,1 )...   0.000 0.000 0.000 0.000             |    .00003710    .00000027    .00000000    .00000000
 TR(2,1 )...  -0.001 0.667 0.000 0.333             |   -.00051391    .66717687    .00000000    .33333333
 TI(2,1 )...   0.000 0.000 0.000 0.000             |   -.00007421   -.00000055    .00000000    .00000000
 TR(3,1 )...   0.000-0.334 0.500 0.333             |    .00025696   -.33358843    .50000000    .33333333
 TI(3,1 )...   0.000 0.000 0.000 0.000             |    .00003710    .00000027    .00000000    .00000000
 TR(4,1 )...   1.000 0.007 0.000-1.000             |    .99999442    .00738115    .00000000   -.99999994
 TI(4,1 )...   0.000 0.001 0.000 0.000             |   -.00000081    .00106588    .00000000    .00000006
New MOLDAT = 0        (data vintage).              |$VINTAGE, 0
New MOLDAT = 1        (data vintage).              |$VINTAGE, 1
Comment card.    NUMDCD = 37.                      |C [BUS1][BUS2][BUS3][BUS4][        R      ][        Zs      ][       vel      ][      Leng     ] 1[]!!
 4.654E-02 1.883E+01 1.579E+07 1.883E+01 6.335E-05|-1BUS2CABUS3CA              4.65361E-02 1.88346E+01 1.57853E+07 1.00000E+03 1    4
 8.066E-04 2.588E+01 1.103E+08 2.588E+01 9.068E-06|-2BUS2CBBUS3CB              8.06561E-04 2.58844E+01 1.10277E+08 1.00000E+03 1    4
 5.988E-04 1.794E+01 1.191E+08 1.794E+01 8.394E-06|-3BUS2CCBUS3CC              5.98804E-04 1.79416E+01 1.19139E+08 1.00000E+03 1    4
 3.054E-04 7.870E+00 1.811E+08 7.870E+00 5.523E-06|-4BUS2SNBUS3SN              3.05397E-04 7.86983E+00 1.81067E+08 1.00000E+03 1    4
 TR(1,1 )...   0.000-0.334-0.500 0.333             |    .00025696   -.33358843   -.50000000    .33333333
 TI(1,1 )...   0.000 0.000 0.000 0.000             |    .00003710    .00000027    .00000000    .00000000
 TR(2,1 )...  -0.001 0.667 0.000 0.333             |   -.00051391    .66717687    .00000000    .33333333
 TI(2,1 )...   0.000 0.000 0.000 0.000             |   -.00007421   -.00000055    .00000000    .00000000
 TR(3,1 )...   0.000-0.334 0.500 0.333             |    .00025696   -.33358843    .50000000    .33333333
 TI(3,1 )...   0.000 0.000 0.000 0.000             |    .00003710    .00000027    .00000000    .00000000
 TR(4,1 )...   1.000 0.007 0.000-1.000             |    .99999442    .00738115    .00000000   -.99999994
 TI(4,1 )...   0.000 0.001 0.000 0.000             |   -.00000081    .00106588    .00000000    .00000006
New MOLDAT = 0        (data vintage).              |$VINTAGE, 0
New MOLDAT = 1        (data vintage).              |$VINTAGE, 1
Comment card.    NUMDCD = 52.                      |C [BUS1][BUS2][BUS3][BUS4][        R      ][        Zs      ][       vel      ][      Leng     ] 1[]!!
 4.654E-02 1.883E+01 1.579E+07 1.883E+01 6.335E-05|-1BUS3CABUS4CA              4.65361E-02 1.88346E+01 1.57853E+07 1.00000E+03 1    4
 8.066E-04 2.588E+01 1.103E+08 2.588E+01 9.068E-06|-2BUS3CBBUS4CB              8.06561E-04 2.58844E+01 1.10277E+08 1.00000E+03 1    4
 5.988E-04 1.794E+01 1.191E+08 1.794E+01 8.394E-06|-3BUS3CCBUS4CC              5.98804E-04 1.79416E+01 1.19139E+08 1.00000E+03 1    4
 3.054E-04 7.870E+00 1.811E+08 7.870E+00 5.523E-06|-4BUS3SNBUS4SN              3.05397E-04 7.86983E+00 1.81067E+08 1.00000E+03 1    4
 TR(1,1 )...   0.000-0.334-0.500 0.333             |    .00025696   -.33358843   -.50000000    .33333333
 TI(1,1 )...   0.000 0.000 0.000 0.000             |    .00003710    .00000027    .00000000    .00000000
 TR(2,1 )...  -0.001 0.667 0.000 0.333             |   -.00051391    .66717687    .00000000    .33333333
 TI(2,1 )...   0.000 0.000 0.000 0.000             |   -.00007421   -.00000055    .00000000    .00000000
 TR(3,1 )...   0.000-0.334 0.500 0.333             |    .00025696   -.33358843    .50000000    .33333333
 TI(3,1 )...   0.000 0.000 0.000 0.000             |    .00003710    .00000027    .00000000    .00000000
 TR(4,1 )...   1.000 0.007 0.000-1.000             |    .99999442    .00738115    .00000000   -.99999994
 TI(4,1 )...   0.000 0.001 0.000 0.000             |   -.00000081    .00106588    .00000000    .00000006
New MOLDAT = 0        (data vintage).              |$VINTAGE, 0
```

```
New MOLDAT = 1     (data vintage).                    |$VINTAGE, 1
Comment card.   NUMDCD = 67.                           |C [BUS1][BUS2][BUS3][BUS4][     R     ][     Zs     ][     vel     ][     Leng     ] 1[]!!
 4.654E-02 1.883E+01 1.579E+07 1.883E+01 6.335E-05|-1BUS4CABUS5CA              4.65361E-02 1.88346E+01 1.57853E+07 1.00000E+03 1   4
 8.066E-04 2.588E+01 1.103E+08 2.588E+01 9.068E-06|-2BUS4CBBUS5CB              8.06561E-04 2.58844E+01 1.10277E+08 1.00000E+03 1   4
 5.988E-04 1.794E+01 1.191E+08 1.794E+01 8.394E-06|-3BUS4CCBUS5CC              5.98804E-04 1.79416E+01 1.19139E+08 1.00000E+03 1   4
 3.054E-04 7.870E+00 1.811E+08 7.870E+00 5.523E-06|-4BUS4SNBUS5SN              3.05397E-04 7.86983E+00 1.81067E+08 1.00000E+03 1   4
TR(1,1 )...   0.000-0.334-0.500 0.333         |   .00025696   -.33358843   -.50000000    .33333333
TI(1,1 )...   0.000 0.000 0.000 0.000         |   .00003710    .00000027    .00000000    .00000000
TR(2,1 )... -0.001 0.667 0.000 0.333          |  -.00051391    .66717687    .00000000    .33333333
TI(2,1 )...   0.000 0.000 0.000 0.000         |  -.00007421   -.00000055    .00000000    .00000000
TR(3,1 )...   0.000-0.334 0.500 0.333         |   .00025696   -.33358843    .50000000    .33333333
TI(3,1 )...   0.000 0.000 0.000 0.000         |   .00003710    .00000027    .00000000    .00000000
TR(4,1 )...   1.000 0.007 0.000-1.000         |   .99999442    .00738115    .00000000   -.99999994
TI(4,1 )...   0.000 0.001 0.000 0.000         |  -.00000081    .00106588    .00000000    .00000006
New MOLDAT = 0     (data vintage).                    |$VINTAGE, 0
New MOLDAT = 1     (data vintage).                    |$VINTAGE, 1
Comment card.   NUMDCD = 82.                           |C [BUS1][BUS2][BUS3][BUS4][     R     ][     Zs     ][     vel     ][     Leng     ] 1[]!!
 4.654E-02 1.883E+01 1.579E+07 1.883E+01 6.335E-05|-1BUS5CABUS6CA              4.65361E-02 1.88346E+01 1.57853E+07 1.00000E+03 1   4
 8.066E-04 2.588E+01 1.103E+08 2.588E+01 9.068E-06|-2BUS5CBBUS6CB              8.06561E-04 2.58844E+01 1.10277E+08 1.00000E+03 1   4
 5.988E-04 1.794E+01 1.191E+08 1.794E+01 8.394E-06|-3BUS5CCBUS6CC              5.98804E-04 1.79416E+01 1.19139E+08 1.00000E+03 1   4
 3.054E-04 7.870E+00 1.811E+08 7.870E+00 5.523E-06|-4BUS5SNBUS6SN              3.05397E-04 7.86983E+00 1.81067E+08 1.00000E+03 1   4
TR(1,1 )...   0.000-0.334-0.500 0.333         |   .00025696   -.33358843   -.50000000    .33333333
TI(1,1 )...   0.000 0.000 0.000 0.000         |   .00003710    .00000027    .00000000    .00000000
TR(2,1 )... -0.001 0.667 0.000 0.333          |  -.00051391    .66717687    .00000000    .33333333
TI(2,1 )...   0.000 0.000 0.000 0.000         |  -.00007421   -.00000055    .00000000    .00000000
TR(3,1 )...   0.000-0.334 0.500 0.333         |   .00025696   -.33358843    .50000000    .33333333
TI(3,1 )...   0.000 0.000 0.000 0.000         |   .00003710    .00000027    .00000000    .00000000
TR(4,1 )...   1.000 0.007 0.000-1.000         |   .99999442    .00738115    .00000000   -.99999994
TI(4,1 )...   0.000 0.001 0.000 0.000         |  -.00000081    .00106588    .00000000    .00000006
New MOLDAT = 0     (data vintage).                    |$VINTAGE, 0
New MOLDAT = 1     (data vintage).                    |$VINTAGE, 1
Comment card.   NUMDCD = 97.                           |C [BUS1][BUS2][BUS3][BUS4][     R     ][     Zs     ][     vel     ][     Leng     ] 1[]!!
 4.654E-02 1.883E+01 1.579E+07 1.883E+01 6.335E-05|-1BUS6CABUS7CA              4.65361E-02 1.88346E+01 1.57853E+07 1.00000E+03 1   4
 8.066E-04 2.588E+01 1.103E+08 2.588E+01 9.068E-06|-2BUS6CBBUS7CB              8.06561E-04 2.58844E+01 1.10277E+08 1.00000E+03 1   4
 5.988E-04 1.794E+01 1.191E+08 1.794E+01 8.394E-06|-3BUS6CCBUS7CC              5.98804E-04 1.79416E+01 1.19139E+08 1.00000E+03 1   4
 3.054E-04 7.870E+00 1.811E+08 7.870E+00 5.523E-06|-4BUS6SNBUS7SN              3.05397E-04 7.86983E+00 1.81067E+08 1.00000E+03 1   4
TR(1,1 )...   0.000-0.334-0.500 0.333         |   .00025696   -.33358843   -.50000000    .33333333
TI(1,1 )...   0.000 0.000 0.000 0.000         |   .00003710    .00000027    .00000000    .00000000
TR(2,1 )... -0.001 0.667 0.000 0.333          |  -.00051391    .66717687    .00000000    .33333333
TI(2,1 )...   0.000 0.000 0.000 0.000         |  -.00007421   -.00000055    .00000000    .00000000
TR(3,1 )...   0.000-0.334 0.500 0.333         |   .00025696   -.33358843    .50000000    .33333333
TI(3,1 )...   0.000 0.000 0.000 0.000         |   .00003710    .00000027    .00000000    .00000000
TR(4,1 )...   1.000 0.007 0.000-1.000         |   .99999442    .00738115    .00000000   -.99999994
TI(4,1 )...   0.000 0.001 0.000 0.000         |  -.00000081    .00106588    .00000000    .00000006
New MOLDAT = 0     (data vintage).                    |$VINTAGE, 0
Comment card.   NUMDCD = 111.                          |C [BUS1][BUS2][BUS3][BUS4][  R ][  L ][  C ]                                   !
Series R-L-C.   2.000E+00  0.000E+00  0.000E+00    |  SOURCEBUS1CA                2.00
Blank card ending branches.  IBR, NTOT = 32   30   |BLANK ending BRANCH
Blank card ending switches.   KSWTCH = 0.          |BLANK ending SWITCH
Comment card.   NUMDCD = 115.                          |C [BUS1][][  AMP.  ][  alpha ][  beta  ]                  [ TSTART ][  TSTOP ]
Source.   1.00E+03 -1.03E+04 -1.53E+06 -1.00E+00  |15SOURCE          1.E3    -1.03E4    -1.53E6                    -1             1
Blank card ends electric network sources.          |BLANK ending SOURCE

List of input elements that are connected to each node.   Only the physical connections of multi-phase lines are shown (capacitive
and inductive coupling are ignored).   Repeated entries indicate parallel connections.   Switches are included,   although sources
(including rotating machinery) are omitted -- except that U.M. usage produces extra,   internally-defined nodes   "UMXXXX".
--------------+------------------------------
From bus name |   Names of all adjacent busses.
--------------+------------------------------
        BUS7CA |TERRA *BUS6CA*
        BUS1CB |TERRA *BUS2CB*
        BUS7CB |TERRA *BUS6CB*
        BUS1CC |TERRA *BUS2CC*
        BUS7CC |TERRA *BUS6CC*
        BUS1SN |TERRA *BUS2SN*
        BUS7SN |TERRA *BUS6SN*
        BUS1CA |BUS2CA*SOURCE*
        BUS2CA |BUS1CA*BUS3CA*
        BUS2CB |BUS1CB*BUS3CB*
        BUS2CC |BUS1CC*BUS3CC*
        BUS2SN |BUS1SN*BUS3SN*
        BUS3CA |BUS2CA*BUS4CA*
        BUS3CB |BUS2CB*BUS4CB*
        BUS3CC |BUS2CC*BUS4CC*
        BUS3SN |BUS2SN*BUS4SN*
        BUS4CA |BUS3CA*BUS5CA*
        BUS4CB |BUS3CB*BUS5CB*
        BUS4CC |BUS3CC*BUS5CC*
        BUS4SN |BUS3SN*BUS5SN*
```

```
BUS5CA |BUS4CA*BUS6CA*
BUS5CB |BUS4CB*BUS6CB*
BUS5CC |BUS4CC*BUS6CC*
BUS5SN |BUS4SN*BUS6SN*
BUS6CA |BUS7CA*BUS5CA*
BUS6CB |BUS7CB*BUS5CB*
BUS6CC |BUS7CC*BUS5CC*
BUS6SN |BUS7SN*BUS5SN*
SOURCE |BUS1CA*
TERRA  |BUS7CA*BUS1CB*BUS7CB*BUS1CC*BUS7CC*BUS1SN*BUS7SN*
--------------+------------------------------
Comment card.   NUMDCD = 118.                  |C [BUS1][BUS2][BUS3][BUS4][BUS5][BUS6][BUS7][BUS8][BUS9][BUSA][BUSB][BUSC][BUSD]
Comment card.   NUMDCD = 119.                  |C  BUS1CABUS7CA
Card of names for time-step loop output.       |  BUS1SNBUS7SN
Blank card ending requests for output variables.  |BLANK ending OUTPUT Specification

Column headings for the   2   EMTP output variables follow.  These are divided among the 5 possible classes as follows ....
  First  2     output variables are electric-network voltage differences (upper voltage minus lower voltage);
  Step       Time        BUS1SN        BUS7SN

     0        0.0          0.0          0.0
    10       .2E-6   28.9967688         0.0
    20       .4E-6   50.2899295         0.0
    30       .6E-6   65.9103791         0.0
    40       .8E-6   77.3536426         0.0
    50       .1E-5   85.7210046         0.0
    60      .12E-5   91.8234357         0.0
    70      .14E-5    96.258124         0.0
    80      .16E-5   99.4648303         0.0
    90      .18E-5   101.767388         0.0
   100       .2E-5   103.404265         0.0
 49950     .999E-3  -.92725112 -.01585723
 49960    .9992E-3  -.90748557 -.01184001
 49970    .9994E-3  -.84773258 -.00891091
 49980    .9996E-3  -.72861822 -.00683176
 49990    .9998E-3  -.61281639 -.00549888
 50000       1.E-3  -.52043729 -.00496233

Extrema of output variables follow.   Order and column positioning are the same as for the preceding time-step loop output.
Variable maxima :  106.446604 .174260114
Times of maxima :       .33E-5  .44628E-3
Variable minima :  -99.895278 -.12594814
Times of minima :    .11146E-3  .73702E-3

Blank card terminating all plot cards.          |BLANK ending PLOT

Memory storage figures for the preceding, now-completed data case.  ------------------------------  Present   Program
A value of  "-9999"  indicates that no figure is available.        25-Feb-14  06.59.27             figure     limit  (name)
    Size List 1.   Number of electric network nodes.                                                    30      8002  (LBUS)
    Size List 2.   Number of electric network branches.                                                 32     15000  (LBRNCH)
    Size List 3.   Number of data values in R, L, C tables.                                              8     15000  (LDATA)
    Size List 4.   Number of electric network sources.                                                   1       340  (LEXCT)
    Size List 5.   Storage for [Y] and triangularized [Y].   No. times =    1     Factors =   113      226     60000  (LYMAT)
    Size List 6.   Number of entries in switch table.                     No. flops = 0                  0      1200  (LSWTCH)
    Size List 7.   Number of distinct ALPHANUMERIC data names plus program SPY variables.               45     13140  (LSIZE7)
    Size List 8.   History points of distributed lines.                                              25938    120000  (LPAST)
    Size List 9.   Number of nonlinear elements.                                                         0      2250  (LNONL)
    Size List 10.  Points of nonlinear characteristics.                                                  0      1900  (LCHAR)
    Size List 11.  Number of Type-59 S.M. outputs.                                                       0       360  (LSMOUT)
    Size List 12.  Total number of EMTP output variables.                                                2       900  (LSIZ12)
    Size List 13.  Working space for batch/SPY plotting.                                             -9999     36400  (LSIZ13)
    Size List 14.  S.M./U.M. connections to TACS.                                                    -9999       255  (LBSTAC)
    Size List 15.  Character storage in bytes for MODELS.                                            -9999    129600  (LCTACS)
    Size List 16.  Total number of Type-59 S.M. masses.                                                  0       384  (LIMASS)
    Size List 17.  Number of Type-59 Synchronous machines.                                               0        45  (LSYN)
    Size List 18.  Branch and switch power/energy outputs.                                               0       254  (MAXPE)
    Size List 19.  Total floating-point TACS table space.                                           235552    240000  (LTACST)
      Tacs table number       1        2        3        4        5        6        7        8
      Present figure          1        1        1       11        0        0        0     1345
      Program limit        2962    13333    14814     1344     4444    37037    51851     8888
```

```
Size List 20.  Non-copied recursive convolution data.                                      0    100000  (LFSEM)
Size List 21.  Total modal/phase [T] matrix storage.                                      96      3000  (LFD)
Size List 22.  Total recursive convolution history.                                        0     15000  (LHIST)
Size List 23.  Giant vectors for renumbering, phasors.                                    43    160000  (LSIZ23)
Size List 24.  Peak phases of compensation for data.                                       0        24  (LCOMP)
Size List 25.  Total table space for all U.M usage.                                    -9999     30000  (LSPCUM)
Size List 26.  Square of max number of coupled phases.                                    16     10000  (LSIZ26)
Size List 28.  MODELS.  Total work space is divided into INTEGER and REAL.  1st, REAL:  -9999    252000  (LRTACS)
                        Second  and last,  statistics for INTEGER work space.              0    252000  (LITACS)
Size List 30.  Taku Noda frequency-dependent circuits.                                     0        60  (LSIZ30)
Timing figures characterizing central processor (CP) solution speed.  ---------------------------  CP sec  Wait sec  Real sec
  Data input time (through blank card ending branches) ....                          0.080     0.000     0.080
  Node renumbering and phasor solution  ....                                              0.000     0.000     0.000
  After phasor solution, but before time-step loop  ....                             0.000     0.000     0.000
  Integration of equations (time-step loop)  ....                                     0.200     0.000     0.200
  Plotting or STATISTICS termination overlays ....                                      0.000     0.000     0.000
                                                                                           ---------------------------
                                                                                 Totals    0.280     0.000     0.280
```

3.7.2.1 One Major Section ($x=6$ km)

Transient Waveform

Figure 3.36 shows results of currents and voltages on a solidly bonded cable when an impulse voltage is applied between phase-b sheath and phase-a and phase-c sheaths as illustrated

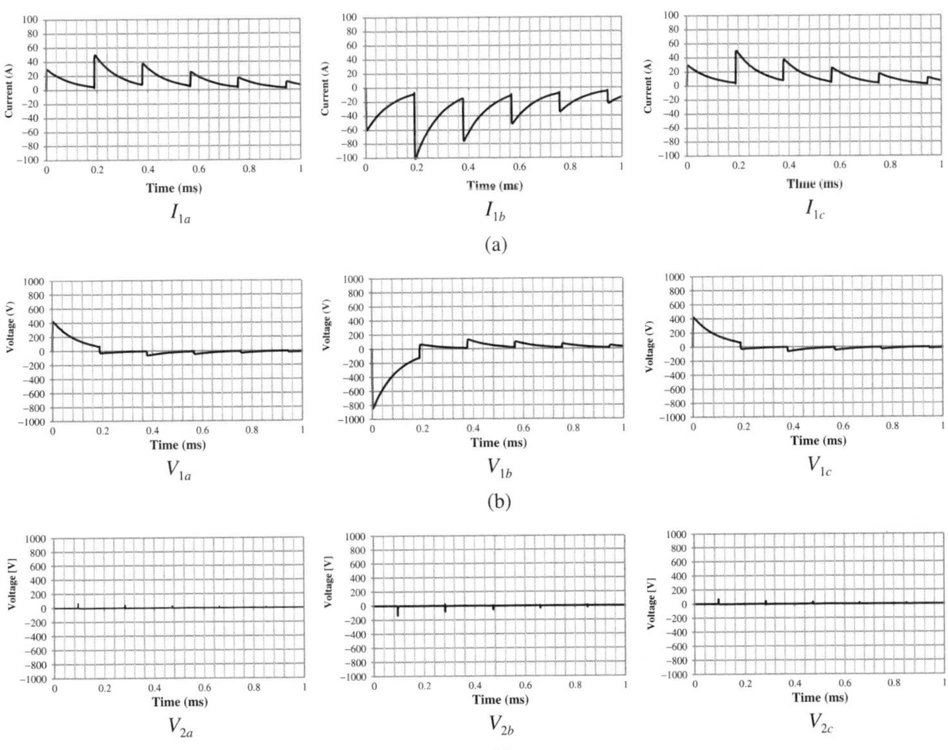

Figure 3.36 Transient currents and voltages associated with the first intersheath mode (mode 1) on a 6 km solidly bonded cable: Case 1N-21: (a) sending-end currents; (b) sending-end voltages; and (c) receiving-end voltages

in Figure 3.29a, and Figure 3.37 shows EMTP simulation results of transient currents and voltages on a cross-bonded cable.

When the cable is solidly bonded, the intersheath modes are clearly observed in current waveforms and in voltage waveforms at the sending end as in Figure 3.31. For example, in the figure, a reflected wave from the receiving end comes back to the sending end at around $t \fallingdotseq 185$ μs. This corresponds to twice the traveling time 2τ. Thus, the propagation velocity is given by:

$$c = x/\tau = 6000/92 \fallingdotseq 64.8\,\mathrm{m/\mu s}$$

The above value is observed to correspond to the first intersheath mode in Table 3.15.

In contrast to Figure 3.36, it is not clear in Figure 3.37, cross-bonded cable, when the first intersheath mode comes back to the sending end. For example, a reflected traveling wave with the same polarity is observed at around $t \fallingdotseq 60$ μs in I_{a1}, the sending-end current. This wave is estimated to be a reflected wave from the first cross-bonding point, that is,

$$2\tau = 60\,\mathrm{\mu s}, x = 2000\,\mathrm{m}$$

$$c = 2000/30 = 66.7\,\mathrm{m/\mu s}$$

The above velocity indicates that the reflected traveling wave is the first intersheath mode from the other end of the first minor section.

At around $t \fallingdotseq 85$ μs, a wave with the opposite polarity is observed. If this wave is assumed to be reflected at the first minor section, then the propagation velocity is:

$$c = 2000/42.5 = 42\,\mathrm{m/\mu s}$$

There exists no such a mode with the propagation velocity of 42 m/μs in Table 3.15. The overall oscillating period of the transient current in Figure 3.37 is estimated as:

$$T \fallingdotseq 220\,\mathrm{\mu s} = 4\tau, \quad \tau = 55\,\mathrm{\mu s}$$

$$c = 6000/55 = 109\,\mathrm{m/\mu s}$$

Again there is no propagation mode with the velocity of 109 m/μs in Table 3.15.

It should be noted that the first intersheath mode is propagating until the end of the first minor section as observed in a solidly bonded cable shown in Figure 3.31. The wave $E/2$ on the phase-a sheath travels into the phase-b sheath, and wave $-E/2$ on phase-c travels into phase-a. Thus, at the beginning of the second minor section, the following modal voltage components appear.

$$(v_e) = [A]^{-1}(V_s) = \frac{1}{3}\begin{bmatrix} 1 & 1 & 1 \\ -1 & 2 & -1 \\ -3/2 & 0 & 3/2 \end{bmatrix}\begin{pmatrix} -E/2 \\ E/2 \\ 0 \end{pmatrix} = \frac{1}{3}\begin{pmatrix} 0 \\ E/2 \\ 3E/4 \end{pmatrix} \quad (3.163)$$

From the above equation, it is clear that the traveling waves propagate as the first and second intersheath modes. Those arrive at the other end of the second minor section, at different propagation times. Then, at the beginning of the third minor section, the traveling wave on phase-b propagates into phase-c and that on phase-a propagates into phase-b. Also, coaxial mode components are generated at the boundaries of minor sections. Thus, all the modal components

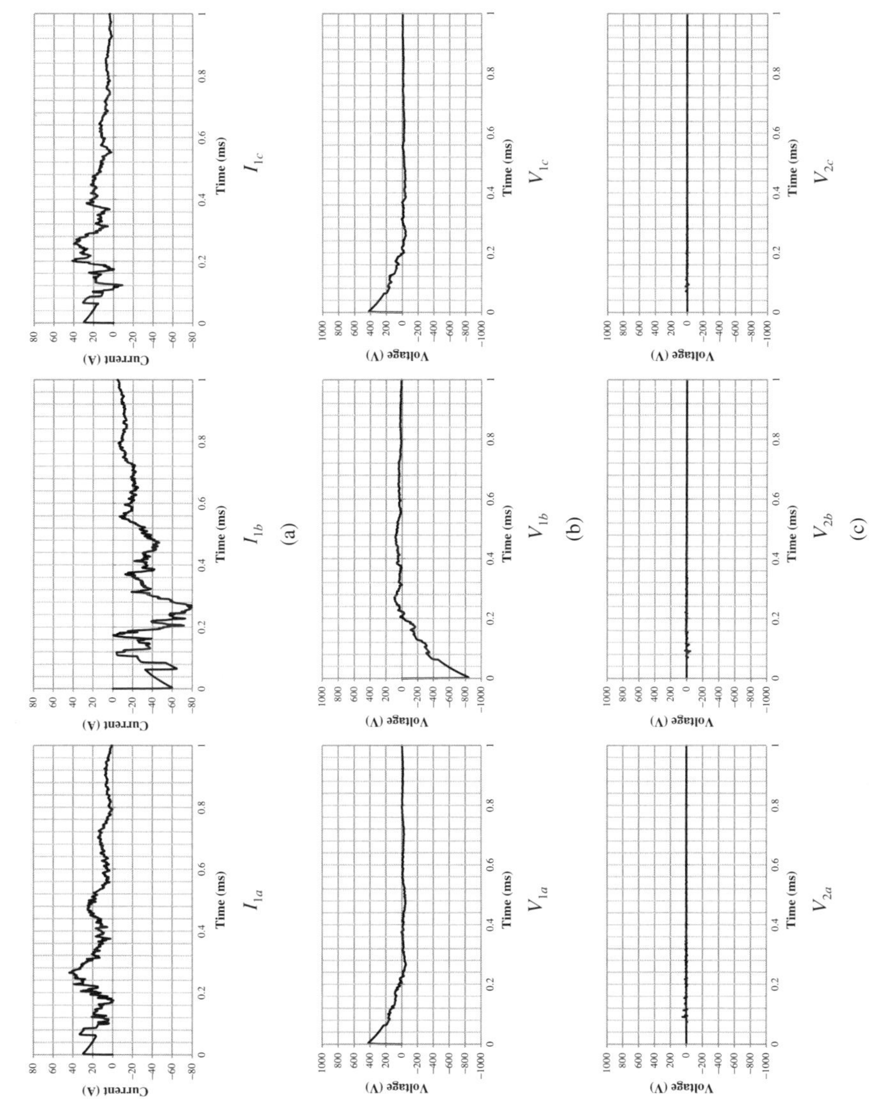

Figure 3.37 Transient currents and voltages associated with the first intersheath mode (mode 1) on a 6 km cross-bonded cable: Case 1X-21: (a) sending-end currents; (b) sending-end voltages; and (c) receiving-end voltages

overlap at every boundary and a waveform containing various frequency components is generated as observed in Figure 3.37 in a cross-bonded cable.

As a result, no clear intersheath mode is observed in the cross-bonded cable. Also, the propagation velocity of the intersheath mode becomes unclear.

In the case of the second intersheath mode (mode 2), quite similar waveforms to those of the first intersheath mode (mode 1) explained above are observed as shown in Figure 3.38, although the maximum current and voltage are greater.

Homogeneous Model of a Cross-bonded Cable

A cross-bonded cable can be represented by an equivalent homogeneous cable composed of three core conductors and one metallic sheath [43]. In this model, the impedance and admittance matrices are given in the following form.

Original Z and Y Matrices:

$$[Z] = \begin{bmatrix} [Z_c] & [Z_m] \\ [Z_m]_t & [Z_s] \end{bmatrix}, \; [Y] = \begin{bmatrix} [Y_c] & -[Y_c] \\ -[Y_c] & [Y_s] \end{bmatrix} \tag{3.164}$$

where

$$[Z_c] = \begin{bmatrix} Z_{11} & Z_{12} & Z_{13} \\ Z_{12} & Z_{22} & Z_{23} \\ Z_{13} & Z_{23} & Z_{33} \end{bmatrix} \quad \text{for cores of phases 1, 2, and 3} \tag{3.165}$$

$[Z_m]$: between cores and sheaths, same form as $[Z_c]$

$[Z_s]$: between three-phase sheaths, same form as $[Z_c]$

$$[Y_c] = \begin{bmatrix} Y_c & 0 & 0 \\ 0 & Y_c & 0 \\ 0 & 0 & Y_c \end{bmatrix}, [Y_s] = \begin{bmatrix} Y_s & 0 & 0 \\ 0 & Y_s & 0 \\ 0 & 0 & Y_s \end{bmatrix} \tag{3.166}$$

Homogeneous Four Conductors:

$$[Z'] = \begin{bmatrix} [Z_c] & (Z'_m)_t \\ (Z'_m) & Z'_s \end{bmatrix}, [Y'] = \begin{bmatrix} [Y_c] & -(Y_c)_t \\ -(Y_c) & Y'_s \end{bmatrix} \; : \; 4 \times 4 \text{ matrix} \tag{3.167}$$

where

$$(Z'_m) = \left(Z'_{m1} \; Z'_{m2} \; Z'_{m3}\right), Z'_s = Z'_{m4}$$

$$Z'_{mi} = \frac{1}{3}\sum_{j=1}^{3} (Z_{ij})_m \; : j = 1 \sim 3, i = 1 \sim 4$$

$$(Y_c) = \left(Y_c \; Y_c \; Y_c\right), Y'_s = 3Y_s$$

Eigenvalue of [P] = [Z'] · [Y']: From (165) and (166), matrix $[P]$ is evaluated analytically assuming flat or trefoil arrangement of a three-phase cable with the approximation of $Z_{12} =$

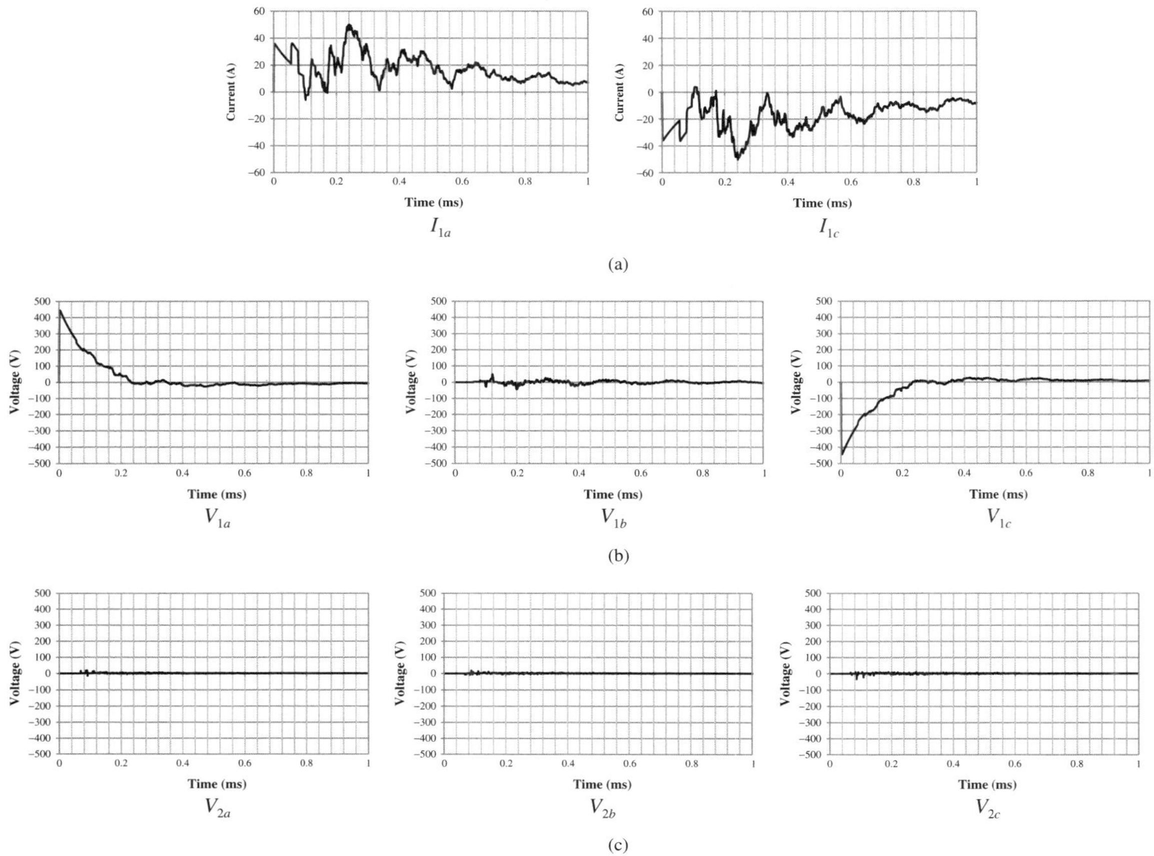

Figure 3.38 Transient currents and voltages associated with the second intersheath mode (mode 2) on a 6 km cross-bonded cable: Case 1X-22: (a) sending-end currents; (b) sending-end voltages; and (c) receiving-end voltages

$Z_{23} = Z_{13} = Z_m$, and $Z_{11} = Z_{22} = Z_{33} = Z_c$.

$$[Z'] = \begin{bmatrix} Z_c & Z_m & Z_m & Z_n \\ Z_m & Z_c & Z_m & Z_n \\ Z_m & Z_m & Z_c & Z_n \\ Z_n & Z_n & Z_n & Z_s \end{bmatrix} \tag{3.168a}$$

$$[Y'] = \begin{bmatrix} Y_c & 0 & 0 & -Y_c \\ 0 & Y_c & 0 & -Y_c \\ 0 & 0 & Y_c & -Y_c \\ -Y_c & -Y_c & -Y_c & Y_s \end{bmatrix} \tag{3.168b}$$

$$[P] = \begin{bmatrix} a & b & b & c \\ b & a & b & c \\ b & b & a & c \\ 0 & 0 & 0 & d \end{bmatrix} \tag{3.168c}$$

Thus, the eigenvalue of P is evaluated as:

$$Q_0 = d = Z_n(Y_s - 3Y_c) \tag{3.169a}$$

$$Q_1 = a - b = (Z_s - Z_m)Y_c \tag{3.169b}$$

$$Q_2 = a = (Z_s - Z_n)Y_c \tag{3.169c}$$

$$Q_3 = (Z_s + Z_m - 2Z_n)Y_c \tag{3.169d}$$

For example, $Z_s = 47.6 + j456$, $Z_m = 46.4 + j355$, $Z_n = 46.5 + j375\,(\Omega/\text{km})$,

$$Y_c = j7.348 \times 10^{-2}, \quad Y_s = j1.275\,(\text{S/km}) \quad \text{at } f = 50\text{ kHz}$$

Then,

$$\gamma_0 = \sqrt{Q_0} = \sqrt{398.5\angle 172.8}\,(\text{km}^{-1}) = 1.253 + j19.91$$

Similarly,

$$\therefore \alpha_0 = 1.253\,\text{N}_\text{p}/\text{km} = 10.82\,\text{dB/km}, \quad c_0' = 15.78\,\text{m}/\mu\text{s} \tag{3.170a}$$

$$\alpha_1 = 0.1445\,\text{dB/km}, \quad c_1' = 115.3\,\text{m}/\mu\text{s} \tag{3.170b}$$

$$\alpha_2 = 0.1442\,\text{dB/km}, \quad c_2' = 128.8\,\text{m}/\mu\text{s} \tag{3.170c}$$

$$\alpha_3 = 64.19 \times 10^{-3}\,\text{dB/km}, \quad c_3' = 148.4\,\text{m}/\mu\text{s} \tag{3.170d}$$

Velocity of 109 m/μs observed in Figure 3.37 corresponds to the above velocities c'_1 and c'_2 of the modified intersheath modes on the homogeneous model of a cross-bonded cable.

Figure 3.39 shows frequency responses of modal attenuation and propagation velocity on the homogeneous cable model. It is observed that there exists no difference for mode 0 between Figure 3.16a and b. Also, no significant difference for the coaxial modes (modes 3–5) is observed.

Figure 3.40 shows a comparison of transient voltages on the phase-a core and sheath between the original (discrete) cross-bonded cable and its homogeneous model when a voltage is applied to the phase-a core at the sending end. It is clear that the homogeneous

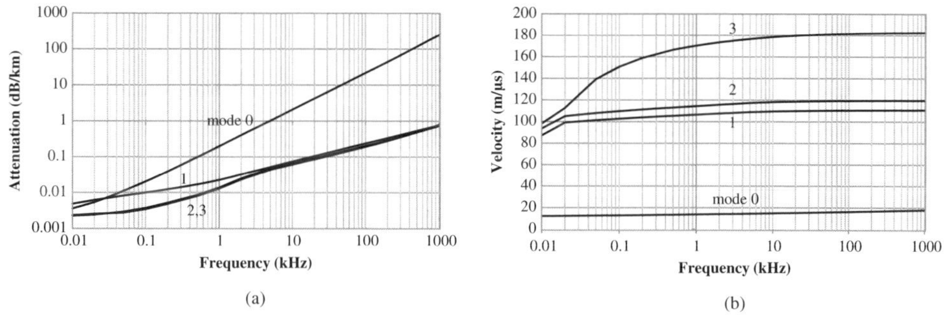

Figure 3.39 Frequency responses of modal (a) attenuation and (b) propagation velocity on the homogeneous model of a cross-bonded cable corresponding to Figure 3.16

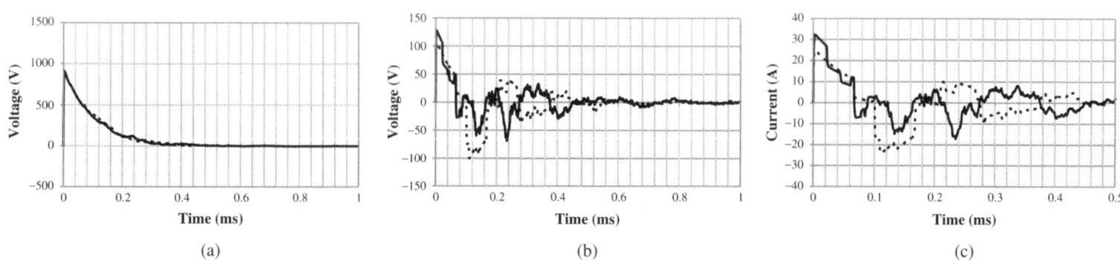

Figure 3.40 Transient currents and voltages on a 6 km cross-bonded cable: Case 1X-1.(a) V_{1ac}, (b) V_{1as}, and (c) $I_{1ac.}$ (—) Original cross-bonded cable; (---) homogeneous model

model gives a reasonable accuracy compared with the original one even in a cross-bonded cable with one major section. This fact has indicated that multiple reflection and refraction in the cross-bonded cable result in intersheath modes quite different from those defined in a solidly bonded cable [43]. The theoretical results in (3.170) agree well with those in Figure 3.16b. It should be noted that the homogeneous model is not applicable transient responses in the intersheath mode circuits illustrated in Figure 3.28 and Figure 3.29, because the three-phase sheaths are reduced to one equivalent sheath.

3.7.2.2 Three Major Sections

Transient Responses on a Cross-bonded Cable

Figure 3.41 shows transient currents and voltages on a cross-bonded cable with three major sections in the case of the first intersheath mode. It is observed that there is no difference up to about 100 μs compared with those on a cable with one major section in Figure 3.37. The time of 60 μs corresponds to the time when a traveling wave is reflected at the first minor section of the intersheath mode as already explained for Figure 3.37.

Transient Waveform on a Homogeneous Model

Figure 3.41 shows transient voltages and currents on a cross-bonded cable with three major sections and Figure 3.42 also shows results for the homogeneous model, when a source voltage

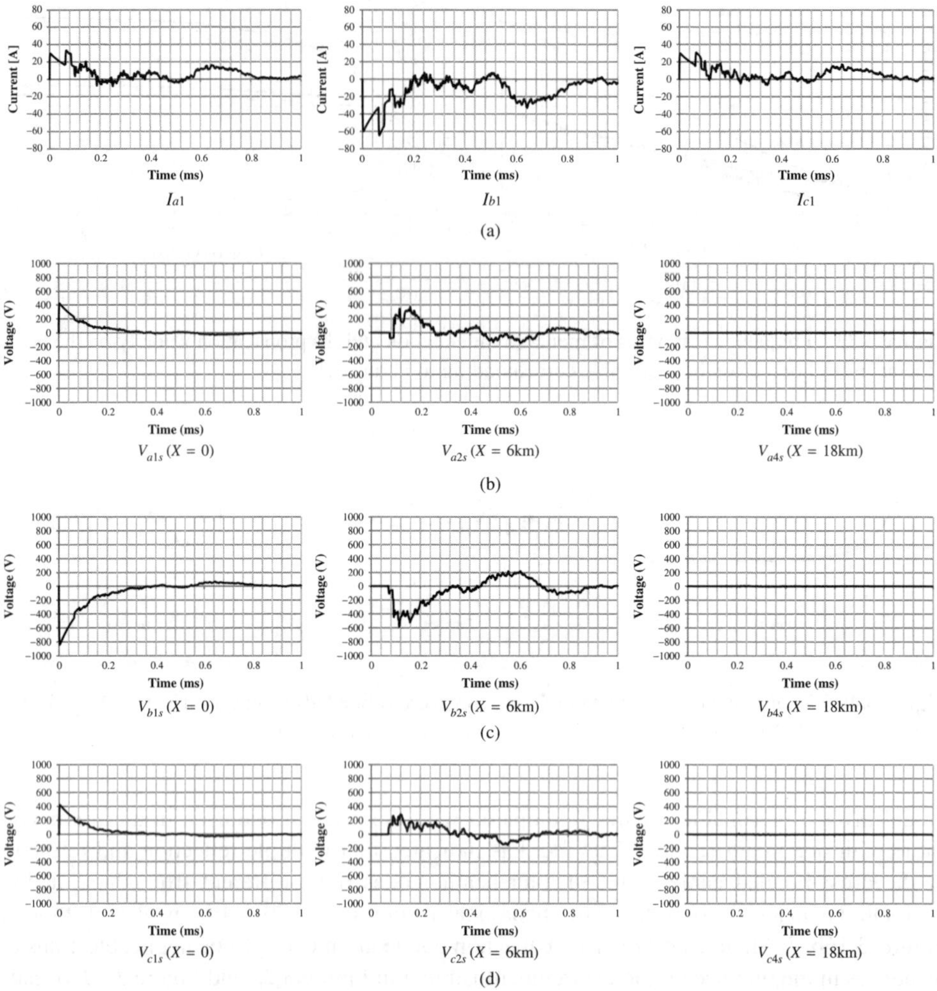

Figure 3.41 Transient currents and voltages associated with the first intersheath mode on a 18 km cross-bonded cable: Case 2X-21: (a) sending-end currents; (b) phase-a voltages; (c) phase-b voltages; and (d) phase-c voltages

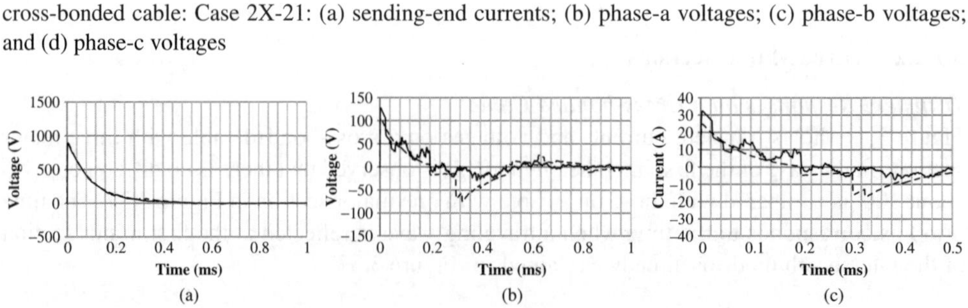

Figure 3.42 Transient currents and voltages on a 18 km cross-bonded cable: Case 2X-1. (a) V_{1ac}, (b) V_{1as}, and (c) I_{1ac}. (—) Original cross-bonded cable; (---) homogeneous model

is applied to the phase-a core. It is observed from the figures that the homogeneous model gives reasonable results compared with those of the original cross-bonded cable.

3.7.3 Earth-return Mode

Figure 3.43 shows transient current and voltage waveforms on phase-a at the sending end. Figure 3.43a is for a solidly bonded cable with a length of 6 km, Figure 3.43b is for a cross-bonded cable composed of one major section with a length of 6 km, and Figure 3.43c is for a cable with three major sections with a total length of 18 km. No difference is observed between the three cases.

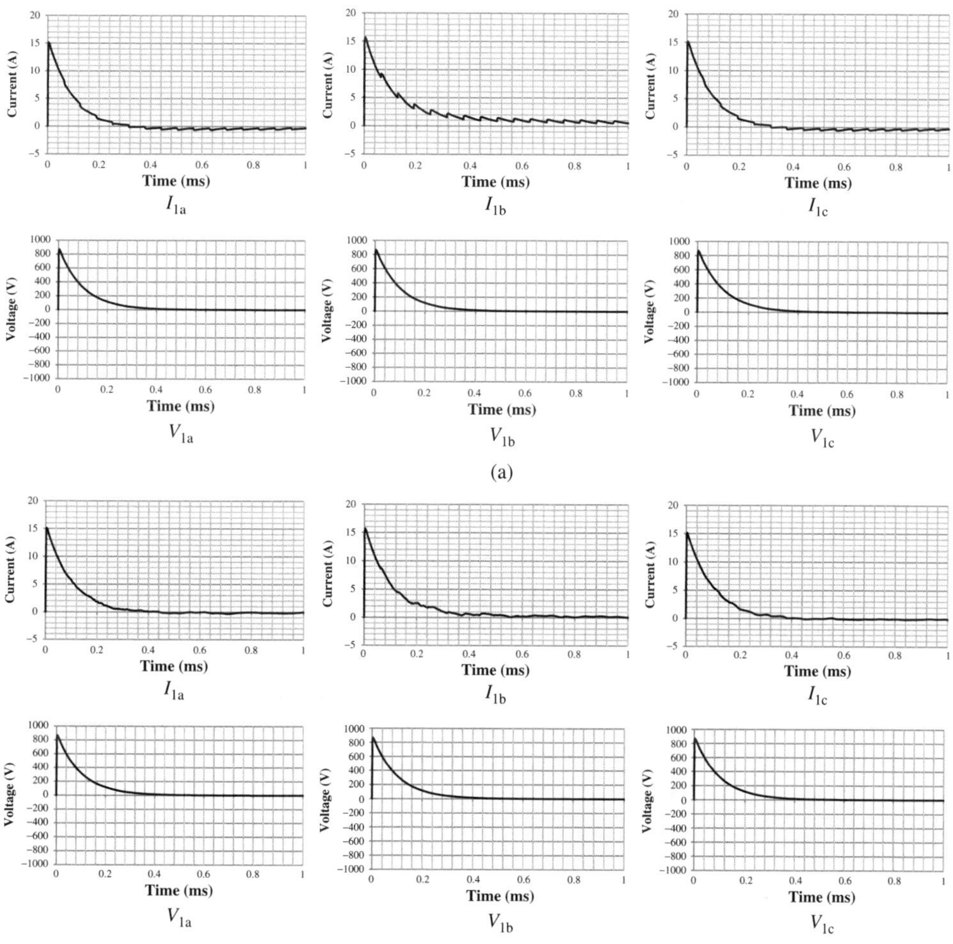

Figure 3.43 Transient currents and voltages of the earth-return mode: (a) solidly bonded cable: Case 1N-3 (6 km); (b) one-major cross-bonded cable: Case 1X-3 (6 km); and (c) three-major cross-bonded cable: Case 3X-3 (18 km)

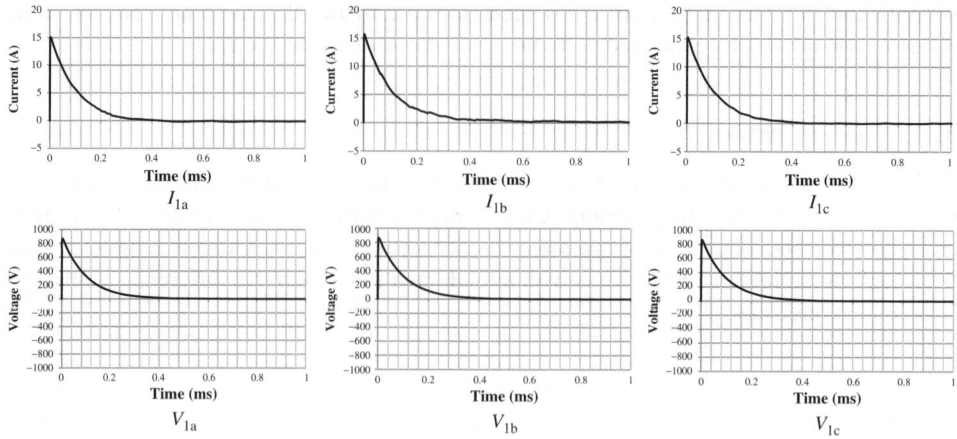

(c)

Figure 3.43 (*Continued*)

3.7.4 Concluding Remarks

Wave propagation and transients associated with intersheath modes on underground cables have been investigated based on EMTP simulations and the characteristics are explained theoretically.

From the investigations, the following points are obtained.

1. Intersheath mode propagation is clearly observed in transient responses of voltages and currents on a solidly bonded cable provided that the other end of the cables and the ground terminal of a voltage source (pulse or impulse generator) are isolated.
2. When the ground terminal is grounded, various modal components (voltages and currents) are produced and thus it becomes difficult to distinguish the intersheath modes.
3. The intersheath mode propagation is also clearly observed when a current source is used, provided that the other end of the cable is short-circuited, although a current and a voltage are replaced by a voltage and a current in the voltage source case as is well-known in a current theory.
4. In the case of a cross-bonded cable, the intersheath mode propagation can be observed only in the wavefront of a current reflected from the other end of the first minor section. After this time period of the current wavefront, various modal components generated at every minor and major section are superposed in the waveform and thus it is almost impossible to distinguish the intersheath mode propagation. In a voltage waveform, the intersheath mode cannot be observed.

References

[1] Wedepohl, L.M. (1963) Application of matrix methods to the solution of travelling wave phenomena in polyphase systems. *Proceedings of the IEE*, **110** (12), 2200–12.

[2] Wedepohl, L.M. and Mohamed, S.E.T. (1969) Multiconductor transmission lines; Theory of natural mode and Fourier integral applied to transient analysis. *Proceedings of the IEE*, **116** (9), 1553–63.

[3] Ametani, A. (1990) *Distributed -Parameter Circuit Theory*, Corona Publishing Co. (in Japanese).
[4] Fortescue, C.L. (1918) Method of symmetrical coordinates applied to the solution of polyphase network. *AIEE Transactions*, **37** (Pt II), 1027–140.
[5] Clark, E. (1943) *Circuit Analysis of AC Power Systems*, Vol. **1**, John Wiley & Sons, Ltd.
[6] Wedepohl, L.M. and Wilcox, D.J (1973) Transient analysis of underground power-transmission systems. *Proceedings of the IEE*, **120**, 253–60.
[7] IEE Japan Working Group (1991) Surge Phenomena on Cables and the Countermeasures. IEE Japan Technical Report 366.
[8] Deri, A., Tevan, G., Semlyen, A., and Castanheira, A. (1981) The complex ground return plane: A simplified model for homogeneous and multi-layer earth return. *IEEE Transactions on Power Apparatus and Systems*, **PAS-100** (8), 3686–963.
[9] Ametani, A., Tanaka, A., Ohsaki, E., and Honaga, Y. (1981) Wave propagation characteristics on an untransposed vertical twin-circuit line. *IEE Japan Transaction B*, **101-B** (11), 675–82.
[10] Ametani, A., Ono, T., Honaga, Y., and Ouchi, Y. (1974) Surge propagation on Japanese 500 kV untransposed transmission line. *Proceedings of the IEE*, **121** (2), 136–8.
[11] Shinozaki, H., Kawakako, T., Hamamoto, J. *et al.* (1971) Abnormal voltages of a core at a crossbonding point. *Journal of the Technology Laboratory of Chugoku Electric Power Company*, **39**, 175–98.
[12] Wedepohl, L.M. and Indulkar, S.S. (1975) Switching overvoltages in short crossbonded cable systems using Fourier Transform. *Proceedings of the IEE*, **122** (11), 1217–21.
[13] Wilcox, D.J. and Lawer, K.J. (1978) Transient phenomena in crossbonded cable systems; analytical results. *Proceedings of the IEE*, **125** (10), 999–1005.
[14] Humpage, W.D., Wong. K.P., and Nguyen, T.T. (1981) Z-transform electromagnetic transient analysis of cross-bonded cable transmission systems. *Proceedings of the IEE*, **128** (2), 55–62.
[15] Nagaoka, N. and Ametani, A. (1983) Transient calculation on cross-bonded cables. *IEEE Transactions on Power Apparatus and Systems*, **PAS-102** (4), 779–86.
[16] Itoh, Y., Nagaoka, N., and Ametani, A (1990) Transient analysis of a crossbonded cable system underneath a bridge. *IEEE Transactions on Power Delivery*, **5** (2), 527–32.
[17] CIGRE WG C4.502 (2013) Power System Technical Performance Issues Related to the Application of Long HVAC Cables. *CIGRE Technical Brochure* **556**.
[18] IEEE (2007) Guide for fault locating techniques on shielded power cable systems. *IEEE Standard* 1234-2007.
[19] CIGRE WG B1.30 (2013) Cable Systems Electrical Characteristics. *CIGRE Technical Brochure* **531**.
[20] Nagaoka, N. and Ametani, A. (1988) A development of a generalized frequency-domain transient program – FTP. *IEEE Transactions on Power Delivery*, **3** (4), 1996–2004.
[21] Nagaoka, N. (1992) Transient analysis of cable systems by means of a frequency transform method. DEng thesis. Doshisha University.
[22] Tegopoulos, J.A. and Kriezis, E.E. (1971) Eddy current distribution in cylindrical shells of infinite length due to axial currents, Part II. Shells of finite thickness. IEEE Transactions on Power Apparatus and Systems, **PAS-90**, 1287–94.
[23] Brown, G.W. and Rocamora, R.G. (1976) Surge propagation in three-phase pipe-type cables, Part I. Unsaturated pipe. *IEEE Transactions on Power Apparatus and Systems*, **PAS-95**, 88–95.
[24] Dugan, R.C., Rocamora, R.G., Brown, G.W. *et al.* (1977) Surge propagation in three-phase pipe-type cables, Part II. Duplication of field tests including the effects of neutral wires and pipe saturation. *IEEE Transactions on Power Apparatus and Systems*, **PAS-96**, 826–33.
[25] Schinzinger, R. and Ametani, A. (1978) Surge propagation characteristics of pipe enclosed underground cables. *IEEE Transactions on Power Apparatus and Systems*, **PAS-97**, 1680–7.
[26] Ametani, A. (1980) A general formulation of impedance and admittance of cables. *IEEE Transactions on Power Apparatus and Systems*, **PAS-99** (3), 902–10.
[27] Dokopoulos, P. and Tampakis, D. (1984) Analysis of field and losses in three-phase gas cable with thick walls: Part I. Field analysis. *IEEE Transactions*, **PAS-103** (9), 2728–34.
[28] Dokopoulos, P. and Tampakis, D. (1985) Analysis of field and losses in three-phase gas cable with thick walls: Part II. Calculation of losses and results. *IEEE Transactions*, **PAS-104** (1), 9–15.
[29] Poltz, J., Kuffel, E., Grzybowski, S., and Raghuveer, M.R. (1982) Eddy-current losses in pipe-type cable systems. *IEEE Transactions on Power Apparatus and Systems*, **PAS-101** (4), 825–32.
[30] Fortin, S., Yang, Y., Ma, J., and Dawalibi, F.P. (2005) Effects of eddy current on the impedance of pipe-type cables with arbitrary pipe thickness. *ICEE 2005*, Paper TD2-09.

[31] Chien, C.H. and Bucknall, R.W.G. (2009) Harmonic calculations of proximity effect on impedance characteristics in subsea power transmission cables. *IEEE Transactions on Power Delivery*, **24** (2), 2150–8.
[32] Gustavsen, B., Bruaset, A., Bremnes, J.J., and Hassel, A. (2009) A finite element approach for calculating electrical parameters of umbilical cables. *IEEE Transactions on. Power Delivery*, **24** (4), 2375–84.
[33] Ishikawa, T., Kawasaki, K., and Okamoto, I. (1976) Eddy current losses in cable sheaths, Part I. *Dainichi Nihon Cable Journal*, **61**, 34–42.
[34] Ishikawa, T., Kawasaki, K., and Okamoto, I. (1977) Eddy current losses in cable sheaths, Part II. *Dainichi Nihon Cable Journal*, **62**, 21–64.
[35] Kawasaki, K., Inami, M., and Ishikawa, T. (1981) Theoretical consideration on eddy current losses on non-magnetic and magnetic pipes for power transmission systems. *IEEE Transactions on Power Apparatus and Systems*, **PAS-100** (2), 474–84.
[36] Mekjian, A. and Sosnowski, M. (1983) Calculation of alternating current losses in steel pipe containing power cables. *IEEE Transactions on Power Apparatus and Systems*, **PAS-102** (2), 382–8.
[37] Pagetli, A. (2012) Cable modeling for electromagnetic transients in power systems. PhD thesis. University of Clermont Ferrand, France and University of Bologna, Italy.
[38] Ametani, A., Kawamura, K., Shendge, A. *et al.* (2013) Wave propagation characteristics on a pipe-type cable in particular reference to the proximity effect. *IEE Japan Transactions on Power and Energy*, **133** (12), 954–60.
[39] Nagaoka, N., Yamamoto, M., and Ametani, A. (1985) Surge propagation characteristics of a POF cable. *Electrical Engineering in Japan*, **105** (5), 67–75.
[40] CIGRE WG C4.502 (2013) Power System Technical Performance Issues Related to the Application of Long HVAC Cables. *CIGRE Technical Brochure* **556**.
[41] Gudmundsdottir, U.S. (2010) Modeling of long high voltage AC cables in transmission systems. PhD thesis. Aalborg University.
[42] Ametani, A. (ed.) (2013) *Power System Transients-Theory and Application*, CRC Press.
[43] Ametani, A., Miyamoto, Y., and Nagaoka, N. (2003) An investigation of a wave propagation characteristic on a cross-bonded cable. *IEE Japan Transaction B*, **123-B** (3), 395–401.

4

Cable Modeling for Transient Simulations

Teruo Ohno[1] and Akihiro Ametani[2]
[1]*Tokyo Electric Power Company, Tokyo, Japan*
[2]*Graduate School of Electrical Engineering, Doshisha University, Kyoto, Japan*

4.1 Sequence Impedances Using a Lumped PI-circuit Model

Before we discuss the cable modeling for electromagnetic transient (EMT) simulations, we first explain sequence impedances (lumped PI-circuit model) of the cable system used in steady-state [root mean square (RMS) value] simulations. The sequence impedances are important for the transient simulations as the initial conditions are set up based on the result of the RMS simulations.

There are formulas to derive sequence impedances of cables, but it has been common practice to measure the sequence impedances of the cables after installation. This is because it was thought that the existing formulas are not accurate enough to be used for the simulation of important lines. The main reason for this inaccuracy is the fact that the existing formulas do not consider the cable as the cable system, including the sheath bonding.

Recent literature tackled this issue and derived new formulas of sequence impedances of cable systems considering the sheath bonding [1, 2, 3, 4]. This section introduces the derivation of these new formulas and compares calculated sequence impedances using these new formulas.

4.1.1 Solidly Bonded Cables

This section derives sequence impedances of solidly bonded cables. We start from the simple one cable case as shown in Figure 4.1 in order to better understand the derivation process. The sheath circuit is solidly grounded at both ends.

Cable System Transients: Theory, Modeling and Simulation, First Edition.
Akihiro Ametani, Teruo Ohno and Naoto Nagaoka.

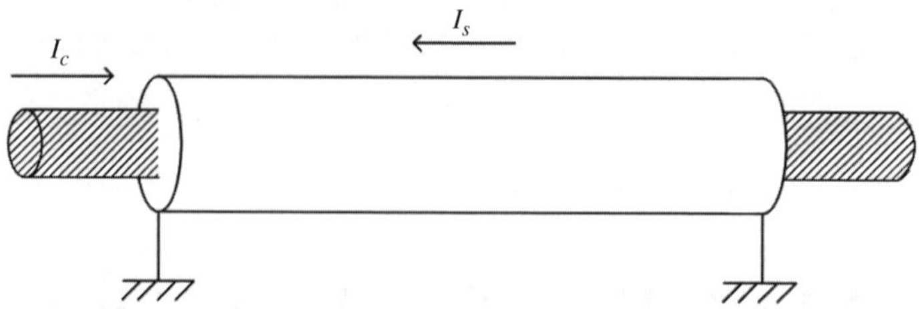

Figure 4.1 One cable case

In this figure, the following set of equations is satisfied.

$$V_c = Z_c I_c - Z_m I_s \tag{4.1a}$$

$$0 = Z_s I_s - Z_m I_c \tag{4.1b}$$

where V_c is the voltage drop in the core, I_c is the core current, I_s is the sheath current, Z_c is the core impedance, Z_s is the sheath impedance, and Z_m is the mutual impedance between the core and the sheath.

From (4.1b), the sheath current can be expressed as:

$$I_s = (Z_m / Z_s) I_c \tag{4.2}$$

Eliminating I_s from (4.1a),

$$V_c = Z_c I_c - Z_m (Z_m / Z_s) I_c = (Z_c - Z_m^2 / Z_s) I_c \tag{4.3}$$

Therefore, the impedance of the cable in Figure 4.1 is found as:

$$Z = Z_c - Z_m^2 / Z_s \tag{4.4}$$

Now, in order to derive sequence impedances, the three-phase single-core coaxial (SC) cable case as shown in Figure 4.2 is considered. Since the application of the sequence impedances assumes a balanced circuit, we assume these cables are laid in a trefoil formation and the following relationship is satisfied.

$$Z_p = Z_{ab} = Z_{bc} = Z_{ca} \tag{4.5}$$

where Z_p is the mutual impedance between phases (balanced), Z_{ab} is the mutual impedance between phase-a cable and phase-b cable, Z_{bc} is the mutual impedance between phase-b cable and phase-c cable, and Z_{ca} is the mutual impedance between phase-c cable and phase-a cable.

In Figure 4.2, $R/2$ is the substation grounding resistance of the sheath circuit at both ends. The value of the substation grounding resistance is normally below 1 Ω and is often around 0.1 Ω. It does not have a significant effect on the sequence impedances when the value as low as 0.1 Ω is achieved. When the value is higher, it is important to consider the substation grounding resistance in a zero-sequence impedance calculation.

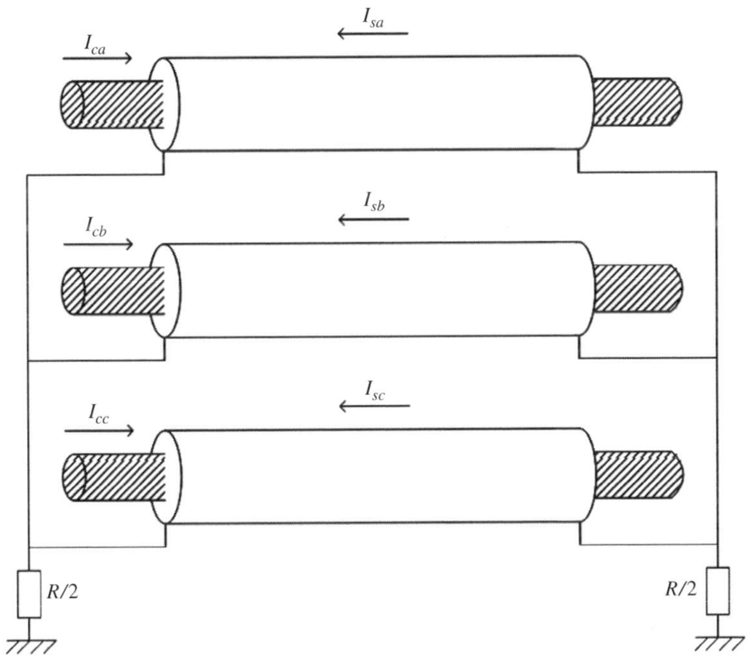

Figure 4.2 Single-core three-phase cable case

First, the core current is assumed to be positive-sequence current in order to derive the positive-sequence impedance. The following set of equations is satisfied with the three-phase SC cable case in Figure 4.2.

$$V_{ca} = Z_c I_{ca} + Z_p(I_{cb} + I_{cc}) - Z_m I_{sa} - Z_p(I_{sb} + I_{sc}) \tag{4.6a}$$

$$0 = Z_s I_{sa} + Z_p(I_{sb} + I_{sc}) - Z_m I_{ca} - Z_p(I_{cb} + I_{cc}) \tag{4.6b}$$

where V_{ca} is the voltage drop in the core of phase-a.

Since the core current is the positive-sequence current, the sheath current also becomes the positive-sequence current.

$$I_{ca} + I_{cb} + I_{cc} = 0, I_{sa} + I_{sb} + I_{sc} = 0 \tag{4.7}$$

Using (4.7), the set of equations in (4.6a) and (4.6b) can be simplified as follows:

$$V_{ca} = (Z_c - Z_p)I_{ca} - (Z_m - Z_p)I_{sa} \tag{4.8a}$$

$$0 = (Z_s - Z_p)I_{sa} - (Z_m - Z_p)I_{ca} \tag{4.8b}$$

Solving (4.8b) for I_{sa},

$$I_{sa} = \frac{Z_m - Z_p}{Z_s - Z_p} I_{ca} \tag{4.9}$$

Eliminating I_{sa} from (4.8a),

$$V_{ca} = (Z_c - Z_p)I_{ca} - (Z_m - Z_p)\frac{Z_m - Z_p}{Z_s - Z_p}I_{ca}$$

$$= \left\{ (Z_c - Z_p) - \frac{(Z_m - Z_p)^2}{Z_s - Z_p} \right\} I_{ca} \quad (4.10)$$

The positive-sequence impedance Z_1 of the three-phase cable in Figure 4.2 is found as:

$$Z_1 = (Z_c - Z_p) - \frac{(Z_m - Z_p)^2}{Z_s - Z_p} \quad (4.11)$$

Secondly, the core current is assumed to be zero-sequence current in order to derive the zero-sequence impedance. This time, the sheath-return circuit through the substation grounding resistance and the earth has to be considered in parallel with the sheath circuit. Therefore, Kirchhoff's voltage law gives:

$$V_{ca} = Z_c I_{ca} + Z_p(I_{cb} + I_{cc}) - Z_m I_{sa} - Z_p(I_{sb} + I_{sc}) \quad (4.12a)$$

$$0 = Z_s I_{sa} + Z_p(I_{sb} + I_{sc}) - Z_m I_{ca} - Z_p(I_{cb} + I_{cc}) + R(I_{sa} + I_{sb} + I_{sc}) \quad (4.12b)$$

Since the core current is the zero-sequence current, the sheath current also becomes the zero-sequence current.

$$I_{ca} = I_{cb} = I_{cc}, I_{sa} = I_{sb} = I_{sc} \quad (4.13)$$

Using (4.13), the set of equations in (4.12a,b) can be simplified as follows:

$$V_{ca} = (Z_c + 2Z_p)I_{ca} - (Z_m + 2Z_p)I_{sa} \quad (4.14a)$$

$$0 = (Z_s + 2Z_p + 3R)I_{sa} - (Z_m + 2Z_p)I_{ca} \quad (4.14b)$$

Solving (4.14b) for I_{sa},

$$I_{sa} = \frac{Z_m + 2Z_p}{Z_s + 2Z_p + 3R}I_{ca} \quad (4.15)$$

Eliminating I_{sa} from (4.14a),

$$V_{ca} = (Z_c + 2Z_p)I_{ca} - (Z_m + 2Z_p)\frac{Z_m + 2Z_p}{Z_s + 2Z_p + 3R}I_{ca}$$

$$= \left\{ (Z_c + 2Z_p) - \frac{(Z_m + 2Z_p)^2}{Z_s + 2Z_p + 3R} \right\} I_{ca} \quad (4.16)$$

The zero-sequence impedance Z_0 of the three-phase cable in Figure 4.2 is found as:

$$Z_0 = (Z_c + 2Z_p) - \frac{(Z_m + 2Z_p)^2}{Z_s + 2Z_p + 3R} \quad (4.17)$$

Note that when the zero-sequence core current flows through the substation grounding resistance, for example, in the case of a cable internal fault, (4.12b) becomes:

$$0 = Z_s I_{sa} + Z_p(I_{sb} + I_{sc}) - Z_m I_{ca} - Z_p(I_{cb} + I_{cc})$$

$$+ R(I_{sa} + I_{sb} + I_{sc}) - R(I_{ca} + I_{cb} + I_{cc}) \quad (4.18)$$

In this case, the zero-sequence impedance Z_0 is found as:

$$Z_0 = (Z_c + 2Z_p) - \frac{(Z_m + 2Z_p)(Z_m + 2Z_p + 3R)}{Z_s + 2Z_p + 3R} \tag{4.19}$$

4.1.2 Cross-bonded Cables

This section derives sequence impedances of cross-bonded cables. As in the last section, we focus on the phase-a cable, which is shown on the top of Figure 4.3. In the figure, the subscript of the sheath currents, I_{sa}, I_{sb}, and I_{sc}, includes two letters. The first letter '*s*' represents that they are sheath currents. The second letter, for example '*a*', shows that this sheath current is paired with the phase-a core current in the first (counting from the left) minor section. As shown in Figure 4.3 core current I_{ca} is paired with, I_{sa}, I_{sc}, and I_{sb}, in the first, second, and third minor sections, respectively.

First, the core current is assumed to be the positive-sequence current in order to derive the positive-sequence impedance. The following equation is satisfied with the three-phase SC cable case in Figure 4.3.

$$\begin{aligned} V_{ca} &= Z_c I_{ca} + Z_p(I_{cb} + I_{cc}) \\ &- \frac{Z_m}{3} I_{sa} - \frac{Z_p}{3}(I_{sb} + I_{sc}) \quad \text{(1st minor section)} \\ &- \frac{Z_m}{3} I_{sb} - \frac{Z_p}{3}(I_{sc} + I_{sa}) \quad \text{(2nd minor section)} \\ &- \frac{Z_m}{3} I_{sc} - \frac{Z_p}{3}(I_{sa} + I_{sb}) \quad \text{(3rd minor section)} \end{aligned} \tag{4.20}$$

When the core current is the positive-sequence current and impedances are balanced, the sheath current does not flow in cross-bonded cables. As a result, (4.20) is reduced to:

$$V_{ca} = Z_c I_{ca} + Z_p(I_{cb} + I_{cc}) = (Z_c - Z_p) I_{ca} \tag{4.21}$$

The positive-sequence impedance Z_1 of the cross-bonded cable is found as:

$$Z_1 = Z_c - Z_p \tag{4.22}$$

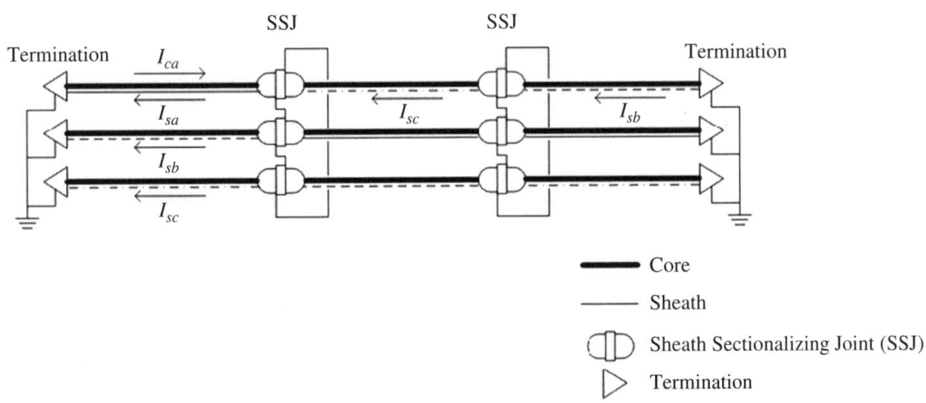

Figure 4.3 Cross-bonded cable case

Secondly, the core current is assumed to be the zero-sequence current in order to derive the zero-sequence impedance. The sheath current also becomes the zero-sequence current, and (4.20) is reduced to:

$$V_{ca} = (Z_c + 2Z_p)I_{ca} - (Z_m + 2Z_p)I_{sa} \tag{4.23}$$

Kirchhoff's voltage law for the sheath circuit gives:

$$\begin{aligned} 0 = {} & Z_s I_{sa} + Z_p(I_{sb} + I_{sc}) + R(I_{sa} + I_{sb} + I_{sc}) \\ & - \frac{Z_m}{3} I_{ca} - \frac{Z_p}{3}(I_{cb} + I_{cc}) \quad \text{(1st minor section)} \\ & - \frac{Z_m}{3} I_{cb} - \frac{Z_p}{3}(I_{cc} + I_{ca}) \quad \text{(2nd minor section)} \\ & - \frac{Z_m}{3} I_{cc} - \frac{Z_p}{3}(I_{ca} + I_{cb}) \quad \text{(3rd minor section)} \end{aligned} \tag{4.24}$$

Since the core current and the sheath current are the zero-sequence current,

$$0 = (Z_s + 2Z_p)I_{sa} + 3RI_{sa} - (Z_m + 2Z_p)I_{ca} \tag{4.25}$$

Solving this equation for I_{sa},

$$I_{sa} = \frac{Z_m + 2Z_p}{Z_s + 2Z_p + 3R} I_{ca} \tag{4.26}$$

Eliminating I_{sa} from (4.23),

$$\begin{aligned} V_{ca} &= (Z_c + 2Z_p)I_{ca} - (Z_m + 2Z_p)\frac{Z_m + 2Z_p}{Z_s + 2Z_p + 3R} I_{ca} \\ &= \left\{ (Z_c + 2Z_p) - \frac{(Z_m + 2Z_p)^2}{Z_s + 2Z_p + 3R} \right\} I_{ca} \end{aligned} \tag{4.27}$$

The zero-sequence impedance Z_0 of the cross-bonded cable is:

$$Z_0 = (Z_c + 2Z_p) - \frac{(Z_m + 2Z_p)^2}{Z_s + 2Z_p + 3R} \tag{4.28}$$

It is found that the zero-sequence impedance of the cross-bonded cable is derived by the same formula as the solidly bonded cable.

4.1.3 Derivation of Sequence Impedance Formulas

Previous sections found formulas to derive sequence impedances of cable systems assuming a balanced circuit. Previous sections did not discuss:

- how to derive formulas when cables are not laid in a balanced formation, especially in a flat formation instead of a trefoil formation;

- how to consider parallel circuits composed of sheath circuits and the earth through earthing joints when cables have multiple major sections.

As for the first item, there exists no transmission line which has a perfect balance even though the definition of sequence impedances assumes a balanced circuit. Normally RMS simulations are performed in a positive-sequence simulation platform, in which sequence impedances are used to express every transmission line including an unbalanced line. Therefore, positive-sequence and zero-sequence impedances are calculated or measured for an unbalanced line, for example, for cables laid in a flat formation, tolerating an error caused by the imbalance.

Formulas of sequence impedances in this section are derived from a setup to measure the sequence impedances of cable systems after the installation [4]. The error caused by the imbalance is thus comparable with the error which has been tolerated when we use the sequence impedances obtained from field measurements.

In addition, formulas of the sequence impedances in this section address the second item by utilizing homogenous nature of cable impedances. A 6×6 impedance matrix of a cross-bonded cable can be reduced to a 4×4 impedance matrix, which makes it possible to derive formulas of sequence impedances without considering the path to the earth through earthing joints [5, 6].

4.1.3.1 Solidly Bonded Cable

One cable system corresponds to a six conductor system composed of three cores and three metallic sheaths. The 6×6 impedance matrix of the cable system is given in the following equation [7]:

$$[Z] = \begin{bmatrix} [Zcc] & [Zcs] \\ [Zcs]_t & [Zss] \end{bmatrix} = \begin{bmatrix} [Zcc] & [Zcs] \\ [Zcs] & [Zss] \end{bmatrix} \tag{4.29}$$

$$[Zcc] = \begin{bmatrix} Z_{cc11} & Z_{cc12} & Z_{cc13} \\ Z_{cc12} & Z_{cc22} & Z_{cc23} \\ Z_{cc13} & Z_{cc23} & Z_{cc33} \end{bmatrix}, \quad [Zss] = \begin{bmatrix} Z_{ss11} & Z_{ss12} & Z_{ss13} \\ Z_{ss12} & Z_{ss22} & Z_{ss23} \\ Z_{ss13} & Z_{ss23} & Z_{ss33} \end{bmatrix},$$

$$[Zcs] = \begin{bmatrix} Z_{cs11} & Z_{cs12} & Z_{cs13} \\ Z_{cs12} & Z_{cs22} & Z_{cs23} \\ Z_{cs13} & Z_{cs23} & Z_{cs33} \end{bmatrix}$$

where c is for core, s is for sheath, m is for the mutual coupling between core and sheath, and t is for transpose.

In (4.29), phase-a is assumed to be laid symmetrical to phase-c against phase-b. The flat configuration and the trefoil configuration, which are typically adopted, satisfy this assumption.

Figure 4.4 shows a sequence current measurement circuit for a solidly bonded cable. The following equations are given from the 6×6 impedance matrix in (4.29) and Figure 4.4.

$$(V_c) = [Zcc](I_c) + [Zcs](I_s) \tag{4.30}$$

$$(V_s) = [Zcs](I_c) + [Zss](I_s) = -2[Rg](I_s) \tag{4.31}$$

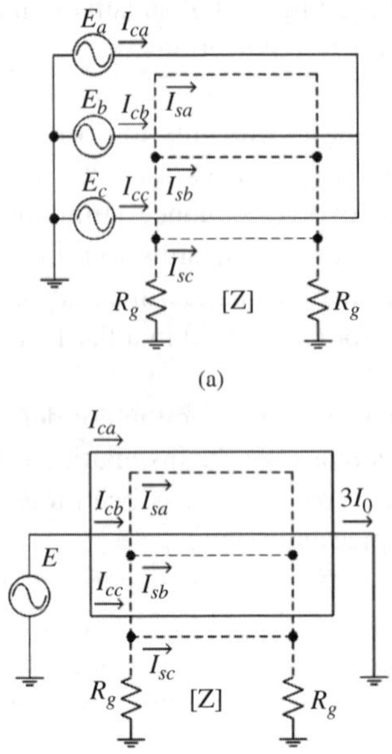

Figure 4.4 Setup for measuring sequence currents for a solidly bonded cable: (a) positive-sequence current; and (b) zero-sequence current. *Source*: © 2011 IPST. Reprinted, with permission, from [1]

Here, $(I_c) = \left(I_{ca}\ I_{cb}\ I_{ca}\right)_t$: core current

$(I_s) = \left(I_{sa}\ I_{sb}\ I_{sa}\right)_t$: sheath current

$[R_g] = R_g \begin{bmatrix} 1 & 1 & 1 \\ 1 & 1 & 1 \\ 1 & 1 & 1 \end{bmatrix}$: sheath grounding resistance at terminations

From (4.31), sheath current (I_s) is found by:

$$(I_s) = -([Zss] + 2[Rg])^{-1}[Zcs](I_c) \tag{4.32}$$

Eliminating sheath current (I_s) in (4.30), core current (I_c) can be derived as:

$$(I_c) = ([Zcc] - [Zcs]([Zss] + 2[Rg])^{-1}[Zcs])^{-1}(V_c) \tag{4.33}$$

In order to find the positive-sequence impedance, it is necessary to eliminate (V_c) and (I_c) in (4.33), using the following equations which can be found from Figure 4.4a.

$$(V_c) = \left[E\ \alpha^2 E\ \alpha E\right]_t,\ (I_c) = \left[I_{ca}\ I_{cb}\ I_{ca}\right]_t \tag{4.34}$$

Core current (I_c) is obtained from (4.33) and (4.34). Using this core current, the positive-sequence current is calculated as $I_1 = (I_{ca} + \alpha I_{cb} + \alpha^2 I_{cc})/3,\ \ \alpha = \exp(j2\pi/3)$.

Once the positive-sequence current is found, the positive-sequence impedance is simply calculated as $Z_1 = E/I_1$.

In order to find the zero-sequence impedance, it is then necessary to eliminate (V_c) and (I_c) in (4.33), using the following equations which can be found from Figure 4.4b.

$$(V_c) = \left[E\ E\ E\right]_t, (I_c) = \left[I_{ca}\ I_{cb}\ I_{ca}\right]_t \tag{4.35}$$

Core current (I_c) is obtained from (4.33) and (4.35). Using this core current, the zero-sequence current is calculated as $I_0 = (I_{ca} + I_{cb} + I_{cc})/3$. Once the zero-sequence current is found, the zero-sequence impedance is simply calculated as $Z_0 = E/I_0$.

4.1.3.2 Cross-bonded Cable

In the case of cross-bonded cables, lengths of minor sections can have imbalances due to the constraint on the location of joints. The imbalances are designed to be as small as possible since they increase sheath currents and raise sheath voltages. When a cable system has multiple major sections, the overall balance is considered to minimize sheath currents. As a result, when a cable system has more than a couple of major sections, sheath currents are generally balanced among three sheath conductors, which allows us to reduce three metallic sheaths to one conductor.

Reducing the sheath conductors, a six conductor system is reduced to a four conductor system composed of three cores and one equivalent metallic sheath as shown in Figure 4.5. The 4×4 reduced impedance matrix can be expressed as:

$$[Z'] = \begin{bmatrix} Z_{cc11} & Z_{cc12} & Z_{cc13} & Z'_{14} \\ Z_{cc12} & Z_{cc22} & Z_{cc23} & Z'_{24} \\ Z_{cc13} & Z_{cc23} & Z_{cc33} & Z'_{14} \\ Z'_{14} & Z'_{24} & Z'_{14} & Z'_{ss} \end{bmatrix} \tag{4.36}$$

Here, $Z'(4,k) = Z'(k,4)\ \ (k = 1, 2, 3)$ is given as an average of $[Z_{cs}]$ and $Z'(4,4)$ is given as an average of $[Z_{ss}]$. They can be calculated from the 6×6 impedance matrix $[Z]$ in (4.29) as:

$$Z'(4,k) = \frac{1}{3}\sum_{j=4}^{6} Z(j,k) \quad (k = 1, \cdots, 4) \tag{4.37}$$

$Z'_{14} = Z'_{34}$ stands in the flat configuration and the trefoil configuration.

In order to find the positive-sequence impedance, we use a setup for measuring positive-sequence currents for a cross-bonded cable as shown in Figure 4.6a. In the figure, the equation $I_{sa} + I_{sb} + I_{sc} = 0$ is satisfied at both ends of the cable line. As a result, the sheath voltage at both ends of the cable line becomes zero since the sheath current that flows through the substation grounding resistance R_g is zero.

In Figure 4.6a, the applied positive-sequence voltage and the measured positive-sequence current are expressed as:

$$(V_1) = \left[E\ \alpha^2 E\ \alpha E\ 0\right]_t, (I_1) = \left[I_{ca}\ I_{cb}\ I_{cc}\ I_s\right]_t \tag{4.38}$$

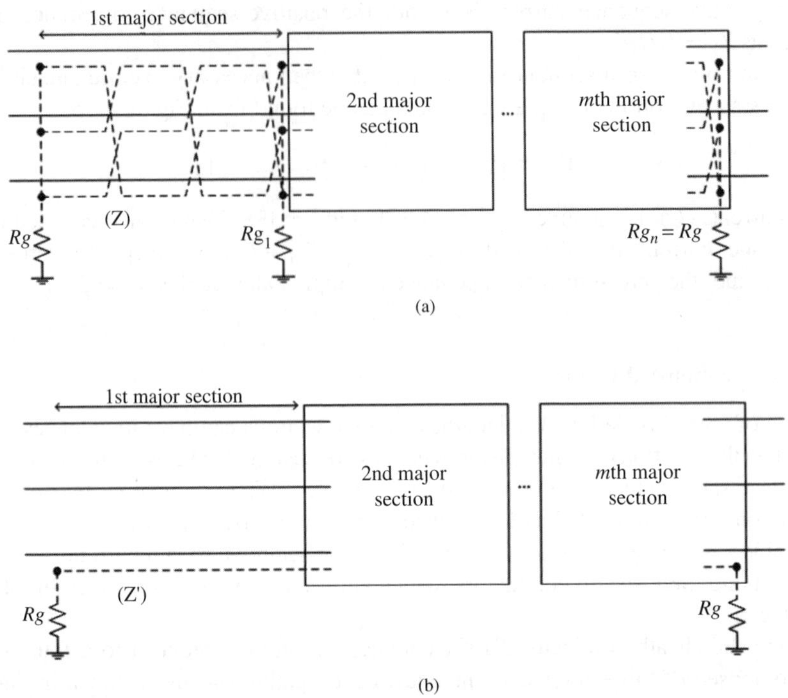

Figure 4.5 Cross-bonded cable and its equivalent model: (a) cross-bonded cable system with "*m*" major sections; and (b) equivalent four conductor system. *Source*: © 2011 IPST. Reprinted, with permission, from [1]

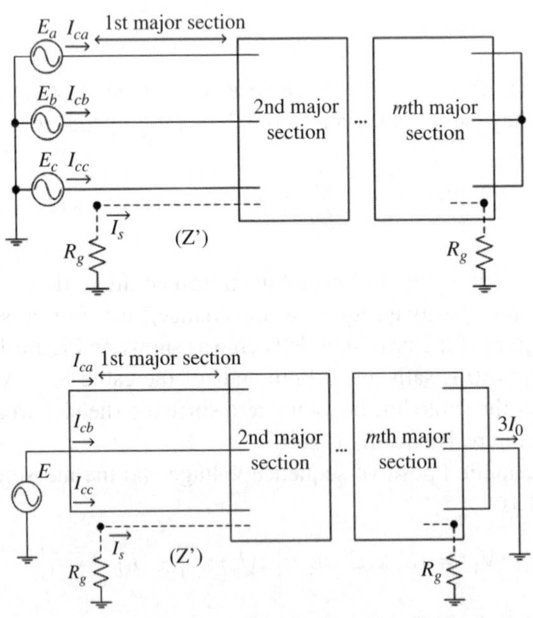

Figure 4.6 Setup for measuring sequence currents for a cross-bonded cable: (a) positive-sequence current; and (b) zero-sequence current. *Source*: © 2011 IPST. Reprinted, with permission, from [1]

Solving (4.38) for I_{ca}, I_{cb}, and I_{cc} yields (4.39).

$$\begin{pmatrix} E \\ \alpha^2 E \\ \alpha E \end{pmatrix} = \begin{bmatrix} Z_{11} & Z_{12} & Z_{13} \\ Z_{12} & Z_{22} & Z_{12} \\ Z_{13} & Z_{12} & Z_{11} \end{bmatrix} \begin{pmatrix} I_{ca} \\ I_{cb} \\ I_{cc} \end{pmatrix}$$

$$\therefore \begin{pmatrix} I_{ca} \\ I_{cb} \\ I_{cc} \end{pmatrix} = \begin{bmatrix} Z_{11} & Z_{12} & Z_{13} \\ Z_{12} & Z_{22} & Z_{12} \\ Z_{13} & Z_{12} & Z_{11} \end{bmatrix}^{-1} \begin{pmatrix} E \\ \alpha^2 E \\ \alpha E \end{pmatrix}$$

$$= \frac{1}{\Delta} \begin{bmatrix} Z_{11}Z_{22} - Z_{12}^2 & Z_{12}\left(Z_{13} - Z_{11}\right) & Z_{12}^2 - Z_{13}Z_{22} \\ Z_{12}(Z_{13} - Z_{11}) & Z_{11}^2 - Z_{13}^2 & Z_{12}(Z_{13} - Z_{11}) \\ Z_{12}^2 - Z_{13}Z_{22} & Z_{12}(Z_{13} - Z_{11}) & Z_{11}Z_{22} - Z_{12}^2 \end{bmatrix} \begin{pmatrix} E \\ \alpha^2 E \\ \alpha E \end{pmatrix} \tag{4.39}$$

Here,

$$Z_{11} = Z_{cc11} - Z_{14}'^2 / Z_{ss}'$$

$$Z_{22} = Z_{cc22} - Z_{24}'^2 / Z_{ss}'$$

$$Z_{12} = Z_{cc12} - Z_{14}' Z_{24}' / Z_{ss}'$$

$$Z_{13} = Z_{cc12} - Z_{14}'^2 / Z_{ss}'$$

The positive-sequence current is derived from (4.39).

$$I_1 = \frac{1}{3}(I_{ca} + \alpha I_{cb} + \alpha^2 I_{cc})$$

$$= \frac{E}{3\Delta_2}\{(Z_{11} - Z_{13})(Z_{11} + Z_{13} + 2Z_{12}) + Z_{22}(2Z_{11} + Z_{13}) - 3Z_{12}^2\} \tag{4.40}$$

where

$$\Delta_2 = (Z_{11} - Z_{13})\{Z_{22}(Z_{11} + Z_{13}) - 2Z_{12}^2\}$$

The positive-sequence impedance is

$$Z_1 = \frac{3\Delta_2}{(Z_{11} - Z_{13})(Z_{11} + Z_{13} + 2Z_{12}) + Z_{22}(2Z_{11} + Z_{13}) - 3Z_{12}^2} \tag{4.41}$$

Next, in order to find the zero-sequence impedance, we use a setup for measuring zero-sequence currents for a cross-bonded cable as shown in Figure 4.6b. In the figure, the applied zero-sequence voltage and the measured zero-sequence current are expressed as:

$$(V_0) = \left[E \ E \ E \ V_s\right]_t \tag{4.42a}$$

$$(I_0) = \left[I_{ca} \ I_{cb} \ I_{cc} \ I_s\right]_t \tag{4.42b}$$

Assuming the grounding resistance at substations R_g, the sheath voltage drop V_s between both ends can be found by:

$$V_s = -2R_g I_s \tag{4.43}$$

From (4.42) and (4.43), the following equation should be solved to find the zero-sequence impedance.

$$\begin{pmatrix} I_{ca} \\ I_{cb} \\ I_{cc} \\ I_s \end{pmatrix} = [Z']^{-1} \begin{pmatrix} E \\ E \\ E \\ -2RgI_s \end{pmatrix} \tag{4.44}$$

The following equations are obtained by solving (4.44).

$$I_{ca} = I_{cc} = (Z_{22} - Z_{12})E/\Delta_0 \tag{4.45a}$$

$$I_{cb} = (Z_{11} - Z_{21})E/\Delta_0 \tag{4.45b}$$

where

$$\begin{aligned}
\Delta_0 &= Z_{11}Z_{22} - Z_{12}Z_{21} \\
Z_{11} &= Z_{cc11} + Z_{cc13} - 2Z_{14}'^2/Z_{SR}' \\
Z_{22} &= Z_{cc22} - Z_{24}'^2/Z_{SR}' \\
Z_{12} &= Z_{cc12} - Z_{14}'Z_{24}'/Z_{SR}', \quad Z_{21} = 2Z_{12} \\
Z_{SR}' &= Z_{ss}' + 2Rg
\end{aligned}$$

The zero-sequence current can be found from (4.45a,b) in the following equation.

$$I_0 = \frac{2I_{ca} + I_{cb}}{3} = \frac{E}{3\Delta_0}(Z_{11} + 2Z_{22} - 2Z_{12} - Z_{21}) \tag{4.46}$$

The zero-sequence impedance is now obtained.

$$Z_0 = \frac{3\Delta_0}{Z_{11} + 2Z_{22} - 2Z_{12} - Z_{21}} \tag{4.47}$$

4.2 Electromagnetic Transients Program (EMTP) Cable Models for Transient Simulations

1. Dommel model [7, 8]
   Multi-phase distributed line model at a fixed frequency defined by a user.
2. PI-circuit model [7]
   Multi-phase lumped parameter circuit model.
3. Semlyen model [9]
   Multi-phase frequency-dependent line model.
   Frequency-dependent transformation matrices cannot be handled.
4. Marti model [10]
   Multi-phase frequency-dependent line model.
   Frequency-dependent transformation matrices cannot be handled.
5. Noda model [11, 12]
   Multi-phase frequency-dependent line model.
   Frequency-dependent transformation matrices are taken into account.

The impedance and admittance of a cable are calculated by the EMTP Cable Constants (CC) [8], and are processed to prepare the cable model for a transient simulation. The cable model data are prepared in the CC for Dommel and PI-circuit models, and are given as an output of the CC. For the frequency-dependent models, all the necessary parameters of a given cable are transferred from the CC to subroutine Semlyen, Marti and Noda Setups of the EMTP. The Semlyen Setup, for example, calculates a step response of the cable based on the data transferred from the CC. Then, it carries out curve fitting of the step response by exponential functions in the time domain. The parameters of the fitted response are transformed into the frequency domain for recursive convolution in a time-domain transient simulation. The output of the Semlyen Setup is the cable model data for the transient simulation. In the Marti model, polynomials of frequency functions are used. Both the Semlyen and Marti cable model data are prepared in the modal domain, and thus those necessitate a transformation matrix for a transient simulation. The weakness of the models is that the transformation matrix is given at a fixed frequency, that is, the matrix is not frequency-dependent. Thus, the accuracy of a transient simulation by the Semlyen and Marti models cannot be guaranteed for a transient where the frequency dependence of the transformation matrix is significant. A typical example is a fault surge and a transient on a long cable.

The Noda model is in the phase domain. Thus, there is no need to worry about the frequency-dependent transformation matrix. However, it requires a number of curve fitting and recursive convolutions, because a phase-domain simulation necessitates handling an $n \times n$ full matrix for an n-phase line or cable. Therefore, it involves a numerical instability problem.

It should be remembered that there are a number of sophisticated methods for dealing with a frequency-dependent line/cable [13–15]. Some of the methods, introduced in Section 4.6, have been implemented into the EMTP-RV [16], PSCAD [17] and XTAP [18]. These latest frequency-dependent models show a higher accuracy than that of the conventional frequency-dependent models explained above.

4.3 Dommel Model

This model was originally called the Schnyder–Bergeron model, because its theory came from Schnyder's paper in 1929 [19] and Bergeron's book [20]. The theory originated in the dynamic hydraulic phenomenon called water hammer, and the details can be found in Parmakian's book [21]. The water hammer theory was first applied to an electrical transient analysis by Frey and Althammer in 1961 [22]. Then, Dommel generalized the theory for a transient analysis in a power system [23, 24], and developed an original version of the EMTP [25]. One of the basic components of this original EMTP is a frequency-independent distributed line model often called the Bergeron model. In this book, the model is called the Dommel model for the reason explained above.

The Dommel model is given as a combination of a loss-less distributed line of its surge impedance "Z_0" and velocity "c" with resistances representing attenuation as illustrated in Figure 4.7. All the parameters of the model are calculated at a fixed frequency which is given in most cases as a dominant transient frequency f_t of a line with length x [7].

$$\text{open-circuited line } f_t = 1/4\tau \tag{4.48a}$$

$$\text{short-circuited line } f_t = 1/2\tau \tag{4.48b}$$

$$\text{impedance-terminated line } f_t \fallingdotseq 1/3\tau \tag{4.48c}$$

where $\tau = x/c$ is the traveling time and c is the propagation velocity at f_t.

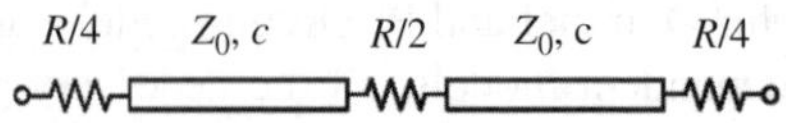

Figure 4.7 Dommel's distributed line model

All the parameters, Z_0, c and R, are obtained from Line Constants or CC routine of the EMTP [26]. In fact, the Dommel model for a transient simulation in the EMTP is automatically output by specifying the output as the Line Constants or CC input data.

In a high frequency region, the parameters are evaluated approximately by hand using the following equation.

(a) Overhead line

$$Z_0 = 60\ln(2h/r), \tag{4.49a}$$

$$c = c_0 = 300\ \mathrm{m}/\mu\mathrm{s} \tag{4.49b}$$

$$R = (1/2\pi r)\sqrt{\omega\mu_0\rho_c/2} \tag{4.49c}$$

where h is the conductor height, r is the radius, and ρ_c is the resistivity.

(b) Underground cable – coaxial mode

$$Z_0 = \left(60/\sqrt{\varepsilon_r}\right)\ln(r_0/r_i) \tag{4.50a}$$

$$c = c_0/\sqrt{\varepsilon_r} \tag{4.50b}$$

$$R = (1/2\pi r_0)\sqrt{\omega\mu_0\rho_c/2} \tag{4.50c}$$

where r_0, r_i are the outer and inner radii of the core, respectively, ρ_c is the core resistivity, and ε_r is the insulator relative permittivity.

Remember that there exists the following relation between the attenuation constant α and the corresponding resistance R.

$$\alpha = R/2Z_0 \tag{4.51}$$

4.4 Semlyen Frequency-dependent Model

In a time-domain transient analysis program such as the EMTP, inclusion of the frequency-dependent effect of a distributed line requires a real-time convolution which is very time consuming and necessitates large computer memories. Remember that computer CPU time and memories in the 1970s and 1980s were the most significant limitation to produce a computer program. To avoid the problems, recursive convolution was proposed [9, 27]. The real-time convolution is defined as:

$$\begin{aligned} f_t &= L^{-1}S(s)\cdot E(s) = L^{-1}[s\cdot\exp\{-\Gamma(s)\cdot x/s\}\cdot E(s)] \\ &= (d/dt)\int_0^t s(\tau)\cdot e(t-\tau)\cdot d\tau \end{aligned}$$

$$= s(0) \cdot e(t) + \int_0^t s'(\tau) \cdot e(t-\tau) \cdot d\tau$$

$$= s(0) \cdot e(nt_0) + \sum_{k=0}^{K-1} s'(\tau) \cdot e(nt_0 - k\tau_0) \cdot \tau_0$$

$$= s(t) * e(t) \tag{4.52}$$

where

$$e(t) = L^{-1}E(s)$$

$$s(t) = L^{-1}S(s), \quad S(s) = \exp\{-\Gamma(s) \cdot x/s\} \tag{4.53}$$

$*$ is the real-time convolution
L^{-1} is the Laplace inverse transform
$K = \tau/\tau_0$, τ_0 is the time step of numerical convolution

The numerical calculation of (4.52) requires a number of multiplications together with large memories.

4.4.1 Semlyen Model

Assume that step response $s(t)$ is expressed by:

$$s(t) = \beta\{1 - \exp(-\alpha t)\}/\alpha = m\{1 - \exp(-\alpha t)\} \tag{4.54}$$

Also, $e(t)$ is assumed to be represented by a quadratic curve.
Substituting the above equation into (4.52),

$$f(t) = (d/dt) \int_0^t (b/a) \cdot \{1 - \exp(-aZ)\} \cdot e(t-\tau) \cdot d\tau$$

$$= as(t - t_0) + be(t) + ce(t - t_0) + de(t - 2t_0) \tag{4.55}$$

where

$$a = \exp(-\alpha t_0),\ m = \beta/\alpha$$

$$b = m\{1 - \alpha/(\alpha t_0)^2 - (3 - \alpha)/(2\alpha t_0)/ + 1\}$$

$$c = m\{-2(1 - \alpha)/(\alpha t_0)^2 + 2/\alpha t_0 - \alpha\}$$

$$d = m\{(1 - \alpha)/(\alpha t_0)^2 - (1 + \alpha)/(2\alpha t_0)\}$$

t_0 is the time step of convolution

The above equation shows that the real-time convolution is calculated as a sum of four multiplications. This is the basic principle of the recursive convolution. In practice, it is impossible to approximate the step function $s(t)$ only by two exponential functions as in (4.54), a few exponential functions are necessary. The accuracy of (4.55) in comparison with its original form, (4.52), is dependent on the number of the exponential functions. Then, the calculation of the coefficients a, b, … in (4.55) becomes hard, and it occasionally involves a numerical instability.

The frequency dependence of a characteristic impedance is included in the same manner as the step response of the propagation constant.

4.4.2 Linear Model

When $s(t)$ is approximated by linear functions as:

$$s(t) = a_i t + b_i \text{ for } t_{i-1} \leq t < t_i \tag{4.56}$$

where $t = 0$ for $i = 1$, $i = 1, 2, \ldots$

Then, (4.52) is rewritten as [27]:

$$f_t = s(0) \cdot e(t) + \sum_{i=0}^{I} A_i \cdot G_i(t) \tag{4.57}$$

where

$$\begin{aligned} A_i &= a_i \tau_0 \\ G_i(t) &= (1/\tau_0) \int_{t_i-1}^{t_i} e(t-\tau) = \sum_{m=i_{m-1}}^{i_m} e(t - m\tau_0) \\ i_m &= t_i/\tau_0 \end{aligned} \tag{4.58}$$

In the above equation, $G_i(t)$ is stored as a past history of $e(t)$, it is obtained by only addition of the present value to the past history. A comparison of (4.57) with (4.52) indicates that the computation time of (4.57) is far smaller than that of (4.52) if I/K is set very small.

4.5 Marti Model

In (4.54), step function $s(t)$ is represented in the time domain. It is quite possible to approximate the original frequency response $S(s)$ in (4.53) by a reactance function explained in a textbook of an a.c. circuit theory. Since step response $s(t)$ is a monotonously increasing smooth function, $S(s)$ is represented approximately by a CR ladder circuit where the reactance function is given in the s-domain by:

$$S(s) = \sum_{i=1}^{m} (a_i/s + b_i) \tag{4.59}$$

Then, the time response $s(t)$ is evaluated by Laplace transform.

$$s(t) = \sum a_i \cdot \exp(-b_i t) \tag{4.60}$$

Instead of performing curve fitting of $s(t)$ in the time domain in the Semleyen model, it is performed in the frequency (s) domain in the Marti model. Having obtained the s(t) as a series of exponential functions, the recursive convolution is carried out simlarly to (4.55).

4.6 Latest Frequency-dependent Models

4.6.1 Vector Fitting

The frequency-dependent model using the vector fitting is described in detail in References [12, 28–30]. This section summarizes the outline of the model and how the model works in transient simulations.

In the frequency-dependent model using the vector fitting, line and cable parameters are first calculated in the frequency domain and then fitted directly in the phase domain as originated by Nakanishi and Ametani in 1986 [28] and followed by Noda *et al.* [12]. Hence, the model does not involve the modal transformation required for the Dommel, Semlyen and Marti models [8–10]. Instead, it requires the parameter fitting process which is called vector fitting.

The Marti model is one of the frequency-dependent models, but its modal transformation matrix is derived at one frequency, assuming that the frequency dependence of the modal transformation matrix is not significant. The assumption is true for horizontal overhead lines but does not hold for untransposed vertical overhead lines and cables [7]. This does not mean that simulation results of a cable line using the Marti model or the Dommel model contain measurable errors. It highly depends on frequency components contained in the transients. The frequency-dependent model using the vector fitting is often recommended as a model that can be used universally for the simulation of various cable system transients.

As described in the previous section, cable parameters are expressed by the propagation constant matrix $S(\omega)$ and the characteristic admittance matrix $Y_0(\omega)$, which are both frequency-dependent. The propagation matrix is the exponential function of the propagation constant $\Gamma(\omega)$. From (3.94) and (3.42) in Chapter 3, these cable parameter matrices are expressed as:

$$S(\omega) = \exp\{-\Gamma(\omega) \cdot x\} \tag{4.61a}$$

$$Y_0(\omega) = Z(\omega)^{-1} \cdot \Gamma(\omega) \tag{4.61b}$$

where $\Gamma(\omega) = \{Y(\omega) \cdot Z(\omega)\}^{1/2}$ and x is the cable length.

The purpose of the vector fitting is to approximate each entry of $S(\omega)$ and $Y_0(\omega)$ in the following form:

$$h(s) \cong \sum_{m=1}^{N} \frac{r_m}{s - a_m} + d + se \tag{4.62}$$

where $h(s)$ is each entry of $S(\omega)$ and $Y_0(\omega)$, N is the number of poles, r_m is the residue for the mth pole, a_m is the mth pole, and d and e are fitting parameters.

Since each entry of the cable parameter matrices is expressed by rational functions in (4.62), convolutions, required for the conversion from the frequency domain to the time domain, become recursive convolutions, which makes the computation significantly efficient [29].

In order to achieve the fitting, the vector fitting changes the fitting problem (4.62) to (4.63) by introducing the new fitting parameter ($\widetilde{r_m}$). As discussed below, (4.63) has an advantage since the set of poles can be updated systematically [28].

$$\left(\sum_{m=1}^{N} \frac{\widetilde{r_m}}{s - a_m} + 1\right) h(s) \cong \sum_{m=1}^{N} \frac{r_m}{s - a_m} + d + se \tag{4.63}$$

First, an initial set of poles (a_m) is given for (4.63). For one frequency point s_k, (4.63) can be written as:

$$\sum_{m=1}^{N} \frac{r_m}{s_k - a_m} + d + s_k e - \sum_{m=1}^{N} \frac{\tilde{r}_m}{s_k - a_m} h(s_k) = h(s_k) \tag{4.64}$$

Here, unknown fitting parameters are ($\tilde{r}_m$), (r_m), d, and e, and known parameters are (a_m) and $h(s_k)$. When we fit parameters using N_ω frequency points, N_ω equations are obtained from (4.64).

$$\sum_{m=1}^{N} \frac{r_m}{s_1 - a_m} + d + s_1 e - \sum_{m=1}^{N} \frac{\tilde{r}_m}{s_1 - a_m} h(s_1) = h(s_1) \tag{4.65a}$$

$$\vdots$$

$$\sum_{m=1}^{N} \frac{r_m}{s_{N_\omega} - a_m} + d + s_{N_\omega} e - \sum_{m=1}^{N} \frac{\tilde{r}_m}{s_{N_\omega} - a_m} h(s_{N_\omega}) = h(s_{N_\omega}) \tag{4.65b}$$

In a matrix form,

$$Ax = b \tag{4.66}$$

where

$$A = \begin{bmatrix} \frac{1}{s_1 - a_1} & \cdots & \frac{1}{s_1 - a_N} & 1 & s_1 & \frac{-h(s_1)}{s_1 - a_1} & \cdots & \frac{-h(s_1)}{s_1 - a_N} \\ & & & \vdots & & & & \\ \frac{1}{s_{N_\omega} - a_1} & \cdots & \frac{1}{s_{N_\omega} - a_N} & 1 & s_{N_\omega} & \frac{-h(s_{N_\omega})}{s_{N_\omega} - a_1} & \cdots & \frac{-h(s_{N_\omega})}{s_{N_\omega} - a_N} \end{bmatrix}$$

$$x = \begin{bmatrix} r_1 & \cdots & r_N & d & e & \tilde{r}_1 & \cdots & \tilde{r}_N \end{bmatrix}^T$$

$$b = \begin{bmatrix} h(s_1) & \cdots & h(s_{N_\omega}) \end{bmatrix}$$

Since the number of frequency points N_ω is always larger than the number of unknown fitting parameters, (4.66) becomes the overdetermined linear system. With the given set of poles (a_m), unknown fitting parameters ($\tilde{r}_m$), (r_m), d, and e are determined by solving the least squares problem given from (4.66).

Equation (4.66) needs to be solved iteratively with an updated set of poles (a_m) until the set of residues ($\tilde{r}_m$) becomes close to zero. Since poles of $h(s)$ have to be equal to zeros of $\left(\sum_{m=1}^{N} \frac{\tilde{r}_m}{s - a_m} + 1\right)$, the set of poles ($a_m$) can be updated by eigenvalues of the following matrix [28–30]. This process is called pole relocation.

$$(a_m) = \mathrm{eig}\begin{bmatrix} a_1 - \tilde{r}_1 & -\tilde{r}_2 & \cdots & -\tilde{r}_N \\ -\tilde{r}_1 & a_2 - \tilde{r}_2 & \cdots & -\tilde{r}_N \\ \vdots & \vdots & \ddots & \vdots \\ -\tilde{r}_1 & -\tilde{r}_2 & \cdots & a_N - \tilde{r}_N \end{bmatrix} \tag{4.67}$$

Once ($\tilde{r}_m$) $\approx$ (0) is satisfied, the final set of poles (a_m) is derived by (4.67). Using this final set of poles, fitting parameters (r_m), d, and e are found from (4.62) using the least squares approximation. This iterative process of the vector fitting is illustrated in Figure 4.8.

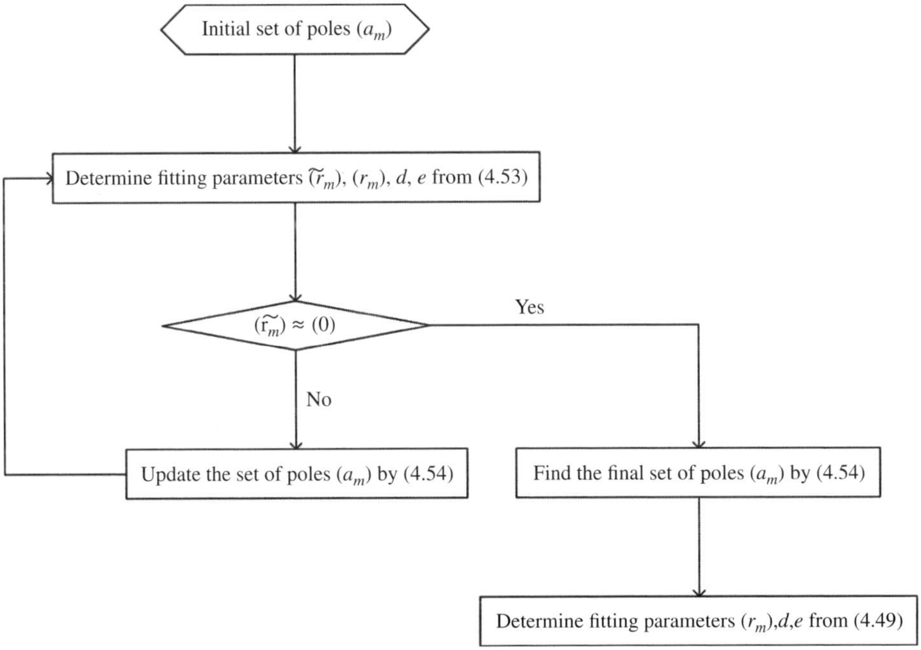

Figure 4.8 Procedure for the vector fitting

4.6.2 Frequency Region Partitioning Algorithm

Another method to find fitting parameters in (4.62) is called the frequency region partitioning algorithm as it divides the fitting frequency range into multiple regions [15, 32, 33]. The frequency region partitioning algorithm identifies poles in (4.62) using the traces of $\Gamma(\omega)$ and $Y_0(\omega)$ since their traces are expected to contain information of all poles evenly. For example, in order to identify poles of $\Gamma(\omega)$, the trace of $\Gamma(\omega)$ is approximated in the following rational function model.

$$f(s) \cong \sum_{i=1}^{L} \Gamma_{ii}(\omega) = \frac{b_0 + b_1 s + \cdots + b_{N-1} s^{N-1}}{1 + a_1 s + \cdots + a_{N-1} s^{N-1} + a_N s^N} \tag{4.68}$$

where $f(s)$ is the trace of $\Gamma(\omega)$, N is the number of poles, and a_m and b_m are fitting parameters.
Equation (4.68) is expressed as:

$$(1 + a_1 s + \cdots + a_{N-1} s^{N-1} + a_N s^N) f(s) = b_0 + b_1 s + \cdots + b_{N-1} s^{N-1} \tag{4.69}$$

When we fit parameters using N_ω frequency points, (4.69) can be expressed in the following matrix form.

$$Ax = b \tag{4.70}$$

where

$$A = \begin{bmatrix} 1 & -s_1 f(s_1) & s_1 & -s_1{}^2 f(s_1) & s_1{}^2 & \cdots & s_1{}^{N-1} & -s_1{}^N f(s_1) \\ & & & \vdots & & & & \\ 1 & -s_{N_\omega} f(s_{N_\omega}) & s_{N_\omega} & -s_{N_\omega}{}^2 f(s_{N_\omega}) & s_{N_\omega}{}^2 & \cdots & s_{N_\omega}{}^{N-1} & -s_{N_\omega}{}^N f(s_{N_\omega}) \end{bmatrix}$$

$$x = \begin{bmatrix} b_0 & a_1 & b_1 & a_2 & b_2 & \cdots & b_{N-1} & a_N \end{bmatrix}^T$$

$$b = \begin{bmatrix} f(s_1) & \cdots & f(s_{N_\omega}) \end{bmatrix}$$

Similarly to the vector fitting, the number of frequency points N_ω is always larger than the number of unknown fitting parameters, (4.66) becomes the overdetermined linear system. The fitting parameters a_m and b_m are found as the solutions of the least squares problem.

The frequency region partitioning algorithm solves the least squares problem by the QR decomposition with Householder transformations after applying the adaptive weighting and the column scaling [15]. Equation (4.70) is solved repeatedly, changing the weighting function, until the solution reaches the desired accuracy.

Once fitting parameters a_m and b_m in (4.68) are found, the transformation to the partial fraction expansion (PFE) model in (4.62) is straightforward. The poles of (4.62) are obtained as the eigenvalues of the companion matrix of the polynomial $1 + a_1 s + \cdots + a_{N-1}s^{N-1} + a_N s^N$, that is

$$\text{Poles (PFE model)} = \text{eig}\left(\frac{1}{a_N} \begin{bmatrix} 0 & 0 & \cdots & 0 & -1 \\ 1 & 0 & \cdots & 0 & -a_1 \\ 0 & 1 & \cdots & 0 & -a_2 \\ \vdots & \vdots & \ddots & \vdots & \vdots \\ 0 & 0 & \cdots & 1 & -a_{N-1} \end{bmatrix} \right) \tag{4.71}$$

In (4.70), the lower limit of the fitting range (s_1) is normally set as low as 0.01 Hz, and the higher limit of the fitting range (s_{N_ω}) is normally set as high as 1 MHz. As a result, matrix A contains both small values and large values, whose gaps can reach eight digits. This makes matrix A ill-conditioned for solving (4.70).

As a countermeasure to this problem, the frequency region partitioning algorithm divides the fitting frequency range into multiple regions when the desired accuracy cannot be achieved [33]. In the initial try, the fitting frequency range is not divided, and (4.70) is solved for the entire frequency range. When the desired accuracy cannot be achieved, the entire frequency range is divided into two, and (4.70) is solved separately for two frequency ranges. If the desired accuracy cannot be achieved in one of two frequency ranges, this frequency range is further divided into two frequency ranges. This division of the frequency range is repeated until the desired accuracy is achieved in all frequency ranges.

When a frequency range is divided into two frequency ranges, these two neighboring frequency ranges should not identify the same poles. In order to avoid it, the boundary of the two neighboring frequency ranges should be drawn at a local minimum of the magnitude of $\Gamma(\omega)$ and $Y_0(\omega)$. The frequency region partitioning algorithm places the boundary at the local minimum closest to the center of the frequency range to be divided.

References

[1] Ohno, T., Ametani, A., and Bak, C. L. (2011) Derivation of theoretical formulas of sequence currents on underground cable systems. Proceedings of the International Conference on Power Systems Transients (IPST), Delft, The Netherlands, Paper # 245.

[2] Ametani, A. and Ohno, T. (2011) Derivation of theoretical formulas of sequence currents on underground cable systems. *IEE Japan Transactions*, **B-131** (3), 277–82.
[3] Ohno, T. (2012) Dynamic study on the 400 kV 60 km Kyndbyværket–Asænsvrket line. PhD thesis. Aalborg University.
[4] CIGRE WG B1.30 (2013) Cable Systems Electrical Characteristics. *CIGRE Technical Brochure* 531.
[5] Nagaoka N. and Ametani, A. (1983) Transient calculations on crossbonded cables. *IEEE Transactions on Power Apparatus and Systems*, **PAS-102** (4), 779–87.
[6] Ametani, A., Miyamoto, Y., and Nagaoka, N. (2003) An investigation of a wave propagation characteristic on a crossbonded cable. *IEE Japan Transactions PE*, **123** (3), 395–401 (in Japanese).
[7] Ametani, A. (1990) *Distributed-Parameter Circuit Theory*, Corona Publishing. Co. (in Japanese).
[8] Dommel H.W. (1986) *EMTP Theory Book*, Bonneville Power Administration.
[9] Semlyen, A. and Dabuleanu, A. (1975) Fast and accurate switching transient calculations on transmission lines with ground return using recursive convolutions. *IEEE Transactions on Power Apparatus and Systems*, **94**, 561–71.
[10] Marti, J. R. (1982) Accurate modeling of frequency-dependent transmission line transients. *IEEE Transactions on Power Apparatus and Systems*, **PAS-101** (1), 147–55.
[11] Noda, T. (1996) Development of a transmission line model considering the skin and corona effects for power system transient analysis. PhD thesis. Doshisha University.
[12] Noda, T., Nagaoka, N., and Ametani, A. (1996) Phase domain modeling of frequency-dependent transmission lines by means of an ARMA model. *IEEE Transactions on Power Delivery*, **11** (1), 401–11.
[13] Morched, A., Gustavsen, B., and Tartibi, M. (1997) A universal line model for accurate calculation of electromagnetic transients on overhead lines and cables. *IEEE*, Paper PE-112-PWRD-0-11-1997.
[14] Gustavsen, B and Semlyen, A. (1998) Combined phase and modal domain calculation of transmission line transients based on vector fitting. *IEEE Transactions on Power Delivery*, **13**, 596–604.
[15] Noda, T. (2005) Identification of a multiphase network equivalent for electromagnetic transient calculations using partitioned frequency response. *IEEE Transactions on Power Delivery*, **20** (2), 1134–42.
[16] EMTP Development Coordination Group and Electric Power Research Institute (1990) EMTP Ver. 2.0 Revised Code Documentation.
[17] Manitoba HVDC Research Center. http://pscad.com/ (accessed 24 January 2015).
[18] Central Research Institute of Electric Power Industry. https://www.xtap.jp/ (accessed 24 January 2015).
[19] Schnyder, O. (1929) Druckstosse in Pumpensteigleitungen. *Schweizerische Bauzeitung*, **92** (22), 271–3/(23), 283–6.
[20] Bergeron, L. (1961) *Water Hammer in Hydraulics and Wave Surges in Electricity*, John Wiley & Sons, Ltd.
[21] Parmakian, J. (1963) *Waterhammer Analysis*, Dover.
[22] Frey, W. and Althammer, P. (1961) The calculation of electromagnetic transients on lines by means of a digital computer. *Brown Boveri Review*, **48** (5/6), 344–55.
[23] Dommel, H.W. (1969) Digital computer solution of electromagnetic transients in single and multi-phase networks. *IEEE Transactions on Power Apparatus and Systems*, **88** (4), 388–99.
[24] Dommel, H.W. and Scott-Meyer, W. (1974) Computation of electromagnetic transients. *Proceedings of the IEEE*, **62**, 983–93.
[25] Dommel, H. W. (1976) *Transient Program Users Manual*, University of British Columbia.
[26] Scott-Meyer, W. (1980) *EMTP Rule Book*, Bonneville Power Administration.
[27] Ametani, A. (1976) A highly efficient method for calculating transmission line transients. *IEEE Transactions on Power Apparatus and Systems*, **95** (5), 1545–51.
[28] Gustavsen, B and Semlyen, A. (1999) Rational approximation of frequency domain responses by vector fitting. *IEEE Transactions on Power Delivery*, **14** (3), 1052–61.
[29] Gustavsen, B. (2006) Improving the pole relocating properties of vector fitting. *IEEE Transactions on Power Delivery*, **21** (3), 1587–92.
[30] Deschrijver, D., Mrozowski, M., Dhaene, T., and De Zutter, D. (2008) Macromodeling of multiport systems using a fast implementation of the vector fitting method. *IEEE Microwave and Wireless Components Letters*, **18** (6), 383–5.
[31] Nakanishi, H. and Ametani, A. (1986) Transient calculation of a transmission line using superposition law. *IEE Proceedings*, **133**, Pt C (5), 263–9.
[32] Noda, T. (2007) A binary frequency-region partitioning algorithm for the identification of a multiphase network equivalent for EMT studies. *IEEE Transactions on Power Delivery*, **22** (2), 1257–8.
[33] Noda, T. (2008) A Transmission Line Model for Electromagnetic Transient Analysis Based on the Frequency Region Partitioning Algorithm. *Electric Power Engineering Research Laboratory Report* H07005.

5

Basic Characteristics of Transients on Single-phase Cables

Akihiro Ametani
Graduate School of Electrical Engineering, Doshisha University, Kyoto, Japan

5.1 Single-core Coaxial (SC) Cable

Basic characteristics of transients on a single-phase SC cable are explained in this section based on experimental and electromagnetic transients program (EMTP) simulation results with various terminating conditions of the core and the sheath. Each case is analyzed theoretically by adopting a traveling-wave theory.

5.1.1 Experimental Observations

5.1.1.1 Experimental Circuit

Figure 5.1a illustrates an experimental setup to measure wave propagation on a 3D2V coaxial cable with a length of 90 m. The cross section is shown in Figure 5.1b, and the geometrical and physical parameters are given in Table 5.1. The cable is laid on a building floor and is folded at the center as in Figure 5.1c so that the sending and receiving ends are placed nearby on an Al plate to measure voltages. This solves problems of power source and oscilloscope grounding, simultaneous measurements of voltages and currents at the sending and receiving ends of the cable, and also voltage reference for the voltage measurement. It is not well understood that there is no power source, no voltage reference and no terminal to ground the power source, the impulse generator and the oscilloscope in a field, for example, in a construction site of a substation and a transmission line. A pulse-like voltage as in Figure 5.1d is applied to the cable core from a pulse generator (PG) of which the source impedance is 68 Ω. The voltage at the core sending end is kept to be about 10 V by adjusting the output of the PG. Experimental conditions and measured voltages are summarized in Table 5.2.

Cable System Transients: Theory, Modeling and Simulation, First Edition.
Akihiro Ametani, Teruo Ohno and Naoto Nagaoka.

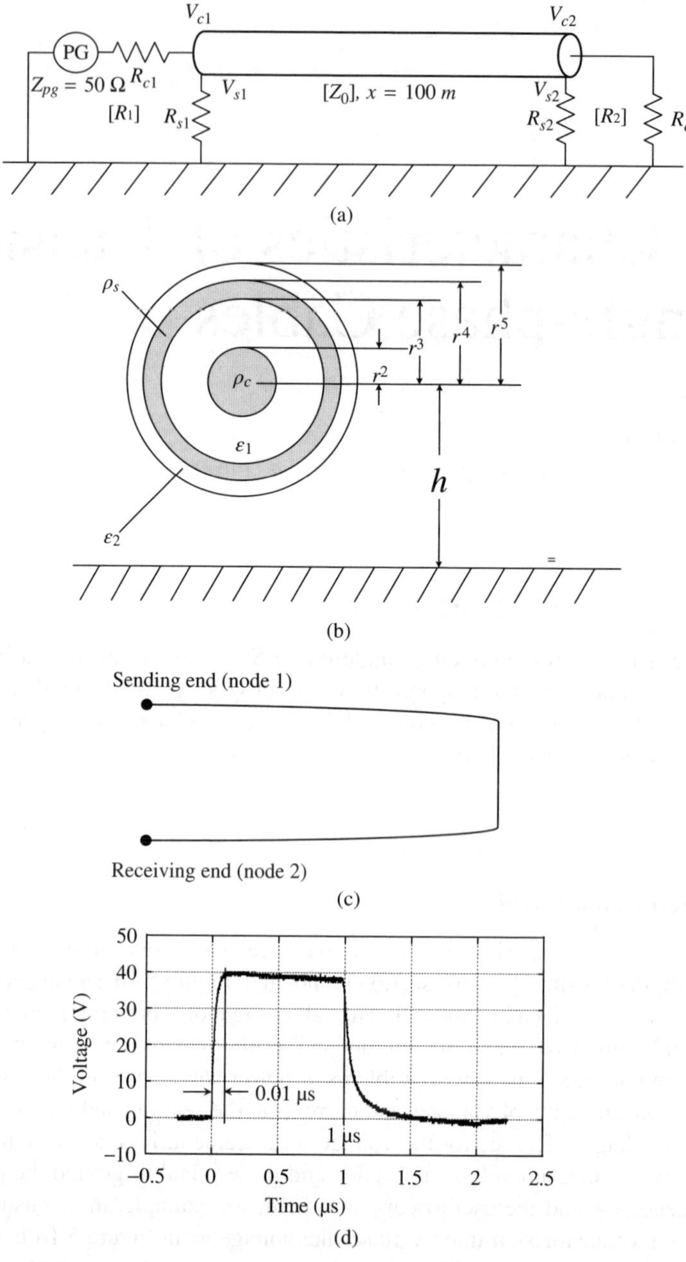

Figure 5.1 Experimental setup: (a) experimental circuit; (b) cable cross-section; (c) cable layout; and (d) applied voltage

Table 5.1 Cable parameters

$r_1 = 0, r_2 = 0.48, r_3 = 1.5, r_4 = 2.3, r_5 = 2.75$ mm
$\rho_c = 1.6 \times 10^{-8},\ \rho_c = 2.0 \times 10^{-8}\ \Omega\text{m}, \varepsilon_1 = 2.3,\ \varepsilon_2 = 5$
$h_1 = r_5 = 2.75\,\text{mm}, \rho_e = 50\,\Omega\text{m}$

Table 5.2 Experimental conditions and measured and simulated voltages at initial peaks

Case *ij-kl*	Terminating resistance (Ω)				Measured voltages of initial peak (V)				Ratio to V_{c1} (pu)					
									Measured			EMTP simulations		
	R_{c1}	R_{c2}	R_{s1}	R_{s2}	V_{c1}	V_{c2}	V_{s1}	V_{s2}	V_{c2}	V_{s1}	V_{s2}	V_{c2}	V_{s1}	V_{s2}
11–11	0	0	0	0	10	0	0	0	0	0	0	0	0	0
11–22	0	0	150	150	10	0	5.8	−3.4	0	0.6	−0.3	0	0.6	−0.4
11–33	0	0	10 k	10 k	10	0	7.4	−2.9	0	0.7	−0.3	0	0.7	−0.3
12–11	0	200	0	0	9.2	13.2	0	0	1.4	0	0	1.4	0	0
12–12	0	200	0	150	9.4	10.8	0	−4	1.1	0	−0.4	1.1	0	−0.4
12–13	0	200	0	10 k	9.2	8.8	0	−6.6	0.9	0	−0.7	0.9	0	−0.7
12–22	0	200	150	150	9.8	8	5.8	2.4	0.8	0.6	0.2	0.7	0.6	0.2
12–23	0	200	150	10 k	9.6	9.2	5.8	3.6	1	0.6	0.4	0.8	0.6	0.3
12–33	0	200	10 k	10 k	9.4	9	7	6.2	1	0.7	0.6	0.8	0.8	0.5
22–11	200	200	0	0	2.4	3.6	0	0	1.5	0	0	1.4	0	0
22–12	200	200	0	150	2.4	2.4	0	−1.4	1	0	−0.6	1.1	0	−0.4
22–13	200	200	0	10 k	2.4	2.4	0	−1.5	1	0	−0.6	0.9	0	−0.7
22–22	200	200	150	150	4	3.4	2.4	0.9	0.9	0.6	0.2	0.7	0.6	0.2
22–23	200	200	150	10 k	4	4	2.4	1.4	1	0.6	0.4	0.8	0.6	0.3
22–33	200	200	10 k	10 k	5	5	3.8	3.3	1	0.8	0.6	0.8	0.7	0.5
23–11	200	10 k	0	0	2.4	4.2	0	0	1.8	0	0	0.8	0	0
23–22	200	10 k	150	150	4	5.4	2.4	1.9	1.4	0.6	0.5	1.1	0.6	0.4
23–33	200	10 k	10 k	10 k	5	9.4	3.8	6.6	1.9	0.8	1.3	1.5	0.7	1.1

i for R_{c1}, *j* for R_{c2}, *k* for R_{s1}, *l* for R_{s2}.
i, *j* = 1: $R_c = 0\ \Omega$; 2: 200 Ω (core matching impedance to the earth); and 3: 10 kΩ.
k, *l* = 1: $R_s = 0\ \Omega$; 2: 150 Ω (sheath matching impedance); and 3: 10 kΩ.

5.1.1.2 Measured Results

Measured waveforms of transient voltages are shown in Figure 5.2 for Case *11-kl*, Figure 5.3 for Case *12-kl*, Figure 5.4 for Case *22-kl*, and Figure 5.5 for Case *23-kl*, where each code of Case *ij-kl* is explained in Table 5.2.

5.1.2 EMTP Simulations

5.1.2.1 Model Circuit

A model circuit for EMTP simulations is the same as in Figure 5.1a and b. Data 5.1 shows the output of the cable parameters. A distributed line model at frequency 1 MHz with no resistance [1, 2] is adopted in transient simulations so as to make a comparison with analytical observations in the following section straightforward and physical interpretation easy.

Data 5.1 Output of the Cable Parameters

```
1. −pase SC cable with 2 conductors. 0 grounded. Total # of conductors = 2
Cable on the surface of Eerth
Cable constants
Phase 1 Boundary radii; 1   0.00000E+00   2   4.80000E-04   3   1.50000E-03   4   2.30000E-03
5   2.75000E−03
                Resistivity (ohm-m): Core   1.67300000E-08,   Seath   2.00000000E-08,   Armor
9.99999994E−09
                Relative permeability;   1.00000000E+00,   1.00000000E+00,   1.00000000E+00
                Insulator:   1;  (Relative)  Permeability   1.0   Permittivity   2.30   Velocity  (m/s)
1.97677E+08
                Insulator:   2;  (Relative)  Permeability   1.0   Permittivity   5.00   Velocity  (m/s)
1.34071E+08
                DC resistance (ohm/m),   2.311.4E−02   2.09414E−03
        Distance from Earth Surface;   1            .0028
        Distance between phases:               .0000
        Earth ; resistivity 1.00000000E+02
    Punched cards: 1)pi-circuit (IPUNCH=1) or distributed line (PUNCH=2);
        IPUNCH=2   total length=9.00000000E+01
        Frequency 1.00000000E+06 Hz
        Total impedance [Zc]
          1.12620E+00   1.00580E+00
          1.15700E+01   1.00234E+01
          1.00580E+00   1.00580E+00
          1.00234E+01   1.00234E+01
    Total admittance [Yc]
          0.00000E+00   0.00000E+00
          7.05581E−04  −7.05581E−04
          0.00000E+00   0.00000E+00
         −7.05581E−04   1.18515E−03
    Characteristic impedance matrix [Zc] in phase varables, in [ohm]:
         191.60584          144.75318
          −9.06543           −7.24441
         144.75318          144.75319
          −7.24441           −7.24441
    Modal transformation matrices follow. These are complex, with the real part displayed above the
imaginary part.
    Voltage transformation matrix [Tv]. This gives the mapping from modal to phase variables: v-phase
[Tv] v-mode
          1.00000          1.00000
           .00000           .00000
          1.00000           .00000
           .00000           .00000
    Modal components
    Mode   Attenuation Velocity       Impedance (Ohm/m)                      Admittance (s/m)
       Charact. imp. (Ohm) Charac. Admit. (s)
    No.     (db/km)     (m/mic.s)     Real                 Imag.              Real                 Imag.
       Real     Imag.                 Real     Imag.
    1    3.01767E+01    90.51       1.00580E+00      1.00234E+01     0.00000E+00 4.79567E−04
    144.753                −7.244                        6.89105E−03  3.4487E−04
    2    1.11604E+01  190.06        1.20399E−01      1.54653E+00     0.00000E+00 7.05581E−04
    46.853            −1.821                              2.13113E−02  8.28302E−04
    Request for flushing of punch buffer.   |SPUN
```

```
C   <++++++> Cards punched by support routine on 05-Aug-05 17.48.11 <+++++>
C CABLE CONSTANTS
C CABLE PARAMETERS
C C TY] [SYS] [NPC] [EAR] [KMO] [ZFL] [YFL] [NPP] [NGD] [DAT] [IYG] [PUN]
C   2    0    1                     1    1                       2
C C NP] [NCR] [IRS] [MAJOR] [RSG]!
C C      1     0    0     500.     .000S
C C N1]
C   2
C C   r1     ][   r2    ][    r3    ][    r4    ][   r5    ]
C     0.0000   0.4800E−3   1.5000E−3   2.3000E−3   2.7500E−3
C C  rohc   ] [   muc    ] [   mul1    ] [   epsl1    ] [   rohs    ] [   mus    ] [   mul2   ]
[   epsl2
C   1.673E−8   1.0000   1.0000   2.3000   2.000E−8   1.0000   1.0000   5.0
C C   Vert1    ] [   Horiz1    ]
```

```
  2.7500E-3   0.0000
C C    rohe     ] ]      Freq.     ]  [DEC]  [PNT]  [DIST] [      IPUN     ]
C        100.00.              1.E6                              90.              2
SVINTAGE, 1
C   Branch cards for constant-parameter distributed:
-1                                         1.00580E+00   1.44753E+02   9.05113E+07-9.00000E+01   1   12
-2                                              1.20399E-01     4.68527E+01   1.90063E+08-9.00000E+01    1
12
          -.00000013       1.00000013
          -.00000010        .00000000
          1.00000013     -1.00000004
           .00000010        .00000007
SVINTAGE, 0
```

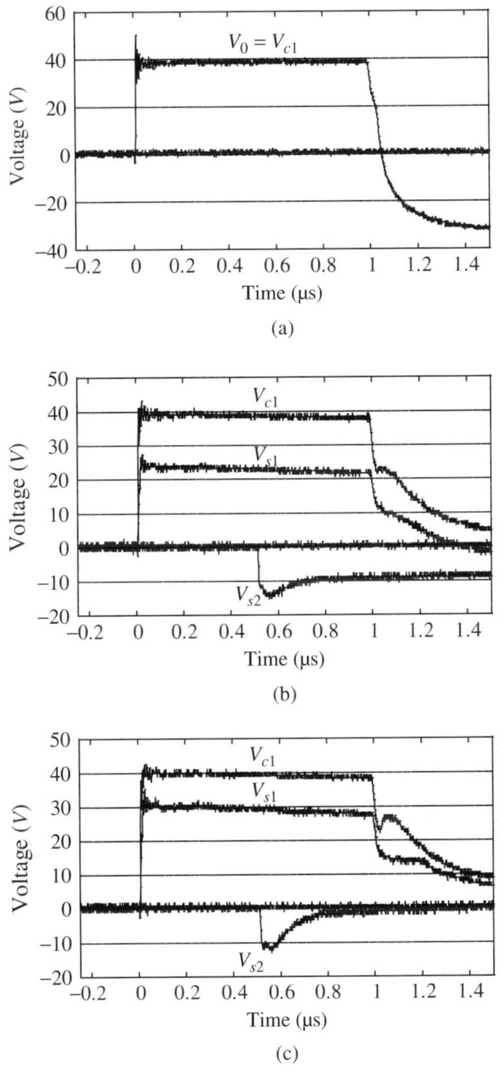

Figure 5.2 Measured results for Case *11-kl*. $R_{c1} = R_{c2} = 0\ \Omega$, $V_{c2} = 0$ V, and V_0 is the applied voltage. (a) 11-11 ($R_{s1} = 0\ \Omega$, $R_{s2} = 0\ \Omega$); (b) 11-22 ($R_{s1} = 150\ \Omega$, $R_{s2} = 150\ \Omega$); and (c) 11-33 ($R_{s1} = 10$ kΩ, $R_{s2} = 10$ kΩ)

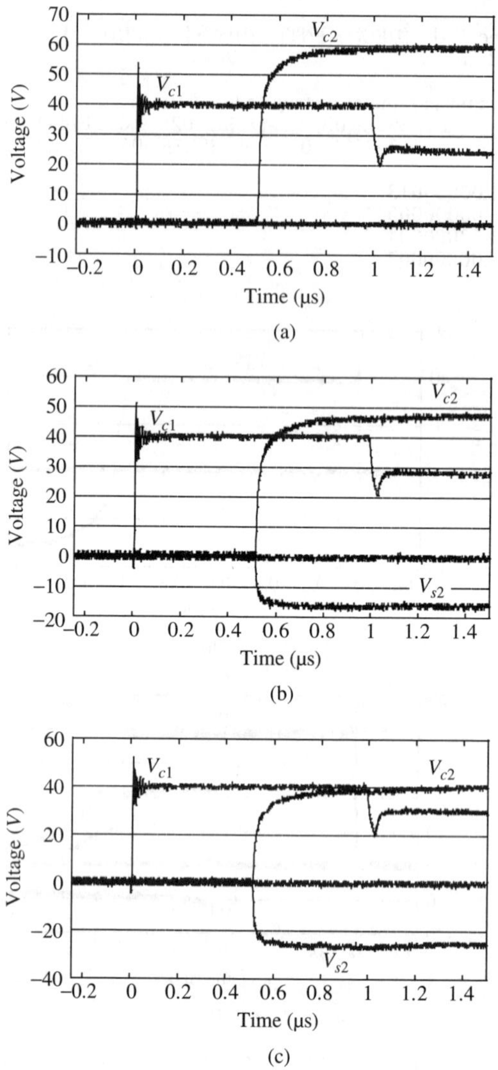

Figure 5.3 Measured results for Case *12-kl*. $R_{c1} = 0\ \Omega$, $R_{c2} = 200\ \Omega$. (a) 12-11 ($R_{s1} = 0\ \Omega$, $R_{s2} = 0\ \Omega$); (b) 12-12 ($R_{s1} = 0\ \Omega$, $R_{s2} = 150\ \Omega$); (c) 12-13 ($R_{s1} = 0\ \Omega$, $R_{s2} = 10\ \text{k}\Omega$); (d) 12-22 ($R_{s1} = 150\ \Omega$, $R_{s2} = 150\ \Omega$); (e) 12-23 ($R_{s1} = 150\ \Omega$, $R_{s2} = 10\ \text{k}\Omega$); and (f) 12-33 ($R_{s1} = 10\ \text{k}\Omega$, $R_{s2} = 10\ \text{k}\Omega$)

5.1.2.2 Simulation Results

Figure 5.6, Figure 5.7, Figure 5.8, and Figure 5.9 show EMTP simulation results corresponding to the measured results in Figure 5.2, Figure 5.3, Figure 5.4, and Figure 5.5, respectively.

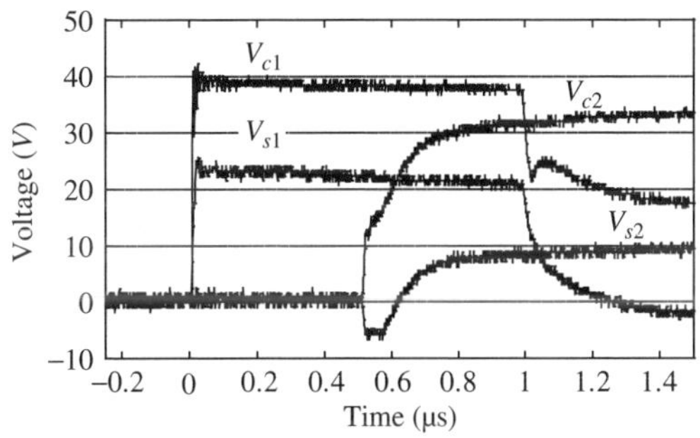

(d)

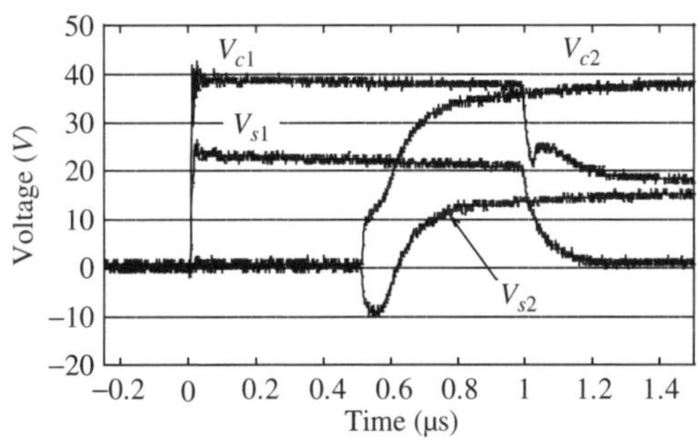

(e)

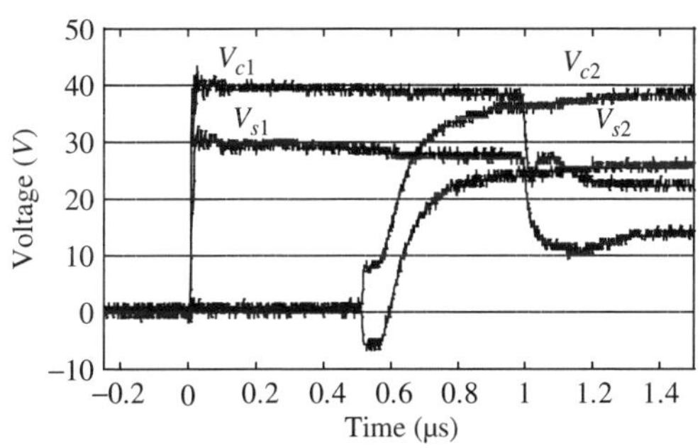

(f)

Figure 5.3 (*Continued*)

The simulation results agree qualitatively well with the measured results. The arrival time of the first wave at the receiving end is observed to be 0.474 μs in the EMTP simulations and 0.48 μs in the measured results. This observation indicates that the parameters calculated at $f = 1$ MHz are reasonable. Normalized values of the initial peaks by that of the core sending voltage V_{c1} are given in Table 5.2 in comparison with the measured results.

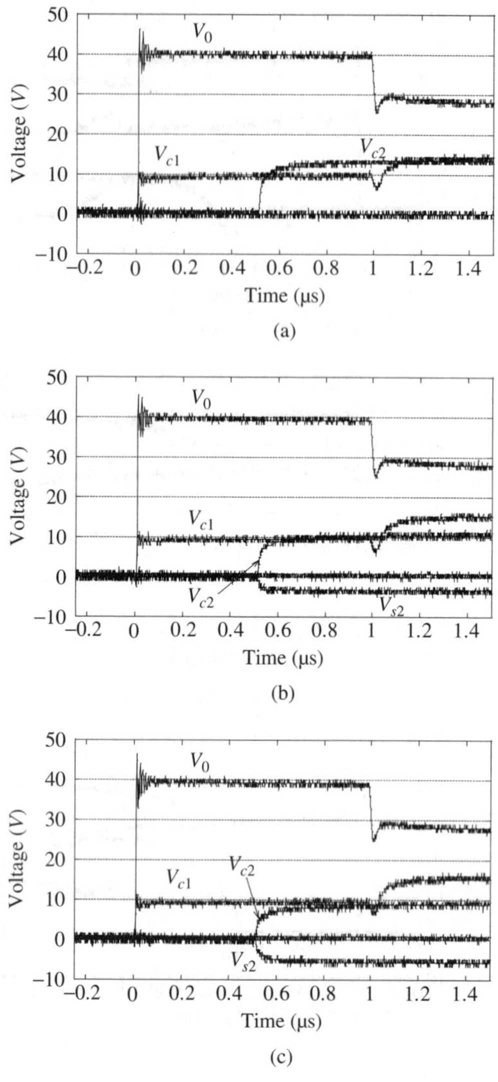

Figure 5.4 Measured results for Case 22-*kl*. $R_{c1} = R_{c2} = 200\ \Omega$. (a) 22-11 ($R_{s1} = 0\ \Omega$, $R_{s2} = 0\ \Omega$); (b) 22-12 ($R_{s1} = 0\ \Omega$, $R_{s2} = 150\ \Omega$); (c) 22-13 ($R_{s1} = 0\ \Omega$, $R_{s2} = 10\ \text{k}\Omega$); (d) 22-22 ($R_{s1} = 150\ \Omega$, $R_{s2} = 150\ \Omega$); (e) 22-23 ($R_{s1} = 150\ \Omega$, $R_{s2} = 10\ \text{k}\Omega$); and (f) 22-33 ($R_{s1} = 10\ \text{k}\Omega$, $R_{s2} = 10\ \text{k}\Omega$)

5.1.3 Theoretical Analysis

5.1.3.1 Refraction Coefficient Matrix

Refraction coefficients at node 1 of Figure 5.10 are given in the following equations [3, 4].

$$\text{To the right}\,[\lambda_f] = 2[Z_0]([R_1] + [Z_0])^{-1} \tag{5.1a}$$

$$\text{To the left}\ [\lambda_b] = 2[R_1]([R_1] + [Z_0])^{-1} \tag{5.1b}$$

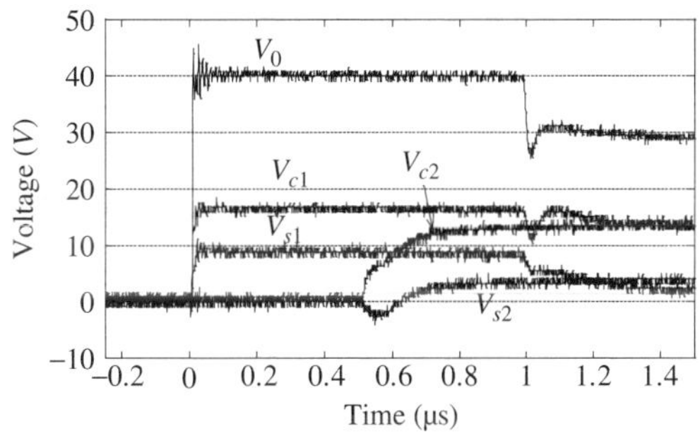

(d)

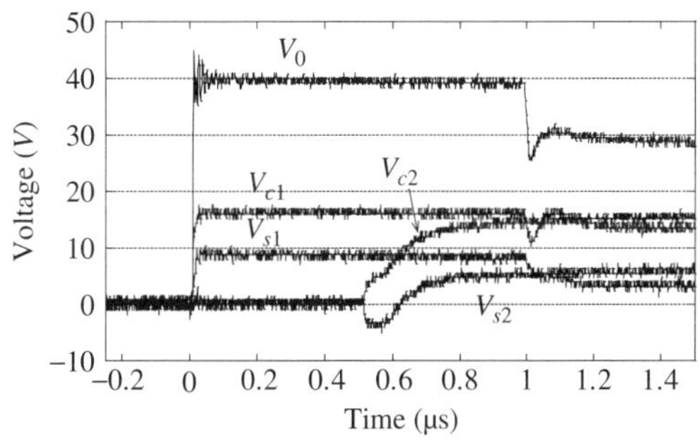

(e)

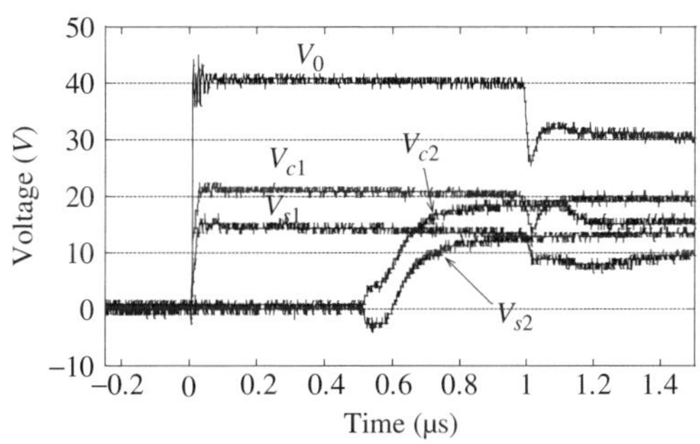

(f)

Figure 5.4 (*Continued*)

where

$$[R_1] = \begin{bmatrix} R_{c1} & 0 \\ 0 & R_{s1} \end{bmatrix} \tag{5.2a}$$

$$[Z_0] = \begin{bmatrix} Z_{0c} & Z_{0m} \\ Z_{0m} & Z_{0s} \end{bmatrix} \tag{5.2b}$$

and Z_{0c}: is the core self surge impedance, Z_{0s} is the sheath self surge impedance, and Z_{0m} is the mutual surge impedance between core and sheath.

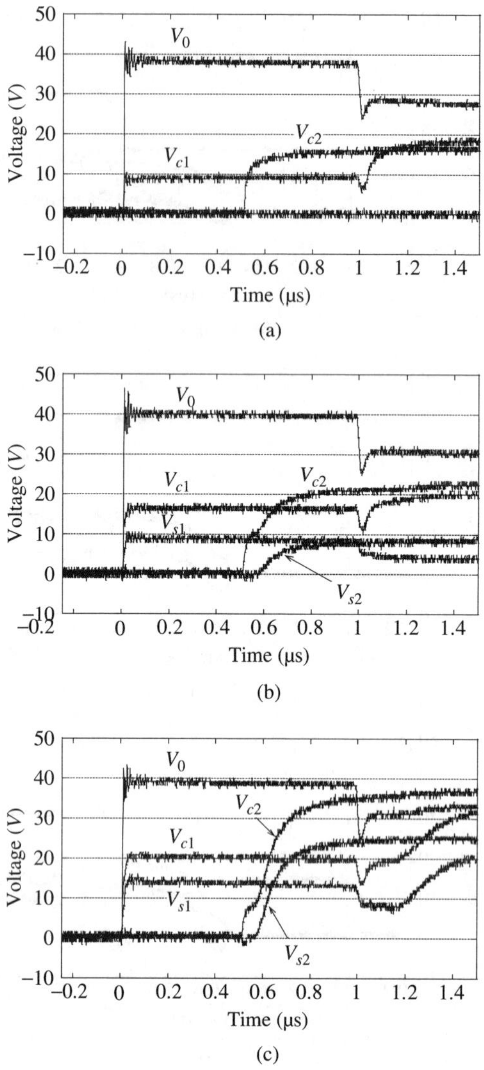

Figure 5.5 Measured results for Case *23-kl*. $R_{c1} = 200\ \Omega$, $R_{c2} = 10\ \text{k}\Omega$. (a) *23-11* ($R_{s1} = 0\ \Omega$, $R_{s2} = 0\ \Omega$); (b) *23-22* ($R_{s1} = 150\ \Omega$, $R_{s2} = 150\ \Omega$); and (c) *23-33* ($R_{s1} = 10\ \text{k}\Omega$, $R_{s2} = 10\ \text{k}\Omega$)

Substituting (5.2a,b) into (5.1a,b), the following results are obtained.

$$[\lambda_{1f}] = (2/\Delta_1)\begin{bmatrix} Z_c\left(Z_{0s}+R_{s1}\right) - Z_{0m}^2 & R_{c1}Z_{0m} \\ R_{s1}Z_{0m} & Z_{0s}(Z_{0c}+R_{c1}) - Z_{0m}^2 \end{bmatrix} \tag{5.3a}$$

$$[\lambda_{1b}] = (2/\Delta_1)\begin{bmatrix} R_{c1}\left(Z_{0s}+R_{s1}\right) - Z_{0m}^2 & -R_{c1}Z_{0m} \\ -R_{s1}Z_{0m} & R_{s1}(Z_{0c}+R_{c1}) \end{bmatrix} \tag{5.3b}$$

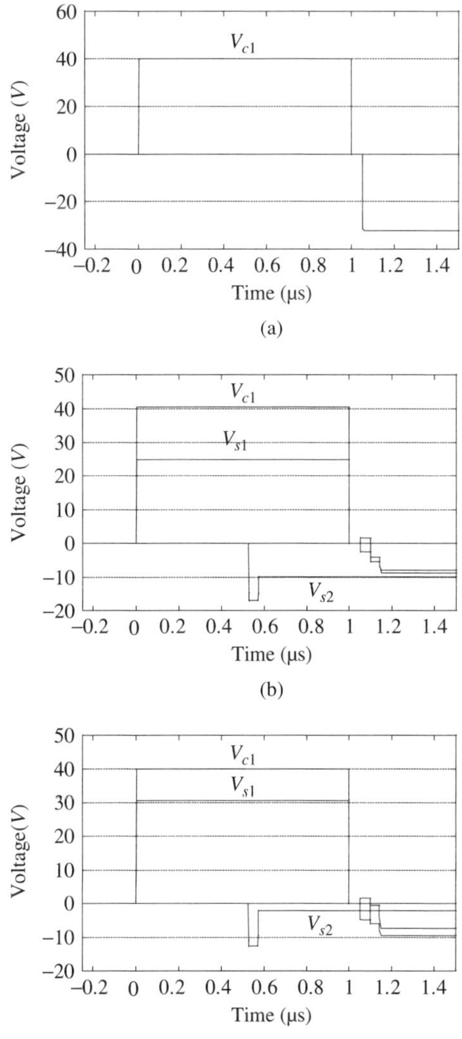

Figure 5.6 Simulated results for Case 11-*kl*. $R_{c1} = R_{c2} = 0\ \Omega$, $V_{c2} = 0$ V. (a) *11-11* ($R_{s1} = 0\ \Omega$, $R_{s2} = 0\ \Omega$); (b) *11-22* ($R_{s1} = 150\ \Omega$, $R_{s2} = 150\ \Omega$); and (c) *11-33* ($R_{s1} = 10$ kΩ, $R_{s2} = 10$ kΩ)

where

$$\Delta_1 = (Z_{0c} + R_{c1})(Z_{0s} + R_{s1}) - Z_{0m}^2$$

It should be noted from (5.1a,b) that the following relation exists between $[\lambda_{1f}]$ and $[\lambda_{1b}]$.

$$[\lambda_{1f}] + [\lambda_{1b}] = 2[U] \quad \text{or} \quad [\lambda_{1b}] = 2[U] - [\lambda_{1f}] \tag{5.4}$$

Also, an internal impedance of a source should be examined if a measured result is analyzed. An applied voltage or a current from a pulse generator (PG) is the output of the PG involving

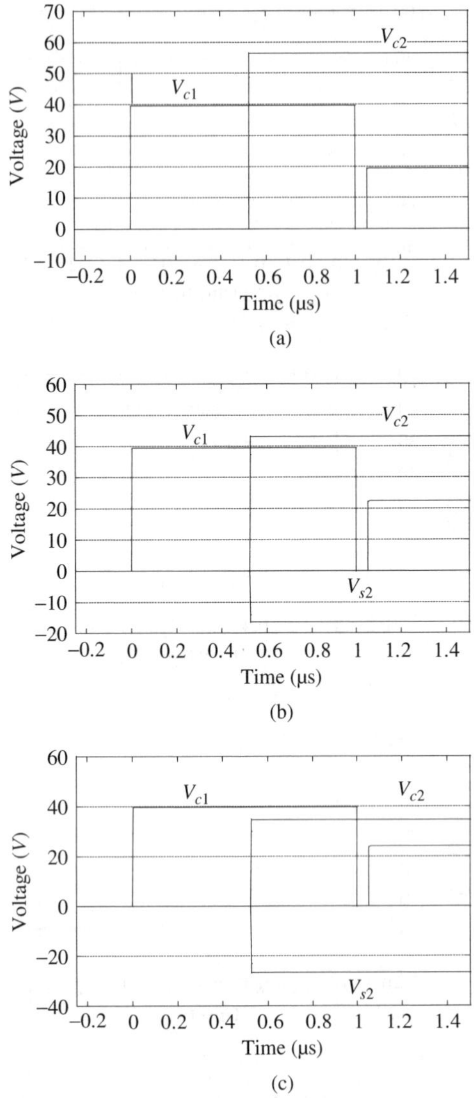

Figure 5.7 Simulated results for Case 12-*kl*. $R_{c1} = 0\ \Omega$, $R_{c2} = 200\ \Omega$. (a) 12-11 ($R_{s1} = 0\ \Omega$, $R_{s2} = 0\ \Omega$); (b) 12-12 ($R_{s1} = 0\ \Omega$, $R_{s2} = 150\ \Omega$); (c) 12-13 ($R_{s1} = 0\ \Omega$, $R_{s2} = 10\ \mathrm{k}\Omega$); (d) 12-22 ($R_{s1} = 150\ \Omega$, $R_{s2} = 150\ \Omega$); (e) 12-23 ($R_{s1} = 150\ \Omega$, $R_{s2} = 10\ \mathrm{k}\Omega$); and (f) 12-33 ($R_{s1} = 10\ \mathrm{k}\Omega$, $R_{s2} = 10\ \mathrm{k}\Omega$)

its internal circuit. A reflected traveling wave from the remote end of the cable in Figure 5.1a sees not only the external resistance R_{c1} and R_{s1} but also the internal source impedance (68 Ω) of the PG in the experiment carried out. If it is a pure voltage source, then the internal source impedance is zero, and infinite for a pure current source. But this never happens in actuality.

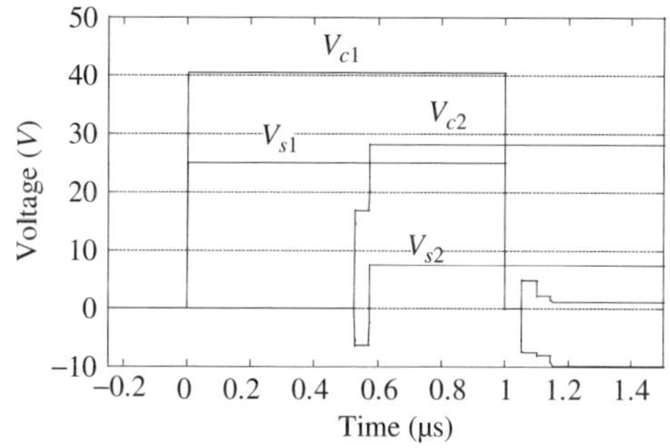

(d)

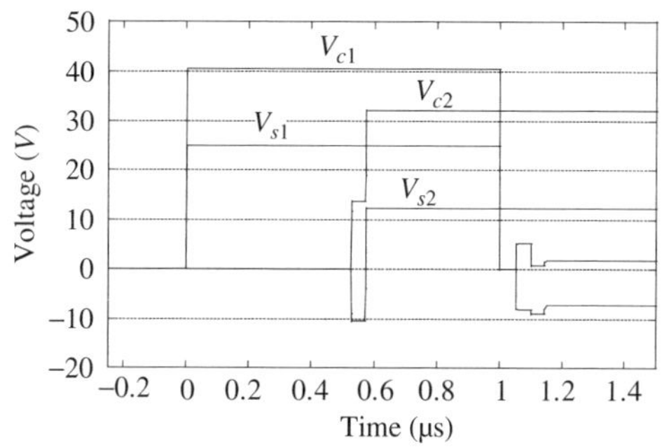

(e)

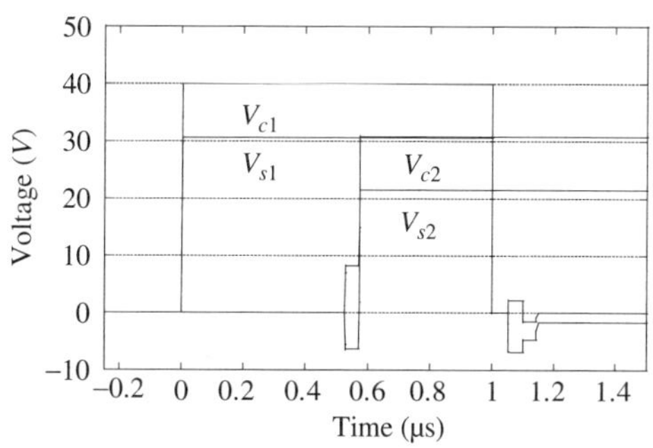

(f)

Figure 5.7 *(Continued)*

Therefore, $[\lambda_{1b}]$ in (5.3) is rewritten taking the source impedance into account.

$$[\lambda_{1b}] = (2/\Delta_1{}') \begin{bmatrix} R_{c1}{}'\left(Z_{0s} + R_{s1}\right) & -R_{c1}{}'Z_{0m} \\ -R_{s1}Z_{0m} & R_{s1}(Z_{0c} + R_{c1}{}') \end{bmatrix} \tag{5.5}$$

where

$\Delta_1' = (Z_{0c} + R'_{c1})(Z_{0s} + R_{s1}) - Z_m^2, R_{c1}{}' = R_{c1} + R_0, R_0 = 68\ \Omega$

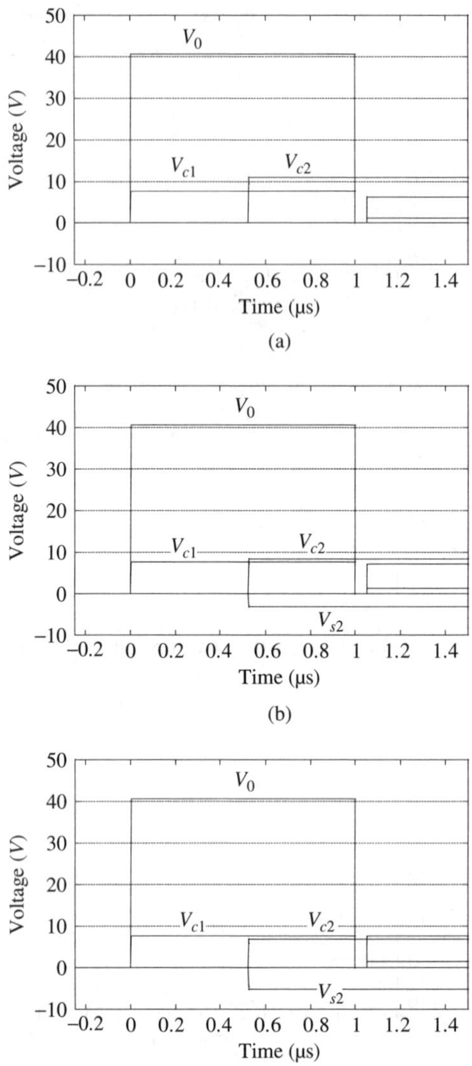

Figure 5.8 Simulated results for Case 22-*kl*. $R_{c1} = R_{c2} = 200\ \Omega$. (a) 22-11 ($R_{s1} = 0\ \Omega$, $R_{s2} = 0\ \Omega$); (b) 22-12 ($R_{s1} = 0\ \Omega$, $R_{s2} = 150\ \Omega$); (c) 22-13 ($R_{s1} = 0\ \Omega$, $R_{s2} = 10\ \text{k}\Omega$); (d) 22-22 ($R_{s1} = 150\ \Omega$, $R_{s2} = 150\ \Omega$); (e) 22-23 ($R_{s1} = 150\ \Omega$, $R_{s2} = 10\ \text{k}\Omega$); and (f) 22-33 ($R_{s1} = 10\ \text{k}\Omega$, $R_{s2} = 10\ \text{k}\Omega$)

At the cable receiving end in Figure 5.10b, only the forward refraction coefficient is necessary.

$$[\lambda_{2f}] = (2/\Delta_2)\begin{bmatrix} R_{c2}\left(Z_{0s} + R_{s2}\right) & -R_{c2}Z_{0m} \\ -R_{s2}Z_{0m} & R_{s2}(Z_{0c} + R_{c1}{}') \end{bmatrix} \tag{5.6}$$

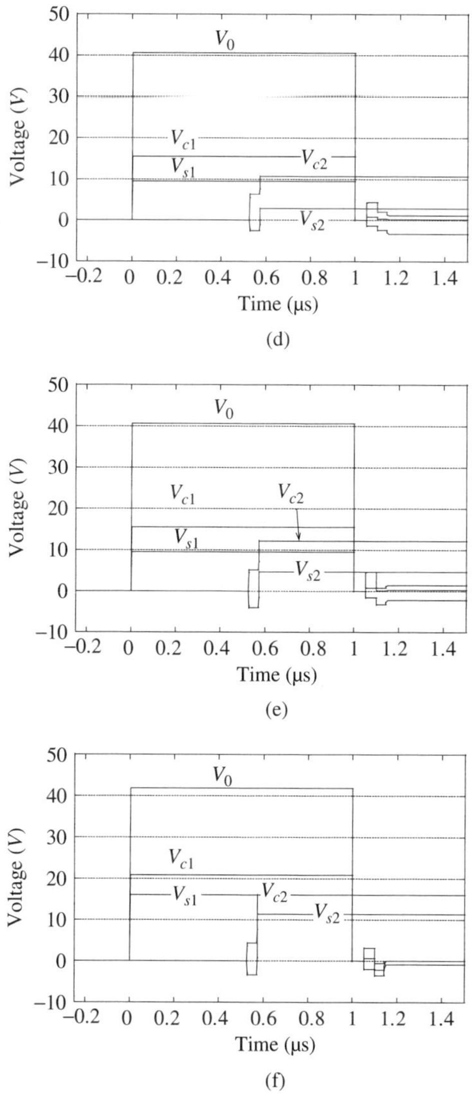

Figure 5.8 (*Continued*)

where

$$\Delta_2 = (Z_{0c} + R_{c2})(Z_{0s} + R_{s2}) - Z_{0m}^2$$

5.1.3.2 Node Voltage

Node voltages (V_1) and (V_2) due to incoming traveling wave (e_0) at node 1 in Figure 5.6a are analytically calculated as a function of traveling time from node 1 to node 2. The traveling

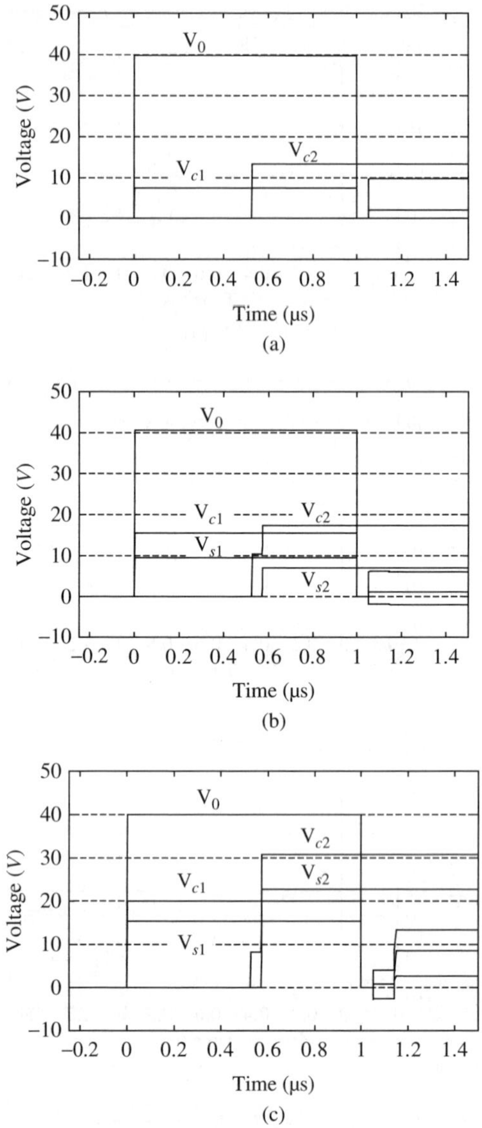

Figure 5.9 Simulated results for Case 23-*kl*. $R_{c1} = 200\ \Omega$, $R_{c2} = 10\ \text{k}\Omega$. (a) 23-11 ($R_{s1} = 0\ \Omega$, $R_{s2} = 0\ \Omega$); (b) 23-22 ($R_{s1} = 150\ \Omega$, $R_{s2} = 150\ \Omega$); and (c) 23-33 ($R_{s1} = 10\ \text{k}\Omega$, $R_{s2} = 10\ \text{k}\Omega$)

time τ is determined by the propagation velocity "c" of the conductor, that is,

$$\tau = x/c \tag{5.7}$$

where x is the conductor length.

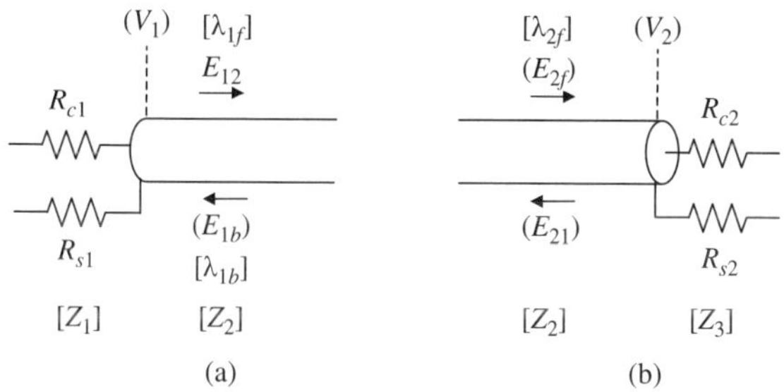

Figure 5.10 Cable ends: (a) sending end; and (b) receiving end

In an "N" conductor system, there exist N modes of propagation, and thus N traveling times correspondingly. In a single-phase coaxial cable, there are two propagation modes:

- coaxial mode: propagation velocity c_1, traveling time $\tau_1 = x/c_1$
- earth-return mode: propagation velocity c_0, traveling time $\tau_0 = x/c_0$

Considering a difference between the traveling times τ_1 and τ_0, the node voltages are analytically calculated in the following procedure.

(1) $t = 0$

$$V_1(t) = [\lambda_{1f}](e_0(t)),$$

$$\text{traveling wave from node 1 to node 2 } : (E_{12}(t)) = (V_1(t))$$

The above traveling wave (E_{12}) is to be transformed into a modal domain to consider the traveling time difference from node 1 to 2.

$$\text{modal domain traveling wave } : (e_{12}(t)) = [A]^{-1}(E_{12}(t)) = \begin{pmatrix} e_{012}(t) \\ e_{112}(t) \end{pmatrix}$$

where e_{012} is the mode 0 (earth-return mode) traveling wave from node 1 to node 2, e_{112} is the mode 1 (coaxial mode) wave, and $[A]^{-1}$ is the voltage transformation matrix from phase to modal components.

(2) $t = \tau_1$
Mode 1 wave e_{112} arrives at node 2 at time τ_1 as forward wave (e^1_{2f}), that is,

$$(e^1_{2f}(t)) = \begin{pmatrix} 0 \\ e_{112}\left(t - \tau_1\right) \end{pmatrix}$$

Transforming (e^1_{2f}) into phasor domain gives forward phasor wave (E^1_{2f}) which generates node 2 voltage (V_2).

$$(E^1_{2f}(t)) = [A](e^1_{2f}(t))$$
$$(V_2(\tau_1)) = [\lambda_{2f}](E^1_{2f}(\tau_1)),$$
$$\text{traveling wave from node 2 to node 1}: (E^1_{21}(\tau_1)) = (V_2(\tau_1)) - (E^1_{2f}(\tau_1))$$

The traveling wave (E^1_{21}) propagates toward node 1 as a modal component (e^1_{21}).

$$(e^1_{21}(t)) = [A]^{-1}(E^1_{21}(t)) = \begin{pmatrix} e^1_{021} \\ e^1_{121} \end{pmatrix}$$

Again, e^1_{021} arrives at node 1 by the traveling time τ_0, while e^1_{121} by τ_1.

(3) $t = \tau_0$
Mode 0 wave e_{012} arrives at node 2 at $t = \tau_0$ as forward wave (e^0_{2f})

$$(e^0_{2f}(t)) = \begin{pmatrix} e_{012}\,(t-\tau_0) \\ 0 \end{pmatrix}, \quad (e^0_{2f}(t)) = [A](e^0_{2f}(t))$$
$$(V_2(\tau_0)) = [\lambda_{2f}](E^0_{2f}(\tau_0)),$$
$$\text{traveling wave from node 2 to node 1}: (E^0_{21}(\tau_0)) = (V_2(\tau_0)) - (E^0_{2f}(\tau_0))$$

The traveling wave (E^0_{21}) propagates toward node 1 in the same manner as (E^1_{21})

$$(e^0_{21}(t)) = [A]^{-1}(E^0_{21}(t)) = \begin{pmatrix} e^0_{021} \\ e^0_{121} \end{pmatrix}$$

(4) $t = 2\tau_1$
Modal traveling wave e^1_{121} generated at node 2 at $t = \tau_1$ arrives at node 1 at $t = 2\tau_1$.

$$(e^1_{1b}(t)) = \begin{pmatrix} 0 \\ e^1_{121}\,(t-\tau_1) \end{pmatrix}, (E^1_{1b}(t)) = [A](e^1_{1b}(t))$$
$$(V_1(t)) = [\lambda_{1f}](e_0(t)) + [\lambda'_{1b}](E^1_{1b}(t)), \quad (E^1_{12}(t)) = (V_1(t)) - (E^1_{1b}(t))$$

It should be noted that the original wave $e_0(t)$ always generates (V_1) at node 1.

(5) $t = \tau_1 + \tau_0$
Mode 0 wave e^1_{021} and the mode 1 wave e^1_{121} arrive at node 1 at the same instance $t = \tau_1 + \tau_0$ and are transformed to the phasor wave (E^2_{1b}).

$$(e^2_{1b}(t)) = \begin{pmatrix} e^1_{021} \\ e^0_{121} \end{pmatrix}, (E^2_{1b}(t)) = [A](e^2_{1b}(t))$$
$$(V_1(t)) = (V_1(2\tau_1)) + [\lambda'_{1b}]E^2_{1b}(t), (E^2_{12}(t)) = (V_1(t)) - (E^2_{1b}(t))$$

(6) $t = 2\tau_0$

Mode 0 wave e_{121}^0 arrives at node 1 at $t = 2\tau_0$ and generates $V_1(t = 2\tau_0)$

$$(e_{1b}^0(t)) = \begin{pmatrix} e_{021}^1 \\ 0 \end{pmatrix}, (E_{1b}^0(t)) = [A](e_{1b}^0(t))$$

$$(V_1(t)) = (V_1(\tau_1 + \tau_0)) + [\lambda'_{1b}]E_{1b}^0(t), (E_{12}^0(t)) = (V_1(t)) - (E_{1b}^0(t))$$

Repeating the process of (1)–(6), node voltages V_1 and V_2 are analytically evaluated, although a hand calculation for $t \geq 2\tau_0$ is rather tedious.

The above procedure already explains the physical meaning of transient voltages on a single-phase cable, and is applicable to a three-phase cable by adding an inter-phase mode of propagation because the coaxial and earth-return modes are nearly the same in each phase cable.

5.1.4 Analytical Evaluation of Parameters

In most cases of cable transients, the dominant transient frequency is high enough to produce a complete coaxial mode of propagation. The critical frequency f_c is given approximately in the following equation as explained in Chapter 3, Section 3.2.1.

$$d = r_4 - r_3 > \sqrt{\rho_s/\omega\mu_s} \quad \text{or} \quad f_c \cong \rho_s/2\pi d^2\mu_s \tag{5.8}$$

where d is the sheath thickness, ρ_s is the sheath resistivity, and μ_s is the sheath permeability.

Above the critical frequency , the propagation velocity and the characteristic impedance of a coaxial cable is given in the following formula.

$$\text{coaxial mode}: \; c_1 = v_0/\sqrt{\varepsilon_1}, z_{01} = (60/\sqrt{\varepsilon_1})\ln(r_3/r_2),$$
$$v_0 = 300\,\text{m}/\mu\text{s}: \; \text{light velocity in free space}$$

characteristic impedance in phase domain, more accurate one in (3.88) and (3.89).

$$Z_{0s} = 60\ln(2h/r_4) \text{ for } h \gg r_4 \tag{5.9a}$$
$$= 60\ln(2r_5/r_4) \text{ on the earth surface for } 2r_5 \gg r_4 \tag{5.9b}$$
$$= \left(60/\sqrt{\varepsilon_2}\right)\ln(r_5/r_4) \text{ for underground} \tag{5.9c}$$
$$Z_{0c} = Z_{0s} + z_{01}, Z_{0m} = Z_{0s}$$

The voltage transformation matrix A is approximated in the frequency region higher than f_c by:

$$[A] = \begin{bmatrix} 1 & 1 \\ 1 & 0 \end{bmatrix} \tag{5.10a}$$

$$[A]^{-1} = \begin{bmatrix} 0 & 1 \\ 1 & -1 \end{bmatrix} \tag{5.10b}$$

The dominant transient frequency is given by:

$$f_t = 1/4\tau \tag{5.11}$$

where $\tau = x/c$ and c is the fastest propagation velocity.

f_t for this experiment and f_c for the tested cable are evaluted as:

$$f_t = 1/4\tau_1 \cong 550\ \text{kHz} > f_c = 4\ \text{kHz} \tag{5.12}$$

where $c_1 = 300/\sqrt{2.3} \cong 198\text{m}/\mu\text{s}$, $x = 90$ m, and $\tau_1 \cong 0.45\ \mu\text{s}$ Thus, the paramters given in (5.9a–c) and (5.10a,b) can be adopted in the analytical calculation.

$$z_{01} = 48\ \Omega \tag{5.13a}$$

$$Z_{0s} = Z_{0m} = 144\ \Omega \tag{5.13b}$$

$$Z_{0c} = 192\ \Omega \tag{5.13c}$$

The above analytical evaluation of cable parameters are not necessary if the parameters are calculated by the Cable Constants/Parameters of the EMTP [1, 5]. However, the analytical parameter evaluation is useful to catch physical insight of a cable characteristic, and also to judge the reliability of the output from the Cable Constants/Parameters because those are not perfect from the viewpoint of theoretical completeness as explained in References [2–8].

The analytical values in (5.10), (5.12) and (5.13) agree well with those calculated by the Cable Parameters [5] given in Data 5.1.

In the evaluation of node voltages in the next section, the following parameters are adopted.

$$[Z_0] = \begin{bmatrix} 192 & 144 \\ 144 & 144 \end{bmatrix}, \quad \tau_1 = 0.45\ \mu\text{s}, \quad \tau_0 \cong 1\mu\text{s}, \quad [A] \text{ and } [A]^{-1} \text{ in}$$

5.1.5 *Analytical Calculation of Transient Voltages*

5.1.5.1 Case *11-11* : $R_{c1} = R_{c2} = R_{s1} = R_{s2} = 0$

$$[\lambda_{1f}] = \begin{bmatrix} 2 & 0 \\ 0 & 2 \end{bmatrix}, \quad [\lambda'_{1b}] = \begin{bmatrix} 1.17 & -1.17 \\ 0 & 0 \end{bmatrix}, \quad [\lambda_{2f}] = \begin{bmatrix} 0 & 0 \\ 0 & 0 \end{bmatrix}$$

In the experimental result in Figure 5.2a, the applied voltage V_0 from the PG is observed to be 9 V. By Thevenen's theorem, the incoming traveling wave $E_0 = V_0/2 = 4.5$ V. For simplicity, the incoming wave E_0 is taken to be 5 V in all the analytical calculations.

(1) $t = 0$

$$\begin{pmatrix} V_{c1} \\ V_{s1} \end{pmatrix} = [\lambda_{1f}] \begin{pmatrix} 5 \\ 0 \end{pmatrix} = \begin{pmatrix} 10 \\ 0 \end{pmatrix} = (E_{12})$$

Because of the voltage transformation matrix in (5.10) and a traveling wave only on the core, there exists only the coaxial mode of propagation. Thus, there is no need of modal decomposition/composition.

(2) $t = \tau_1 \cong 0.45$ μs

$$(E_{2f}(t)) = E_{12}(t-\tau_1) = \begin{pmatrix}10\\0\end{pmatrix}, (V_2(t)) = [\lambda_{2f}](E_{2f}(t)) = \begin{pmatrix}0\\0\end{pmatrix},$$

$$(E_{21}(t)) = (V_2(t)) - (E_{2f}(t)) = \begin{pmatrix}-10\\0\end{pmatrix}$$

(3) $t = 2\tau_1 \cong 0.9$ μs

$$(V_1(t)) = [\lambda_{1f}]\begin{pmatrix}5\\0\end{pmatrix} + [\lambda_{1b}]\begin{pmatrix}-10\\0\end{pmatrix} = \begin{pmatrix}10 - 1.17 \times 10\\0\end{pmatrix} = \begin{pmatrix}-1.7\\0\end{pmatrix}$$

(4) $t > 1$ μs

For the pulse width of the applied voltage V_0 is 1 μs as shown in Figure 5.1d, the incoming wave E_0 becomes 0 at $t = 1$ μs. Thus, V_1 for $t > 1$ μs is given by:

$$(V_1(t)) = [\lambda_{1f}]\begin{pmatrix}0\\0\end{pmatrix} + [\lambda_{1b}]\begin{pmatrix}-10\\0\end{pmatrix} = \begin{pmatrix}-11.7\\0\end{pmatrix}$$

Figure 5.11a shows that the above analytical results agree well with EMTP simulation results in Figure 5.6a and with the measured result in Figure 5.2a qualitatively. Each step in the above analytical calculation explains a transient waveform calculated by the EMTP with the lossless line model of which the parameters are evaluated by Cable Contstans/Parameters. Also, the analytical evaluation makes it possible to check the line/Cable Parameters calculated by the EMTP subroutines.

It should be clear from the above analysis that only the coaxial mode contributes the transient voltage under the condition of $R_{s1} = R_{s2} = R_{c2} = 0$. If the applied voltage waveshape in Figure 5.1d is taken into account, a better agreement with the measured result is achieved. The frequency-dependent effect of the coaxial mode propagation is evaluated by a difference of $V_1(t)$ for $t = 2\tau_1$ from $V_1(t)$ for $0 \le t \le 2\tau_1$ considering the amplitude of the applied voltage in Figure 5.2.

5.1.5.2 Case *11-33* : $R_{c1} = R_{c2} = 0$, $R_{s1} = R_{s2} = 10$ kΩ (Nearly Open-circuited)

$$[\lambda_{1f}] = \begin{bmatrix}2 & 0\\1.5 & 0.2\end{bmatrix}, \quad [\lambda'_{1b}] = \begin{bmatrix}0.51 & 0\\-1.12 & 2\end{bmatrix}, \quad [\lambda_{2f}] = \begin{bmatrix}0 & 0\\-1.5 & 2\end{bmatrix}$$

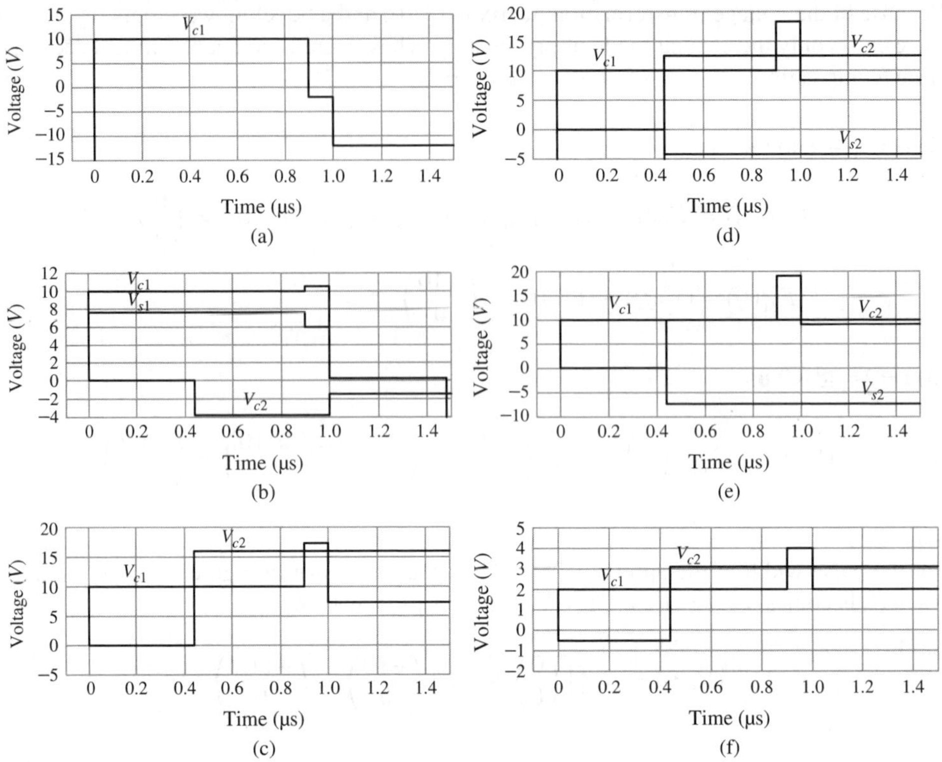

Figure 5.11 Analytical results: (a) 11-11 ($R_{c1} = R_{c2} = 0\ \Omega$, $R_{s1} = 0\ \Omega$, $R_{s2} = 0\ \Omega$); (b) 11-33 ($R_{c1} = R_{c2} = 0\ \Omega$, $R_{s1} = 10\ \text{k}\Omega$, $R_{s2} = 10\ \text{k}\Omega$); (c) 12-11 ($R_{c1} = 0\ \Omega$, $R_{c2} = 200\ \Omega$, $R_{s1} = 0\ \Omega$, $R_{s2} = 0\ \Omega$); (d) 12-12 ($R_{c1} = 0\ \Omega$, $R_{c2} = 200\ \Omega$, $R_{s1} = 0\ \Omega$, $R_{s2} = 150\ \Omega$); (e) 12-13 ($R_{c1} = 0\ \Omega$, $R_{c2} = 200\ \Omega$, $R_{s1} = 0\ \Omega$, $R_{s2} = 10\ \text{k}\Omega$); and (f) 22-11 ($R_{c1} = 200\ \Omega$, $R_{c2} = 200\ \Omega$, $R_{s1} = 0\ \Omega$, $R_{s2} = 0\ \Omega$)

(1) $t = 0$

$$\begin{pmatrix} V_{c1} \\ V_{s1} \end{pmatrix} = [\lambda_{1f}] \begin{pmatrix} 5 \\ 0 \end{pmatrix} = \begin{pmatrix} 10 \\ 7.5 \end{pmatrix} = (E_{12}),$$

$$(e_{12}) = [\text{A}]^{-1}(E_{12}) = \begin{pmatrix} 7.5 \\ 2.5 \end{pmatrix} \begin{matrix} : \text{earth-return mode (mode 0)} \\ : \text{coaxial mode (mode 1)} \end{matrix}$$

(E_{12}) shows that there exist traveling waves both on the core and on the sheath. This means that the traveling waves propagate partially as the coaxial and the earth-return modes. Thus, we need to make modal decomposition and obtain modal traveling waves (e_{12}) from node 1 to node 2.

The coaxial mode arrives at node 2 at $t = \tau_1 \cong 0.45$ μs, while the earth-return mode at $t = \tau_0 \cong 1$ μs.

(2) $t = \tau_1 \cong 0.45$ μs: *Only Coaxial Mode*

$$e^1_{2f}(t) = (e^1_{12}(t-\tau_1)) = \begin{pmatrix} 0 \\ 2.5 \end{pmatrix}, \quad (E^1_{2f}) = [\mathrm{A}]e^1_{2f} = \begin{pmatrix} 2.5 \\ 0 \end{pmatrix}$$

$$\begin{pmatrix} V_{c2} \\ V_{s2} \end{pmatrix} = [\lambda_{2f}](E^1_{2f}) = \begin{pmatrix} 0 \\ -3.75 \end{pmatrix}$$

$$(E^1_{21}) = (V_2) - (E^1_{2f}) = \begin{pmatrix} -2.5 \\ -3.75 \end{pmatrix}, \quad (e^1_{21}) = [\mathrm{A}]^{-1}(E^1_{21}) = \begin{pmatrix} -3.75 \\ 1.25 \end{pmatrix} \begin{matrix} : \text{mode } 0 \\ : \text{mode } 1 \end{matrix}$$

(e^1_{21}) shows two modal traveling waves. The mode 1 component arrives at node 1 at $t = 2\tau_1$, while the mode 0 at $t = \tau_1 + \tau_0$.

(3) $t = \tau_0 = 1$ μs: *Mode 0 of* $(\mathbf{e}_{12})$ *at Step* (1)

$$(e^0_{2f}(t)) = (e^0_{12}(t-\tau_0)) = \begin{pmatrix} 7.5 \\ 0 \end{pmatrix}, \quad (E^0_{2f}) = [\mathrm{A}](e^0_{2f}) = \begin{pmatrix} 7.5 \\ 7.5 \end{pmatrix} \begin{matrix} : \text{core} \\ : \text{sheath} \end{matrix}$$

$$\begin{pmatrix} V_{c2}\,(t=\tau_0) \\ V_{s2}(t=\tau_0) \end{pmatrix} = [\lambda_{2f}]\begin{pmatrix} 7.5 \\ 7.5 \end{pmatrix} + (V_2(t=\tau_1)) = \begin{pmatrix} 0 \\ 3.75 - 3.75 \end{pmatrix} = \begin{pmatrix} 0 \\ 0 \end{pmatrix}$$

E^0_{2f} shows a traveling wave of 7.5 V both on the core and the sheath. This is due to the earth-return mode (e^0_{2f}). As is well-known, the earth-return (or so-called zero-sequence) mode propagates along all the conductors (phases) with the same voltage and polarity. The earth-return mode voltage cancels out the sheath voltage −3.75 V at $t = \tau_1$ generated by the coaxial mode in this analysis. This phenomenon is again well-known in a three-phase overhead line with one phase energization, that is, an aerial mode produces a negative voltage on the energized phase at the receiving end first, and later the positive earth-return mode arrives and cancels out the negative voltage. Finally, the receiving-end voltage becomes positive [3]. In the present case, the terminating resistance of the energized phase (core) at the receiving end is zero, no voltage appears on the core.

$$(E^0_{21}) = (V_2) - (E^0_{2f}) = -(E^0_{2f}) = \begin{pmatrix} -7.5 \\ -7.5 \end{pmatrix}, \quad (e^0_{21}) = [A]^{-1}(E^0_{21}) = \begin{pmatrix} -7.5 \\ 0 \end{pmatrix}$$

Since the forward traveling wave (e^0_{21}) contains only the earth-return mode, the reflected wave to node 1 involves only the earth-return mode for the terminating condition.

(4) $\mathrm{t} = 2\tau_1 = 0.9$ μs: *Coaxial Mode* $\boldsymbol{e^1_{121}} = \mathbf{1.25}$ *V*

$$(e^1_{1b}(t)) = (e^1_{21}(t-\tau_1)) = \begin{pmatrix} 0 \\ 1.25 \end{pmatrix}, \quad (E^1_{1b}) = [\mathrm{A}](e^1_{1b}) = \begin{pmatrix} 1.25 \\ 0 \end{pmatrix}$$

A coaxial mode propagates along the core with reference to the sheath as observed in (E_{1b}).

$$\begin{pmatrix} V_{c1} \\ V_{s1} \end{pmatrix} = [\lambda'_{1b}](E^1_{1b}) + [\lambda_{1f}](E_0) = [\lambda'_{1b}]\begin{pmatrix} 1.25 \\ 0 \end{pmatrix} + (V_1(t=0))$$

$$= \begin{pmatrix} 0.64 + 10 \\ -1.4 + 7.5 \end{pmatrix} = \begin{pmatrix} 10.64 \\ 6.1 \end{pmatrix}$$

(5) $t = 1$ µs : *Original Wave* $\mathbf{E_0}$ *Becomes Zero Because of the Pulse Width of the Source Voltage*

$$\begin{pmatrix} V_{c1} \\ V_{s1} \end{pmatrix} = [\lambda'_{1b}](E^1_{1b}) = \begin{pmatrix} 0.64 \\ -1.4 \end{pmatrix}$$

(6) $t = \tau_1 + \tau_0 = 1.45$ µs: *Earth-return Mode* $e^1_{021} = -3.75$ *V*

$$(e^2_{1b}(t)) = (e^1_{21}(t - \tau_0)) = \begin{pmatrix} -3.75 \\ 0 \end{pmatrix}, \quad (E^2_{1b}) = [\mathrm{A}](e^2_{1b}) = \begin{pmatrix} -3.75 \\ -3.75 \end{pmatrix}$$

$$\begin{pmatrix} V_{c1} \\ V_{s1} \end{pmatrix} = [\lambda'_{1b}](E^2_{1b}) + (V_1(t = 1\ \mu\mathrm{s})) = \begin{pmatrix} -1.9 + 0.64 \\ -3.3 - 1.4 \end{pmatrix} = \begin{pmatrix} -1.26 \\ -4.7 \end{pmatrix}$$

(7) $t = 2\tau_0 = 2$ µs: *Earth-return Mode* $e^0_{021} = -7.5$ *V*

$$(e^0_{1b}(t)) = (e^0_{21}(t - \tau_0)) = \begin{pmatrix} -7.5 \\ 0 \end{pmatrix}, \quad (E^0_{1b}) = [\mathrm{A}](e^0_{1b}) = \begin{pmatrix} -7.5 \\ -7.5 \end{pmatrix}$$

$$\begin{pmatrix} V_{c1} \\ V_{s1} \end{pmatrix} = [\lambda'_{1b}](E^0_{1b}) + (V_1(t = 1.45\ \mu\mathrm{s})) = \begin{pmatrix} -3.85 - 1.26 \\ -6.6 - 4.7 \end{pmatrix} = \begin{pmatrix} -5.08 \\ -11.3 \end{pmatrix}$$

Figure 5.11b shows the above result. It is observed that the results agrees well with the EMTP simulataion results in Figure 5.6c, and qualitatively well with the measured result in Figure 5.2c.

5.1.5.3 Case *12-11* : $R_{c1} = 0\ \Omega, R_{c2} = 200\ \Omega, R_{s1} = R_{s2} = 0\ \Omega$

$$[\lambda_{1f}] = \begin{bmatrix} 2 & 0 \\ 0 & 2 \end{bmatrix}, \quad [\lambda'_{1b}] = \begin{bmatrix} 1.172 & -1.172 \\ 0 & 2 \end{bmatrix}, \quad [\lambda_{2f}] = \begin{bmatrix} 1.163 & -1.163 \\ 0 & 2 \end{bmatrix}$$

(1) $t = 0$

$$\begin{pmatrix} V_{c1} \\ V_{s1} \end{pmatrix} = [\lambda_{1f}] \begin{pmatrix} 5 \\ 10 \end{pmatrix} = \begin{pmatrix} 10 \\ 0 \end{pmatrix} = (E_{12})$$

No modal decomposition is required.

(2) $t = \tau_1 = 0.45$ µs

$$(E_{2f}(t)) = (E_{12}(t - \tau_1)) = \begin{pmatrix} 10 \\ 0 \end{pmatrix}, \quad (V_2(t)) = [\lambda_{2f}](E_{2f}(t)) = \begin{pmatrix} 16.13 \\ 0 \end{pmatrix}$$

$$(E_{21}) = (V_2) - (E_{2f}) = \begin{pmatrix} 6.13 \\ 0 \end{pmatrix}$$

(3) $t = 2\tau_1 = 0.9$ μs

$$(E_{1b}(t)) = (E_{21}(t-\tau_1)) = \begin{pmatrix} 6.13 \\ 0 \end{pmatrix},$$

$$(V_1(t)) = [\lambda'_{1b}](E_{1b}) + [\lambda_{1f}](E_0) = \begin{pmatrix} 7.19 + 10 \\ 0 \end{pmatrix} = \begin{pmatrix} 17.19 \\ 0 \end{pmatrix}$$

(4) $t = \tau_0 = 1$ μs $: E_0 = 0$

$$(V_1(t)) = (E_{1b}) = \begin{pmatrix} 7.19 \\ 0 \end{pmatrix}$$

5.1.5.4 Case *12-12* : $R_{c1} = 0\ \Omega, R_{c2} = 200\ \Omega, R_{s1} = 0, R_{s2} = 150\ \Omega$

$$[\lambda_{1f}] = \begin{bmatrix} 2 & 0 \\ 0 & 2 \end{bmatrix}, \quad [\lambda'_{1b}] = \begin{bmatrix} 1.172 & -1.172 \\ 0 & 2 \end{bmatrix}, \quad [\lambda_{2f}] = \begin{bmatrix} 1.244 & -0.609 \\ -0.457 & 1.244 \end{bmatrix}$$

(1) $t = 0$

$$\begin{pmatrix} V_{c1} \\ V_{s1} \end{pmatrix} = [\lambda_{1f}] \begin{pmatrix} 5 \\ 0 \end{pmatrix} = \begin{pmatrix} 10 \\ 0 \end{pmatrix} = (E_{12})$$

There is only coaxial mode, and thus no modal decomposition is required, until $t \geq \tau_1$.

(2) $t = \tau_1 = 0.45$ μs

$$(E_{2f}(t)) = (E_{21}(t-\tau_1)) = \begin{pmatrix} 10 \\ 0 \end{pmatrix}, \quad (V_2(t)) = [\lambda_{2f}](E_{2f}(t)) = \begin{pmatrix} 12.44 \\ -4.57 \end{pmatrix}$$

Because the core and the sheath at the receiving end show a voltage being not zero, the reflected traveling wave (E_{12}) involves both the coaxial and the earth-return mode of propagation. Thus, modal decomposition is necessary.

$$(E_{12}) = (V_2) - (E_{2f}) = \begin{pmatrix} 2.44 \\ -4.57 \end{pmatrix}, \quad (e_{21}) = [A]^{-1}(E_{21}) = \begin{pmatrix} -4.57 \\ 7.01 \end{pmatrix}$$

(3) $t = 2\tau_1 = 0.9$ μs

$$(e^1_{1b}(t)) = (e^1_{21}(t-\tau_1)) = \begin{pmatrix} 0 \\ 7.01 \end{pmatrix}, \quad (E^1_{1b}) = [A](e^1_{1b}) = \begin{pmatrix} 7.01 \\ 0 \end{pmatrix}$$

$$(V_1(t)) = [\lambda'_{1b}](E_{1b}) + (V_1(t=0)) = \begin{pmatrix} 8.22 + 10 \\ 0 \end{pmatrix} = \begin{pmatrix} 18.22 \\ 0 \end{pmatrix}$$

(4) $t = \tau_0 = 1$ μs $: E_0 = 0$

$$(V_1(t)) = (E_{1b}) = \begin{pmatrix} 8.22 \\ 0 \end{pmatrix}$$

(5) $t = \tau_0 + \tau_0 = 1.45\ \mu s$

$$(e_{1b}^1(t)) = (e_{21}^0(t-\tau_0)) = \begin{pmatrix} -4.57 \\ 0 \end{pmatrix},\quad (E_b^0) = [A](e_b^0) = \begin{pmatrix} -4.57 \\ -4.57 \end{pmatrix}$$

$$(V_1(t)) = [\lambda'_{1b}](E_{1b}) + (V_1(t = 1\ \mu s)) = \begin{pmatrix} 8.22 \\ 0 \end{pmatrix}$$

5.1.5.5 Case *12-13* : $R_{c1} = 0, R_{c2} = 200\ \Omega, R_{s1} = 0\ \Omega, R_{s2} = 10\ k\Omega$

$$[\lambda_{1f}] = \begin{bmatrix} 2 & 0 \\ 0 & 2 \end{bmatrix},\quad [\lambda'_{1b}] = \begin{bmatrix} 1.172 & -1.172 \\ 0 & 0 \end{bmatrix},\quad [\lambda_{2f}] = \begin{bmatrix} 1.026 & -0.015 \\ -0.728 & 1.982 \end{bmatrix}$$

There is only coaxial mode, and thus no modal decomposition is required until $t \geq \tau_1$, but it is required after $t \leq \tau_1$, similarly to the previous cases.

(1) $t = 0$

$$\begin{pmatrix} V_{c1} \\ V_{s1} \end{pmatrix} = [\lambda_{1f}] \begin{pmatrix} 5 \\ 0 \end{pmatrix} = \begin{pmatrix} 10 \\ 0 \end{pmatrix} = (E_{12})$$

(2) $t = \tau_1 = 0.45\ \mu s$

$$(E_{2f}(t)) = (E_{21}(t-\tau_1)) = \begin{pmatrix} 10 \\ 0 \end{pmatrix},\quad (V_2(t)) = [\lambda_{2f}](E_{2f}(t)) = \begin{pmatrix} 10.26 \\ -7.28 \end{pmatrix}$$

$$(E_{21}) = (V_2) - (E_{2f}) = \begin{pmatrix} 0.26 \\ -7.28 \end{pmatrix},\quad (e_{21}) = [\mathrm{A}]^{-1}(E_{21}) = \begin{pmatrix} -7.28 \\ 7.54 \end{pmatrix}$$

(3) $t = 2\tau_1 = 0.9\ \mu s$

$$(\mathrm{e}_{1\mathrm{b}}^1(t)) = (\mathrm{e}_{21}^1(t-\tau_1)) = \begin{pmatrix} 0 \\ 7.54 \end{pmatrix},\quad (\mathrm{E}_{1\mathrm{b}}^1) = [A](\mathrm{e}_{1\mathrm{b}}^1) = \begin{pmatrix} 7.54 \\ 0 \end{pmatrix}$$

$$(V_1(t)) = [\lambda'_{1b}](E_{1b}) + (\mathrm{V}_1(t = 0)) = \begin{pmatrix} 8.84 + 10 \\ 0 \end{pmatrix} = \begin{pmatrix} 18.84 \\ 0 \end{pmatrix}$$

(4) $t = \tau_0 = 1\ \mu s : E_0 = 0$

$$(V_1(t)) = (E_{1b}) = \begin{pmatrix} 8.84 \\ 0 \end{pmatrix}$$

(5) $t = \tau_0 + \tau_0 = 1.45\ \mu s$

$$(e_{1b}^0(t)) = (e_{21}^0(t-\tau_0)) = \begin{pmatrix} -7.28 \\ 0 \end{pmatrix},\quad (E_b^0) = [A](e_b^0) = \begin{pmatrix} -7.28 \\ -7.28 \end{pmatrix}$$

$$(V_1(t)) = [\lambda'_{1b}](E_{1b}) + (V_1(t = 1\ \mu s)) = \begin{pmatrix} 8.84 \\ 0 \end{pmatrix}$$

5.1.5.6 Case *22-11* : $R_{c1} = R_{c2} = 200\ \Omega$, $R_{s1} = R_{s2} = 0\ \Omega$

$$[\lambda_{1f}] = \begin{bmatrix} 0.387 & 0.613 \\ 0 & 2 \end{bmatrix}, \quad [\lambda'_{1b}] = \begin{bmatrix} 1.696 & -1.696 \\ 0 & 0 \end{bmatrix}$$

$$[\lambda_{2f}] = \begin{bmatrix} 1.613 & -1.613 \\ 0 & 0 \end{bmatrix}$$

(1) $t = 0$

$$\begin{pmatrix} V_{c1} \\ V_{s1} \end{pmatrix} = [\lambda_{1f}] \begin{pmatrix} 5 \\ 0 \end{pmatrix} = \begin{pmatrix} 1.94 \\ 0 \end{pmatrix} = (E_{12})$$

There is only coaxial mode, and thus no modal decomposition is required.

(2) $t = \tau_1 = 0.45\ \mu s$

$$(E_{2f}(t)) = (E_{12}(t - \tau_1)) = \begin{pmatrix} 1.94 \\ 0 \end{pmatrix}, \quad (V_2(t)) = [\lambda_{2f}](E_{2f}(t)) = \begin{pmatrix} 3.13 \\ 0 \end{pmatrix}$$

$$(E_{21}) = (V_2) - (E_{2f}) = \begin{pmatrix} 1.19 \\ 0 \end{pmatrix}$$

(3) $t = 2\tau_1 = 0.9\ \mu s$

$$(E^1_{1b}(t)) = (E^1_{21}(t - \tau_1)) = \begin{pmatrix} 1.19 \\ 0 \end{pmatrix}$$

$$(V_1(t)) = [\lambda'_{1b}](E_{1b}) + [\lambda_{2f}](E_0) = \begin{pmatrix} 2.02 + 1.94 \\ 0 \end{pmatrix} = \begin{pmatrix} 3.96 \\ 0 \end{pmatrix}$$

(4) $t = \tau_0 = 1\ \mu s : E_0 = 0$

$$(V_1(t)) = (E_{1b}) = \begin{pmatrix} 2.02 \\ 0 \end{pmatrix}$$

Figure 5.11c–f shows node voltage waveforms analytically calculated in the above. The analytical results agree well with EMTP simulation results in Figure 5.6, Figure 5.7, Figure 5.8, and Figure 5.9. It should be noted that the modal composition/decomposition is required when a phase traveling wave appears both on the core and the sheath.

5.1.6 Concluding Remarks

Transient voltages of coaxial cable called "3D2V" have been measured with various terminating conditions of the core and the sheath at the sending and receiving ends. EMTP simulations of the transient voltages are also carried out in comparison with the measured results. Also, analytical calculations of the transient voltages have been made based on a traveling wave theory and a modal theory. The analytical calculations make it easy to understand the wave propagation characteristics of a cable and to explain a transient phenomenon physically.

In a high frequency transient, only coaxial mode propagation contributes to a node voltage if a traveling wave appears only on a core, for example, if a voltage is applied to the core at the sending end of a coaxial cable and the sending-end sheath is short circuited to ground, that is, the grounding resistance of the sheath is zero. However, even the coaxial mode propagation produces an earth-return mode component if the terminating impedances of the core and the sheath are not zero at the other end.

The experimental and analytical results of transient voltages in this section are expected to be useful to understand transient characteristics of a coaxial cable.

5.2 Pipe-enclosed Type (PT) Cable – Effect of Eccentricity

The significance of the proximity effect on conductor impedance is well-known, and there are a number of papers, which derive a theoretical formula of the impedance and the admittance [9–15] and discuss the impedance variation due to the proximity effect based on numerical simulations [16–18]. Also, power losses in a steady state are discussed based on measured results [19–22].

Recently, it has been pointed out that the proximity effect is also significant in a transient state for a surge waveform is noticeably distorted by an increase of a conductor resistance due to the proximity effect [23–25]. Unfortunately, there exists almost no measured data investigating the proximity effect for a transient [24].

In a PT cable, the proximity is represented by the eccentricity of the inner conductor from the pipe center.

In this section, the effect of the eccentricity on transients in a PT cable is investigated based on a theoretical analysis, EMTP and finite-difference time-domain (FDTD) [26] simulations. The effect of the inner conductor eccentricity on the frequency responses of the impedance, the characteristic impedance, the attenuation and the propagation velocity have been already explained in Section 3.6.3. From EMTP simulation results, the transient characteristics in the PT cable are discussed as a function of the eccentricity of the inner conductor. Also, an FDTD simulation [27] is carried out to confirm the observation of the proximity effect on transient (time) responses made by the EMTP simulations.

5.2.1 Model Circuit for the EMTP Simulation

Figure 5.12 illustrates a model circuit for EMTP and FDTD simulations. Figure 5.12a is the model circuit, and Figure 5.12b is the cross section of a PT cable with a single inner conductor of which the distance '*d*' between the centers of the pipe and the inner conductor is changed from $d=4.5$ to 0 cm (Cases 1–5) to discuss the effect of the inner conductor eccentricity. The cable and the parameters are the same as those in Section 3.6.3 (Figure 3.23).

As a voltage source, the following two waveforms are used: (1) a step voltage with the amplitude V_0 = 100 V; and (2) a pulse voltage with the rise time T_f = 15 ns and the amplitude V_0 = 100 V as in Figure 5.12c. The voltage is applied at the sending end between the inner conductor and the pipe as in Figure 5.12a. The sending end of the pipe is short-circuited to the ground terminal of the voltage source. The inner conductor and the pipe are short-circuited

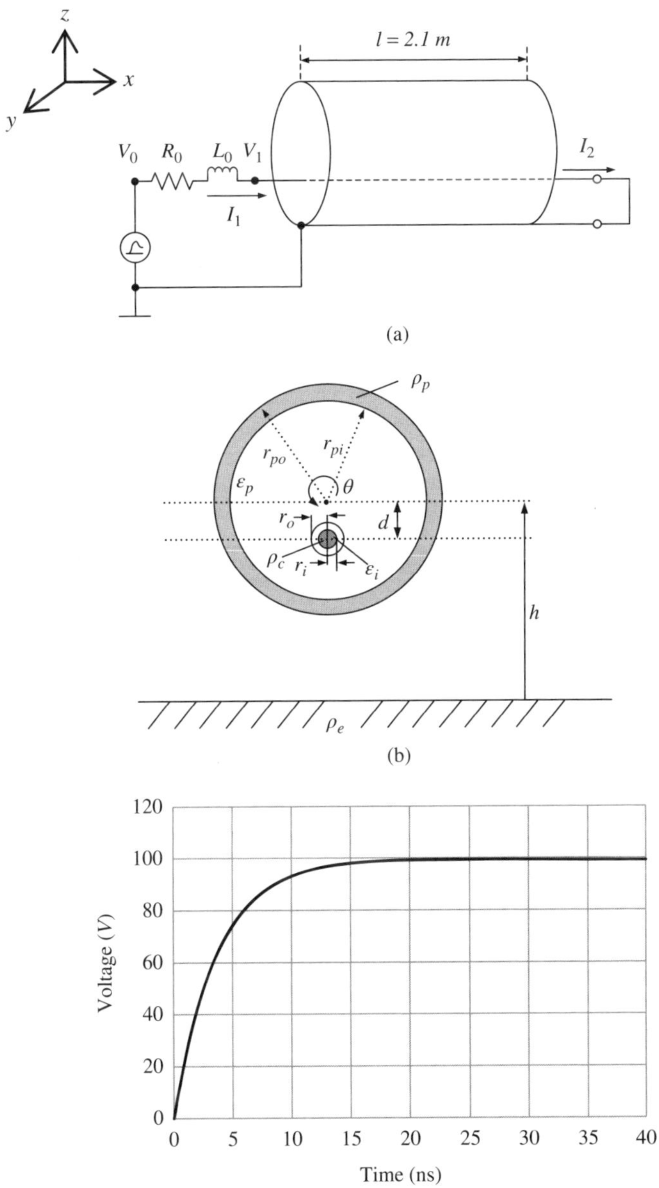

Figure 5.12 Simulation model: (a) a model circuit of a PT cable for a transient simulation; (b) cross section of the PT cable (pipe: $h = 7.3$ cm, $r_{pi} = 5$ cm, $r_{po} = 5.3$ cm, $\rho_p = 2.0 \times 10^{-8}$ Ωm; inner conductor: $r_i = 3.1$ mm, $r_0 = 4.1$ mm, $\varepsilon_r = 1.0$, $\rho_c = 2.0 \times 10^{-8}$ Ωm; Al plate: 10×5 m, $\rho_e = 2.8 \times 10^{-8}$ Ωm; cable length: $x = 2.1$ m; Case 1: $d = 4.5$ cm; Case 2: $d = 4.0$ cm; Case 3: $d = 2.5$ cm; Case 4: $d = 1.25$ cm; Case 5: $d = 0$); and (c) applied voltage waveform with $T_f = 15$ ns

at the receiving end. The sending-end voltage V_1 and the currents I_1 and I_2 at both ends are simulated as a function of the eccentricity $d/r_{pi} = 0$–0.9 (Cases 1–5).

5.2.2 *Simulation Results for Step-function Voltage Source*

5.2.2.1 Effect of the Eccentricity

Figure 5.13a and b shows EMTP simulation results, V_1 and I_1, when a step voltage with an amplitude of 100 V is applied in the case of $L_0 = 0$. The sending-end voltage shows a typical decreasing waveform on a distributed parameter line where the receiving end is short-circuited. That is, at every 2 τ (where τ is the travel time), negative voltage reflection from the short-circuited end comes back to the sending end, and the voltage V_1 decrease at every 2 τ.

At the same time, the current I_1 increases. The frequency f_1 of this change is given by:

$$f_1 = 1/T_1, \quad T_1 = 2\,\tau, \tau = l/c$$

For $l = 2.1$ m and $c \cong 300$ m/μs as explained in Section 3.6.3 (Figure 3.26),

$$\tau = 7 \text{ ns}, \quad T_1 = 2\,\tau = 14 \text{ ns}, \; f_1 = 1/T_1 \cong 35.7 \text{ MHz}$$

The currents I_1 and I_2 increase as time passes, and converge to the steady-state current I_s.

$$I_s = V_0/R_0 = 2 \text{ A}.$$

The effect of the eccentricity is very clear in the simulation results corresponding to the investigations in Section 3.6.3. For example, the first peak voltage for $d/r_{pi} = 0.9$ is 58 V, while that for $d/r_{pi} \leq 0.5$ is about 77 V, which is greater by 1.3 times. This corresponds to the greater surge impedance for $d/r_{pi} < 0.5$ in Figure 3.25. The current for $d/r_{pi} = 0.9$ becomes greater by about 1.7 times than that for $d/r_{pi} \leq 0.5$ correspondingly.

The first peak of the transient voltage and the current at the sending end are determined by the coaxial mode surge impedance z_{0c} of the conductor and the source resistance R_0, that is,

$$I_1 \cong V_0/(R_0 + z_{0c}) \tag{5.14a}$$

$$V_1 = z_{0c} I_1 \tag{5.14b}$$

$$d/r_{pi} < 0.5 \;:\; z_{0c} = 167\ \Omega, \quad I_1 = 0.46 \text{ A}, \quad V_1 = 77 \text{ V}$$

$$= 0.9 \;:\; z_{0c} = 67\ \Omega, \quad I_1 = 0.85 \text{ A}, \quad V_1 = 57 \text{ V}$$

It is clear that the above analysis agrees well with the simulation results in Figure 5.13a and b.

The analytical observations have clearly shown the significance of the proximity effect as a function of the eccentricity of an inner conductor against the pipe enclosure in the PT cable case, and thus the eccentricity should be considered in a transient simulation. Such a simulation is easily carried out by the EMTP, and also related impedances, admittances and propagation constants are obtained by the Cable Constants/Parameters of the EMTP [1, 5, 8].

Remember that the proximity effect becomes significant for eccentricity $d/r_{pi} > 1/4$. Otherwise, the eccentricity of the inner conductor of a PT cable can be neglected, and thus a coaxial cable approach is good enough.

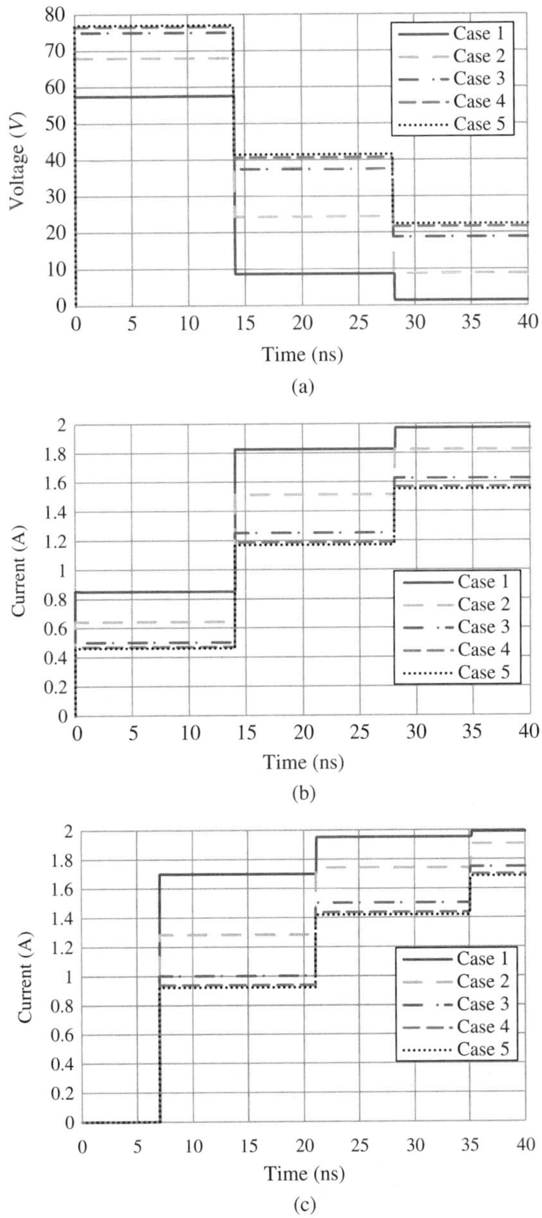

Figure 5.13 Simulation results for step-function voltage source with $L_0 = 0$: (a) sending-end voltage V_1; (b) sending-end current I_1; and (c) receiving-end current I_2

5.2.2.2 Effect of Lead Wire Inductance

In any transient measurement, there are lead wires to apply a source voltage and to measure a voltage and a current. The lead wires affect the measured responses significantly especially in a fast transient. The effect is discussed in this section.

Figure 5.14 shows simulation results for $L_0 = 0.5\ \mu\text{H}$ which represents the inductance of a current lead wire from the voltage source to the conductor, 1 μH/m, when an experiment is carried out. It is observed in Figure 5.13 and Figure 5.14 that the lead wire inductance affects the rise time of the voltage and current waveforms. The reason for this is readily explained as the time constant τ of an RL series circuit. From Figure 3.24 in Section 3.6.3.1, the conductor inductance of the coaxial mode at 5 MH is given approximately as:

$$d/r_{pi} < 0.2\ :\ L_c = \Delta L \times l = 0.55 \times 2.1 = 1.16\ \mu\text{H}$$

$$d/r_{pi} < 0.9\ :\ L_c = \Delta L \times l = 0.22 \times 2.1 = 0.46\ \mu\text{H}$$

The total inductance is given as $L = L_0 + L_c$. The resistance $R_0 = 50\ \Omega$ is far greater than the conductor resistance. Thus, the time constant τ becomes:

$$\tau \cong L/R_0 \tag{5.15}$$

$$\text{(a)}\ \ L_0 = 0\ :\ \tau_0 = 1.16 \times 10^{-6}/50 = 23.2\ \text{ns} \quad \text{for} \quad d/r_{pi} < 0.2$$

$$= 0.46 \times 10^{-6}/50 = 9.2\ \text{ns} \quad \text{for} \quad d/r_{pi} = 0.9$$

$$\text{(b)}\ \ L_0 = 0.5\ \mu\text{H}\ :\ \tau_1 = 1.16 \times 10^{-6}/50 = 33.2\ \text{ns} \quad \text{for} \quad d/r_{pi} < 0.2$$

$$= 0.96 \times 10^{-6}/50 = 19.2\ \text{ns} \quad \text{for} \quad d/r_{pi} = 0.9$$

Considering the steady-state current of 2 A, the above time constants agree qualitatively with the simulation results of the currents in Figure 5.13 and Figure 5.14.

5.2.2.3 Effect of Applied Voltage Waveform

Figure 5.15 and Figure 5.16 show simulation results of V_1 and I_1 in the model circuit in Figure 5.12a when a source voltage with the rise time $T_f = 15$ ns in Figure 5.12c is applied. The waveform in Figure 5.12c corresponds to an output voltage of a pulse generator often used in an experimental observation of a transient. Figure 5.15 is for the lead wire inductance $L_0 = 0$, and Figure 5.16 for $L_0 = 0.5\ \mu\text{H}$. Figure 5.15 clearly indicates that the transient voltage and current are very much distorted due to the applied voltage waveform in comparison with those in Figure 5.13 for a step-function voltage source. It is not easy to discuss the basic characteristic of traveling wave reflection and refraction from the physical viewpoint in Figure 5.15, while it is very clear in Figure 5.13. This difference should be remembered when an experimental observation is to be carried out, and it is suggested to make predictive calculations by using a simulation tool such as the EMTP with various parameters. A comparison of Figure 5.16 for $L_0 = 0.5\ \mu\text{H}$ with Figure 5.15 for $L_0 = 0$ clearly indicates the significance of a lead wire inductance in a transient on a short length conductor. As already discussed in the previous section, the lead wire inductance even with 0.5 μH corresponding to wire length of 0.5 m is comparative to the inductance of a tested conductor, and thus a measured waveform is not clear if it is the response of the test conductor or it is the overall response of the

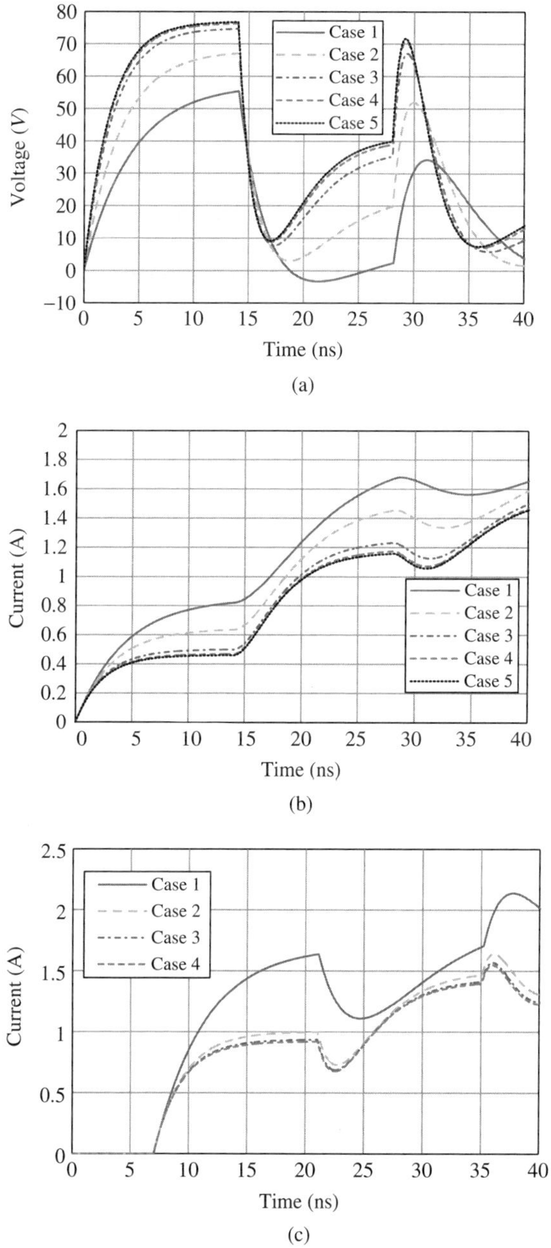

Figure 5.14 Simulation results for step-function voltage and $L_0 = 0.5$ μH: (a) sending-end voltage V_1; (b) sending-end current I_1; and (c) receiving-end current I_2

test conductor and measuring circuits. A comparison of Figure 5.16 with Figure 5.14 shows the effect of the applied source voltage with lead wire length of 0.5 m. It is observed that the oscillating period T of the voltage V_1 is:

$$\text{Figure5.14} : T = 14.5\ \text{ns}, \quad f = 68.9\ \text{MHz}$$

$$\text{Figure5.16} : T = 17.0\ \text{ns}, \quad f = 58.8\ \text{MHz}$$

The above result indicates a difficulty of a physical interpretation of a transient waveform when the applied source voltage is not a step function. Again, to avoid an erroneous interpretation of an obtained simulation/measured result, predictive calculations with various parameters are recommended.

5.2.3 FDTD Simulation

5.2.3.1 Model Circuit

The model circuit and parameters for an FDTD simulation are the same as in Figure 5.12. The FDTD simulation is carried out with cell size $\Delta x = 2.5$ mm, $\Delta y = 10$ mm, $\Delta z = 2.5$ mm and the analytical space $x = 0.35$ m, $y = 3.3$ m, $z = 0.275$ m. The applied voltage is the same as in Figure 5.12c. VSTL is used as the simulation tool [27].

5.2.3.2 Simulation Results

Figure 5.17 shows simulation results corresponding to Figure 5.16. It is observed that the FDTD simulation results agree satisfactorily with the EMTP simulation results. The peak voltage at around $t = 15$ ns in Figure 5.17 is a little higher than that in Figure 5.16, and the voltage decrease after 15 ns is smaller in Figure 5.17. The currents at 15 and 30 ns are greater in Figure 5.17, the FDTD simulation. The reason for this is smaller attenuation of the inner conductor, that is, no resistance of the inner conductor, because the FDTD (VSTL) assumes a perfect conductor. However, Figure 5.17 shows the proximity effect clearly, and the observation made by the EMTP is confirmed by the FDTD simulation.

5.2.4 Theoretical Analysis

5.2.4.1 Steady-state Solution in a Frequency Domain for a Long Time Period

In a two-conductor system in Figure 5.12 voltages and currents at the sending and receiving ends are related by adopting an F-parameter (four-terminal parameter) theory in the following form.

$$(V_s) = [A] \cdot (V_r) + [B](I_r) \tag{5.16a}$$

$$(I_s) = [C] \cdot (V_r) + [D](I_r) \tag{5.16b}$$

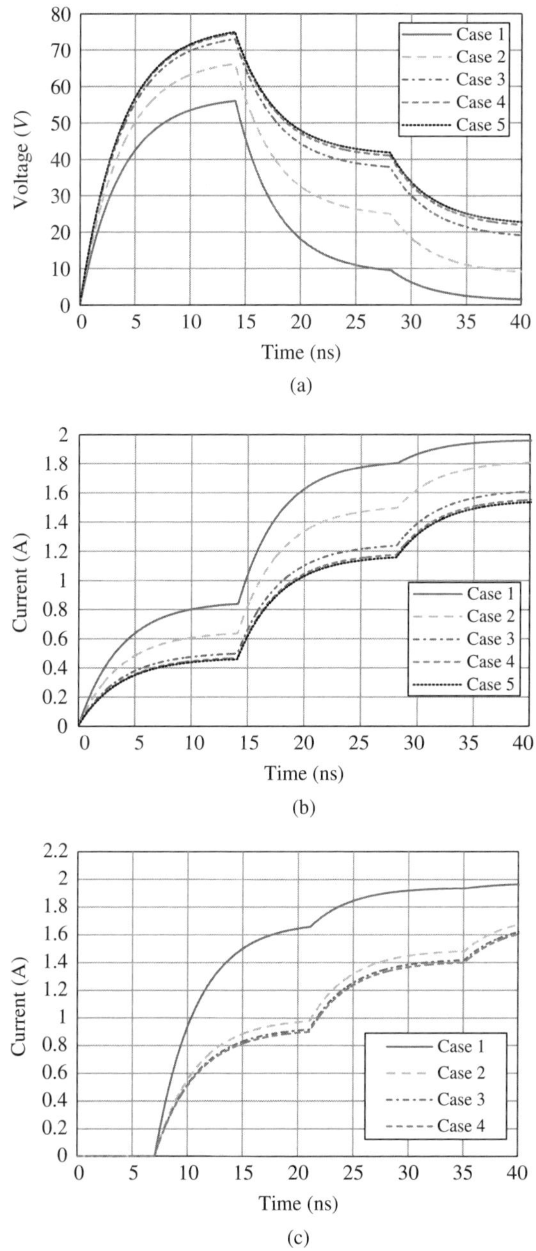

Figure 5.15 Simulation results for voltage source of $T_f = 15$ ns and $L_0 = 0$: (a) sending-end voltage V_1; (b) sending-end current I_1; and (c) receiving-end current I_2

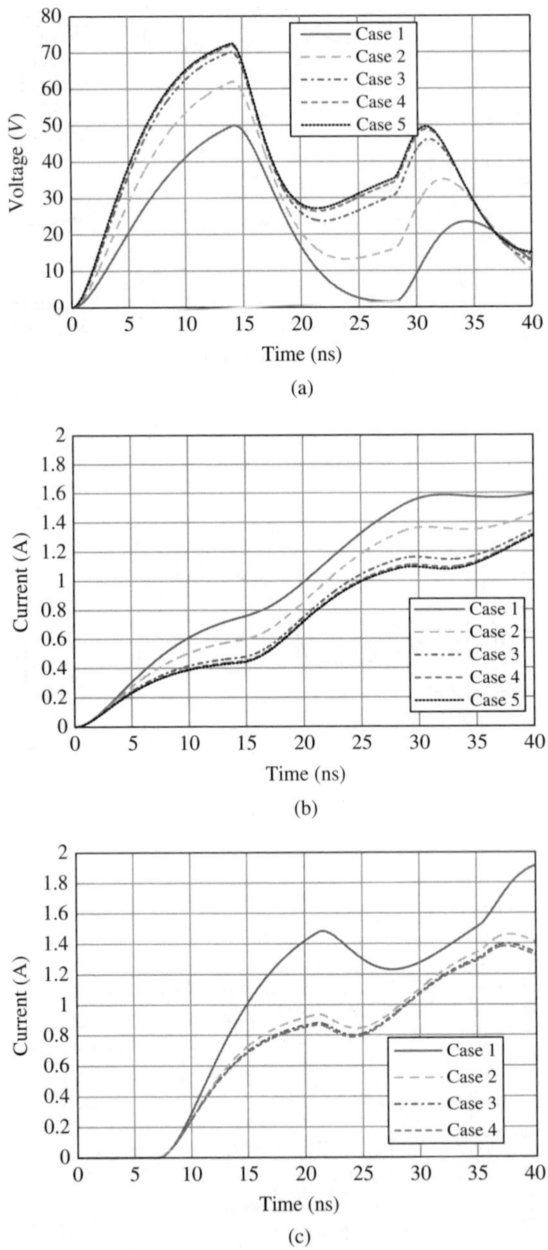

Figure 5.16 Simulation results for $T_f = 15$ ns and $L_0 = 0.5$ μH; (a) sending-end voltage V_1; (b) sending-end current I_1; and (c) receiving-end current I_2

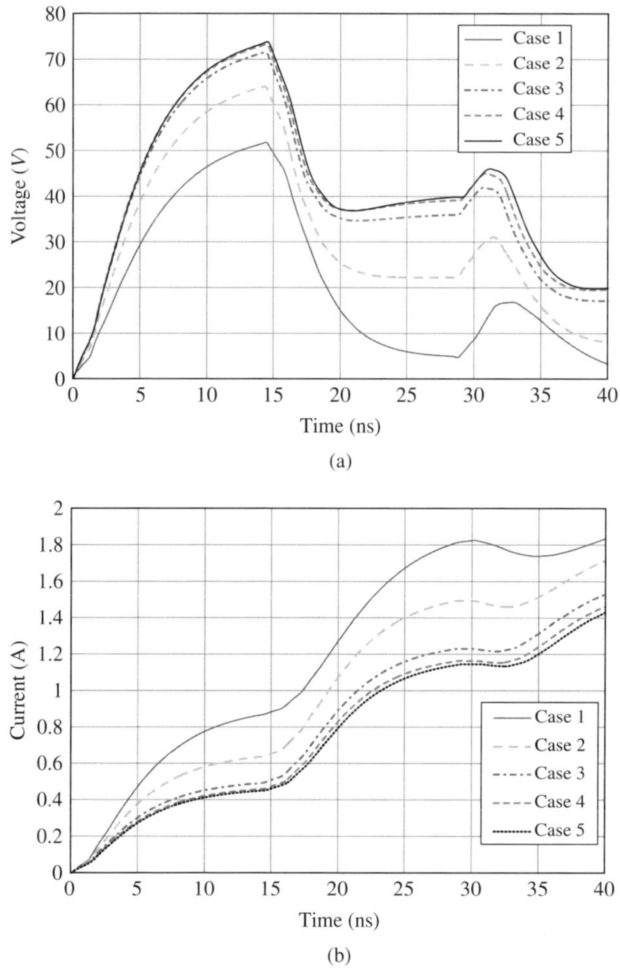

Figure 5.17 FDTD simulation results corresponding to Figure 5.16: (a) sending-end voltage V_1; and (b) sending-end current I_1

where

$$[A] = \cos([\Gamma] \cdot l) \tag{5.17a}$$

$$[B] = \sinh([\Gamma] \cdot l) \cdot [Z_0] \tag{5.17b}$$

$$[C] = [Y_0] \cdot \sinh([\Gamma] \cdot l) \tag{5.17c}$$

$$[D] = [Y_0] \cdot \cosh([\Gamma] \cdot l) \cdot [Z_0] \tag{5.17d}$$

For the inner conductor (conductor 1) is short-circuited to the pipe (conductor 2) at the receiving end and the pipe is grounded at the sending end, the following boundary conditions are given.

$$V_{s1} = 0 \tag{5.18a}$$

$$V_{r1} = V_{r2} = V_r \tag{5.18b}$$

$$I_{r2} = -I_{r1} = -I_2 \tag{5.18c}$$

Thus, the voltage and current vectors are:

$$(V_s) = \begin{pmatrix} V_1 \\ 0 \end{pmatrix} \tag{5.19a}$$

$$(V_r) = \begin{pmatrix} V_r \\ V_r \end{pmatrix} \tag{5.19b}$$

$$(I_r) = \begin{pmatrix} I_2 \\ -I_2 \end{pmatrix} \tag{5.19c}$$

$$(I_s) = \begin{pmatrix} I_1 \\ I_p \end{pmatrix} \tag{5.19d}$$

Because the conductor length l is very short, the well-known L equivalent lumped circuit can be used to represent the PT cable in Figure 5.12.

$$[A] = [U] \tag{5.20a}$$

$$[B] = [Z] \tag{5.20b}$$

$$[C] = [Y] \tag{5.20c}$$

$$[D] = [U] + [Z] \cdot [Y] = [U] + [\Gamma]^2 \cdot l^2 \tag{5.20d}$$

In the above equations, U is a unit matrix, and Z and Y are the impedance and admittance which are given as the output of the Cable Constants/Parameters of the EMTP. Γ is the propagation constant in a phase domain, and easily evaluated from the modal propagation constant γ by:

$$\Gamma = A \cdot \gamma \cdot A^{-1} \tag{5.21}$$

where A is the voltage transformation matrix.

γ and A are also given as the output of the Cable Constants/Parameters. In the PT cable in Figure 5.12b, Γ becomes:

$$\Gamma = \begin{bmatrix} \gamma_0^2 & \gamma_1^2 - \gamma_0^2 \\ 0 & \gamma_1^2 \end{bmatrix} \tag{5.22}$$

where γ_0 is the mode 0 (earth-return) propagation constant and γ_1 is the mode 1 (coaxial) propagation constant.

Because the conductor length l is very small, the following approximation is satisfied.

$$[U] \gg [\Gamma]^2 \cdot l^2$$

Thus, D in (5.20d) is expressed by:

$$[D] \cong [U] \tag{5.23}$$

Solving (5.16a,b) under the conditions given in (5.19a–d), (5.20a–d), (5.23) and $Y_{11} \cong Y_{12}$,

$$V_1 = z_c \cdot V_0/\{z_c + R_0(1 + 2\gamma_0^2 - \gamma_1^2)\} \cong z_c \cdot V_0/(z_c + R_0) \tag{5.24a}$$

$$I_1 = (1 + 2\gamma_0^2 - \gamma_1^2)V_1/z_c \cong V_0/(z_c + R_0) \tag{5.24b}$$

$$I_2 = V_1/(Z_{11} + Z_{22} - 2Z_{12}) \cong V_1/z_c = V_0/(z_c + R_0) \cong I_1 \tag{5.24c}$$

where $z_c = Z_{11} - Z_{12}$ is the coaxial mode impedance.

The above analytical solutions are effective only for a long time period, but not for the time period of traveling wave reflection/refraction, because the solutions are based on the lumped parameter equivalence in a steady state. However, the solution supports the observations in Section 5.2.2.1 and (5.14a,b), that is, the surge impedance z_{0c} for the time period of traveling wave reflection/refraction and the impedance Z_c for a longer time period. When the lead wire inductance is to be included in an analysis, R_0 in (5.24) is replaced by $R_0 + j\omega L_0$. By Laplace inverse transform of the equation, transient voltages and currents for a long time period are easily obtained.

5.2.4.2 Traveling Wave Solutions for a Short Time Period

For traveling wave solutions of V_1, V_2 and I_2 in a circuit illustrated in Figure 5.12a, the incoming (original) traveling wave e_0 from the source is given according to Thevenin's theorem.

$$e_0 = V_0/2 \tag{5.25}$$

The source circuit surge impedance matrix Z_s and that of the PT cable are:

$$[Z_s] = \begin{bmatrix} R_0 & 0 \\ 0 & 0 \end{bmatrix} \tag{5.26a}$$

$$[Z_0] = \begin{bmatrix} Z_{01} & Z_{02} \\ Z_{02} & Z_{02} \end{bmatrix} \tag{5.26b}$$

The refraction coefficient matrix from the source side to the cable is:

$$[\lambda_{1f}] = 2[Z_0]([Z_s] + [Z_0])^{-1} = \{R_0/(R_0 + z_{0c})\} \begin{bmatrix} z_{0c} & R_0 \\ 0 & R_0 + z_{0c} \end{bmatrix} \tag{5.27}$$

where $z_{0c} = Z_{01} - Z_{02}$ is the coaxial mode surge impedance.

At the receiving end, the inner conductor and the pipe are short-circuited. It is necessary to adopt a rotation matrix to obtain the receiving-end voltage and current as explained in Section 3.5 [3, 28]. The reflected traveling wave becomes the same as the arriving wave with negative polarity. The sending-end voltage at $t = 0$ is evaluated from λ_{1f} in the above equation and e_0 in (5.25).

$$\begin{pmatrix} V_{s1} \\ V_{s2} \end{pmatrix} = [\lambda_{1f}] \begin{pmatrix} e_0 \\ 0 \end{pmatrix} = \{2e_0/(R_0 + z_{0c})\} \begin{pmatrix} z_{0c} \\ 0 \end{pmatrix} \tag{5.28}$$

Thus,

$$V_1 = V_{s1} = z_{0c}V_0/(R_0 + z_{0c}), \quad V_{s2} = 0 \quad \text{for} \quad t = 0$$

$$I_1 = V_1/z_{0c} = V_0/(R_0 + z_{0c}) \tag{5.29}$$

At $t = \tau$:

$$I_2 = 2I_1 = 2V_0/(R_0 + z_{0c}) \quad \text{for} \quad t \leq 2\tau \tag{5.30}$$

The above solution agrees with (5.14a).

5.2.5 Concluding Remarks

This section has investigated the proximity effect on the surge propagation on an inner conductor of a PT cable as a function of the eccentricity of the conductor against the pipe inner surface. Based on the investigations, the following points are obtained.

1. A transient voltage calculated by the EMTP becomes smaller and a transient current becomes greater as the eccentricity becomes larger. It should be noted that the lead wire inductance comparable with the inductance of a conductor and the rise time of an applied voltage affect transient voltages and currents very significantly when a tested conductor is short, and thus it becomes difficult to discuss the effect of the proximity from measured results of a transient response.
2. FDTD simulation results agree satisfactorily with the EMTP simulation results, and show clearly the effect of the eccentricity during a transient. Thus, the significance of the eccentricity, that is, the proximity effect on the transient, has been confirmed by the FDTD simulations.

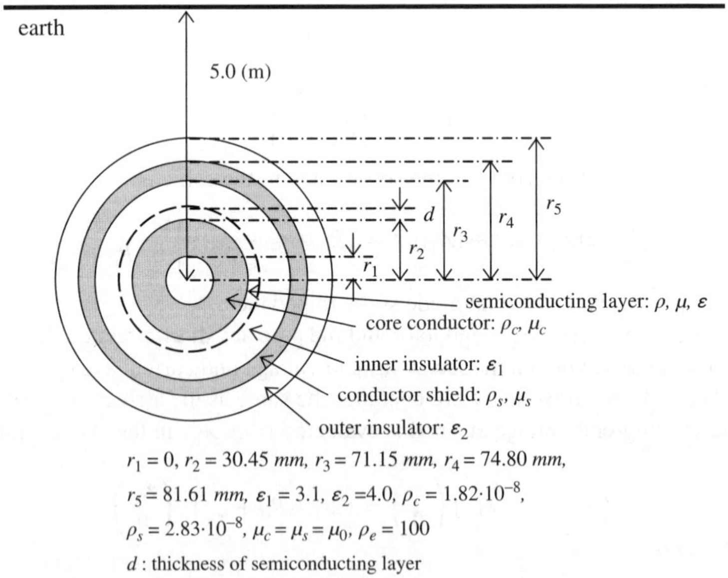

Figure 5.18 An underground 275 kV XLPE cable: $r_1 = 0$, $r_2 = 30.45$ mm, $r_3 = 71.15$ mm, $r_4 = 74.80$ mm, $r_5 = 81.61$ mm, $\varepsilon_1 = 3.1$, $\varepsilon_2 = 4.0$, $\rho_c = 1.82 \times 10^{-8}$, $\rho_s = 2.83 \times 10^{-8}$, $\mu_c = \mu_s = \mu_0$, $\rho_e = 100$, and d is the thickness of the semiconducting layer

5.3 Effect of a Semiconducting Layer on a Transient

Impedance and admittance formulas of a semiconducting layer have been described in Section 2.4 of Chapter 2, and the effect on wave propagation has been explained in Section 2.4.5. In this section, the effect of the semiconducting layer on a transient is discussed.

5.3.1 *Step Function Voltage Applied to a 2 km Cable*

Figure 5.18 illustrates an underground 275 kV XLPE cable with a length of 2 km. Figure 5.19 shows transient voltages at the receiving end of phase-a, when a step-function voltage is applied to the sending-end core of phase-a.

The cable is represented by a multiphase distributed line model with constant parameters at frequency $f = 1/4\tau = 21$ kHz (where τ is the traveling time of coaxial mode) in an EMTP simulation. The core voltage clearly shows a greater attenuation of the peak voltage and more retardation (longer oscillating period) by the semiconducting layer in Cases 1 and 2 than those in Case 0, no semiconducting layer, but the effect of the resistivity is not observed between

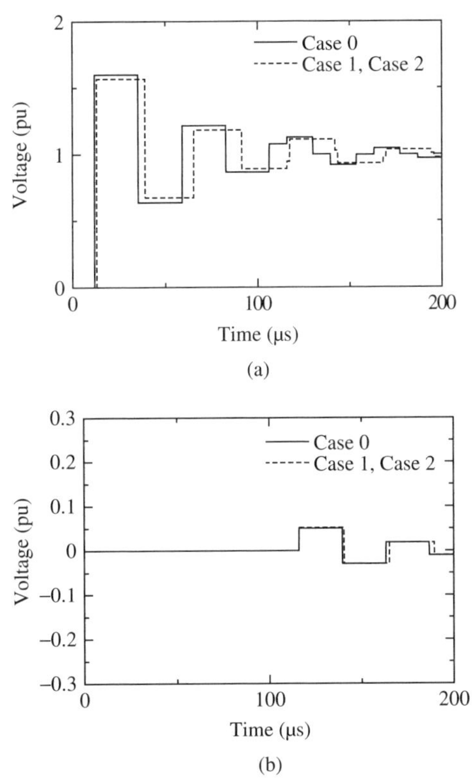

Figure 5.19 Switching transient voltages at phase-a receiving end on a 275 kV cable with a length of 2 km. Case 0: no semiconducting layer; Case 1: $\rho = 0.01$ Ωm; and Case 2: $\rho = 1$ Ωm. (a) Core voltage and (b) sheath voltage

Cases 1 and 2. In contrast, the conductor shield voltage becomes greater in Cases 1 and 2 than that in the case of no semiconducting layer.

5.3.2 5×70 μs *Impulse Voltage Applied to a 40 km Cable*

Figure 5.20a shows transient core voltages at the phase-a receiving end on a 275 kV cable with a length of 40 km, when an impulse voltage of 5 × 70 μs with an amplitude of 1 pu is applied to the phase-a conductor at the sending end, assuming lightning on an overhead line connected to the underground cable. The effect of the semiconducting layer on the core voltage is clearly observed on such a long distance cable in the case of the impulse voltage application. The greater the resistivity, the smaller the peak voltage.

Figure 5.20b shows the distance characteristic of the maximum core voltages along the 40 km cable. It is observed that the peak voltage is reduced to 0.5 pu in Case 3, $\rho = 1\,\Omega$m at a distance of 5 km from the sending end, while it is about 0.7 pu in Case 0, no semiconducting layer.

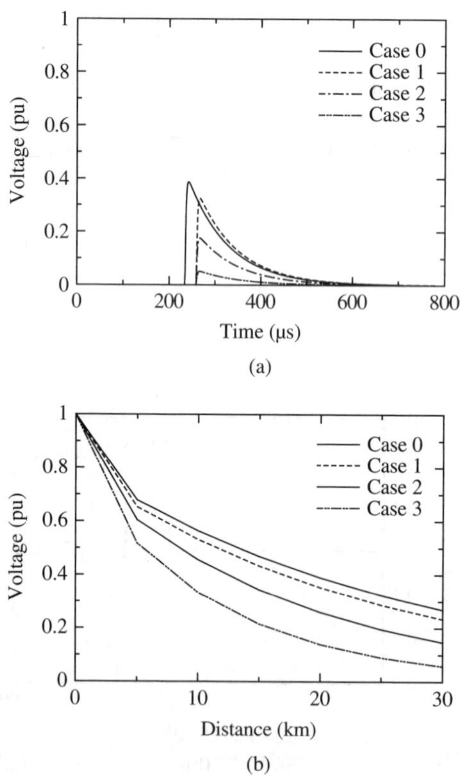

Figure 5.20 Lightning transient voltages on a 275 kV cable with a length of 40 km. Case 0: no semiconducting layer; Case 1: $\rho = 1\ \Omega$m; Case 2: $\rho = 0.03\ \Omega$m; and Case 3: $\rho = 0.01\ \Omega$m. (a) Phase-a core voltage at the receiving end and (b) maximum core voltage vs. distance

This indicates that the semiconducting layer and its resistivity significantly affect the maximum core voltage, when an impulse voltage is applied to the core of a long cable.

References

[1] Scott-Meyer, W. (1982) *EMTP Rule Book*, Bonneville Power Administration.

[2] Dommel, H.W. (1986) *EMTP Theory Book*, Bonneville Power Administration.

[3] Ametani, A. (1990) *Distributed-Parameter Circuit Theory*, Corona Publishing Co. (in Japanese).

[4] Ametani, A. (1973) Modified traveling wave technique to solve electrical transients on lumped and distributed constant circuits. *Proceedings of the IEE*, **120** (4), 497–504.

[5] Ametani, A. (1994) *Cable Parameter Rule Book*, Bonneville Power Administration.

[6] Ametani, A. (1980) Wave propagation characteristics of cables. *IEEE Transactions on Power Apparatus and Systems*, **PAS-99** (2), 499–505.

[7] Ametani, A. (1983) A study of cable transient calculations. *The Science and Engineering Review of Doshisha University*, **24** (2), 110–127.

[8] Ametani, A. (1980) General formulation of impedances and admittances of cable. *IEEE Transactions on Power Apparatus and Systems*, **PAS-99** (3), 902–10.

[9] Tegopoulos, J.A. and Krezis, E.E. (1971) Eddy current distribution in cylindrical shells of infinite length due to axial currents, Part II. Shells of finite thickness. *IEEE Transactions on Power Apparatus and Systems*, **PAS-90**, 1287–94.

[10] Brown, G.W. and Racamora, R.G. (1976) Surge propagation in three-phase pipe-type cables, Part I. Unsaturated pipe. *IEEE Transactions on Power Apparatus and Systems*, **PAS-95**, 88–95.

[11] Dugan, R.C., Racamora, R.G., and Brown, G.W. (1977) Surge propagation in three-phase pipe-type cables, Part II. Duplication of field tests including the effects of neutral wires and pipe saturation. *IEEE Transactions on Power Apparatus and Systems*, **PAS-96**, 826–33.

[12] Schinsinger, R. and Ametani, A. (1978) Surge propagation characteristics of pipe enclosed underground cables. *IEEE Transactions on Power Apparatus and Systems*, **PAS-97**, 1680–7.

[13] Dokopoulos, P. and Tampkins, D. (1984) Analysis of field and losses in three-phase gas cables with thick walls: Part I. Field analysis. *IEEE Transactions on Power Apparatus and Systems*, **PAS-103** (9), 2728–34.

[14] Dokopoulos, P. and Tampkins, D. (1985) Analysis of field and losses in three-phase gas cables with thick walls: Part II. Calculation of losses and results. *IEEE Transactions on Power Apparatus and Systems*, **PAS-104** (1), 9–15.

[15] Poltz, J., Kuffel, E., Grzybowski, S., and Raghuveer, M. R. (1982) Eddy-current losses in pipe-type cable systems. *IEEE Transactions on Power Apparatus and Systems*, **PAS-101** (4), 825–32.

[16] Fortin, S., Yang, Y., Ma, J., and Dawalibi, F.P. (2005) Effects of eddy current on the impedance of pipe-type cable with arbitrary pipe thickness. *ICEE 2005*, Paper TD2-09.

[17] Chien, C.H. and Bucknall, R.W.G. (2009) Harmonic calculations of proximity effect on impedance characteristics in subsea power transmission cables. *IEEE Transactions on Power Delivery*, **24** (2), 2150–8.

[18] Gustavsen, B., Bruaset, A., Bremnes, J.J., and Hassel, H. (2009) A finite element approach for calculating electrical parameters of umbilical cables. *IEEE Transactions on Power Delivery*, **24** (4), 2375–84.

[19] Ishikawa, T., Kawasaki, K., and Okamoto, I. (1976) Eddy current losses in cable sheaths (1). *Dainichi Nihon Cable Journal*, **61**, 34–42.

[20] Ishikawa, T., Kawasaki, K. and Okamoto, I. (1975) Eddy current losses in cable sheaths (2). *Dainichi Nihon Cable Journal*, **62**, 21–64.

[21] Kawasaki, K., Inami, M. and Ishikawa, T. (1981) Theoretical consideration on eddy current losses on non-magnetic and magnetic pipes for power transmission systems. *IEEE Transactions on Power Apparatus and Systems*, **PAS-100** (2), 474–84.

[22] Mekjian, A. and Sosnowski, M. (1983) Calculation of alternating current losses in steel pipe containing power cables. *IEEE Transactions on Power Apparatus and Systems*, **PAS-102** (2), 382–8.

[23] Pagnetti, A. (2012) Cable modeling for electromagnetic transients in power systems. PhD thesis. University of Clermont Ferrand, France and University of Bologna, Italy.

[24] Gudmundsdottir, U.S. (2010) Modeling of long high voltage AC cables in transmission systems. PhD thesis. Aalborg University.

[25] CIGRE WG C4.502 (2013) Power System Technical Performance Issues Related to the Application of Long HVAC Cables. *CIGRE Technical Brochure* 556.
[26] CIGRE WG C4.501 (2013) Guideline for Numerical Electromagnetic Analysis Method and Its Application to Surge Phenomena. *CIGRE Technical Brochure* 543.
[27] CRIEPI (2007) Visual Surge Test Laboratory (VSTL). http://criepi.or.jp (accessed April 2014).
[28] Ametani, A., Kato, R., Nishinaga, H., and Okai, M. (1995) A study on transient induced voltages on a MAGLEV train coil system. *IEEE Transactions on Power Delivery*, **10** (3), 1657–62.

6

Transient on Three-phase Cables in a Real System

Akihiro Ametani
Graduate School of Electrical Engineering, Doshisha University, Kyoto, Japan

In this chapter, field test results on various three-phase cables are explained, and a comparison with simulation results is carried out. Surge characteristics and the effect of various parameters are investigated based on the field test and simulation results. Also, electromagnetic transients program (EMTP) simulation results by various cable models prepared in the EMTP are discussed.

6.1 Cross-bonded Cable

6.1.1 Field Test on an 110 kV Oil-filled (OF) Cable

Figure 6.1 illustrates a field test circuit on an 110 kV OF cable with a core cross section of 610 mm^2 and Al sheath [1].

An impulse voltage 7.3 kV with wavefront duration $T_f \fallingdotseq 0$ and wave tail $T_f = 40$ μs is applied to the middle-phase core at the sending end. Figure 6.2 shows measured results and Figure 6.3 gives the simulation results [2, 3] by a frequency-domain transient analysis program (FTP) [2, 4]. A satisfactory agreement is observed in the results, although the first negative voltage in the measured result, Figure 6.2b, is missed in the oscilloscope. In the FTP calculation, the dielectric loss of a corrosion-proof layer is considered in Figure 6.3. Figure 6.4 shows a FTP simulation result of the applied phase core voltage when neglecting the dielectric loss. An oscillating spike-like voltage is observed in Figure 6.4 in comparison with the field test and simulation results in Figure 6.2a and Figure 6.3a.

6.1.2 Effect of Cross-bonding

Figure 6.5 illustrates a cross section of a cable and its configuration, and Table 6.1 shows cable data. The length of one minor section is 500 m and the total length is 12 km. Thus, the system

Cable System Transients: Theory, Modeling and Simulation, First Edition.
Akihiro Ametani, Teruo Ohno and Naoto Nagaoka.

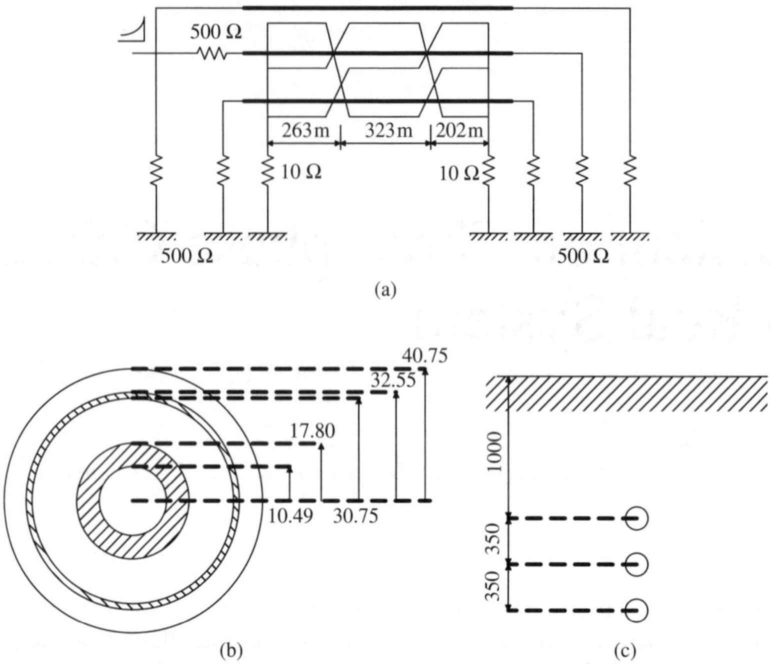

Figure 6.1 A field test circuit of an 110 kV OF cable: (a) field test circuit; (b) cross section of 110 kV OF cable; and (c) cable arrangement

Table 6.1 Parameters of a cable in Figure 6.5

$r_1 = 1.03$ (cm)	$\varepsilon_1 = 3.5$
$r_2 = 1.90$ (cm)	$\varepsilon_2 = 4.0$
$r_3 = 3.45$ (cm)	$\rho_1 = 1.70 \times 10^{-8}$ (Ωm)
$r_4 = 3.85$ (cm)	$\rho_2 = 2.10 \times 10^{-7}$ (Ωm)
$r_5 = 4.25$ (cm)	$\mu_1 = 1.0$
$d = 1.0$ (m)	$\mu_2 = 1.0$
$s = 0.35$ (m)	$\rho_e = 50$ (Ωm)

is composed of eight major sections. A unit step voltage is applied with no source impedance between the outer core and earth. The other cores are open-circuited, and the sheaths are connected together and grounded through a 0.5 Ω resistance at both ends and grounded through a 10 Ω resistance at the boundary of major sections. Earth resistivity is assumed to be 50 Ωm.

Figure 6.6a and b shows the receiving-end core voltages for cross-bonded cable and normal-bonded cable, respectively. Table 6.2 shows the maximum overvoltages. It is clear that when neglecting the cross-bonding, the waveshape becomes a square-like wave and its oscillating period is four times the propagation time of the coaxial mode. When considering the cross-bonding of the cable, the waveshape is quite different from the square-like wave and the maximum sheath voltage is rather high, because the cross-bonding behaves as

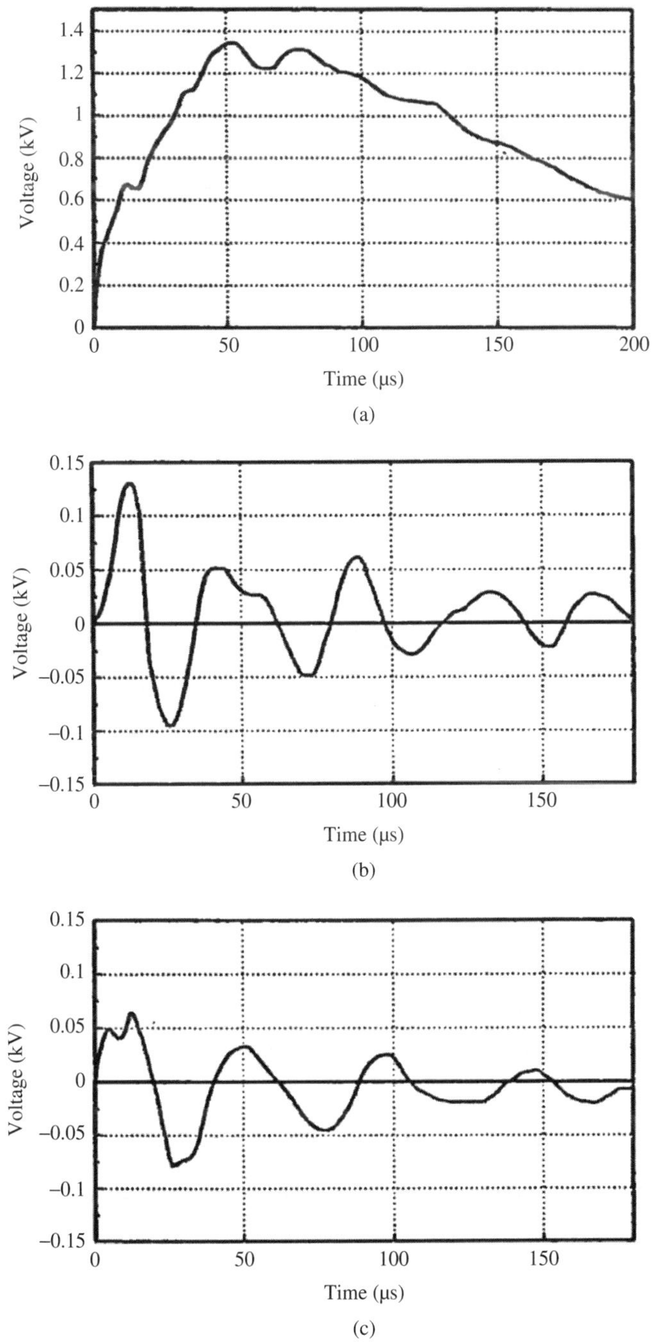

Figure 6.2 Field test results on the 110 kV OF cable in Figure 6.1: (a) sending-end core 2 voltage; (b) second-point sheath 2 voltage; and (c) second-point sheath 3 voltage

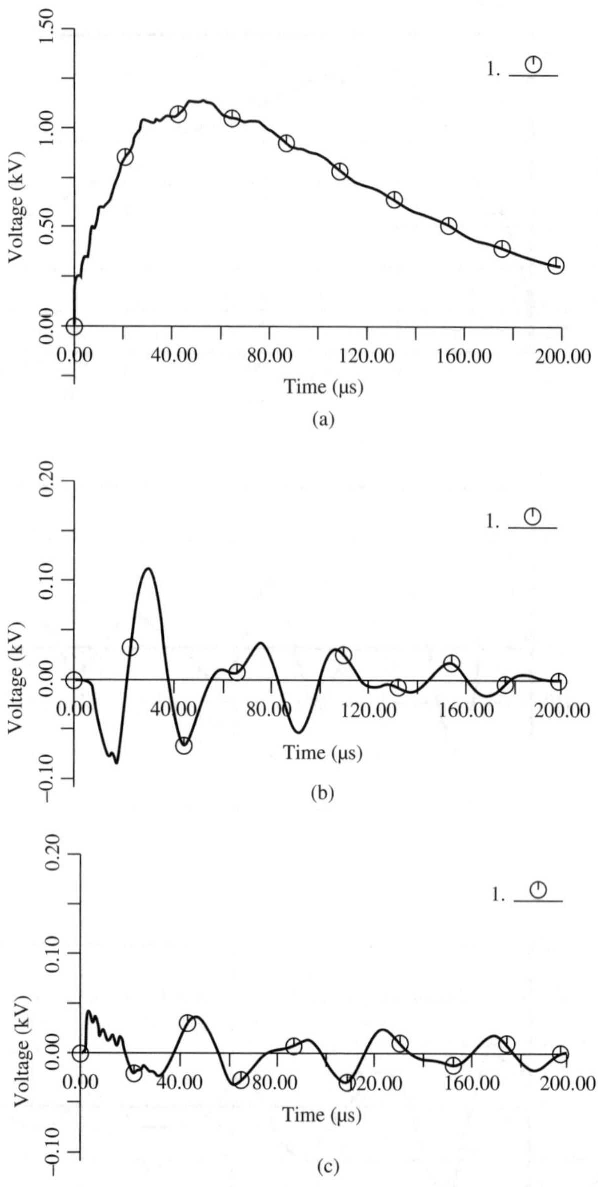

Figure 6.3 FTP simulation results: (a) sending-end core 2 voltage; (b) second-point sheath 2 voltage; and (c) second-point sheath 3 voltage

discontinuous points for traveling waves. Therefore, it is impossible to calculate transients in a cross-bonded cable neglecting the cross-bonding.

6.1.3 Effect of Various Parameters

Table 6.3 shows the maximum voltages at the sending end, first and second cross-bonding points and receiving end when the source impedance is changed. It is clear that as the source

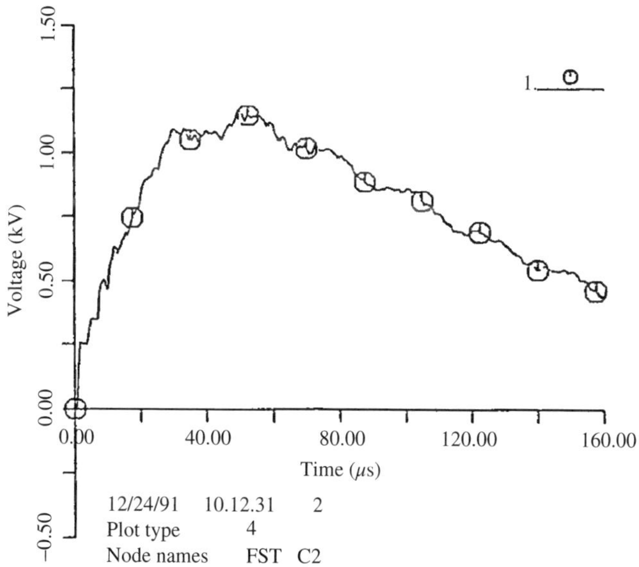

Figure 6.4 FTP simulation result when neglecting dielectric loss

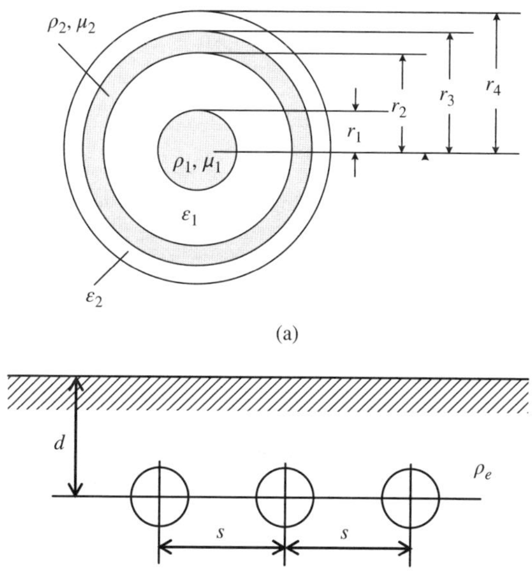

Figure 6.5 An underground SC cable: (a) cross section; and (b) cable configuration

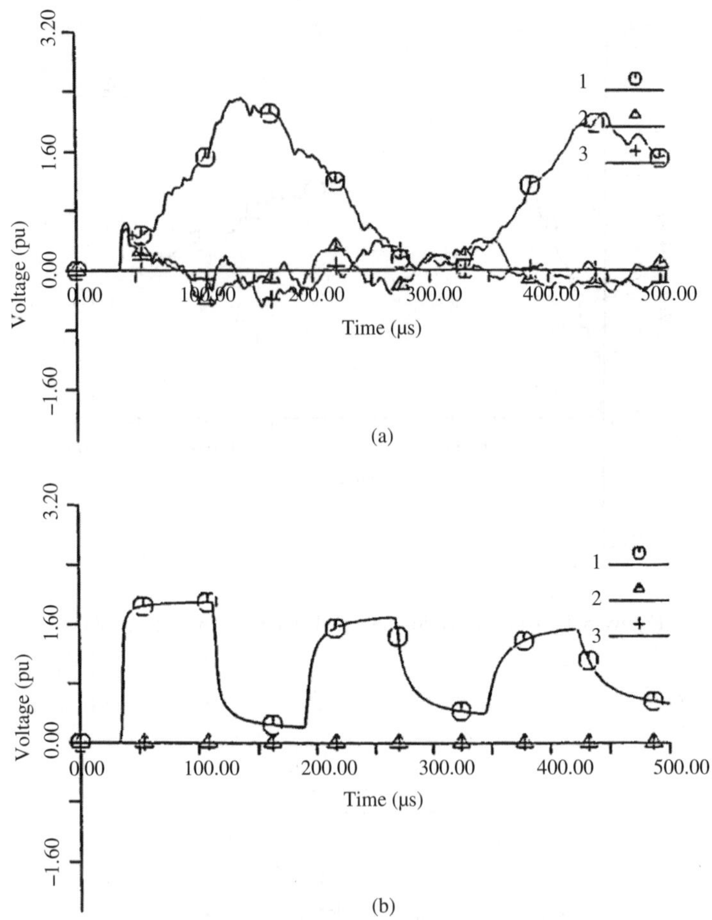

Figure 6.6 Effect of cross-bonding when applying a source voltage to the outer core: (a) cross-bonded; and (b) normal-bonded

impedance increases, the maximum induced voltages at the cross-bonding points and receiving end decrease.

Table 6.4 shows the maximum voltages when changing the sheath grounding resistance. From the table, it can be seen that the resistance at both ends affects the sheath maximum voltages, but the resistance at the intermediate points causes a less significant effect on the maximum voltages. Therefore, it is concluded that a calculation of a cable transient neglecting the sheath grounding resistance except at both ends gives a reasonable accuracy.

Figure 6.7 shows calculated results when a source voltage is applied to the center core. In this case, the time when maximum voltage appears is slightly earlier than in the case of the source being applied to the outer core as in Figure 6.6a. This can be explained in the following manner. Table 6.5 shows the characteristic impedance, voltage transformation matrix, attenuation constants and velocities of each mode. Ignoring the sheath grounding resistances except

Table 6.2 Effect of cross-bonding on maximum voltages (per unit)

		Case a	Case b			Case a	Case b
Sending end	C_1	1.00	1.00	2nd point	C_1	1.23	1.50
	C_2	0.427	0.0235		C_2	0.402	0.0227
	C_3	0.616	0.0235		C_3	0.518	0.0234
	S	0.0176	0.0235		S_1	0.166	0.0226
1st point	C_1	1.13	1.30		S_2	0.130	0.0229
	C_2	0.444	0.0236		S_3	0.200	0.0227
	C_3	0.603	0.0239	3rd point	C_1	1.32	1.60
	S_1	0.190	0.0239		C_2	0.397	0.0156
	S_2	0.190	0.0228		C_3	0.410	0.0136
	S_3	0.141	0.0234		S	0.0094	0.0119
				Receiving end	C_1	2.35	1.81
					C_2	0.468	0.0007
					C_3	0.661	0.0003
					S	0.0002	0.0000

Case a, cross-bonded; Case b, normal-bonded.

Table 6.3 Effect of a source impedance on maximum voltages (per unit)

		0 Ω	400 Ω			0 Ω	400 Ω
Sending end	C_1	1.00	1.00	2nd point	C_1	1.23	1.00
	C_2	0.427	0.026		C_2	0.402	0.0199
	C_3	0.616	0.0454		C_3	0.518	0.0373
	S	0.0176	0.0011		S_1	0.166	0.0134
1st point	C_1	1.13	1.90		S_2	0.130	0.0065
	C_2	0.444	0.0233		S_3	0.200	0.0165
	C_3	0.603	0.0408	3rd point	C_1	1.32	1.00
	S_1	0.190	0.0125		C_2	0.397	0.0191
	S_2	0.190	0.0145		C_3	0.410	0.0365
	S_3	0.141	0.0107		S	0.0094	0.0008
				Receiving end	C_1	2.35	1.00
					C_2	0.468	0.0274
					C_3	0.661	0.0505
					S	0.0002	0.0000

at both ends, the cross-bonded cable is assumed to be a homogenous line approximately as explained in Section 3.5.3 of Chapter 3. When applying a source voltage to the center core, initial induced voltage is the same on each phase. These induced voltages are transformed to modal voltages using the voltage transformation matrix $[A]$ given in Table 6.5. When applying the source voltage to the center core, the mode 1 voltage is zero, and only modes 0, 2, and 3 voltages propagate to the other end. Assuming the sheath voltages at both ends are zero, and neglecting the frequency-dependent effect, the analytical voltage waveshapes can easily evaluated. Figure 6.8 shows the calculated results by a theoretical analysis. From the figure, it is clear that the difference comes from the initial induced voltages due to the different mutual

Table 6.4 Effect of sheath grounding resistances on maximum voltages (per unit)

R_{gend}		0.5 Ω	0.5 Ω	0.5 Ω	1 Ω	10 Ω
R_{gint}		1 Ω	20 Ω	∞	10 Ω	10 Ω
Sending end	C_1	1.00	1.00	1.00	1.00	1.00
	C_2	0.427	0.425	0.426	0.418	0.363
	C_3	0.616	0.616	0.616	0.603	0.463
	S	0.179	0.0175	0.0173	0.0333	0.176
1st point	C_1	1.13	1.14	1.14	1.14	1.17
	C_2	0.439	0.446	0.449	0.429	0.325
	C_3	0.603	0.604	0.604	0.594	0.502
	S_1	0.190	0.190	0.190	0.187	0.179
	S_2	0.190	0.190	0.190	0.187	0.250
	S_3	0.141	0.141	0.141	0.151	0.237
2 nd point	C_1	1.23	1.23	1.23	1.24	1.29
	C_2	0.395	0.405	0.409	0.392	0.330
	C_3	0.516	0.518	0.520	0.513	0.464
	S_1	0.166	0.166	0.166	0.165	0.202
	S_2	0.130	0.130	0.131	0.128	0.167
	S_3	0.200	0.200	0.200	0.198	0.204
3rd point	C_1	1.32	1.32	1.33	1.31	1.38
	C_2	0.390	0.400	0.403	0.395	0.377
	C_3	0.405	0.412	0.416	0.409	0.396
	S	0.0002	0.0119	0.0160	0.0153	0.0829
Receiving end	C_1	2.35	2.35	2.35	2.33	2.09
	C_2	0.467	0.468	0.468	0.456	0.363
	C_3	0.661	0.661	0.661	0.652	0.573
	S	0.0002	0.0002	0.0006	0.0004	0.0024

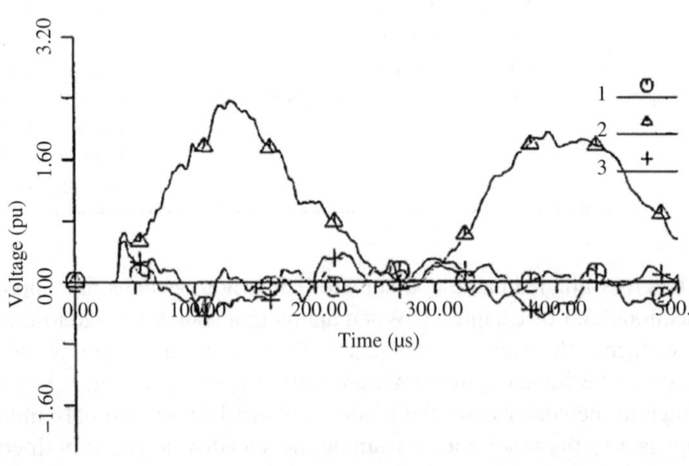

Figure 6.7 A calculated result when applying a source voltage to the center core

Table 6.5 Cable parameters for a theoretical analysis (f = 10 kHz)

Characteristic impedance $[Z_0]$ (Ω)

$$\begin{bmatrix} 49.38 & 6.13 & 2.42 & 12.67 \\ 6.13 & 47.05 & 6.13 & 13.13 \\ 2.42 & 6.13 & 49.38 & 12.67 \\ 12.67 & 13.13 & 12.67 & 12.81 \end{bmatrix}$$

Voltage transformation matrix $[A]$

$$\begin{bmatrix} 1 & 1 & -1/2 & 1 \\ 1 & 0 & 1 & 1 \\ 1 & -1 & -1/2 & 1 \\ 1 & 0 & 0 & 0 \end{bmatrix}$$

Mode	Attenuation (dB/km)	Velocity (m/μs)
0	3.26	11.6
1	0.0678	65.23
2	0.0793	75.75
3	0.130	153.90

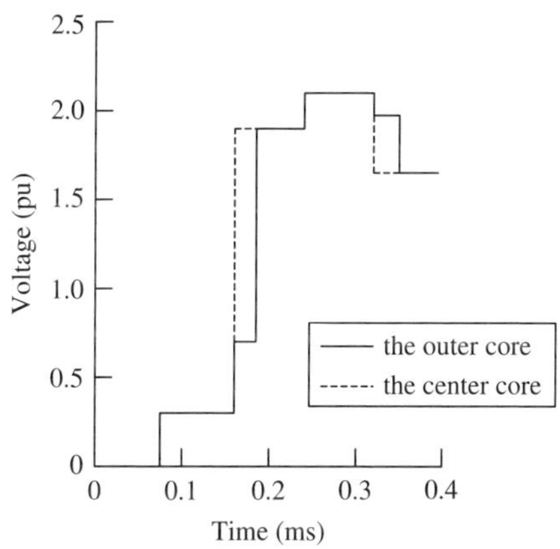

Figure 6.8 Calculated results by accurate and analytical methods

impedance. Also, Figure 6.8 indicates that a qualitative waveshape is easily obtained even in a cross-bonded cable using the theoretical analysis.

6.1.4 *Homogeneous Model (See Section 3.5.3)*

Figure 6.9 shows the calculated results by the homogeneous model explained in Section 3.5.3. It is clear that the results by the homogeneous model show good agreement with those by the accurate method shown in Figure 6.6a. The CPU time was reduced to one third in the homogeneous model.

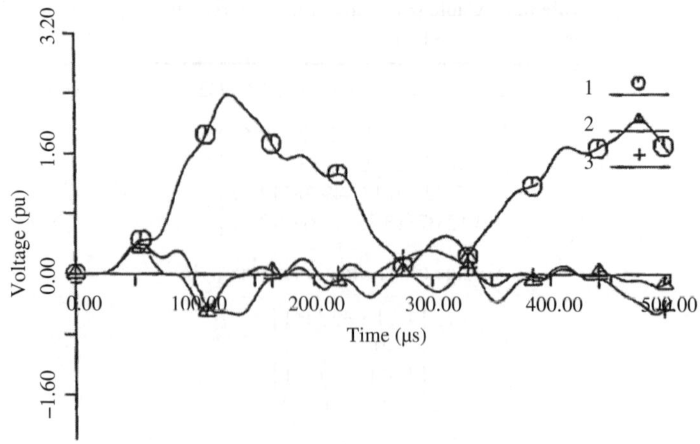

Figure 6.9 Calculated results by the homogeneous model

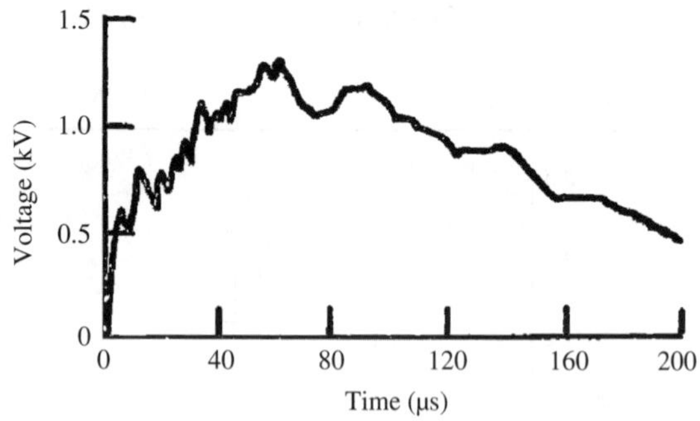

Figure 6.10 A calculated result by a PAI-circuit model corresponding to Figure 6.2a

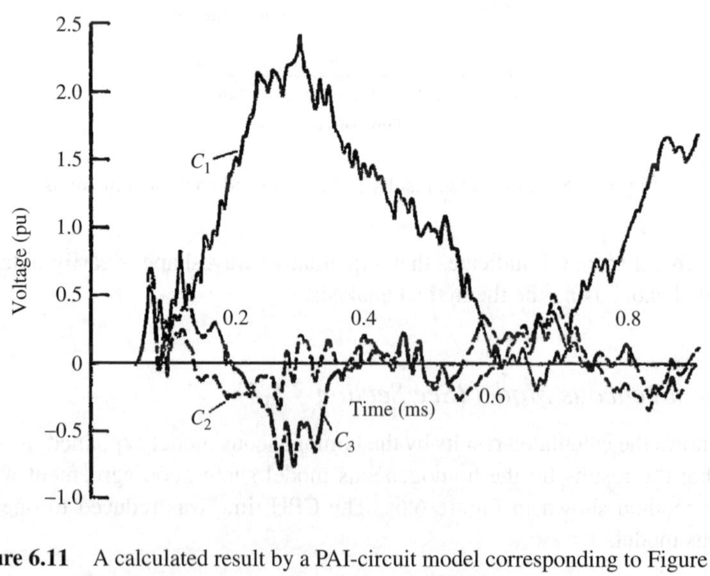

Figure 6.11 A calculated result by a PAI-circuit model corresponding to Figure 6.6a

6.1.5 PAI-circuit Model

Figure 6.10 shows a calculated result by a PAI-circuit model in which each minor section in Figure 6.1 is represented by one PAI circuit. The core voltage in the figure agrees satisfactorily with that in Figure 6.2a and Figure 6.3a. Thus, it can be said that the PAI-circuit model is applicable to a transient calculation of a cross-bonded cable even with one major section.

Figure 6.11 shows a calculated result by the PAI-circuit model corresponding to Figure 6.6a. Figure 6.12 shows a comparison of sheath voltages calculated by the FTP and the PAI-circuit model. The PAI-circuit model can give a reasonable result even for the sheath voltage.

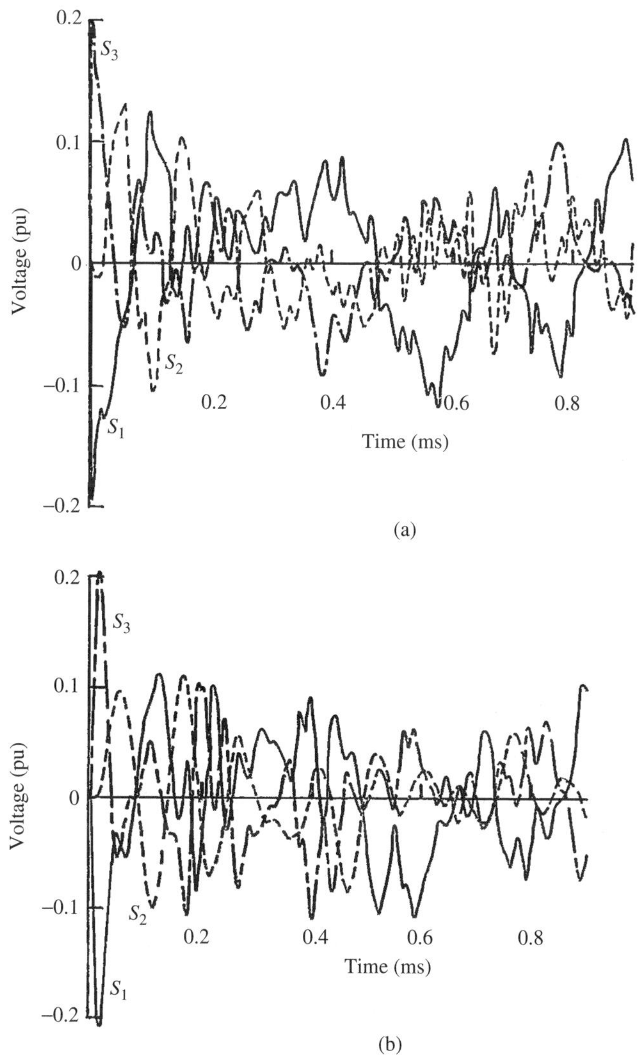

Figure 6.12 Comparison of sheath voltages at the first cross-bonding point: (a) FTP simulation; and (b) PAI-circuit model

6.2 Tunnel-installed 275 kV Cable

Most of the newly constructed cable transmission systems in Japan are installed within a tunnel. Many publications have investigated the cable propagation characteristic of directly buried cable [2, 3, 5–8], but there are few papers on the propagation characteristics of cable installed within a tunnel.

This section investigates the propagation characteristic of a 275 kV cable system based on field measurement [2, 9] and numerical analysis by EMTP. The difference of the propagation characteristics between a directly buried cable and a cable installed in a tunnel is studied. The effects of cable configuration on the characteristic impedance, attenuation and propagation velocity are also investigated. From the results, a method for modeling the cable installed within a tunnel is explained.

6.2.1 Cable Configuration

Figure 6.13a illustrates the configuration of a tunnel-installed 275 kV cable. Figure 6.13b shows a directly buried cable for a comparison. Table 6.6 gives the geometrical parameters of the cables.

The cable system installed within a tunnel consists of seven conductors, that is, three core conductors, three sheath conductors and the tunnel (the seventh conductor). The cable impedance and admittance matrices are calculated as 7×7 matrices. Then, the 7×7 matrices

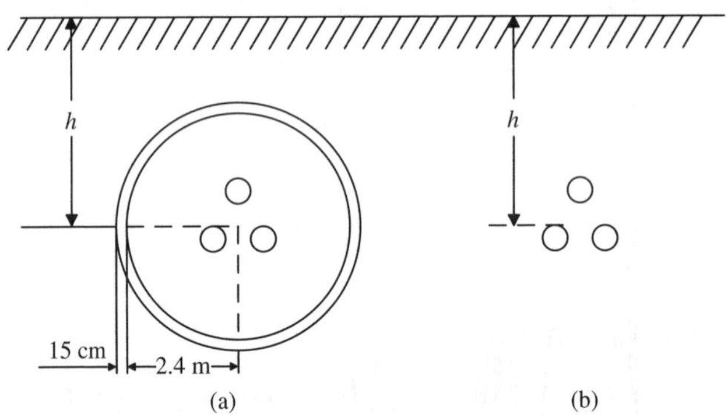

Figure 6.13 A 275 kV cable: (a) installed in a tunnel; and (b) directly buried

Table 6.6 Geometrical parameters of a 275 kV cable

	Case 1	Case 2	Case 3	Case 4	Case 5	Case 6
Installation	Tunnel	Tunnel	Tunnel	Tunnel	Tunnel	Buried
Tunnel length (m)	2.4	2.4	2.4	2.4	1.05	–
Burial depth (m)	10	–	–	–	–	10
Tunnel wall thickness (cm)	15	–	–	–	–	–
Distance from wall (m)	2.4	2.4	2.4	20	20	–
Tunnel wall material	Iron	Iron	Earth	Earth	Earth	–

are reduced to 6×6 matrices (assuming the wall conductor is grounded to the earth) by applying a matrix reduction method that is identical to the ground wire reduction in an overhead line. If the thickness of the tunnel wall is assumed infinite, the matrices are obtained directly as 6×6 matrices. In this case, the calculations required for cable constants are greatly reduced, because evaluation of the earth-return impedance and matrix reduction is avoided.

6.2.2 *Effect of Geometrical Parameters on Wave Propagation*

6.2.2.1 Effect of Earth Surface and Wall Thickness

As the material of the tunnel wall differs from the outer material, that is, soil, the effect of wall thickness is investigated. Table 6.7 shows calculated results of the characteristic impedances and modal propagation constants. Case 1 in Table 6.6 and Table 6.7 corresponds to the case in which the wall thickness is 15 cm and the depth of the tunnel is 10 m. Case 2 is the case in which the wall thickness is infinite. Characteristic impedances Z_0, Z_s and Z_m in Table 6.7 are the core self-impedance, sheath self-impedance and sheath mutual-impedance, respectively.

There are no differences in the propagation characteristic between Cases 1 and 2 in Table 6.7. This is readily explained by the conductor penetration depth. In a frequency region higher than 1 kHz, the penetration depth is less than the thickness of the wall. Thus, both the electric and the magnetic fields in the tunnel are shielded by the tunnel wall, and are not affected by the materials other than the wall. Therefore, the wall thickness causes no effect on the propagation characteristic of the cable.

The same is applicable to the depth of the tunnel from the earth surface, and the tunnel depth has no influence on the propagation characteristic.

6.2.2.2 Effect of Cable Configuration

Case 3 assumes that the material of the wall is a soil with 33 Ωm resistivity. Table 6.7 shows that the characteristic impedances at 1 MHz are decreased by 20% compared with those at 1 kHz. A comparison of Case 3 with Cases 1 and 2 indicates that the material of the tunnel wall, that is, its resistivity, significantly affects the propagation characteristic. The frequency dependence of the characteristic impedance is more noticeable in Case 3 than those in Cases 1 and 2. The coaxial-mode characteristic impedance ($Z_c - Z_s$) is 30 Ω and is not affected by the wall resistivity.

The attenuation and propagation velocity of the coaxial mode and the sheath mode are not influenced by the resistivity of the tunnel wall, because the coaxial and sheath modes are only dependent on the cross section and the physical configuration of the core and sheath. The attenuation and propagation velocity of the earth-return mode are heavily dependent on the frequency. The difference of those parameters at frequencies of 1 kHz and 1 MHz is only 4% in Cases 1 and 2, that is, the case in which the tunnel wall is assumed to be iron. The difference reaches 30% in Case 3, that is, for the soil wall.

6.2.2.3 Effect of Cable Arrangement

The effect of the cable arrangement will be investigated in this section. Cables are positioned at a distance 20 cm apart from the tunnel wall.

Table 6.7 Wave propagation characteristics on a 275 kV cable

	Frequency	Case 1		Case 2		Case 3		Case 4		Case 5		Case 6	
		1 kHz	1 MHz	1 kHz	1 MHz	1 kHz	1 MHz	1 kHz	1 MHz	1 kHz	1 MHz	1 kHz	1 MHz
Z_0 (Ω)	Z_c	253	251	253	251	336	267	223 ± 20	178 ± 17	219 ± 18	172 ± 15	51	45
	Z_s	222	221	222	221	305	237	192 ± 19	147 ± 18	188 ± 17	142 ± 15	20	15
	Z_m	167	166	167	166	250	182	146 ± 8	101 ± 8	141 ± 7	95 ± 7	11	7
α (dB/km)	Coaxial	0.013	0.413	0.013	0.413	0.013	0.413	0.013	0.413	0.013	0.413	0.0131	0.413
	Sheath transm.	0.00242	0.0722	0.00242	0.0722	0.00242	0.0722	0.003	0.65 ± 0.2	0.003	0.96 ± 0.4	0.0147	2.87
	Earth return	0.00055	0.0169	0.00055	0.0167	0.0159	9.97	0.0267	16.4	0.0278	25.5	0.355	464
c (m/μs)	Coaxial	161	166	161	166	161	166	161	166	161	166	161	166
	Sheath transm.	287	289	287	289	287	289	252 ± 4	256 ± 5	248 ± 6	252_n6	46.3	46.7
	Earth return	298	299	298	299	205	275	120	164	116	162	9.97	14.9

Calculated results for Case 4 are given in Table 6.7. It is observed that the characteristic impedance of the sheath is decreased by 30% in Case 4 compared with that in Case 3. This is because the capacitance between the cable and the tunnel wall becomes greater in Case 4. Also, the symmentrical system configuration decreases each component of the characteristic impedance by about 10%. The attenuations of the sheath and earth-return modes are increased in a high frequency region. The propagation velocities of both modes are decreased compared with those in Case 3, and the difference is more noticeable on the earth-return mode.

6.2.2.4 Effect of Tunnel Radius

Case 5 in Table 6.7 shows the results when the tunnel inner radius is 1.05 m and the distance from the cable to the tunnel wall is 20 cm (equal to Case 4). A comparison of Case 5 with Cases 3 and 4 shows a minor difference in all the propagation constants. It is clear that the effect of the tunnel radius is smaller than that of the distance from the cable to the tunnel wall. Therefore, a cylindrical tunnel approximation to a noncylindrical tunnel such as a square tunnel causes no significant effect on the propagation characteristic, as long as the distance between the cable and the tunnel wall is kept the same.

6.2.2.5 Difference from a Directly Buried Cable

The characteristic impedance of a directly buried cable is quite small, because a capacitance between the sheath and the soil is determined by the small thickness of the corrosion proof layer. The capacitance of the cable installed within a tunnel is determined by the distance from the tunnel wall, and thus the characteristic impedance of the cable system becomes greater than that of the directly buried cable. The attenuation of the directly buried cable increases rapidly as frequency increases. The propagation velocities of the sheath and earth-return modes are quite small in comparison with those of the cable system installed within a tunnel. However, there is no difference in the coaxial-mode velocity.

6.2.3 Field Test on 275 kV XLPE Cable

6.2.3.1 Field Test Circuit

A field test was carried out on a 275 kV XLPE cross-bonded cable system. The cable system consisted of five major sections. Figure 6.14 shows the first major section of the test circuit, including a source circuit of the test. A pulse voltage of amplitude 500 V and width $T_f = 1$ μs was applied to a cable core. The sheath voltages are measured at the first and second cross-bonding joints (nodes b and c in Figure 6.14). The cable system behind the first major section is represented by 30 Ω resistance that corresponds to the coaxial-mode characteristic impedance because the maximum observation time in the field test is less than the time at which the reflected wave comes back from the second major section. The first minor section in Figure 6.14 is installed in an underground tunnel as illustrated in Figure 6.15a, and the second and third sections are shown in Figure 6.15b.

In a simulation, the rectangular tunnel is approximated by a cylindrical tunnel as illustrated by a dotted line in Figure 6.15, based on the investigation in the previous section. The circuit conditions are given in Table 6.8.

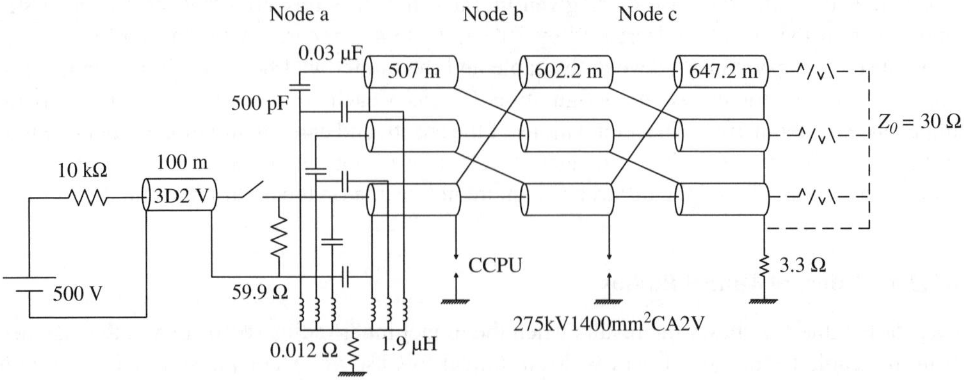

Figure 6.14 A field test circuit of a tunnel-installed cable

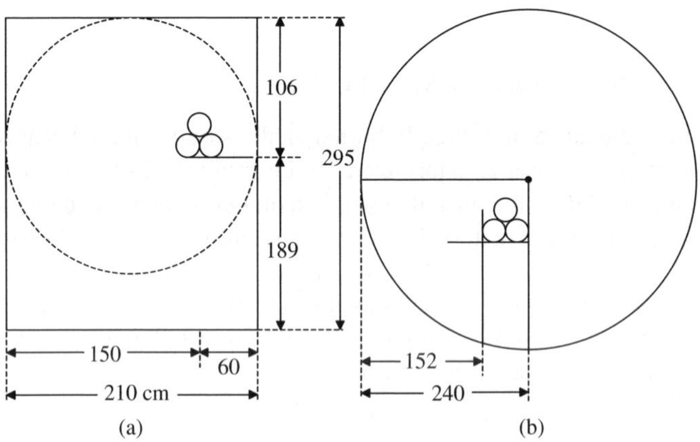

Figure 6.15 Cross section of tunnels: (a) first minor section; and (b) second/third minor sections

Table 6.8 Circuit conditions for EMTP simulations

Case	Type of installation	External medium	Distance to wall (cm)	Bonding lead (μH)	Arrester equivalent capacity (pF)
A	Tunnel	Earth	6		4000
B	Direct buried	Earth		–	4000
C	Tunnel	Iron	–	–	4000
D	Tunnel	Earth	16	–	4000
E	Tunnel	Earth	6	2	4000
F	Tunnel	Earth	6	–	1000

6.2.3.2 Comparison of Calculated Results with Field Test

Field test and simulation results (Case A in Table 6.8) of surge voltages on the test circuit in Figure 6.14 are given in Figure 6.16 and Figure 6.17, respectively. In the calculation, the cable is represented by Dommel's distributed-parameter line model of EMTP. Its parameters are obtained at 25 kHz, which is the dominant transient frequency of the coaxial mode.

High frequency oscillations caused by the multiple reflection at the substation are observed in Figure 6.16a at about $t = 0$. Such oscillations are not observed in calculated results, because the substation is modeled by a 500 pF lumped capacitor. The amplitude of the sinusoidal oscillation observed before $t = 6$ μs in Figure 6.16a is about 15 V in the field measurement, and 10 V in the calculated result. The periods of the oscillations are about 1.5μs, and no significant difference is observed between the field test and calculated results. The oscillation is estimated to be caused by the resonance of the surge bypass capacitor (0.03 μF) and the inductor (1.9 μH) representing the grounding lead wire.

A negative pulse is observed in the time range of $6 < t < 7.5$ μs in Figure 6.16a and Figure 6.17a. The pulse is caused by the reflection at the first cross-bonding point. The first minor section length is 507 m and the coaxial-mode traveling velocity is 166 m/μs; this agrees with the time of the negative pulse. Its amplitude is −55 V in the calculated result which agrees with the field measurement. However, more wave deformation is observed in the calculated result.

Figure 6.16b and Figure 6.17b show the sheath voltages at the first cross-bonding joint. If we neglect the high frequency oscillation of the field test, the calculated maximum voltage of 40 V, and the waveshape at the cross-bonding joint agree well with the field test result.

Figure 6.16c and Figure 6.17c show the sheath voltages at the second joint. The maximum voltage of 40 V is observed at $t = 7$ μs in both the field test and the calculated results.

The maximum voltage at the first cross-bonding joint is greater than that at the second joint, and thus the overvoltage at the first joint is the most important in an insulation design. A theoretical analysis to estimate the sheath overvoltage is carried out.

The reflection coefficient at the cross-bonding point is given by:

$$[K] = (Z_t - Z_0)(Z_t + Z_0)^{-1} \tag{6.1}$$

$$[Z_t] = [R]_t[Z_0][R] \tag{6.2}$$

where $[Z_0]$ is the characteristic impedance
and

$$[R] = \begin{bmatrix} 1 & 0 & 0 & 0 & 0 & 0 \\ 0 & 1 & 0 & 0 & 0 & 0 \\ 0 & 0 & 1 & 0 & 0 & 0 \\ 0 & 0 & 0 & 0 & 0 & 1 \\ 0 & 0 & 0 & 1 & 0 & 0 \\ 0 & 0 & 0 & 0 & 1 & 0 \end{bmatrix} \begin{matrix} \text{core 1} \\ \text{core 2} \\ \text{core 3} \\ \text{sheath1} \\ \text{sheath 2} \\ \text{sheath3} \end{matrix}$$

and t is for a transposed matrix.

The voltage at the first cross-bonding joint is obtained by the following equation.

$$(V_2) = ([K] + [U])(V_1) \tag{6.3}$$

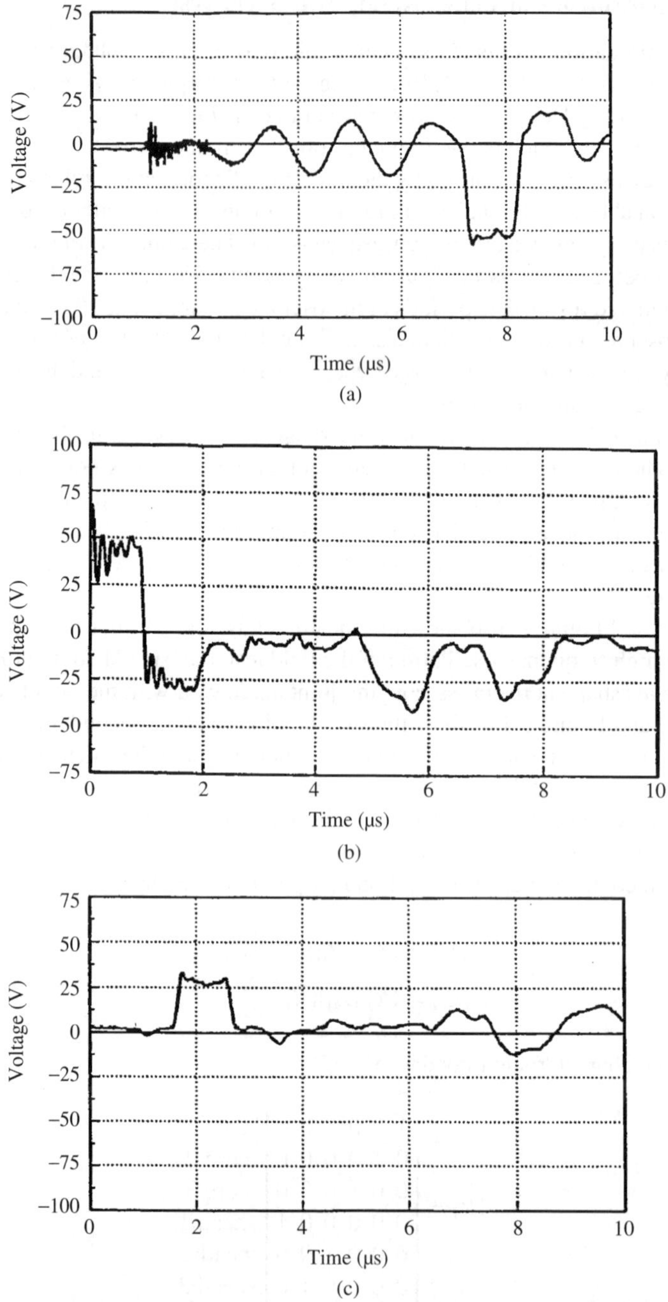

Figure 6.16 Field test results on the 275 kV cable: (a) induced core voltage at node a; (b) sheath voltage at first cross-bonding joint (node b); and (c) sheath voltage at the second cross-bonding joint (node c)

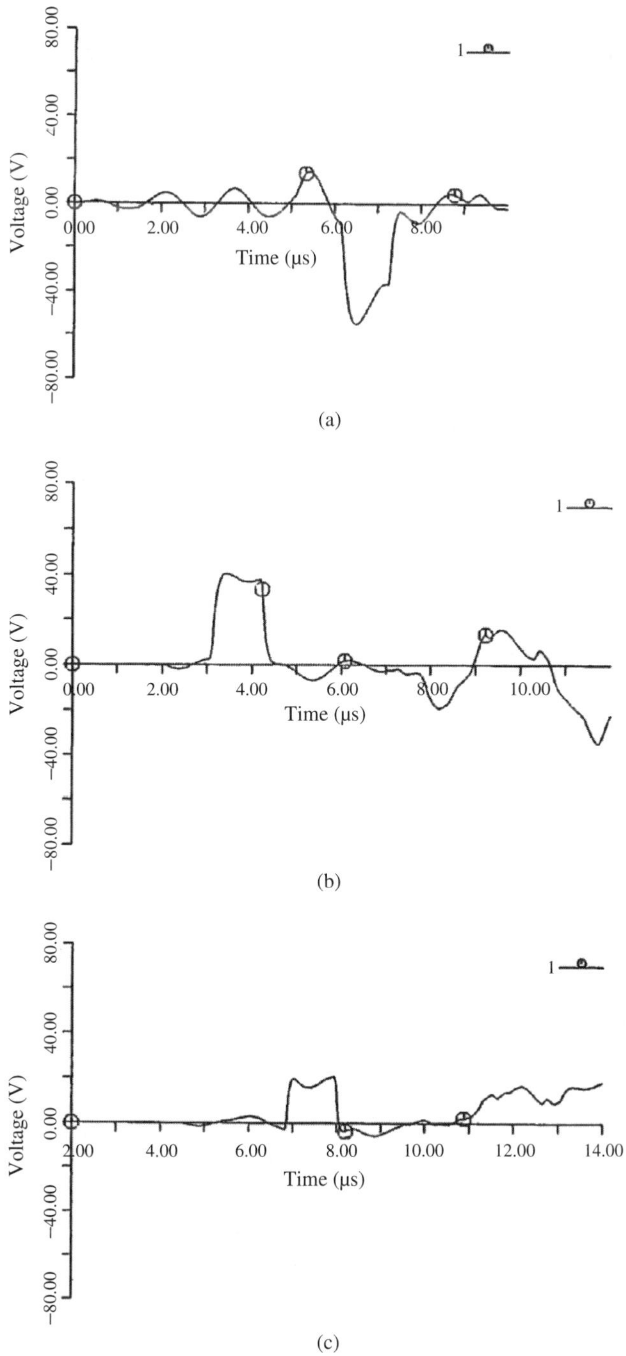

Figure 6.17 EMTP simulation results on the 275 kV cable: (a) induced core voltage at node a; (b) sheath voltage at first cross-bonding joint (node b); and (c) sheath voltage at the second cross-bonding joint (node c)

If the cable is positioned at the tunnel center, that is, if the characteristic impedance matrix is assumed to be symmetrical, the first three columns of the reflection coefficient matrix are expressed as follows:

$$[K] = k \cdot \begin{bmatrix} 0 & 1 & -1 \\ -1 & 0 & 1 \\ 1 & -1 & 0 \\ -2 & 2 & 0 \\ 2 & -2 & 2 \\ 2 & 0 & -2 \end{bmatrix} \begin{matrix} \text{core 1} \\ \text{core 2} \\ \text{core 3} \\ \text{sheath1} \\ \text{sheath 2} \\ \text{sheath3} \end{matrix}$$

$$k = (Z_s - Z_m)/\{4(Z_c - Z_s) + 3(Z_s - Z_m)\} \tag{6.4}$$

Substituting the average value of the characteristic impedance given in Table 6.7, Case 5, into (6.4), k becomes 0.19. Substituting the existing voltage $(V_1) = (300, 0, 0, 0, 0, 0)_t$ into (6.3), an induced sheath voltage of 57 V is obtained. Only a 5 V difference between the theoretical and calculated results is observed. Thus, (6.3) and (6.4) give satisfactory accuracy as a theoretical approach.

6.2.3.3 Comparison with Underground Cable

Figure 6.18 shows a calculated result (Case B in Table 6.8) of a sheath voltage at the first cross-bonding joint, assuming the cable is directly buried in the earth at a depth of 10 m from the earth surface. From the figure, it is clear that the induced voltage is caused by the coaxial-mode traveling wave. The waveshape looks a square wave, and its maximum sheath voltage is less than one third of that of a cable installed within a tunnel. The deviation comes from the difference of the characteristic impedance. Substituting the characteristic impedance in Table 6.7 into (6.2) and (6.3), the induced voltage is estimated to be 17 V, which is identical to the measured value.

6.2.3.4 Effect of Cable Parameter

Effect of Tunnel Wall Material

Figure 6.19a shows a calculated result, in which the wall material is assumed to be steel with a resistivity of 10^{-7} Ωm and a relative permeability of 100. The maximum voltage is estimated to be 50 V, which is the same as that of Case A. The analytical result obtained from (6.3) by substituting the characteristic impedance in Table 6.7 for Case 2 (iron) agrees with that for the case of the soil wall. However, the waveshape after the first traveling wave coming from the sending end differs from that of Case A. The oscillation with a period of 1.5 μs is caused by the earth-return mode, that is, the small attenuation of the earth-return mode due to the iron wall increases the oscillation.

Effect of Cable Configuration

Figure 6.19b shows the sheath voltage waveform at the first cross-bonding joint, when the first section of the cable system is positioned 16 cm away from the tunnel wall. The maximum

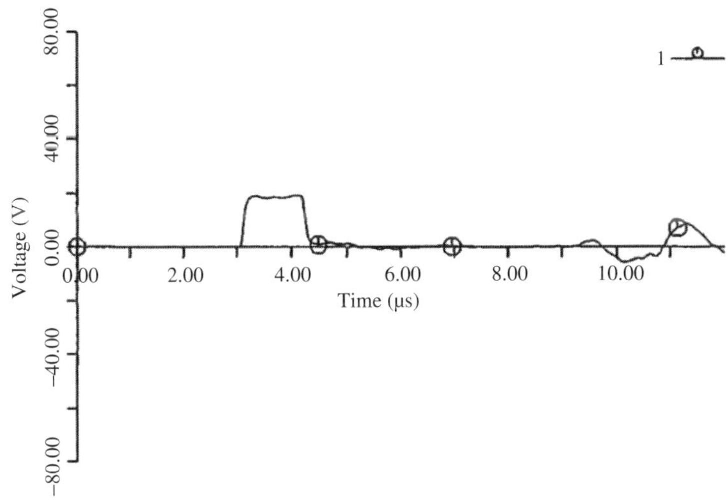

Figure 6.18 Simulation results on a directly buried cable

voltage is almost identical to that of Case A, but a difference is observed at $t = 8$ μs. The difference is estimated to be caused by the earth-return mode traveling wave, because the distance between the cable and the tunnel affects the earth-return mode as explained in Section 6.2.2.

Effect of Bonding Lead Wire

Figure 6.19c shows the sheath voltage waveform when the bonding wire inductance is taken to be 2 μH. An oscillation with a frequency of 2 MHz is observed at the time when the first traveling wave arrives. The impedance seen from the cable at the cross-bonding joint increases as the bonding lead wire is added. This results in an increase of the reflection and refraction coefficients which cause the steep wavefront and voltage increase.

Effect of Equivalent Capacitance of CCPU

Figure 6.19d shows a calculated result when capacitance of 1000 pF equivalent to a CCPU (sheath protection unit) is installed (Case F). In comparison with the calculated result in Figure 6.17b assuming 4000 pF capacitance, the wavefront becomes steeper. Because the equivalent capacitance bypasses high frequency components, the decrease in the capacitance causes the increase in the steepness. Further, the oscillation after 4.5 μs is more noticeable in Case F than in Case A. This observation leads to the fact that the capacitance also affects the propagation characteristics of the sheath and earth-return modes.

6.2.4 Concluding Remarks

1. No difference of the coaxial-mode propagation characteristic is observed between a directly buried cable and a cable installed within a tunnel. However, the characteristics of the sheath and earth-return modes are heavily affected by the installation. The attenuation of the modes

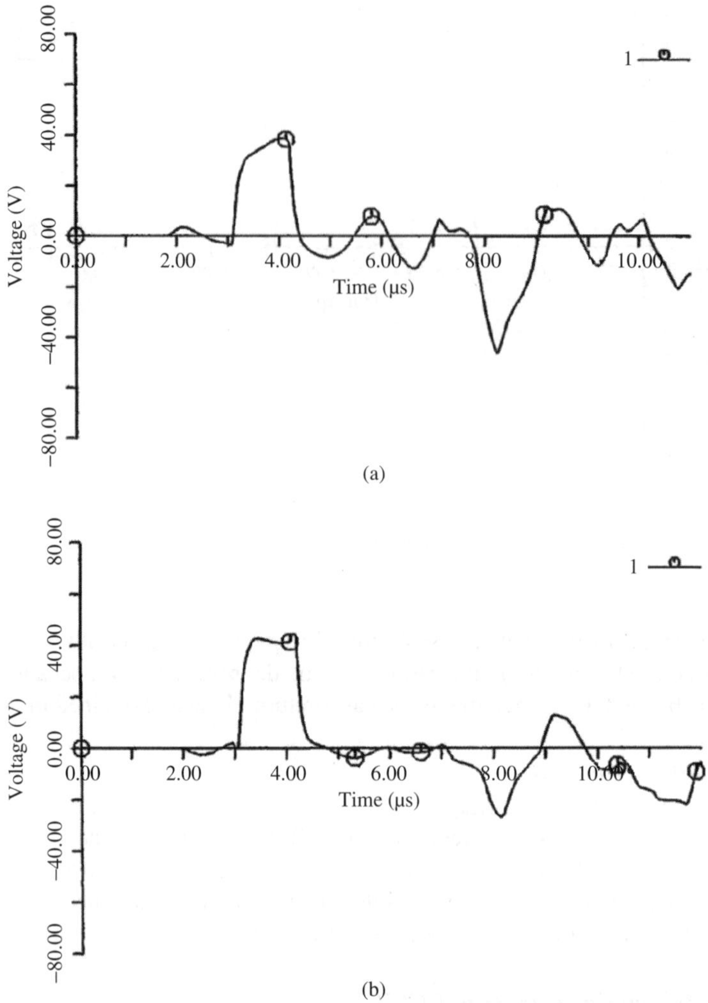

Figure 6.19 Effect of various parameters on a transient sheath voltage: (a) effect of the wall material; (b) effect of cable configuration; (c) effect of bonding wire; and (d) effect of CCPU capacitance

on the cable installed within a tunnel is smaller and the traveling velocity is greater than those of a directly buried cable.

2. The frequency dependence of the earth-return mode increases as the resistivity of the tunnel wall increases. However, the effect of wall thickness is minor. Thus, the tunnel is modeled as a cylindrical hole.
3. The surge characteristic of the tunnel-installed cable heavily depends on the distance between the cable and the tunnel wall surface, but is almost independent of the tunnel radius. Thus, a cable installed within a tunnel, of which the cross section is not cylindrical, is represented by a cable enclosed within a cylindrical tunnel with an infinite wall thickness. The radius of the cylindrical tunnel is approximately determined in the manner

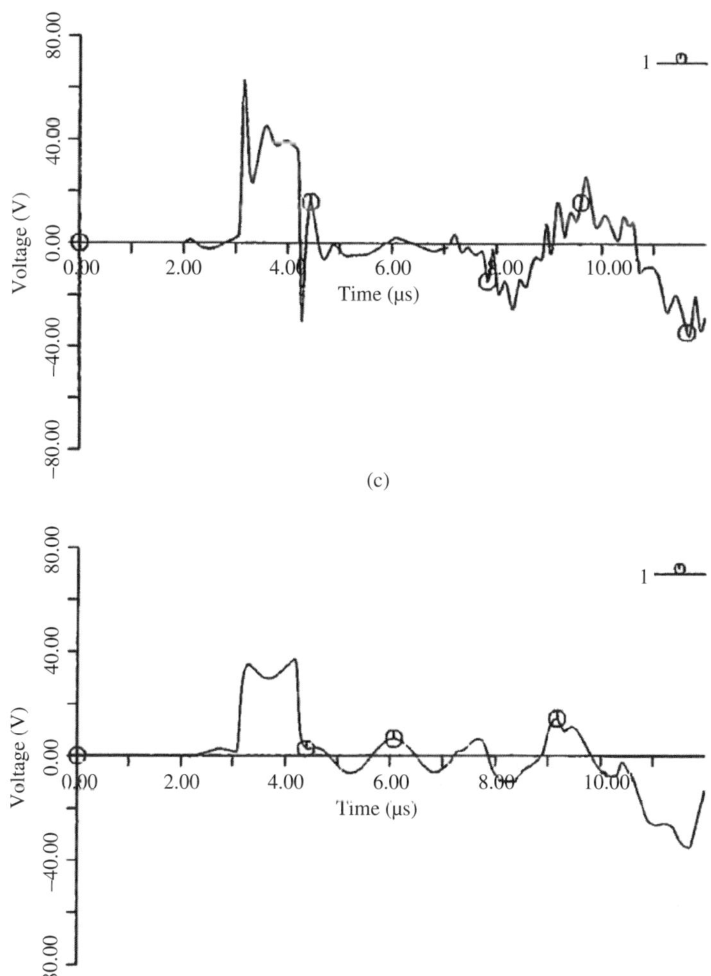

(c)

(d)

Figure 6.19 (*Continued*)

that the shortest distance between the cable and the tunnel surface is maintained, and the distances from the cable to two arbitrary points of the cylindrical surface are nearly the same as those to the original tunnel surface.

4. The inductance of the bonding lead wire causes spike voltages at the wavefront of the sheath voltage.
5. The capacitances of a sheath protection unit (CCPU) cause a significant effect on the sheath voltage, and thus the capacitance has to be included in a transient calculation.

In practice, the depth of a tunnel in which a cable system is installed is not uniform along the cable. It is almost impossible to evaluate a transient with detailed allowance for the tunnel depth. However, a cable installed within a tunnel can be modeled by a uniform distributed parameter line, as is clear from the observation made in this section.

6.3 Cable Installed Underneath a Bridge

This section deals with a transient analysis of a cross-bonded cable along an iron bridge, which is connected with overhead lines at both ends [2, 10]. First, transients on a system composed of cables are investigated for the following cases: (a) all the cable being underground; (b) all the cable being overhead; (c) the cable being partially overhead; and (d) the cable being partially overhead with parallel steel conductors. The first case is that of an ordinary cross-bonded cable. The second case is that of underneath a bridge. The third case is that a part of a cross-bonded cable system is installed underneath a bridge and the remaining part is underground. The last case is the more detailed approximation of a cable installed underneath a bridge.

Secondly, transients on a cross-bonded cable connected with overhead lines at both ends, which is a common practice in Japan, are studied.

A numerical evaluation of transients in this section is carried out using EMTP.

6.3.1 Model System

Figure 6.20 illustrates a model system of EMTP simulations. Both sides of the cable are underground. The middle part is installed underneath a bridge for a length of about 1.7 km and is cross-bonded. The sheaths are grounded through 10 Ω resistance at both ends.

Figure 6.21 shows the cable configurations and the cross section of the cable. The underground part of the cable is horizontally arranged in Figure 6.21a, and the cable has a vertical configuration along the bridge in Figure 6.21b.

In EMTP simulations, the cable is represented by a distributed parameter line model with constant resistances representing attenuation of a surge on the cable at a certain frequency (Dommel model).

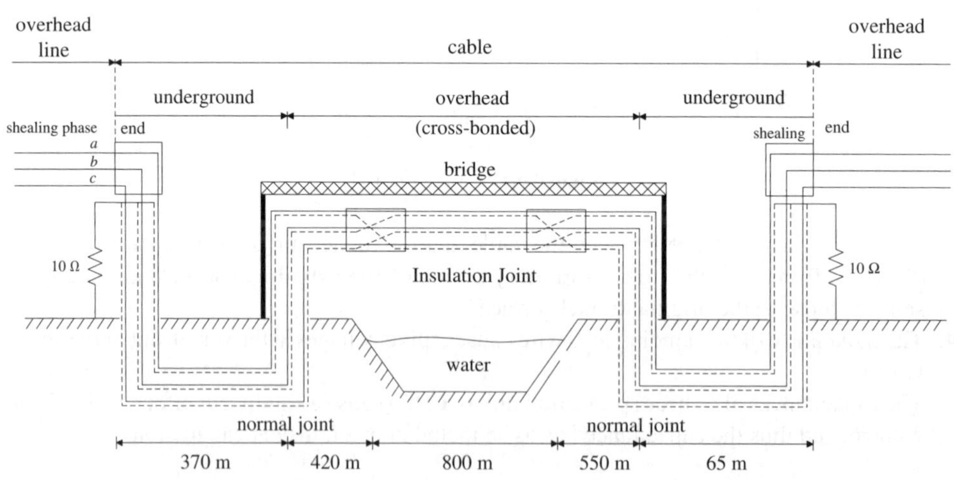

Figure 6.20 A model system of a cable installed underneath a bridge

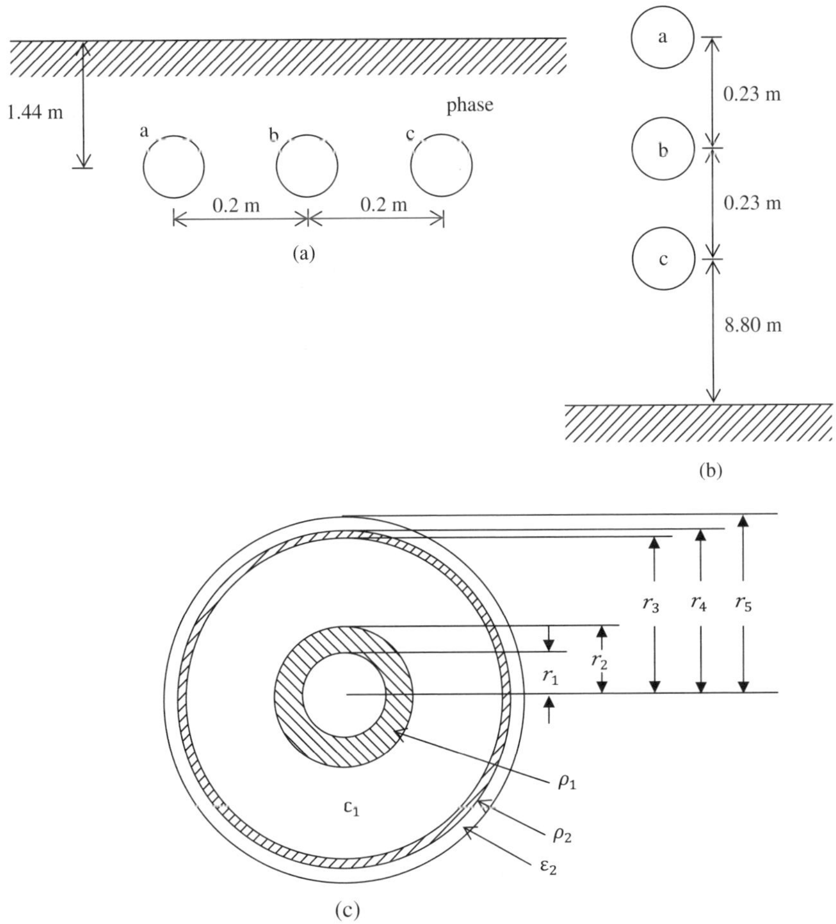

Figure 6.21 Cable configuration: (a) underground part; (b) bridge part; and (c) cross section. $r_1 = 7.80$, $r_2 = 13.59$ mm, $r_3 = 26.64$ mm, $r_4 = 28.24$ mm, $r_5 = 32.74$ mm, $\varepsilon_1 = 3.7$, $\varepsilon_2 = 7.0$, $\rho_1 = 2.468 \times 10^{-8}$ Ωm, $\rho_2 = 2.8 \times 10^{-8}$ Ωm

6.3.2 *Effect of an Overhead Cable and a Bridge*

In this section, the effect of an overhead cable on transients in a composite system of underground and overhead cables, as illustrated in Figure 6.20, will be investigated. The cable system is partially underground and the remaining part is overhead, that is, installed underneath the bridge. Also, the effect of iron frameworks of the bridge on the transients will be studied. To make clear the transient characteristic of the system, overhead lines are disconnected and a zero impedance source is applied to the phase-a core.

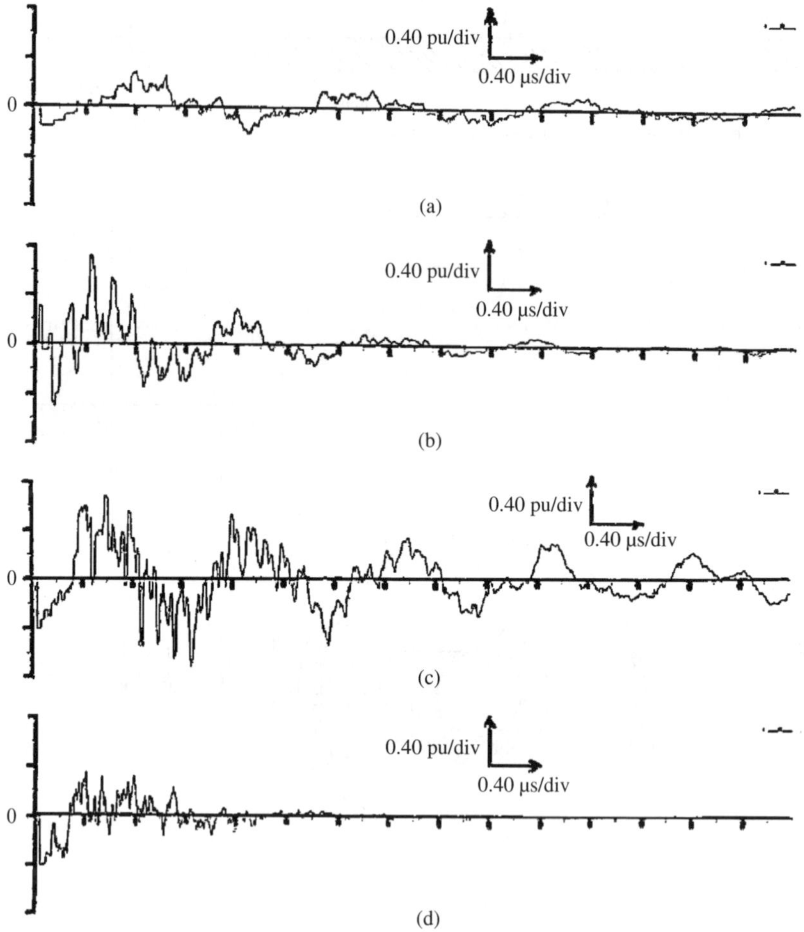

Figure 6.22 Transient sheath voltages at the first cross-bonding point on phase-a: (a) underground (Case A); (b) overhead (Case B); (c) underground-overhead-underground (Case D); (d) same as (c) but with the overhead cable with parallel conductors (Case E)

6.3.2.1 Homogeneous System

As the first step of the investigation, a transient on a homogeneous underground cable (Case A) with the configuration of Figure 6.21a is compared with that of an overhead cable (Case B) of Figure 6.21b.

Calculated results of transient sheath voltages at the first cross-bonded point are shown in Figure 6.22a for the underground cable and in Figure 6.22b for the overhead cable. The maximum sheath overvoltages and the time of the occurrence at the first and second cross-bonded points are given in Table 6.9 together with the other cases investigated later.

A comparison of Figure 6.22a with Figure 6.22b shows that the first wave is negative up to 35 μs in the underground cable case (Figure 6.22a = Case A), while it is positive and oscillates in the overhead cable case. The maximum overvoltage is 0.28 pu in the underground

Table 6.9 Maximum sheath voltages for various cable conditions

Case			First cross-bonded point			Second cross-bonded point		
			Phase-a	Phase-b	Phase-c	Phase-a	Phase-b	Phase-c
A	V_{max}	(pu)	0.28	0.18	−0.16	0.24	0.14	0.20
	t	(μs)	81.00	91.60	106.55	157.55	162.10	101.05
B	V_{max}	(pu)	0.73	0.70	0.51	−0.40	−0.59	0.58
	t	(μs)	42.08	5.25	24.50	37.25	53.80	48.25
C	V_{max}	(pu)	0.73	0.70	0.51	−0.39	−0.58	0.59
	t	(μs)	42.75	5.30	25.25	37.25	52.15	48.25
D	V_{max}	(pu)	−0.71	0.45	−0.46	0.72	0.55	0.57
	t	(μs)	128.10	7.95	103.10	82.40	122.65	56.80
E	V_{max}	(pu)	−0.42	0.34	0.39	−0.32	−0.44	−0.38
	t	(μs)	6.8	55.3	80.10	29.70	34.50	9.60

Case A, underground; Case B, overhead; Case C, overhead, horizontal; Case D, underground-overhead-underground; and Case E, same as Case D but with the overhead cable with parallel conductors.

cable case (Case A) and is 0.73 pu in the overhead case (Case B) as shown in Table 6.9. The dominant frequency of the oscillation is about 5.8 kHz in Case A, while it is about 8.8 kHz in Case B. That is, the maximum voltage and the oscillating frequency are much higher in the overhead cable case than those in the underground cable case. This is due to lower attenuation, faster velocity and higher characteristic impedance of the sheath in the propagation characteristic of an overhead cable [11, 12]. An analytical evaluation of the first waveform is given in Appendix 6.A.

As is clear from the analytical results in Appendix 6.A,the reason why the maximum sheath voltage in the overhead cable case is higher by about 2.5 times than that in the underground case is that the sheath characteristic impedance of the overhead cable is quite different from that of the underground cable, and thus, the refraction coefficient at the cross-bonded point is quite different. That is, the difference of the refraction coefficient between the overhead and underground cable causes significantly different sheath overvoltages.

For the case that the overhead cable is not vertically arranged but horizontally arranged, calculated results of the sheath maximum overvoltages are given in Table 6.9 as Case C. It is clear from a comparison of Case C with Case B that the configuration of the overhead cable does not cause a significant effect on the sheath overvoltages.

6.3.2.2 Partially Overhead Cable

As shown in Figure 6.20, a part of the cable is installed underneath a bridge. This part of the cable can be simulated as an overhead cable (Case D). The remaining part is underground as shown in Figure 6.20. A calculated waveform of the phase-a sheath voltage at the first cross-bonded point is illustrated in Figure 6.22c and the maximum sheath voltages and the time of the occurrence are given in Table 6.9 as Case D. The waveform until 35 μs is qualitatively quite similar to that of Figure 6.22a, though its magnitude is about twice that of Figure 6.22a. In fact, the sheath maximum voltages are not much different from those of Case B (overhead cable) which are much greater than those of Case A. Also, the dominant frequency of the waveform is 8.6 kHz which is about the same as that of Case B.

From the above, it is concluded that the surge on a partially overhead cable shows a similar characteristic to a totally overhead cable except for the beginning of the surge waveform.

6.3.2.3 Effect of Iron Frameworks of a Bridge

A cable underneath a bridge has mutual coupling with the iron frameworks of the bridge. The frameworks are modeled as parallel conductors to the overhead cable as illustrated in Figure 6.23a. The conductors are grounded through 10 Ω resistance at both ends of the bridge. The configuration of the conductor system is shown in Figure 6.23b.

A calculated waveform of the phase-a sheath voltage at the first cross-bonded point is shown in Figure 6.22d, and the maximum overvoltages are given in Table 6.9 as Case E. From the figure, it is observed that the initial negative peak at about t = 5 μs is almost the same as that of Figure 6.22c, Case D, but after t = 30 μs, the voltage is reduced to about half that of Case D. After t = 200 μs, the voltage is reduced to less than 0.1 pu. This is a marked difference from any other cases in Figure 6.22. Table 6.9 shows that the maximum overvoltages in Case E are less by about 20–50% than those in Case D.

The above characteristics are due to the mutual coupling between the cable and the conductors simulating the iron frameworks of a bridge. Thus, it can be said that the iron frameworks of a bridge cause a significant reduction of sheath voltages. The rate of reduction reaches more than 50% on the maximum voltages and the oscillation of the voltage waveforms are dying out much faster than the case with no mutual coupling. But, it should be noticed that the maximum sheath overvoltages, even in the case of considering the mutual coupling between the cable and nearby iron frameworks of a bridge, are higher than those in the underground cable case.

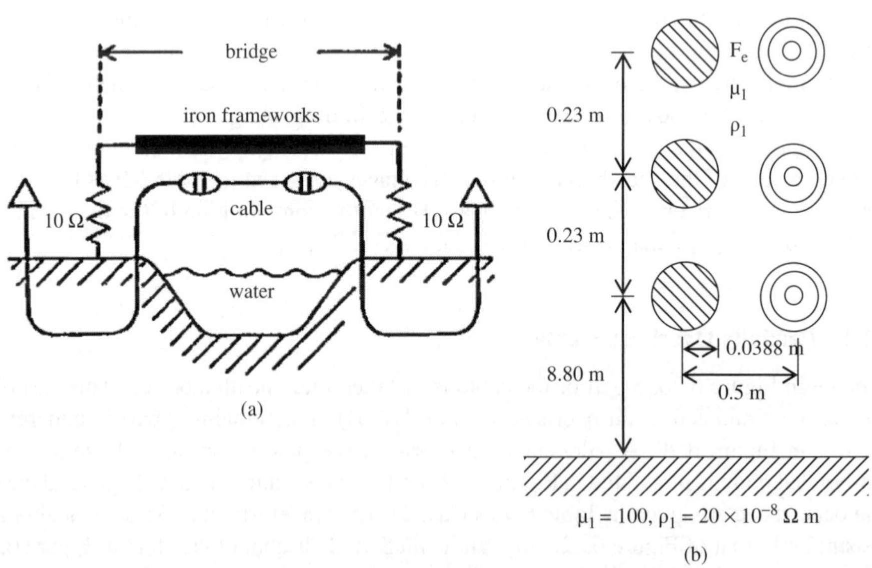

Figure 6.23 A model of iron frameworks of a bridge: (a) model circuit; and (b) configuration of conductors

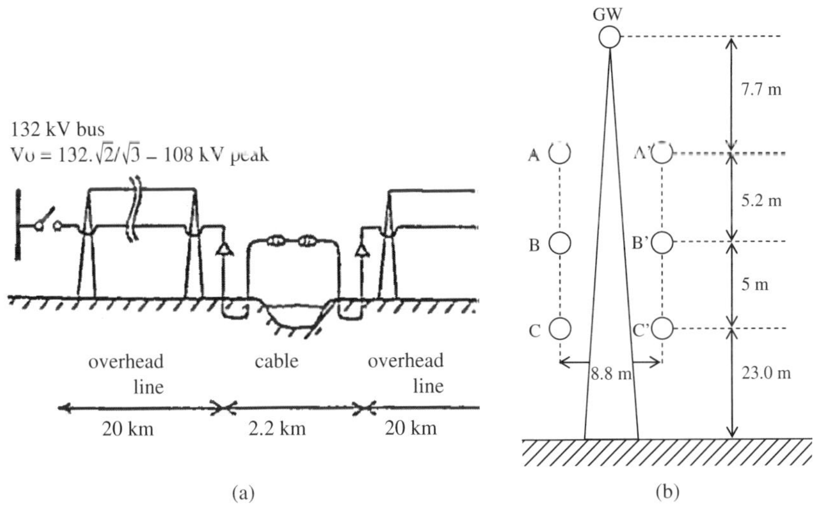

Figure 6.24 A model system of overhead lines and a cable: (a) model system configuration; and (b) overhead line configuration

6.3.3 *Effect of Overhead Lines on a Cable Transient*

In general, a cable is not operated alone, but is connected with overhead lines and/or substations. Also, the sheath overvoltage of a cross-bonded cable is proportional to an incoming surge to the core of the cable. This surge is heavily dependent on the system connected to the cable. In this section, the effect of an overhead line, which is assumed to be connected to the cable system investigated in the previous section, on sheath overvoltages of the cable system will be studied.

Figure 6.24 shows a model system studied in this section. The overhead line connected at the far end of the cable is open-circuited at the remote end. A sinusoidal source voltage is applied to phase-a of the left-hand end of the overhead line at the time of the voltage peak. The overhead line is represented by the Dommel model with constant resistance in the EMTP.

6.3.3.1 Double Circuit Operation of Overhead Line

Calculated waveforms of transient sheath voltages at the first cross-bonded point are shown in Figure 6.25. The waveforms are rather different from those in Figure 6.22 which is for the case of no overhead line. In Figure 6.22, the transient voltage oscillates around the neutral (zero-potential) axis and the oscillation is dying out after 200 μs. In Figure 6.25, the transient sheath voltage consists of three different oscillating components.

$$T_1 = 1700\ \mu\text{s}\ (f_1 = 600\ \text{Hz})$$

$$T_2 = 73\ \mu\text{s}\ (f_2 = 13.8\ \text{kHz})$$

$$T_3 = \text{some}\ \mu\text{s to } 20\ \mu\text{s}\ (f_3 = 50\text{–}500\ \text{kHz})$$

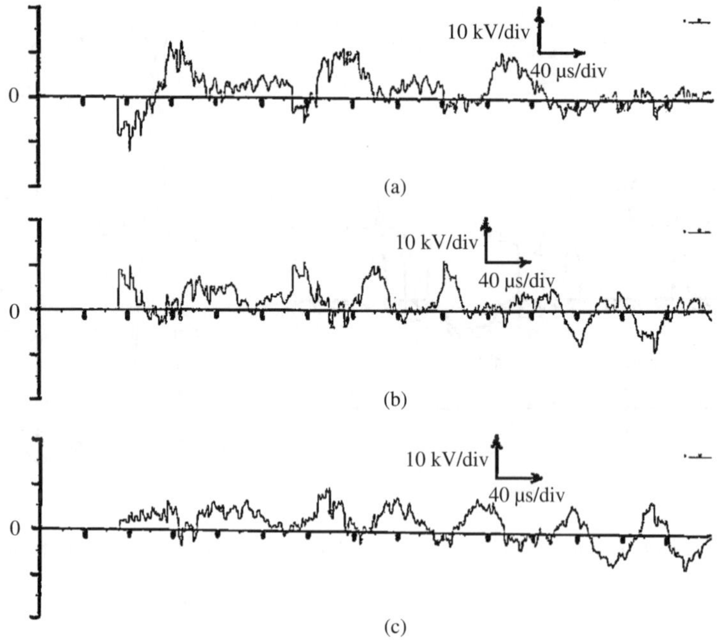

Figure 6.25 Transient sheath voltages at the first cross-bonding point for double circuit operation: (a) phase-a; (b) phase-b; and (c) phase-c

The highest frequency component (f_3) is due to the existance of the cross-bonded cable where the length of one minor section is about 500 m. The second frequency component (f_2) seems to be caused by the cable system where the total length is 2.2 km. The lowest frequency component (f_1) may be caused by the whole sysmtem of Figure 6.24a. Also, the attenuation of higher frequency components is much smaller in Figure 6.25 than that in Figure 6.22. The maximum sheath overvoltages at the first cross-bonded point are given in Table 6.10 as Case A. The maximum overvoltages range from 0.45 to 0.71 pu in Case D of Table 6.9, while these are from 0.09 to 0.117 pu in Case A of Table 6.10. That is, the maximum overvoltages in the cable system connected with the overhead lines are about 20% of those in the cable system alone. This is quite reasonable, because the incoming surge voltage V_c to the cable terminal is reduced by the overhead line in the following equation.

$$V_c = V_0 \cdot 2Z_2/(Z_1 + Z_2)$$

where $V_0$is the s ourve voltage applied to the overhead line terminal (= 1 pu),$Z_1$is theoverhead line surge impedance, and $Z_2$is the cable surge impedance.

Assuming $Z_1 = 400\ \Omega$, and $Z_2 = 40\ \Omega$, V_c becomes about 0.2 pu, that is, 20% of the source voltage applied to the cable terminal in the case of the cable system alone.

From the above observation, it is concluded that the maximum sheath voltage, when the overhead lines are connected to the cable system, are reduced to about 20% of those in the case of the cable system alone.

Table 6.10 Maximum sheath voltages of a cross-bonded cable connected with an overhead line

Case			First cross-bonded point			Second cross-bonded point		
			Phase-a	Phase-b	Phase-c	Phase-a	Phase-b	Phase-c
A	V_{max}	(kV)	12.6	11.1	9.6	9.7	7.9	9.5
	t	(μs)	129.60	361.95	259.30	436.55	320.35	121.90
B	V_{max}	(kV)	8.3	8.3	6.9	5.8	4.2	6.5
	t	(μs)	119.30	361.85	257.10	294.45	91.10	269.70
C	V_{max}	(kV)	−10.43	−9.71	−10.05	−6.92	−7.46	−6.33
	t	(μs)	88.75	99.75	103.55	96.80	101.60	86.70

Case A, double-circuit operation; Case B, single-circuit operation; and Case C, same as Case A but considering the effect of iron frameworks.

6.3.3.2 Frequency-dependent Effect of an Overhead Line

It may be interesting to see the effect of the frequency dependence of overhead line parameters. To take into account the frequency-dependent effect of the overhead line, the Semlyen line model of an overhead line in the EMTP is adopted. Calculated results of the transient sheath voltage are shown in Figure 6.26.

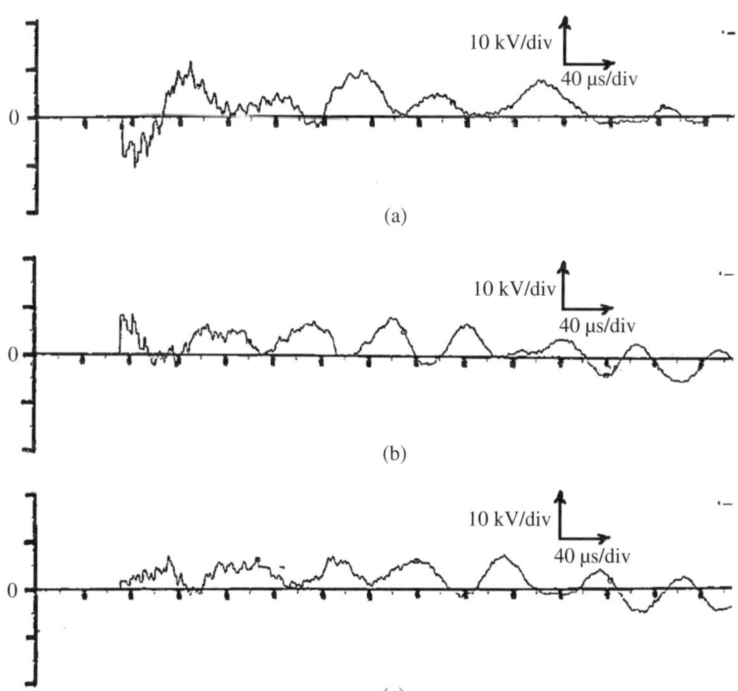

Figure 6.26 Transient sheath voltages when considering the overhead line frequency-dependent effect: (a) phase-a; (b) phase-b; and (c) phase-c

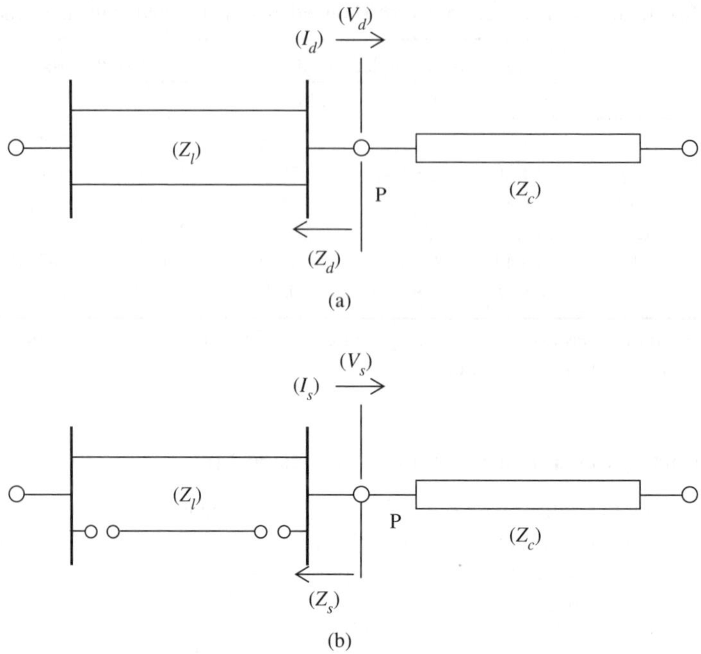

Figure 6.27 Types of circuit operation: (a) double-circuit operation; and (b) single-circuit operation

A comparison of Figure 6.26 with Figure 6.25 indicates that the higher frequency components of oscillations in the transient voltage are damped by the frequency-dependent effect and the voltage waveform is smoothed. But, the maximum overvoltages show a minor difference because they appear in the initial stage of the transient where the frequency-dependent effect is not pronounced. Thus, it is concluded that the frequency-dependent effect of an overhead line is not significant as far as the maximum sheath overvoltages are concerned.

6.3.3.3 Effect of Single-circuit Operation of Overhead Lines

In general, an overhead line with the configuration illustrated in Figure 6.24 is operated as a double-circuit line as illustrated in Figure 6.27a. However, it is possible to be operated as a single-circuit line as in Figure 6.27b due to a fault on the other line. It may be interesting to observe a difference of transient sheath overvoltages between the two types of operation.

Calculated waveforms are shown in Figure 6.28 and the maximum sheath overvoltages are given in Table 6.10 as Case B. A comparison of Figure 6.28 with Figure 6.26 indicates that the voltage waveforms are qualitatively the same until about $t = 360$ μs, but the peak values are smaller in the single-circuit operation case. The maximum overvoltages in the case of single-circuit operation are smaller by about 30% than those in the double-circuit operation case. This is due to the fact that the refraction coefficient at the boundary point between the overhead line and the cable is smaller in the single-circuit operation case. An analytical explanation for this is given in Appendix 6B.

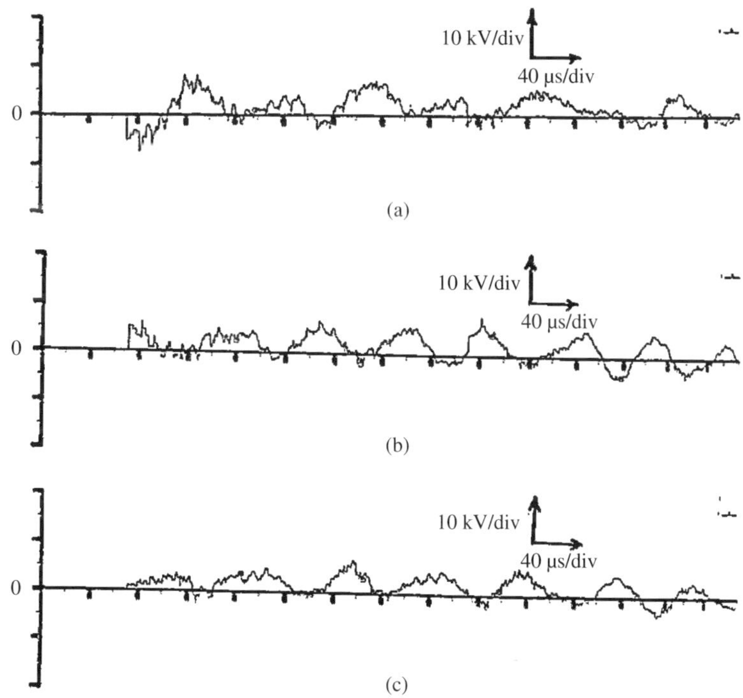

Figure 6.28 Transient sheath voltages for single-circuit operation: (a) phase-a; (b) phase-b; and (c) phase-c

From the above observation, it is concluded that the single-circuit operation of a double-circuit line causes lower sheath overvoltages, and therefore it is not necessary to worry about the single-circuit operation, which may occur during a fault, from the viewpoint of cable sheath overvoltages.

6.3.3.4 Effect of Iron Frameworks of a Bridge

In Section 6.3.2.3, it has been made clear that the iron frameworks of a bridge cause a significant reduction of sheath overvoltages in a cable system. It is interesting to observe if this is also true for a cable system connected with overhead lines.

Figure 6.29 shows calculated waveforms of transient sheath voltages when considering the mutual coupling between the cable and the iron frameworks of the bridge. The maximum overvoltages are given in Table 6.10 as Case C. It is observed that the positive part of the voltage is greatly damped by considering the iron frameworks, and the waveforms in Figure 6.29 are quite different from those in Figure 6.25. But, the maximum overvoltages are not necessarily reduced as is observed from Table 6.10, Cases A and C. At the first cross-bonded point, phase-a sheath voltage is decreased by about 17%, but phase-c voltage is increased by about 4%. At the second cross-bonded point, the rates of reduction are about 28% on phase-a, 5% on phase-b and 34% on phase-c. An interesting fact is that the maximum overvoltages have a positive polarity

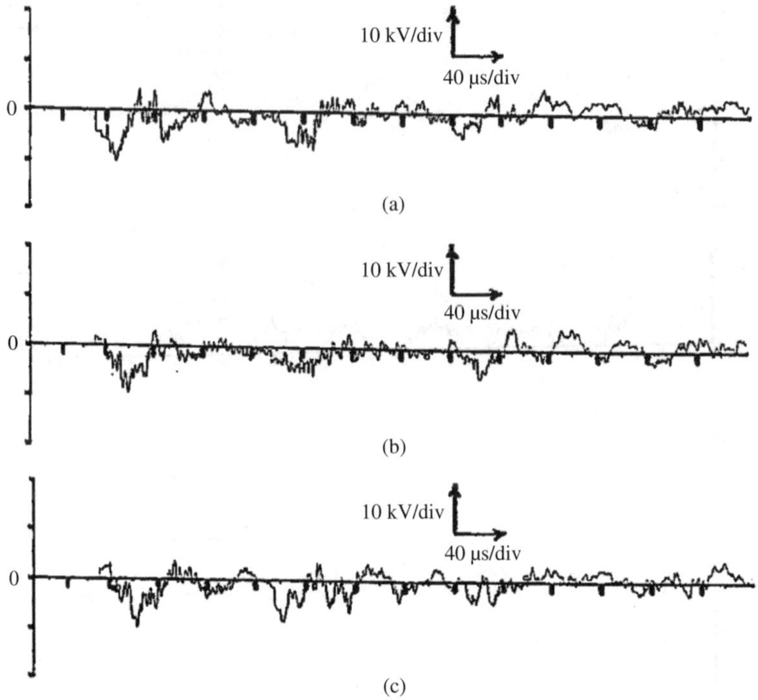

Figure 6.29 Transient sheath voltages when considering mutual coupling with iron frameworks of a bridge: (a) phase-a; (b) phase-b; and (c) phase-c

in Case A, but in Case C, those are negative. The time when the maximum overvoltages appear is much faster in Case C than in Case A.

Thus, it can be said that the iron frameworks cause an effect of reducing the sheath overvoltage as same as Section 6.3.2.3, but the effect is less pronounced when overhead lines are connected to the cable system.

6.4 Cable Modeling in EMTP Simulations

EMTP simulation results on a cable explained in Section 3.5.3 of Chapter 3 are discussed in this section [13, 14].

6.4.1 Marti's and Dommel's Cable Models

Figure 6.30 shows simulation results on normal-bonded and cross-bonded cables by Marti's and Dommel's cable models explained in Chapter 4. It is clear that there exists a significant difference between the normal- and cross-bonded cables as already explained. It should be noted that no significant difference is observed between the results obtained by the Marti's and Dommel's cable models. Thus, it can be said that the Dommel's constant frequency distributed line model is good enough to simulate a transient on a cross-bonded cable.

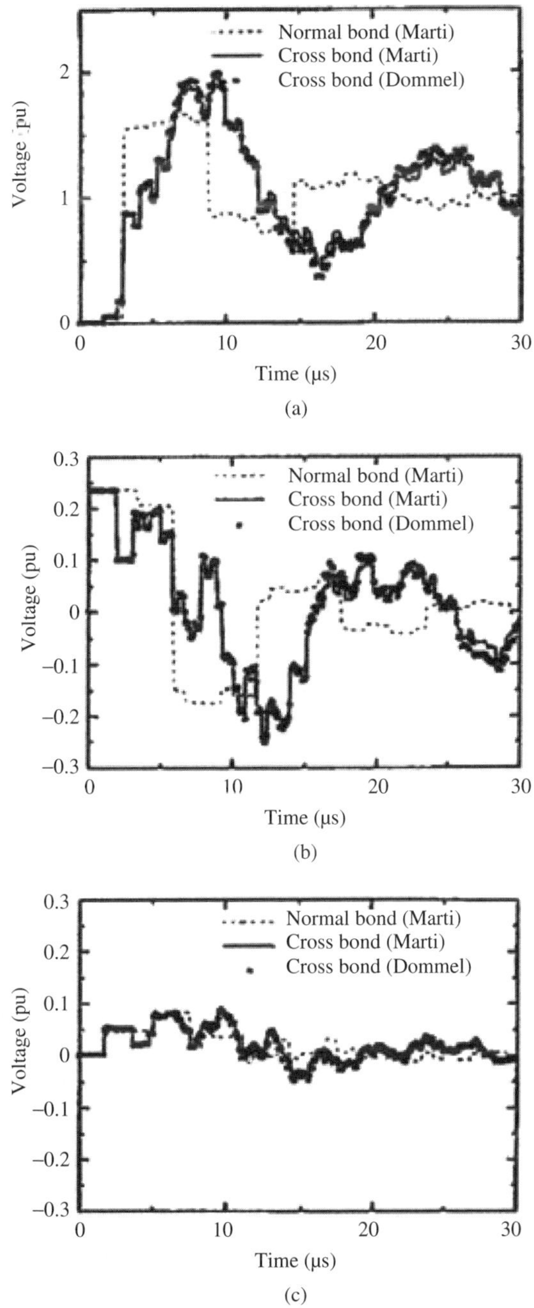

Figure 6.30 Calculated results of transient on a tunnel-installed cable by the Marti frequency-dependent line model: (a) receiving-end core; (b) sending-end sheath; and (c) receiving-end sheath

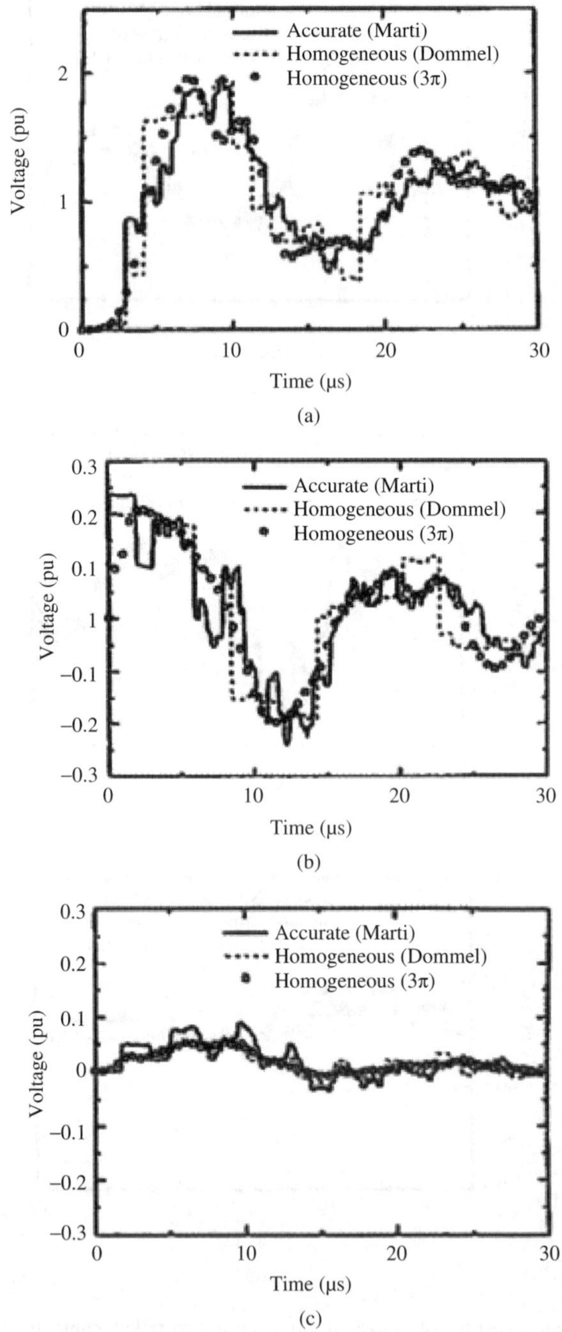

Figure 6.31 Calculated results of transient voltages by a homogeneous cable model: (a) receiving-end core; (b) sending-end sheath; and (c) receiving-end sheath

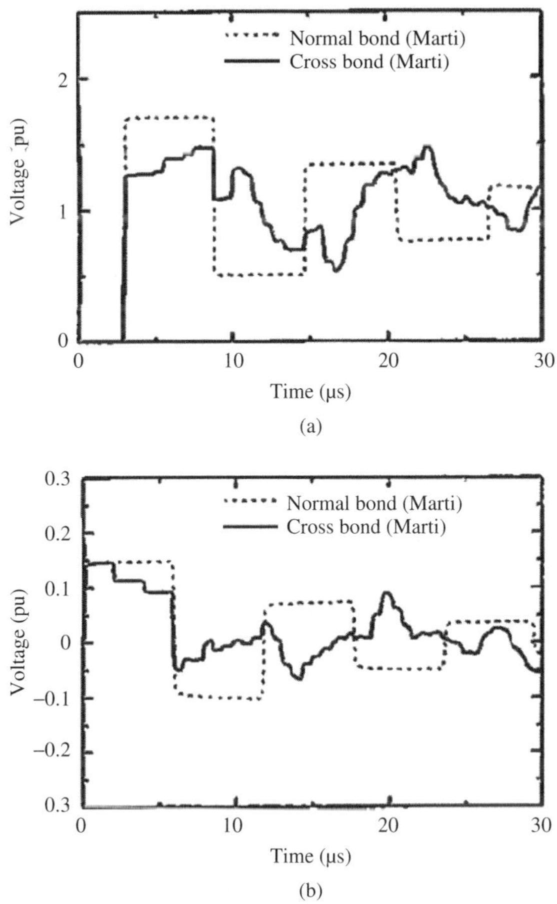

Figure 6.32 Calculated results of transient voltages on an underground cable by the Marti line model: (a) receiving-end core; and (b) sending-end sheath

6.4.2 Homogeneous Cable Model (See Section 3.5.3)

Figure 6.31 shows simulation results by the homogeneous model of a cross-bonded cable consisting of three major sections. Dommel's cable model using the homogeneous model shows a difference from the results by Marti's model with no assumption of the homogeneous model as explained. However, it is observed that the simulation results by a PAI-circuit homogeneous cable model agree satisfactorily with the accurate result by Marti's cable model. In fact, the homogeneous model of a cross-bonded cable was derived for a PAI-circuit line model as described in the *EMTP Rule Book* [15].

6.4.3 Effect of Tunnel-installed Cable

Figure 6.32 shows a comparison of transient waveforms on a tunnel-installed cable and on a buried cable simulated by Marti's cable model. A difference is clearly observed in the figure as already explained in Section 6.2.

6.5 Pipe-enclosed Type (PT) Cable

6.5.1 Field Test on a 275 kV Pressure Oil-filled (POF) Cable

Figure 6.33 illustrates a 275 kV, 250 mm^2 POF cable to connect a generator station and a substation with a length of 3.382 km [2, 16]. Table 6.11 gives the specification of the 275 kV POF cable. Figure 6.34 shows the field test circuit.

Two test cases are set up. As the first case, the pipe is grounded and an impulse voltage with $T_f = 2$ μs and $T_t = 40$ μs is applied between one core conductor and the ground. In the second case, the pipe is isolated from the ground and an impulse voltage of the same waveform is applied between the pipe and the ground. The impulse voltage is applied to a core through the input resistance R_s of 22 Ω, and the parallel resistance R_0 of the impulse generator and the sending and receiving ends of the other two core conductors are open-circuited in the first case. To apply the impulse voltage to the pipe in the second case, the pipe is isolated from the ground and all the core and sheath conductors are connected to the pipe. Both resistances R_s and R_0 are set equal to 11 Ω. The receiving ends of all the core conductors are connected together and isolated from the ground. The internal impedance of the impulse generator is simulated rigidly to take into account the reflection of a traveling wave at the input terminal.

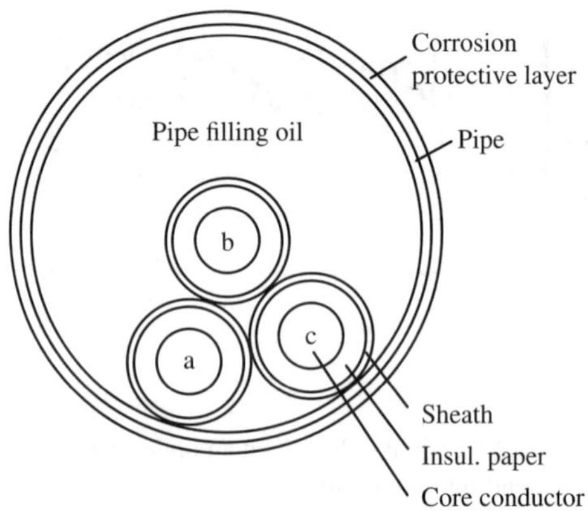

Figure 6.33 Cross section of a 275 kV POF cable

Table 6.11 Specification of a 275 kV POF cable

Core conductor radius	31.10	mm	Core conductor resistivity	0.017	μΩm
Insulating layer thickness	20.45	mm	Sheath resistivity	0.73	μΩm
Sheath thickness	0.15	mm	Pipe resistivity	0.1	μΩm
Pipe inner diameter	150.85	mm	Earth resistivity	100	Ωm
Pipe outer diameter	159.25	mm	Pipe specific permeability	100	
Thickness of corrosion protective layer	4.5	mm	Insulating layer specific dielectric constant	3.83	
Laying depth	3.22	m	Pipe filing oil specific dielectric constant	2.3	
Cable length	3.384	m	Corrosion protective layer specific dielectric constant	3.5	

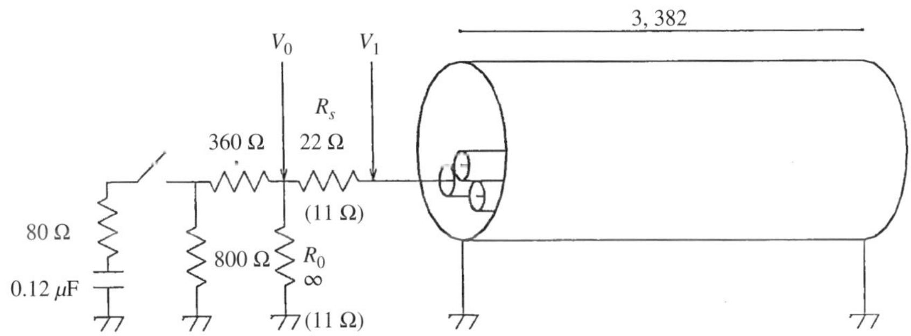

Figure 6.34 Field test circuit of a POF cable

6.5.2 Measured Results

6.5.2.1 Pipe Grounded

The measured results in the grounded pipe case are shown in Figure 6.35. Applied voltage $V_0$and sending-end voltage V_1 of the core are shown in Figure 6.35a. The sending-end voltage of the core reaches the peak value −0.5 pu (1 pu ≒ 400 V) at 2 μs and the first wave reflected from the receiving end is observed at 47 μs. Since the cable length is 3382 m, the propagation velocity becomes 144 m/μs. The peak value of the reflected wave is −0.63 pu and it appears at 73 μs. The wavefront is considerably distorted.

The applied-phase voltage V_2 of the core at the receiving end is shown in Figure 6.35b. In this case too, the wavefront is considerbly distorted and it reaches the peak voltage of −0.66 pu at 21 μs after the traveling wave has reached the receiving end. Since the receiving end is open-circuited and its reflection coefficient is equal to unity, the peak voltage of the traveling wave arriving at the receiving end for the first time is equal to −0.33 pu, which is equal to one half of the observed peak voltage. This voltage gives an attenuation constant of 1.1 dB/km, which is equal to the attenuation constants of coaxial propagation modes 2, 3 and 4 at a frequency of 10 kHz. This frequency is approximately equal to $1/4\tau$, τ being the propagation time corresponding to the dominant transient frequency.

The receiving-end induced voltage of a core is shown in Figure 6.35c. The peak induced voltage is observed at 8.1μs after the arrival of the traveling wave voltage and is 0.056 pu in phase-b and 0.083 pu in phase-c. The second peak voltage appears at 47μs and is −0.056 pu in both phase-b and phase-c.

6.5.2.2 Pipe Isolated

The sending-end pipe voltage in the isolated pipe case is shown in Figure 6.36. A traveling wave reflected from the receiving end is observed at 233 μs and therefore the propagation velocity is 29 m/μs, which is much smaller than that in the grounded pipe case. A comparison of waveforms between Figure 6.35a and Figure 6.36 also indicates that the waveform distortion and attenuation in the isolated pipe case are much more significant than in the ground pipe case. This is because the major mode in the isolated pipe case is the earth-return mode and its propagation velocity is very small and its attenuation constant is very large.

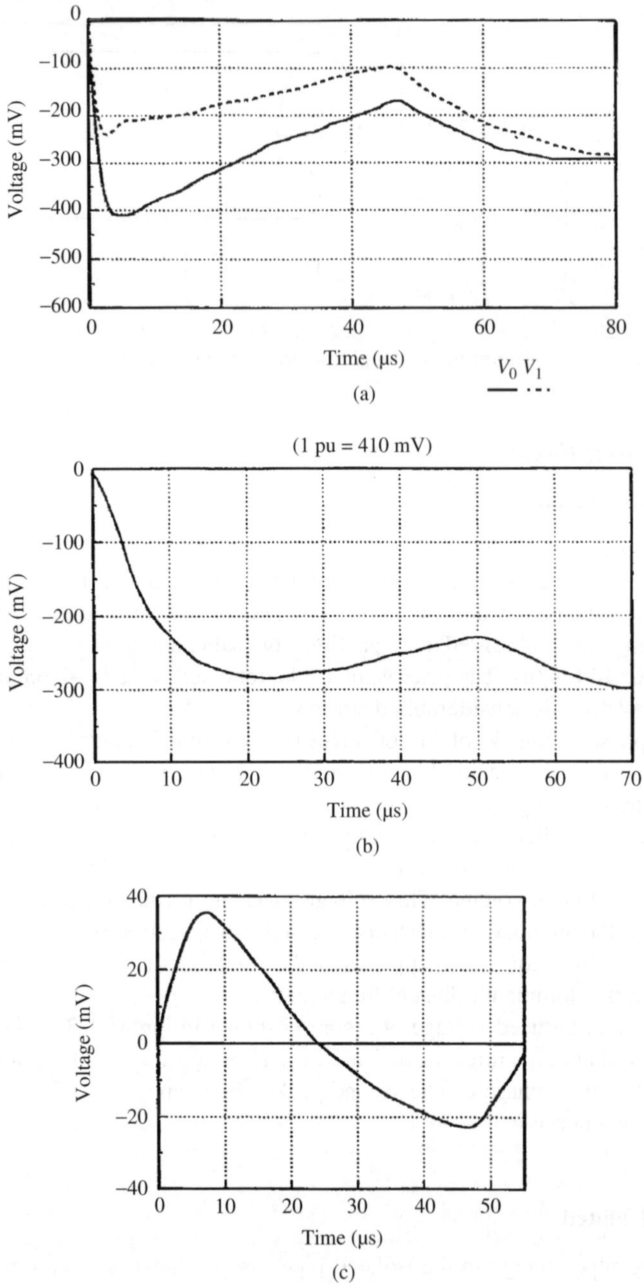

Figure 6.35 Field test results of voltages for core energization: (a) sending-end voltage; (b) receiving-end voltage of voltage-impressed phase; and (c) receiving-end voltage of induced phase

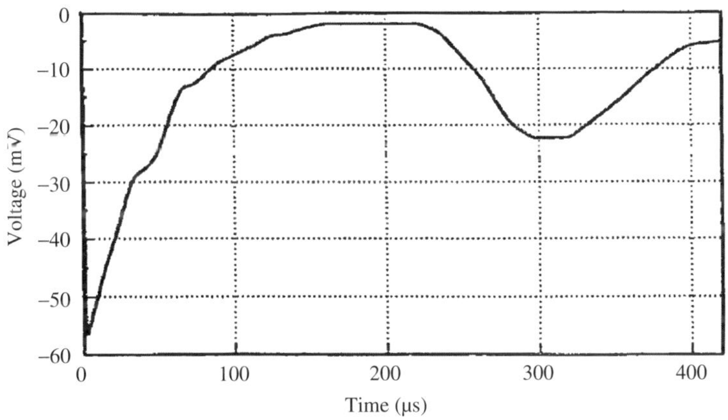

Figure 6.36 Field test results for pipe energization

6.5.3 FTP Simulation

The frequency dependence of the POF cable impedance is taken into account by using the FTP [2, 4]. In this method, the steady-state solution in the frequency domain is analyzed first and it is transformed to the time domain by the numerical Laplace inverse transform [17]. The steady-state solution in the frequency domain is derived by a distributed circuit theory with the modal theory as explained in Chapter 3. The number of sampling points used for the numerical Laplace inverse transformation is taken to be 256.

6.5.3.1 Pipe Grounded

The calculated results in the case of the grounded pipe are shown in Figure 6.37. The applied voltage $V_0$and the sending-end core voltage V_1 are shown in Figure 6.37a. The core voltage reaches the peak amplitude of −0.46 pu at 5 μs. The difference between the measured and calculated voltage is 8%. The traveling wave reflected from the receiving end is observed at 45 μs, which gives a propagation velocity of 150 m/μs between the core and the sheath. This is 4.2% different from the measured velocity. The peak amplitude of the wave reflected from the receiving end is −0.62 pu and it appears at 71 μs. The receiving-end voltage of the core is shown in Figure 6.37b. The peak amplitude of −0.65 pu appears at 19 μs after the arrival of the traveling wave and it is different by 1.5% from the measured value. The induced-phase voltage at the receiving-end core is shown in Figure 6.37c. The first peak amplitude of 0.067 pu is observed at t = 11 μs. It is different by 20% from the measured result.

6.5.3.2 Pipe Isolated

The sending-end pipe voltage in the isolated pipe case is shown in Figure 6.38. The wave reflected from the receiving end is observed at 460 μs and its peak amplitude is 0.18 pu. It gives a propagation velocity of 15 m/μs in the isolated pipe case.

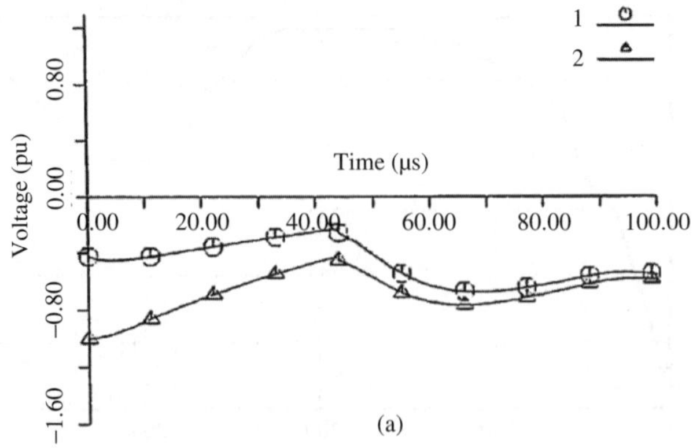

(a)

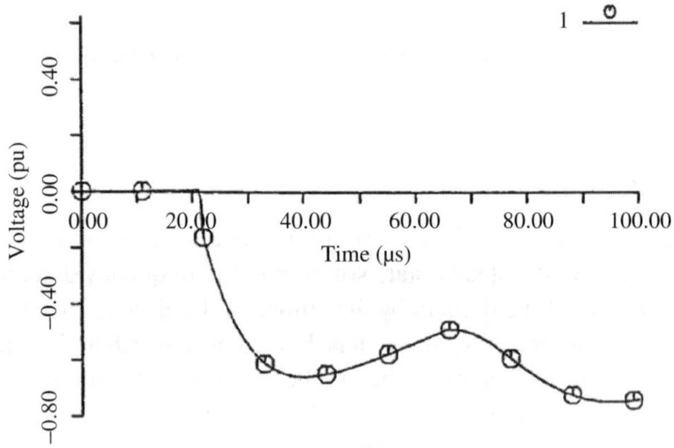

(b)

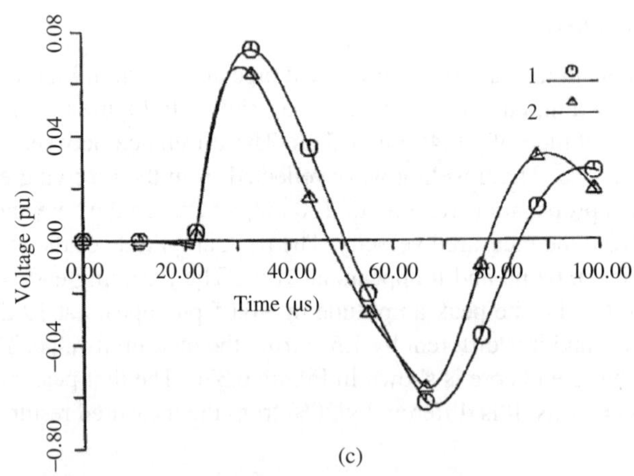

(c)

Figure 6.37 Calculated results for core energization: (a) sending-end voltage; (b) receiving-end voltage of voltage-applied phase; and (c) receiving-end voltage of induced phase

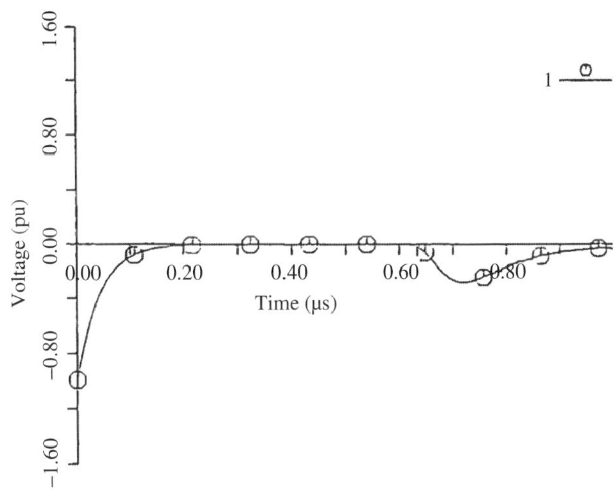

Figure 6.38 Calculated results for pipe energization

Table 6.12 Field test and calculated results

	Measured		Calculated	
V_{1ca} (pu)	−0.5	(2 μs)	−0.46	(2 μs)
V_{2ca} (pu)	−0.63	(τ + 21 μs)	−0.65	(τ + 19 μs)
V_{2cb} (pu)	−0.056	(τ + 8 μs)	−0.061	(τ + 9 μs)
V_{2cc} (pu)	0.083	(τ + 8 μs)	0.067	(τ + 11 μs)
v (m/μs)	144		150	—[a]
(m/μs)	29		15	—[b]

τ, Propagation time.
[a]Voltage applied to the core conductor.
[b]Voltage applied to the pipe.

The aforementioned results are summarized in Table 6.12, which shows good agreement between the measured and the calculated results. However, they are considerably different for the isolated pipe case. Here, the earth-return circuit has a significant effect. The calculation error is estimated to be caused by the earth resistivity and corrosion proof layer constants, as well as their non-uniform distribution in the longitudinal direction.

6.5.3.3 Effect of Sheath Thickness

Calculated results for the sheath thickness of 1 mm are shown in Figure 6.39. The peak amplitude of the core to which the impulse voltage is applied, is almost independent whether the sheath thickness is 1 or 0.15 mm. However, the wavefront is not distorted. The propagation velocity is 154 m/μs and it is also almost independent of the sheath thickness. However, it should be noted that the attenuation in the high frequency range is small when the sheath is thick. The small induced voltage also indicates that the shielding effect is significant when the

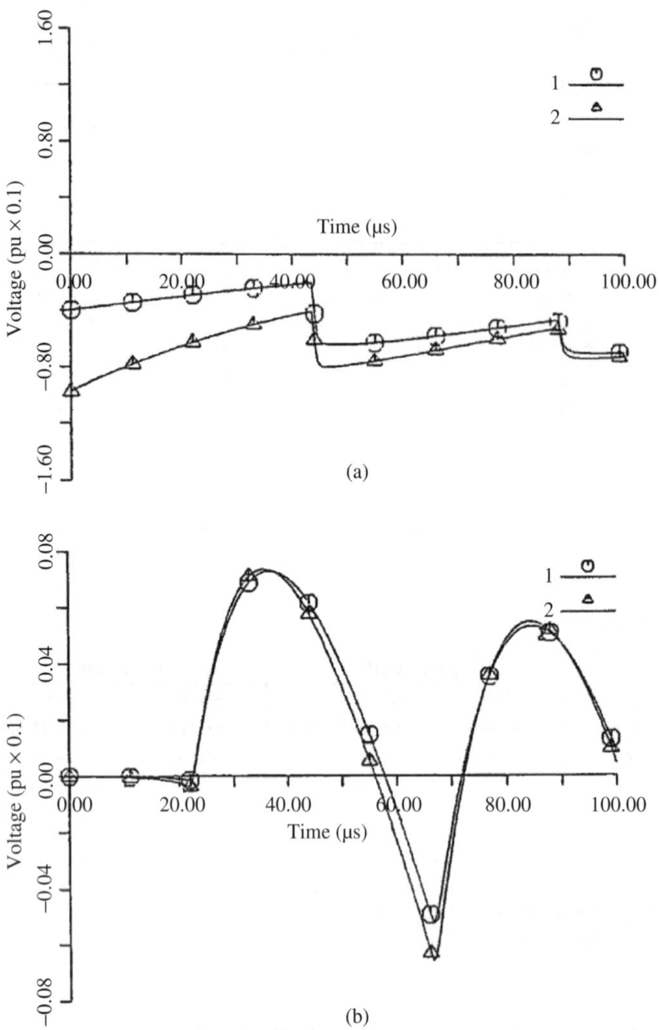

Figure 6.39 Calculated results for the sheath thickness: (a) sending-end voltage; and (b) receiving-end voltage of induced phase

sheath is thick. From the above discussion, it is clear that the sheath thickness influences the wavefront distortion, attenuation and induced voltage significantly.

6.5.3.4 Effect of Inner Cable Arrangement

An inner cable arrangement is shown in Figure 6.40 and Figure 6.41 shows the calculated results for the inner cable arrangement. A comparison between waveforms in Figure 6.37 and Figure 6.41 shows that the inner cable arrangement has little effect. The induced voltage of phase-c in Figure 6.40a is smaller than that of phase-b as illustrated in Figure 6.41b. The

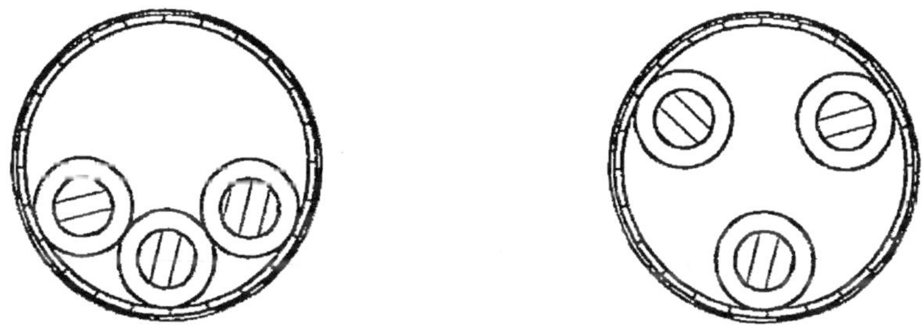

Figure 6.40 Inner cable arrangement: (a) cradle type; and (b) symmetric triangular type

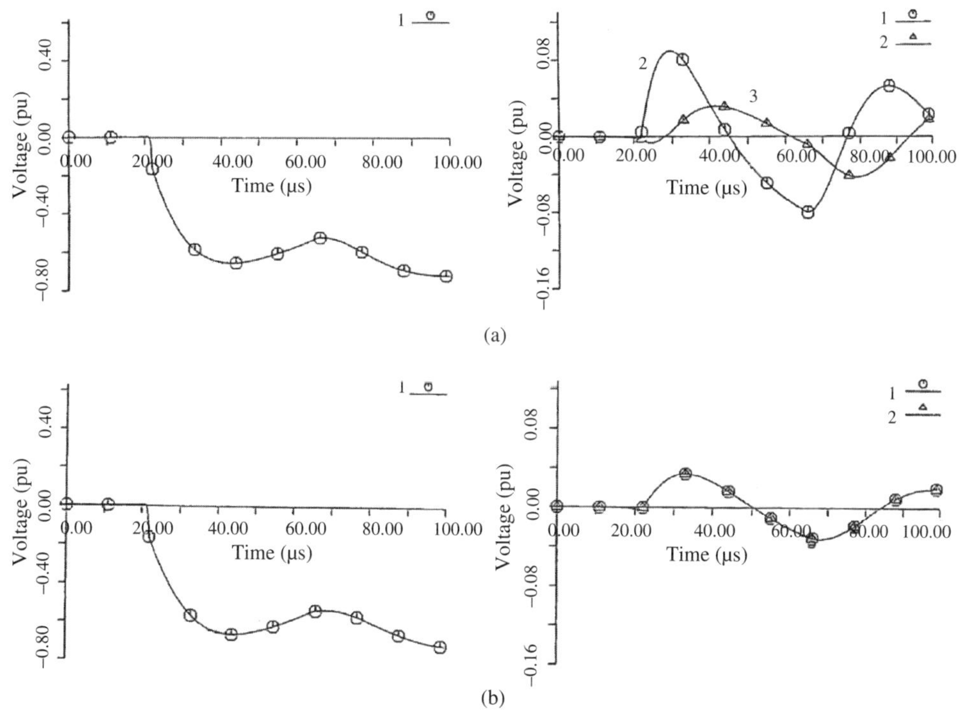

Figure 6.41 Effect of the inner conductor arrangement: (a) cradle; and (b) symmetric. (Left) Voltage-impressed phase and (right) induced phase

induced voltage of phase-c is small partly because the distance between the phase-a conductor to which the impulse voltage is applied, and the phase-c conductor in which a voltage is induced, is large and partly because the phase-b cable has some shielding effect.

In Figure 6.40b where the three cables are arranged symmetrically, the induced voltages are reduced because the distances between the conductors are increased. Since the three cables are arranged symmetrically, the same voltage is induced in the three cables.

From the observations, it is clear that the cable arrangement inside the pipe has little effect on the voltage on the core to which the impulse voltage is applied. However, the cable arrangement in Figure 6.40b has a slightly greater effect on the induced voltages than the other cable arrangement.

6.6 Gas-insulated Substation (GIS) – Overhead Cables

A gas-insulated bus (GIB) system in a GIS is a kind of overhead cable. If the GIB is a three-phase single-core coaxial (SC) cable, it is dealt with as an overhead SC cable. If three-phase cores are installed within a pipe conductor, it is handled as an overhead PT cable. In this section, switching surges on the GIB and GIS are investigated [18, 19].

A significant difference of GIS surges from those in an ordinary cable is that the GIS surges involve very fast transients, that is, very high frequency components [19] due to multiple reflection and refraction of traveling waves in a number of short gas-insulated cables (GIB). Such high frequency components cause electromagnetic disturbances to adjacent control circuits in the substation [19, 20]. Thus, the very fast transients in the GIS have become an important subject in the field of electromagnetic compatibility (EMC) [21]. In fact, an EMC related standard in power stations and substations in Japan was recently revised based on the electromagnetic disturbance experiences [22, 23].

6.6.1 *Basic Characteristic of an Overhead Cable*

It has been explained that there exists no significant difference between the earth-return impedance of an underground cable and an overhead line (cable) in Chapter 2. However, the admittance of the overhead line is, in general, far smaller than that of the underground cable. The potential coefficient P_{ud} of the metalic sheath of the underground cable is given in the following equation.

$$P_{ud} = \ln(r_0/r_i)/2\pi\varepsilon_0\varepsilon_r \tag{6.5}$$

In the overhead line case, P_{oh} is given by:

$$P_{oh} = \ln(2h/r_0)/2\pi\varepsilon_0 \tag{6.6}$$

where r_0 and r_i are the outer and inner radii of the sheath outer insulator (jacket), respectively, h is the conductor height, and ε_r is the relative permittivity of the jacket.

Because r_0 and r_i are in the order of some centimeters and ε_r is 3–6 in a cable jacket, while h is 1–2 m and r_0 is some 10 cm in a GIB, the following relation is satisfied.

$$P_{ud} \ll P_{oh} \tag{6.7}$$

Thus, the overhead line admittance Y_{oh} is far smaller than the admittance Y_{ud} of the underground cable. Then, the propagation constant Γ shows the following characteristic.

$$\text{Underground} \quad \Gamma_{ud} = \sqrt{Z_{ud} \cdot Y_{ud}} = \alpha_{ud} + j\beta_{ud}, \quad Z_{oud} = \sqrt{Z_{ud}/Y_{ud}}$$

$$\text{Overhead} \quad \Gamma_{oh} = \sqrt{Z_{oh} \cdot Y_{oh}} = \alpha_{oh} + j\beta_{oh}, \quad Z_{ooh} = \sqrt{Z_{oh}/Y_{oh}}$$

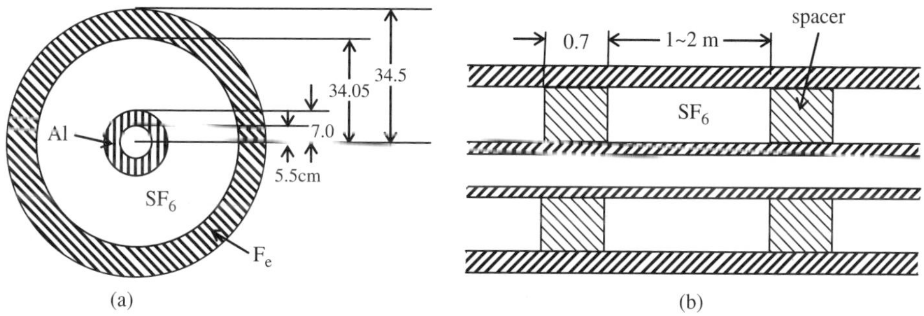

Figure 6.42 Configuration of a GIB: (a) radial cross-section; and (b) longitudinal cross-section

For $Z_{ud} \fallingdotseq Z_{oh}$ and $Y_{ud} \gg Y_{oh}$, the following relation is clear.

$$\alpha_{ud} \gg \alpha_{oh} \tag{6.8a}$$

$$c_{ud} \ll c_{oh}\ (\beta_{ud} \gg \beta_{oh}) \tag{6.8b}$$

$$|Z_{oud}| \leq |Z_{ooh}| \tag{6.8c}$$

where α is the attenuation constant, β is the phase constant, c is the propagation velocity, and Z_0 is the characteristic impedance.

From the above investigation of the earth-return mode, it should be clear that the earth-return mode attenuation is far smaller and the velocity is much greater, near to the velocity of light in free space, in the overhead cable. As already explained, the coaxial mode characteristic is nearly the same at a high frequency in the overhead and underground cables. It should be noted that the coaxial mode propagation velocity is nearly the same as that of the earth-return mode in a GIB, because the relative permittivity of SF_6 gas is nearly the same as that of free space. This fact occasionally results in a numerical instability of eigenvalue and vector calculations.

6.6.2 *Effect of Spacer in a Bus*

In a GIB, the inner conductor (core) is suspended as illustrated in Figure 6.42 by a spacer with relative permittivity of 3–6 depending on its material. Because there are a number of short length buses for connecting circuit breakers, disconnectors, arresters, and so on, in a GIS, the number of spacers is large, and they affect a transient response significantly

6.6.2.1 Propagation Parameters with a Spacer

The velocity v_{coax} and the characteristic impedace z_{ocoax} of coaxial mode propagation at a very high frequency is given in the following equation.

$$v_{coax} = v_0/\sqrt{\varepsilon_1} \tag{6.9a}$$

$$z_{ocoax} = (60/\sqrt{\varepsilon_1}) \cdot \ln(r_{out}/r_{in}) \tag{6.9b}$$

where $\upsilon_0 = 300$ m/μs is the light velocity in free space, $\varepsilon_1 = 1.009$ is the relative permittivity of an SF_6 gas, and r_{in} and r_{out} are the inner and outer radii of the insulator, respectively.

In the case of spacer with $\varepsilon_2 = 6$, the propagation velocity υ_2 becomes 122 m/μs, while for a SF_6 gas $\upsilon_1 = 298$ m/μs. The low velocity υ_2 produces a large traveling time. Also, the characteristic impedance of the spacer part becomes smaller. The low velocity and the small impedance result in a transient voltage different from that neglecting the spacer. In practice, the shape and the structure of a spaser are not simple as shown in Figure 6.42 and Figure 6.43, and there are bushings. Therefore it should be noted that these deform a surge waveform, and an actual waveform measured in a field might differ from the simulated waveforms in this section.

6.6.2.2 EMTP Simulation Results

Transient voltage waveforms in a model circuit of Figure 6.43 for length$x = 150$ m calculated by EMTP are shown in Figure 6.44, and the maximum voltage with the time of its appearance is summarized in Table 6.13. It is observed in the table that the core voltage at the receiving-end GA2 increases as the number of spacers increases. This is readily explained as a result of

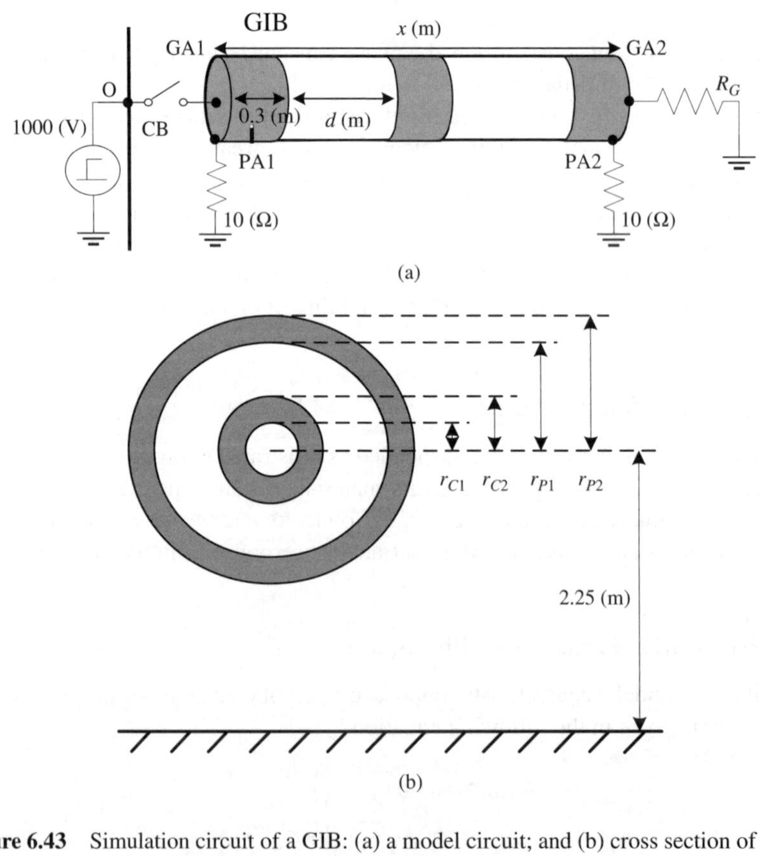

Figure 6.43 Simulation circuit of a GIB: (a) a model circuit; and (b) cross section of a GIB

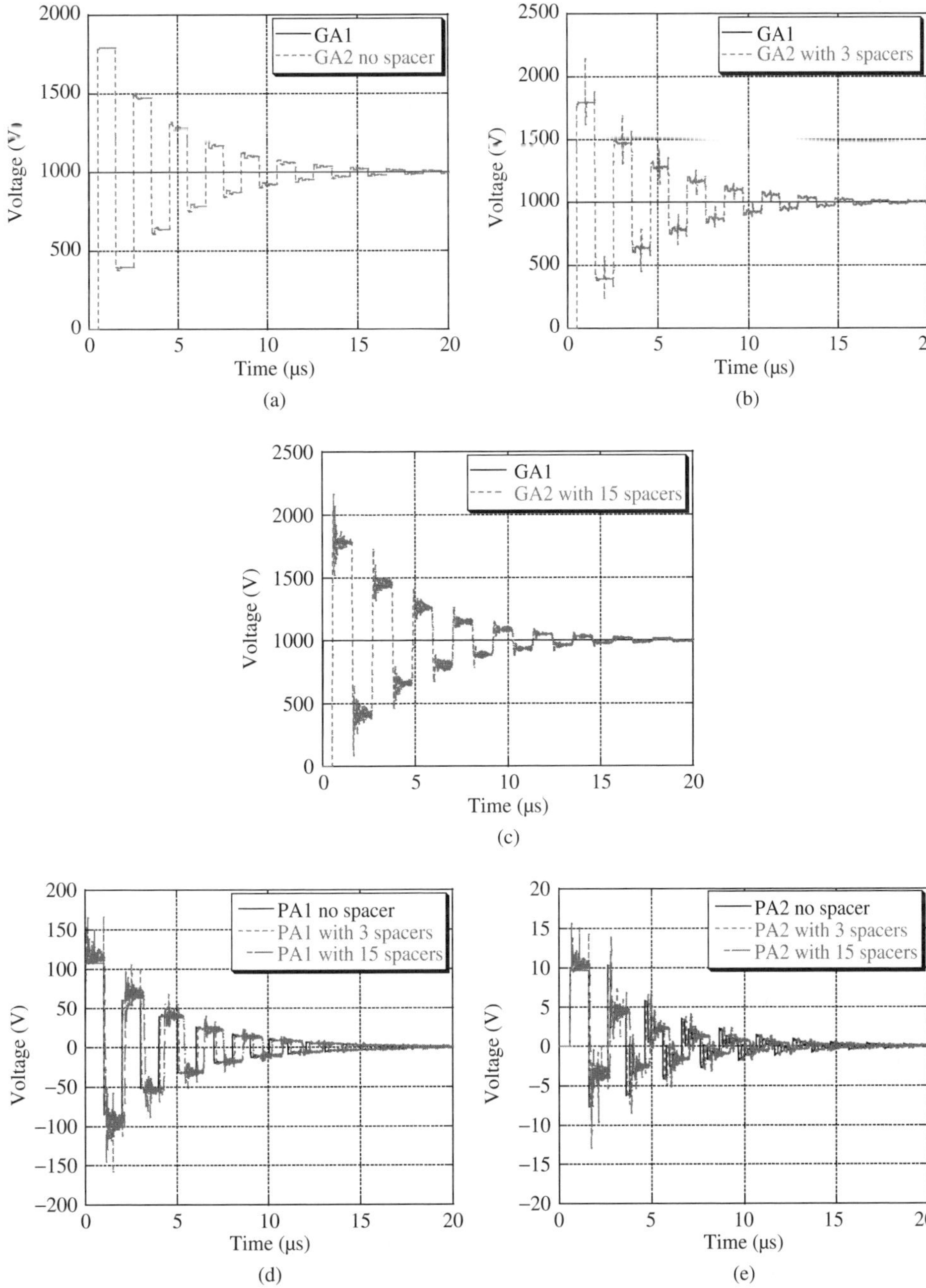

Figure 6.44 Simulation results of transient voltages for $x = 150$ m: (a) no spacer; (b) 3 spacers ($d = 75$ m); and (c) 15 spacers ($d = 10$ m). In (a)–(c), core voltage at both ends (GA1, GA2). (d) Pipe voltage at the sending end (PA1) and (e) pipe voltage at the receiving end (PA2)

Table 6.13 Simulation results for $x = 150$ m

R_G (Ω)	No. of spacer	V_{max} (V) / Time (μs)			f (kHz)
		GA2	PA1	PA2	GA2
∞	No	1793 / 0.592	107.9 / 0.001	10.3 / 2.604	496.7
	3	2141 / 1.009	228.8 / 0.507	20.0 / 0.593	492.4
	15	2228 / 0.603	229.0 / 0.077	17.9 / 0.668	475.8
65.6	No	849 / 0.662	107.9 / 0.001	−101.7 / 0.504	662.2
	3	1077 / 1.010	228.8 / 0.507	−120.5 / 1.010	656.5
	15	1104 / 0.603	229.0 / 0.077	−123.8 / 0.603	634.3

multiple reflection and refraction of traveling waves at the spacer which becomes a transition point to the traveling waves. A similar trend is observed on the tank (outer encolosure of the GIB), but the effect of the number of spacers is not clear.

6.6.2.3 Case of Short-length Bus

There are number of short buses in a GIS. Therefore it is necessary to investigate a transient voltage in the case of a short-length bus. Figure 6.45 shows transient voltage waveforms in the case of $x = 7.5$ m with two and three spacers, and the maximum voltages are given in Table 6.14. A similar trend of the core and pipe voltages to those for $x = 150$ m in Table 6.13 is observed. It is interesting to note that the pipe voltage at the sending end is independent of the bus length x. A difference between no spacer and spacers become more noticeable in the short-length bus.

6.6.2.4 Equivalent Permittivity

Figure 6.46 illustrates an experimental circuit of a single-phase GIB [18]. In the experiment, the spacer has a width of 0.7 m and the relative permittivity $\varepsilon_2 = 6.0$ as illustrated in Figure 6.42.

In such a case, an accurate approach is to represent the GIB as a combination of a very short distributed line (or a lumped capacitance) representing the spacer part and a distributed line corresponding to the GIB as explained in Sections 6.6.2.1–6.6.2.3. However, the approach requires a very short time step and very large memories for a numerical simulation. An alternative simple approach is to adopt an equivalent permittivity ε_r defined in the following equation.

$$\varepsilon_r \fallingdotseq (l_1\varepsilon_1 + l_2 \cdot \varepsilon_2)/(l_1 + l_2) \tag{6.10}$$

where l_1 is the length of the GIB, l_2 is the length (width) of the spacer, $\varepsilon_1 \fallingdotseq 1.0$ is the relative permittivity of SF_6 gas, and ε_2 is the relative permittivity of the spacer.

For example in Figure 6.42, $l_2 = 0.7$ m, and $\varepsilon_2 = 6.0$, the following equivalent permittivity is obtained.

(a) $l_1 = 2$ m : $\varepsilon_r = 2.3$
(b) $l_1 = 1$ m : $\varepsilon_r = 3.0$

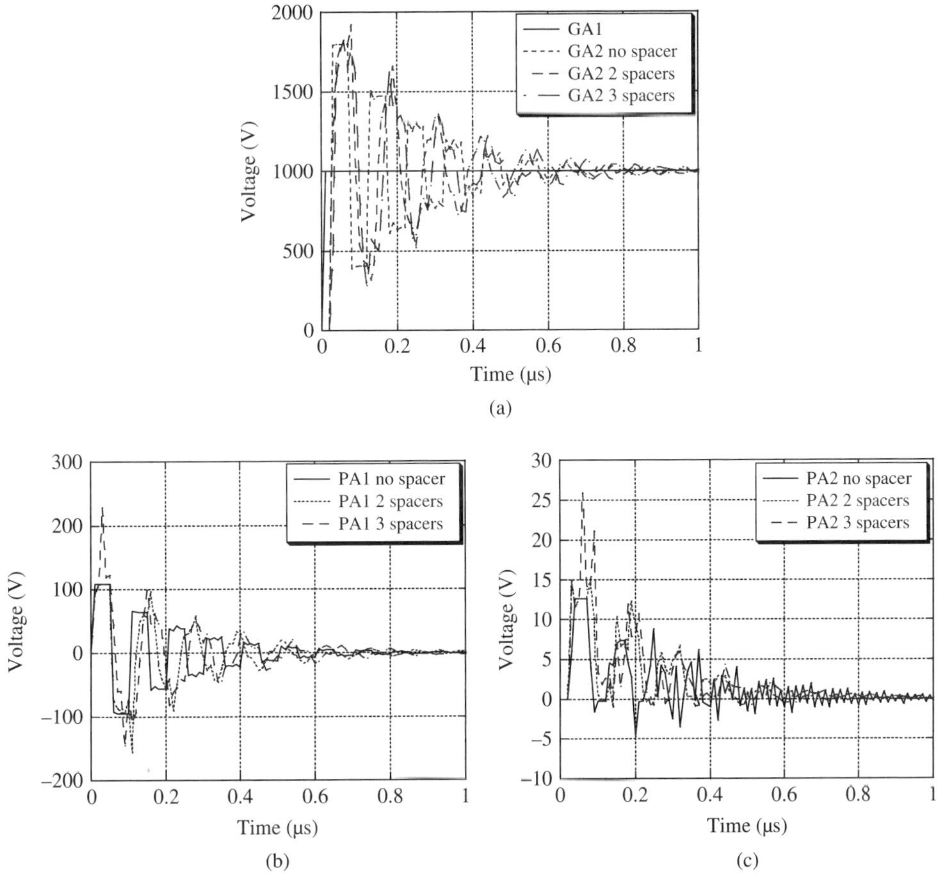

Figure 6.45 Simulation result of transient voltages for $x = 7.5$ m: (a) core voltage at both ends (GA1, GA2); (b) pipe voltage at the sending end (PA1); and (c) pipe voltage at the receiving end (PA2)

Table 6.14 Simulation results for $x = 7.5$ m

		V_{max} (V) / Time (μs)			f (MHz)
R_G (Ω)	No. of spacer	GA2	PA1	PA2	GA2
∞	No	1797 / 0.031	107.9 / 0.001	17.4 / 0.135	9.93
	2	2017 / 0.077	228.9 / 0.054	26.7 / 0.033	8.91
	3	2135 / 0.054	229.2 / 0.030	26.6 / 0.033	8.47
65.6	No	970 / 0.100	107.9 / 0.001	−101.7 / 0.027	13.2
	2	1016 / 0.082	228.9 / 0.054	−108.0 / 0.079	11.9
	3	1081 / 0.055	229.2 / 0.030	−118.0 / 0.055	11.3

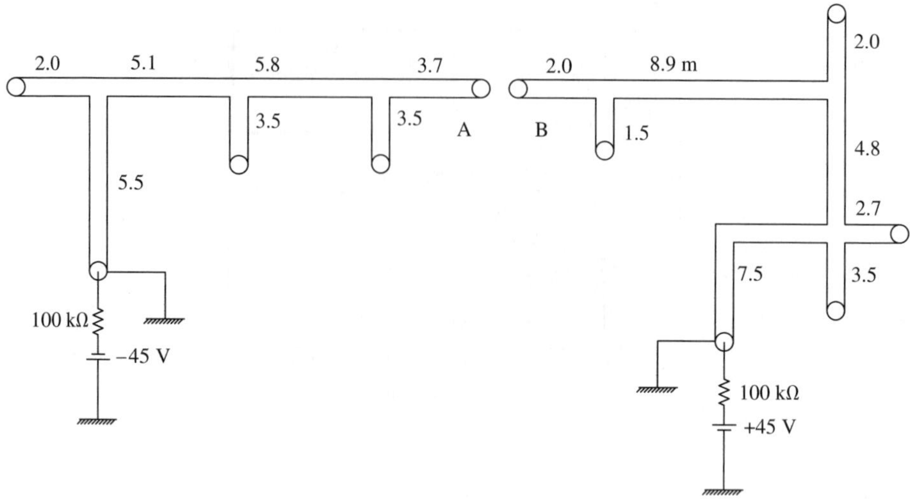

Figure 6.46 An experimental circuit

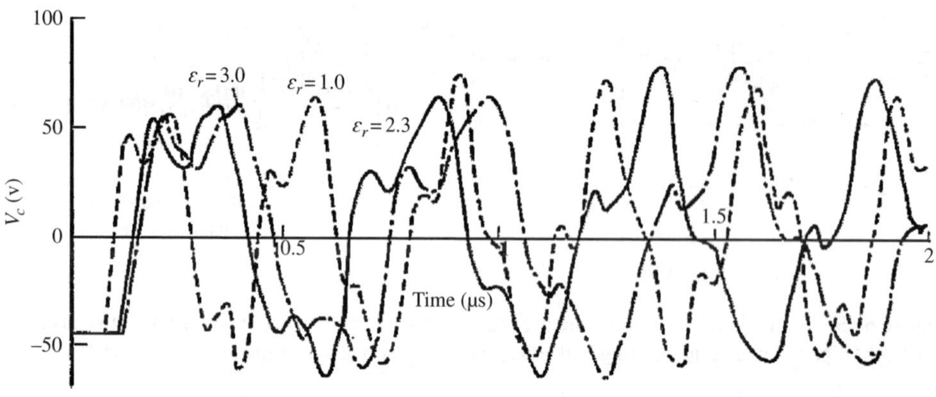

Figure 6.47 Calculated results for various ε_r

Then, the propagation velocity is evaluated at $f = 1$ MHz as:

(a) $\varepsilon_r = 2.3$: $c_1 = 197.0$ m/μs
(b) $\varepsilon_r = 3.0$: $c_2 = 172.5$ m/μs
(c) gas only $\varepsilon_r = 1.0$: $c_3 = 297.2$ m/μs

Figure 6.47 shows simulation results and Figure 6.48 is the measured result, when terminals A and B are short circuited under the condition that the parts A and B of the GIBs are charged by ±45 V, respectively.

Case (a) ($\varepsilon_r = 2.3$) agrees best with the measured result among the three cases, if the peak voltage and the oscillating period in Figure 6.47 are compared with those in Figure 6.48. Thus,

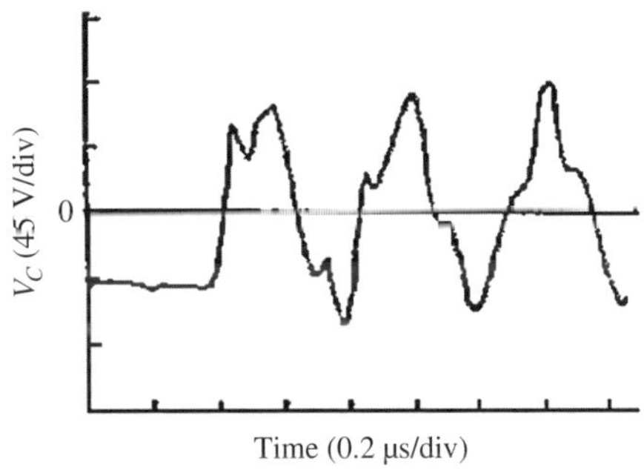

Figure 6.48 A measured result of the experimental circuit in Figure 6.46

the equivalent permittivity in (6.10) is the simplest way to consider the effect of spacers in a GIB for a transient analysis.

6.6.3 Three-phase Underground Gas-insulated Line

Most gas-insulated cables are overhead and are used in a GIS. Figure 6.49 illustrates a field test circuit of an underground gas-insulated line to connect an overhead line and a switching

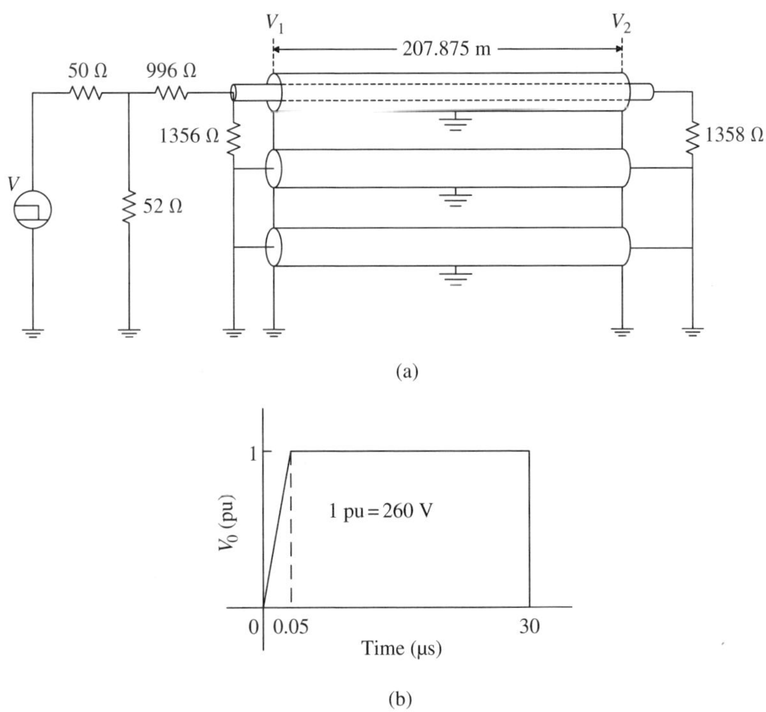

Figure 6.49 (a) A field test circuit and (b) applied voltage

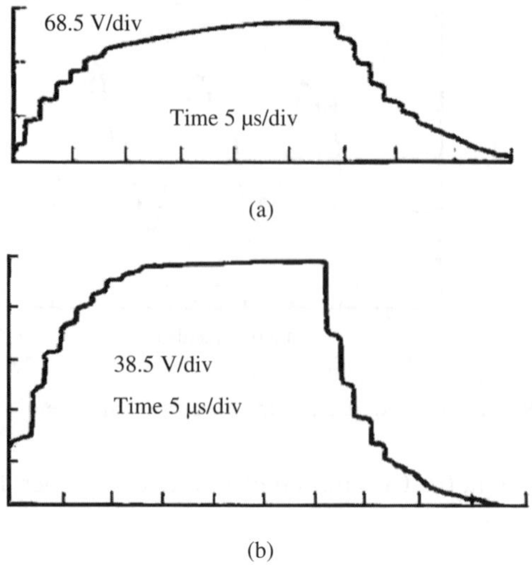

Figure 6.50 Field test results: (a) sending-end voltage V_1; and (b) receiving-end voltage V_2

station. Figure 6.50 shows field test results. Figure 6.51 is simulation results, which agree well with the field test results. Figure 6.52 shows the effect of the relative permittivity of an insulator between a core and a metallic sheath. It should be clear that the gas-insulated line shows a faster voltage rise with a higher amplitude because of a lower permittivity.

6.6.4 Switching Surges in a 500 kV GIS

6.6.4.1 A Model GIS

Figure 6.53 illustrates a part of a 500 kV GIS in Japan. The cross section of each bus is the same as in Figure 6.43b. Table 6.15 gives the parameters of the GIB. The system configuration varies depending on circuit breaker (CB) and disconnector (DS) conditions of "on" and "off". When CB2 is off, the system is composed of the circuits from BUS1 to CB2 with an infinite a.c.source connected to the left of BUS1. This configuration is called "system A". When CB2 is in a closed condition, the system is composed of the circuits from BUS1 to BUS4 which is called "system B". In system B, disconnector DS1 is operating. When considering a residual charge in system B, CB4 is closed first with DS1 being off, and CB4 is made off so that there exists a residual charge in the circuit composed of DS1S and CB1. The total length from node BUS1 to CB2 is about 25 m. A transformer connected to a GIB is represented by its stray capacitance of 2 nF to the earth. The resistance of 300 Ω at node BUS3 represents an outgoing overhead line. Only a single phase circuit is considered in an EMTP simulation because mutual coupling between three-phase buses causes no significant effect on a core voltage of a CB or DS operating phase.

The a.c. source voltage $e_0(t)$ at BUS1 is given by:

$$e_0(t) = \cos(\omega t),\ \ E_0 = 500\sqrt{2/3} = 408.25\ \text{kV},\ \ f = 60\ \text{Hz}$$

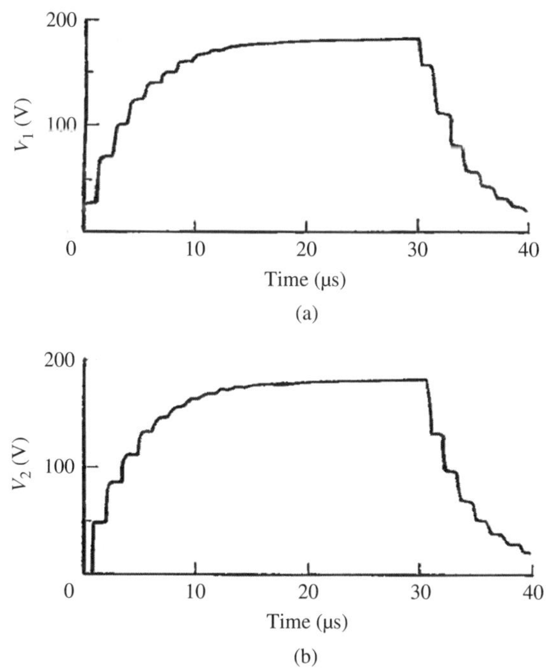

Figure 6.51 Calculated results: (a) V_1; and (b) V_2

Table 6.15 Parameters of a GIB

GIB		Spacer
$r_{c1} = 0.00$ mm	$\rho_c = 1.80 \times 10^{-8}$ Ωm	(Thickness 30cm)
$r_{c2} = 125$ mm	$\rho_s = 3.78 \times 10^{-8}$ Ωm	$\varepsilon_s = 6$
$r_{p1} = 460$ mm	$\mu_c = \mu_s = 1.0$	
$r_{p2} = 480$ mm	$\varepsilon_1 = 1.009, \varepsilon_2 = 1.000$	

The source at BUS4 is in opposite polarity to the above.

A spacer of width 30 cm with relative permittivity $\varepsilon_s = 6$ is installed in the buses in Figure 6.43.

6.6.4.2 EMTP Simulation Results

Table 6.16 summarizes simulation conditions and results. Figure 6.54 and Figure 6.55 show transient voltage waveforms. It is observed in Table 6.16 that the spacer results in a higher overvoltage with a lower oscillating frequency. This is very noticeable in the case of CB switching on with a residual voltage at the BUS4 side: the maximum voltage at node DS1SC is 1.777 pu with no spacer, and 1.946 pu with spacers, and the oscillating frequnecy 18 MHz with no spacer, and 12 MHz with spacers. In the case of spacers, the overvoltage tends to increase at the nodes nearby the operating DS and CB, and to decrease at the source side.

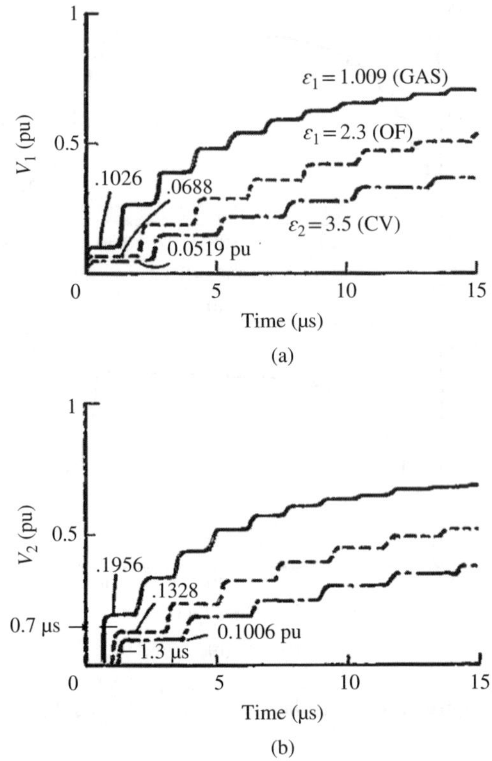

Figure 6.52 Effect of insulator permittivity on transient voltage waveforms: (a) V_1; and (b) V_2

6.6.5 Basic Characteristics of Switching Surges Induced to a Control Cable

6.6.5.1 A Model Circuit

Figure 6.56 illustrates a model circuit to investigate a basic characteristic of induced surges from a GIB to a control cable. The parallel length x_c of the GIB and the control cable is varied. For example, x_c is from 0 to 10 m at an interval of 2.5 m when x_p is taken to be 10 m. The applied voltage is normalized to be 1000 V with a step function, the severest case. The GIB core is either open-circuited or matching resistance ($R_G = 77.8\ \Omega$) terminated at the far end. The GIB tank is grounded by a 10 Ω resistance at the both ends. The control cable core is terminated by the matching resistance (22.3 Ω) at both ends. The control cable sheath is grounded by a 10 Ω resistance at both ends.

A voltage transformer (VT: 1000 V/33.3 V) is installed in node GA2 and CC2.

6.6.5.2 EMTP Simulation Results

Simulation results for $x_p = 10$ m are given in Table 6.17 and case $x_p = 10$ m, $x_c = 5$ m is shown in Figure 6.57, Figure 6.58, and Figure 6.59.

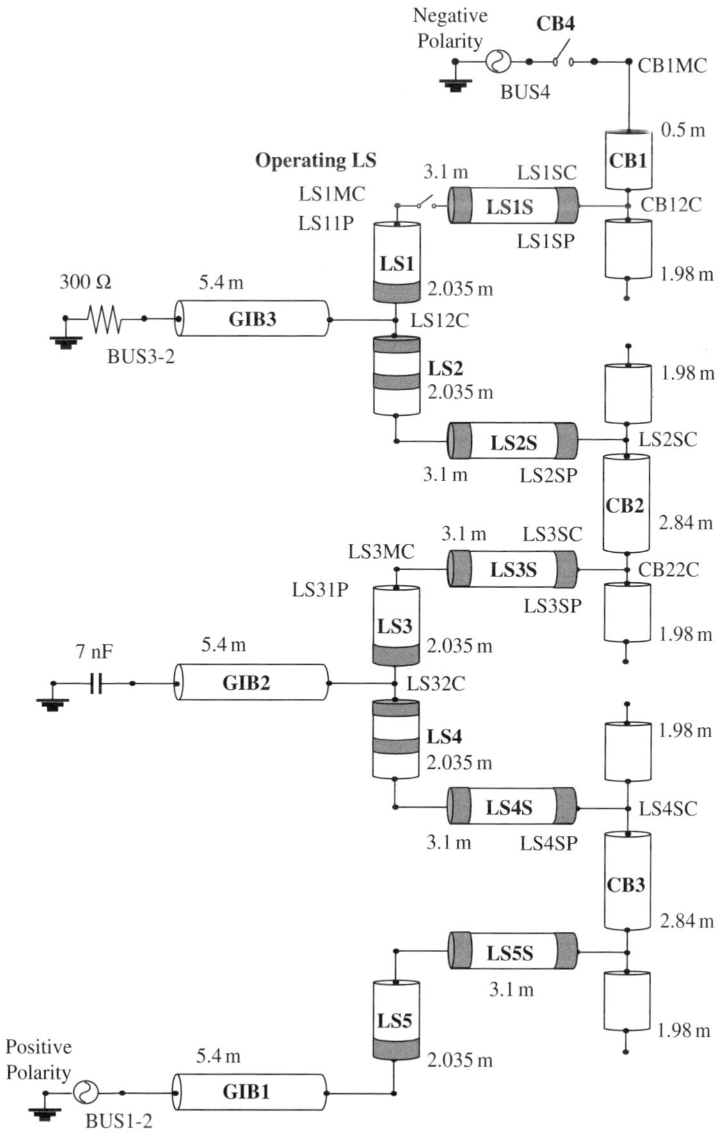

Figure 6.53 Simulation circuit of a 500 kV GIS

It is observed in Table 6.17 that the induced voltage to the control cable tends to increase as the parallel length x_c increases as expected. However, in the case of the open-circuited GIB ($R_G = \infty$), there exists a condition that x_c gives the highest induced voltage. This is estimated due to a positive reflection of a traveling wave at the open-circuited end and electrostatic coupling between the GIB and the control cable. It should be noted that currents on the GIB core and pipe become quite large when a VT is connected to the cable at the node of an operating CB or DS as in Table 6.17 for $x_c = 10$ m. This is caused by VT stray capacitances. The

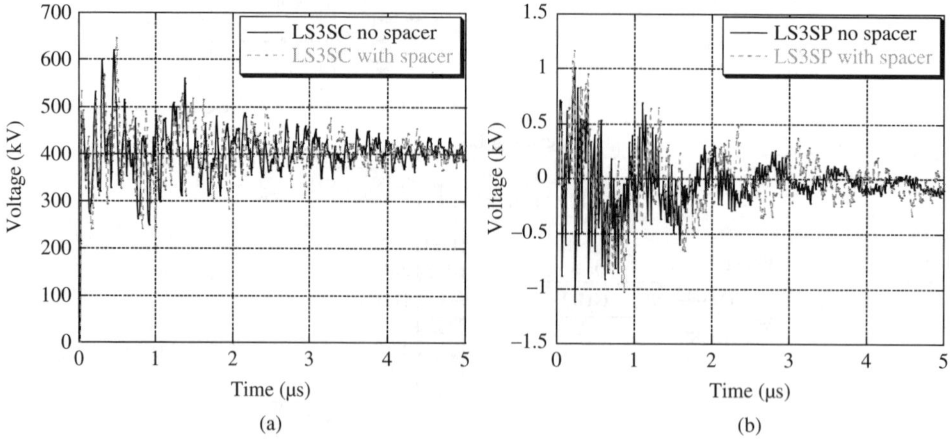

Figure 6.54 Simulation results of core and pipe voltages in system A with LS3 switching on: (a) node LS3SC; and (b) node LS3SP

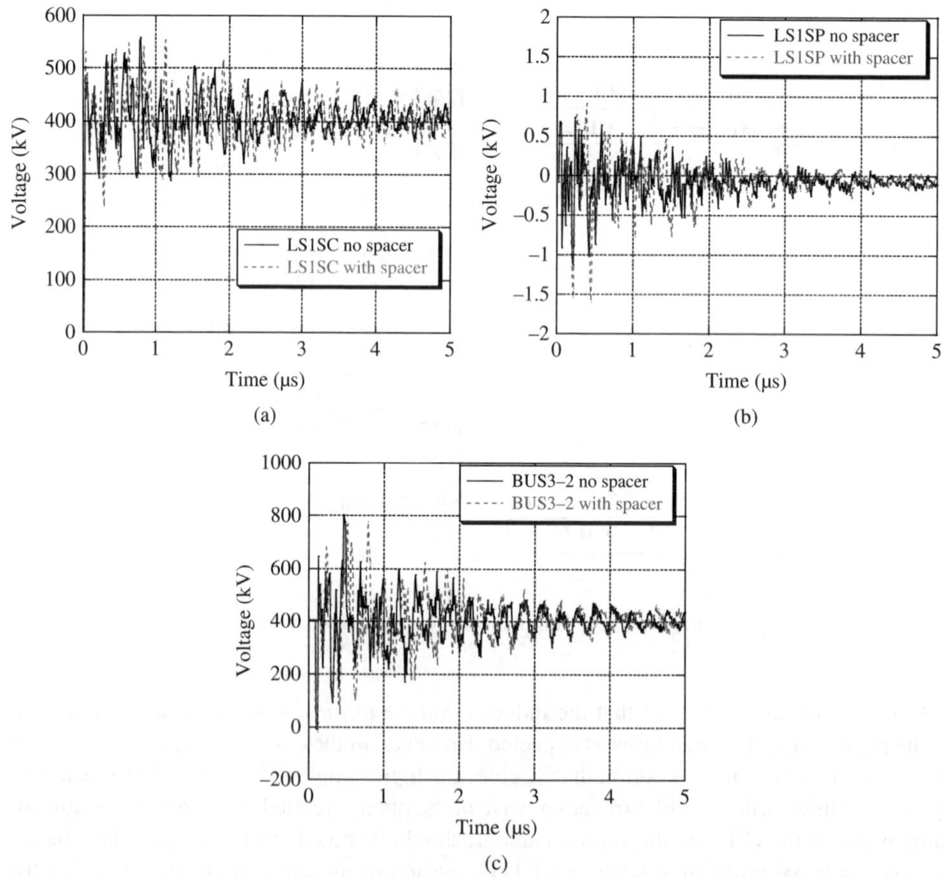

Figure 6.55 Simulation results of core and pipe voltages in system B with LS1 switching on: (a) node LS1SC; (b) node LS1SP; and (c) node BUS3-2

Table 6.16 Simulation results for GIS (1 pu = 408.25 kV)

System	Operating LS	Spacer	Node	V_{max} (pu) / Time (μs)	f (MHz)
A (CB2 OFF)	LS3	NO	LS3SC	1.611 / 0.459	14.0
			LS3MC	1.400 / 0.482	14.0
			LS4SC	1.384 / 0.267	13.0
		YES	LS3SC	1.686 / 0.501	12.0
			LS3MC	1.465 / 0.515	12.0
			LS4SC	1.372 / 0.547	11.0
B (CB2 ON)	LS1 (CB4 OFF)	NO	LS1SC	1.383 / 0.792	13.0
			LS1MC	1.333 / 0.570	13.0
			LS2SC	1.442 / 0.367	12.0
			LS3SC	1.366 / 0.375	14.0
			LS3MC	1.277 / 0.481	14.0
			LS4SC	1.315 / 0.696	13.0
		YES	LS1SC	1.468 / 0.455	10.0
			LS1MC	1.351 / 0.444	10.0
			LS2SC	1.433 / 0.396	9.0
			LS3SC	1.352 / 0.393	10.0
			LS3MC	1.263 / 0.532	10.0
			LS4SC	1.300 / 0.745	9.0
	LS1 (CB4 ON)	NO	LS1SC	1.777 / 0.794	18.0
			LS1MC	1.676 / 0.572	18.0
			LS2SC	1.895 / 0.369	17.0
			LS3SC	1.742 / 0.377	19.0
			LS3MC	1.563 / 0.483	20.0
			LS4SC	1.640 / 0.698	18.0
		YES	LS1SC	1.946 / 0.457	12.0
			LS1MC	1.712 / 0.446	12.0
			LS2SC	1.876 / 0.398	11.0
			LS3SC	1.718 / 0.395	12.0
			LS3MC	1.535 / 0.534	13.0
			LS4SC	1.610 / 0.747	11.0

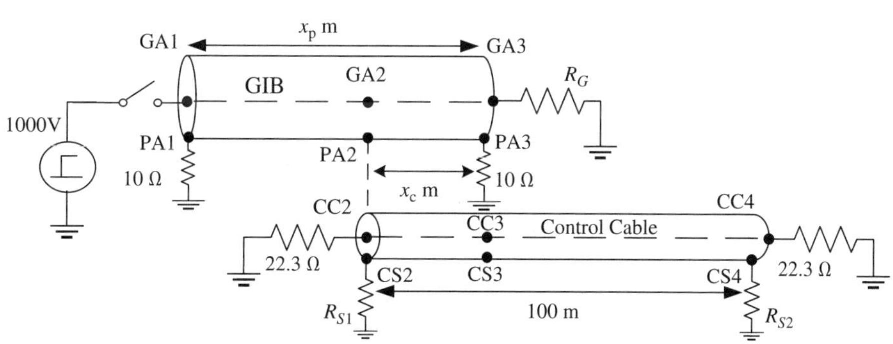

Figure 6.56 A model circuit composed of a GIB and a control cable (VT connected between node GA2 and CC2)

Table 6.17 Simulation results of switching surges induced to a control cable

$x_p = 10$ m and $R_G = \infty$

		Max. current (A)						Max. voltage (V)											
X_p	X_c	GA1			PA1			CC2			CC4			CS2			CS4		
(m)	(m)	Ind.	VT	both	Ind.	VT	both	Ind.	VT	both	Ind.	VT	both	Ind.	VT	both	Ind.	VT	both
10	0	11.5	25.8	25.8	10.8	24.2	24.2	0.00	63.48	63.48	0.00	32.75	32.75	0.00	18.84	18.84	0.00	−13.65	−13.65
	2.5	11.5	18.9	18.9	10.8	16.8	16.9	−2.14	66.94	66.73	1.91	34.58	32.65	−3.60	19.87	27.36	5.30	−14.31	−20.73
	5	11.5	19.2	19.2	10.8	17.0	17.1	−2.66	70.51	69.92	2.15	36.35	36.87	−4.11	20.93	29.81	5.89	−15.25	−24.28
	7.5	11.5	23.5	23.5	10.8	15.6	15.6	−2.29	69.86	69.29	1.60	35.96	36.10	−4.71	20.74	30.69	4.06	−15.22	−22.12
	10	11.5	699	699	10.8	10.8	10.8	−2.35	33.36	33.36	2.58	18.04	20.70	−3.79	9.90	12.42	5.11	−7.03	−10.50

$x_p = 10$ m and $R_G = 77.8\ \Omega$

		Max. current (A)						Max. voltage (V)											
X_p	X_c	GA1			PA1			CC2			CC4			CS2			CS4		
(m)	(m)	Ind.	VT	both	Ind.	VT	both	Ind.	VT	both	Ind.	VT	both	Ind.	VT	both	Ind.	VT	both
10	0	12.4	25.8	25.8	10.8	24.3	24.3	0.00	34.67	34.67	0.00	17.93	17.93	0.00	10.29	10.29	0.00	−7.36	−7.36
	2.5	12.5	18.9	18.9	10.8	16.8	16.9	1.71	46.04	45.77	−1.80	23.78	22.39	3.69	13.67	19.89	−3.76	−9.84	−13.11
	5	12.5	23.1	21.7	10.8	17.0	17.1	2.55	52.83	52.20	−3.05	27.25	25.70	5.47	15.68	23.39	−3.79	−11.38	−18.59
	7.5	12.6	23.5	23.5	10.8	15.6	15.6	2.88	58.12	57.31	−3.16	29.92	29.64	5.69	17.26	27.71	−3.97	−12.65	−19.30
	10	12.6	699	699	10.8	10.8	10.8	4.46	33.36	33.36	−4.42	18.04	16.31	6.37	9.90	15.31	−6.49	−7.03	−13.48

Ind., induced; VT, a voltage transferred via a voltage transformer.

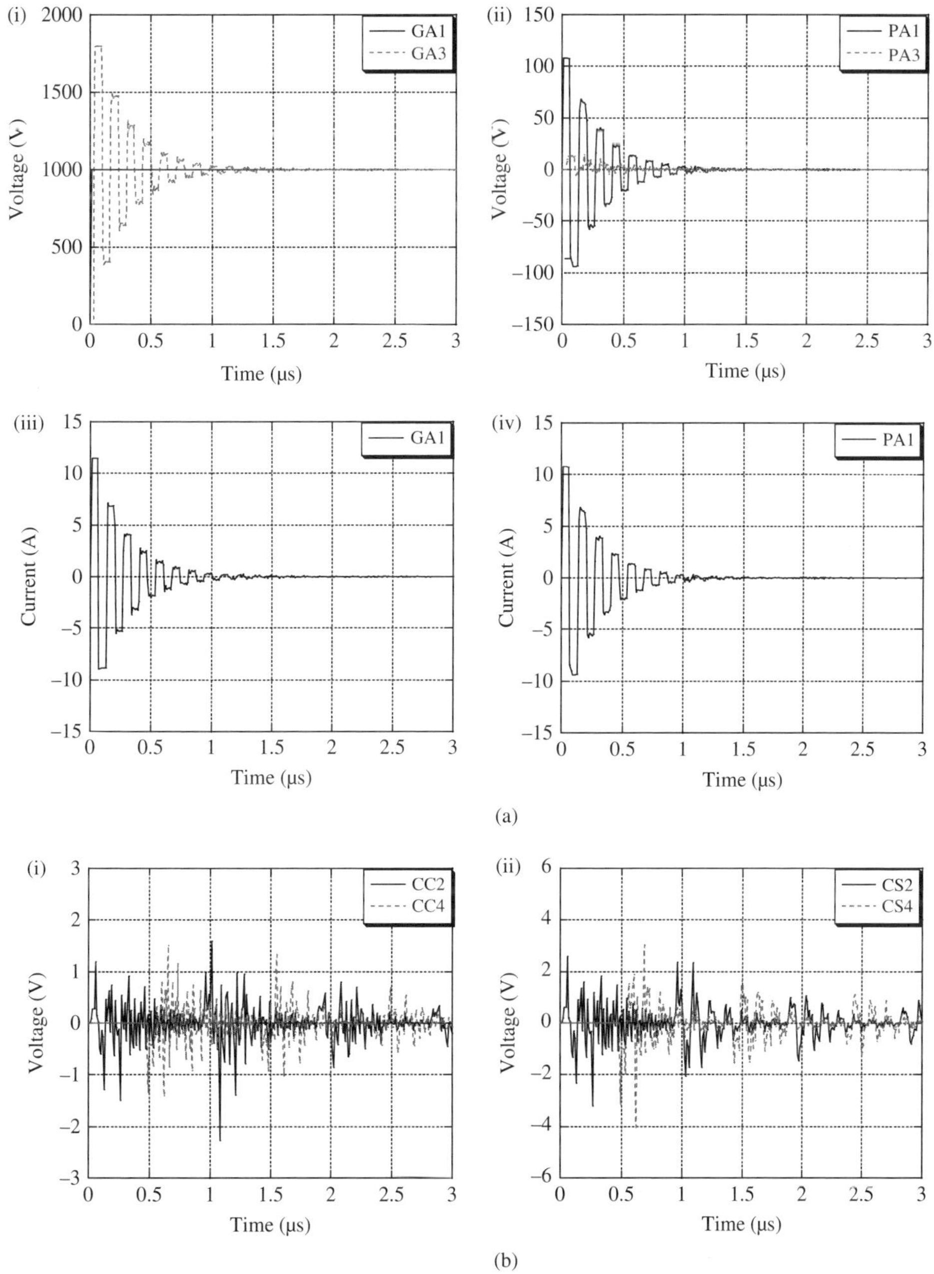

Figure 6.57 Simulation results for induced voltages (x_p = 10, x_c = 5 m): (a) GIB: (i) core voltage; (ii) pipe voltage; (iii) core current; (iv) pipe current; and (b) control cable: (i) core voltage; (ii) sheath voltage

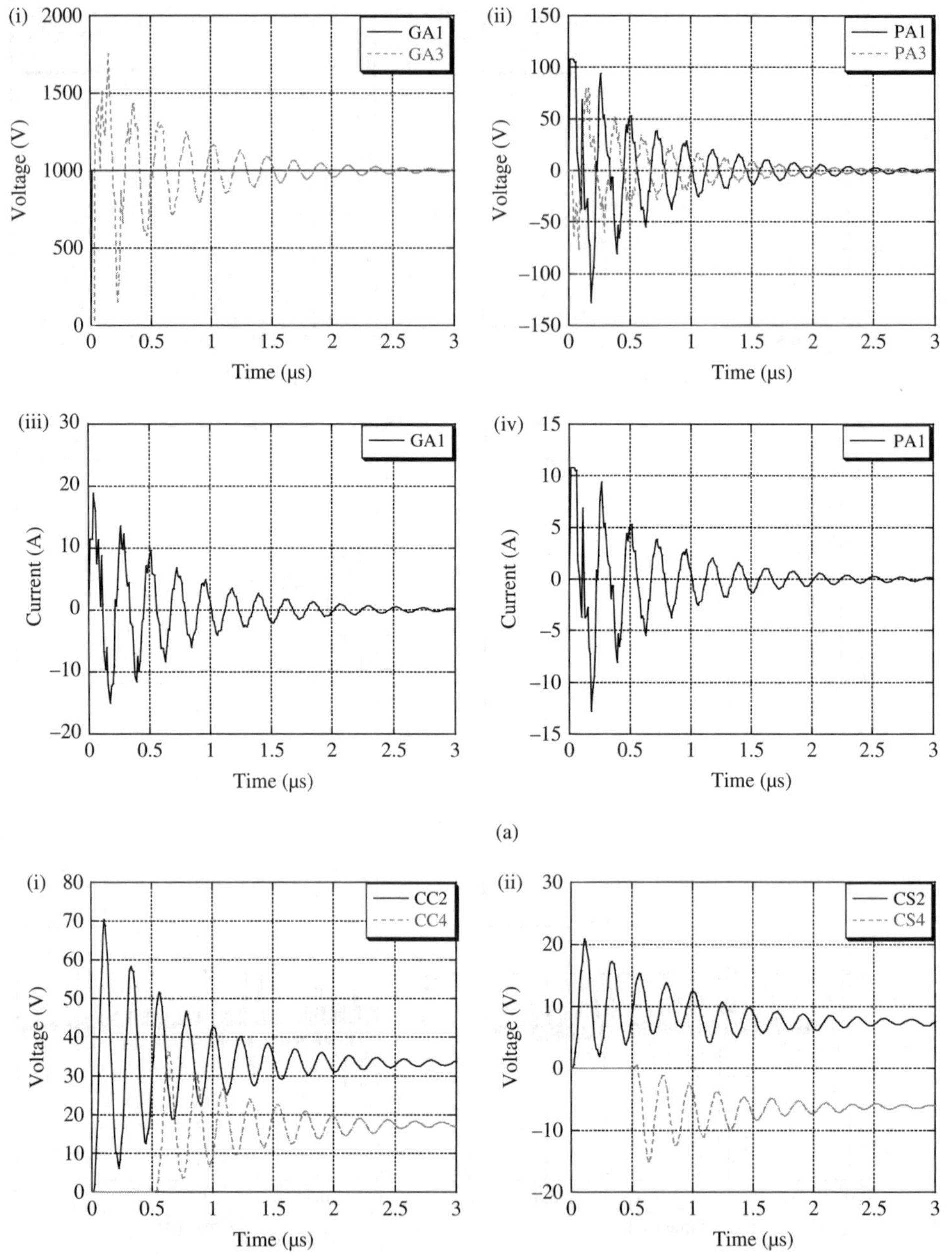

Figure 6.58 Simulation results for VT transferred voltages ($x_p = 10,\ x_c = 5$ m): (a) GIB: (i) core voltage; (ii) pipe voltage; (iii) core current; (iv) pipe current; and (b) control cable: (i) core voltage; (ii) sheath voltage

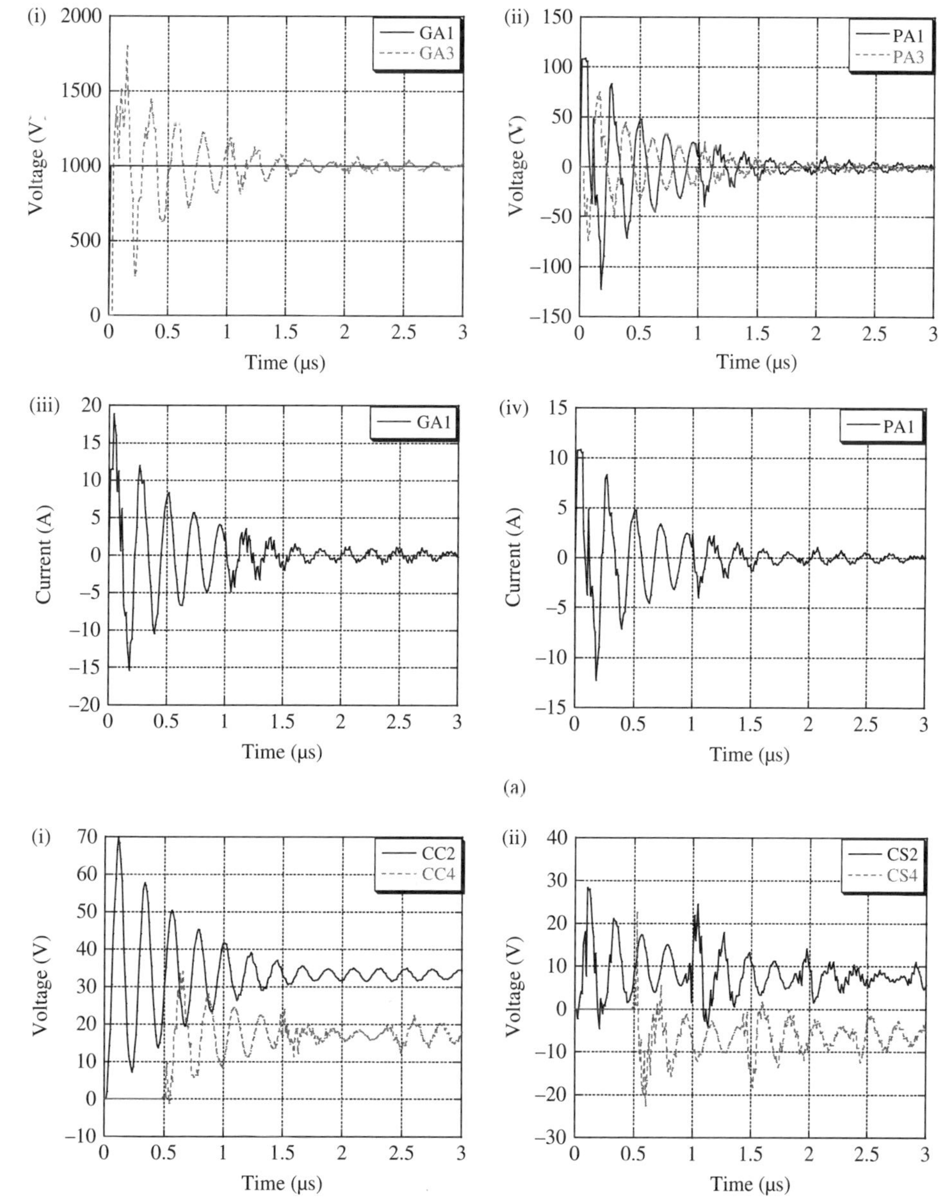

Figure 6.59 Simulation results for both voltages ($x_p = 10$, $x_c = 5$ m): (a) GIB: (i) core voltage; (ii) pipe voltage; (iii) core current; (iv) pipe current; and (b) control cable: (i) core voltage; (ii) sheath voltage

Table 6.18 Simulation results of switching surges induced to a control cable with three spacers ($x_p = 10\,\text{m}$ and $R_G = \infty$)

X_p (m)	X_c (m)	Max. current (A)						Max. voltage (V)											
		GA1			PA1			CC2			CC4			CS2			CS4		
		Ind.	VT	both	Ind.	VT	both	Ind.	VT	both	Ind.	VT	both	Ind.	VT	both	Ind.	VT	both
10	0	24.2	24.2	24.2	22.8	22.8	22.8	0.00	62.23	62.23	0.00	32.11	32.11	0.00	18.47	18.47	0.00	−13.40	−13.40
	2.5	24.2	24.2	24.2	22.8	22.8	22.8	2.53	67.95	67.59	−2.66	35.07	33.24	4.33	20.17	26.91	−6.33	−14.60	−21.70
	5	24.4	31.9	31.9	22.9	26.9	27.0	3.13	70.04	69.52	−2.91	36.14	40.06	5.48	20.79	36.48	−6.92	−15.08	−30.67
	7.5	24.4	35.5	29.2	22.9	22.8	22.8	3.84	66.89	65.99	−3.10	34.48	32.37	4.80	19.86	28.19	−6.97	−14.46	−22.53
	10	24.4	6859	712	22.9	22.9	22.8	3.46	33.56	33.36	−3.40	18.04	19.98	5.94	9.96	14.54	−7.28	−7.06	−12.79

Ind., induced; VT, a voltage transferred via a voltage transformer.

total capacitance C_{VT} of the VT to ground is about 300 pF which is equivalent to a short distributed line of the surge impedance $Z_C = \Delta t/C_{VT} = 0.5\text{ns}/0.3\text{nF} = 1.5\ \Omega$ with the remote end open-circuited. Therefore the current I at $t = 0$ reaches $1000/1.5 = 660$ A. In practice, there exist some resistances from the CB or DS to the VT and such a large current is not observed. However, it is suggested not to install a VT near a CB and a DS.

6.6.5.3 Effect of Spacers

As in Figure 6.56, a spacer of width 30 cm with relative permittivity $\varepsilon_s = 6$ is installed every 4.5 m. Simulation results are given in Table 6.18 corresponding to Table 6.17 ($x_p = 10$ m and $R_G = \infty$). Because the spacer is represented by a capacitance to ground and the first one is next to an operating CB or DS (node GA1), the transient current on the GIB core becomes twice that with no spacer. This results in a greater induced voltage to a control cable, while a VT transferred voltage is not significantly different from that with no spacer, because the GIB core voltage is not much different from that with no spacer.

Appendix 6.A

Figure 6.A.1 shows the diagram of a cross-bonded point. In the figure, Z_0 is the cable surge impedance matrix which is given in Table 6.A.1. Z_t is the impedance matrix seen from node P to the right which is given by:

$$[Z_t] = [A]^{-1}[Z_0][A] \tag{6.A.1}$$

where

$$[A] = \begin{bmatrix} [U] & [0] \\ [0] & [B] \end{bmatrix}, [U] = \begin{bmatrix} 1 & 0 & 0 \\ 0 & 1 & 0 \\ 0 & 0 & 1 \end{bmatrix}, [0] = \begin{bmatrix} 0 & 0 & 0 \\ 0 & 0 & 0 \\ 0 & 0 & 0 \end{bmatrix}, [B] = \begin{bmatrix} 0 & 1 & 0 \\ 0 & 0 & 1 \\ 1 & 0 & 0 \end{bmatrix}$$

Refraction coefficient matrix λ is defined by:

$$[\lambda] = 2[Z_t]([Z_0] + [Z_t])^{-1} \tag{6.A.2}$$

Numerical results of λ are given in Table 6.A.2.

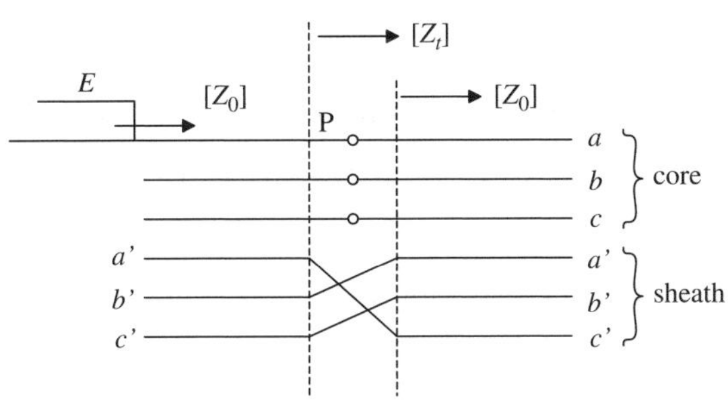

Figure 6.A.1 Cross-bonding point

Table 6.A.1 Cable surge impedance matrix at $f = 20$ kHz

Underground cable (Figure 6.21a)

	a	b	c	a″	b″	c′
a	42.62	8.64	6.91	21.14	8.64	6.91
b	8.64	41.97	8.64	8.64	20.49	8.64
c	6.91	8.64	42.62	6.91	8.64	21.14
a″	21.14	8.64	6.91	21.13	8.65	6.91
b′	8.64	20.49	8.64	8.65	20.48	8.65
c′	6.91	8.64	21.14	6.91	8.65	21.13

Overhead cable (vertical configuration)

	a	b	c	a′	b′	c′
a	442.93	295.36	255.51	421.41	295.36	255.51
b	295.36	440.70	298.54	295.36	419.18	298.54
c	255.51	298.54	438.59	255.51	298.54	417.07
a′	421.41	295.36	255.51	421.37	295.36	255.51
b′	295.36	419.18	298.54	295.36	419.14	298.54
c′	255.51	298.54	417.07	255.51	298.54	417.03

Table 6.A.2 Numerical results of the first column of a refraction coefficient matrix

Underground cable (Figure 6.21a)

$[\lambda_1(k,\ 1)] = [1.00,\ 0.11,\ -0.10,\ -0.21,\ 0.21,\ 0.01]_t$

Overhead cable (vertical configuration)

$[\lambda_2(k,\ 1)] = [1.03,\ 0.31,\ -0.24,\ -0.52,\ 0.58,\ 0.05]_t$

When an incoming wave E of a step function arrives at node P on phase-a, the voltages on each phase at node P are given by:

$$(V) = [\lambda](E) \tag{6.A.3}$$

where

$$(E) = \begin{pmatrix} E \\ (0) \end{pmatrix} \tag{6.A.4}$$

Substituting (6.A.4) and the values in Table 6.A.2 into (6.A.3), the following results are obtained.

$$(V)_{under} = (1.0,\ 0.11,\ -0.10,\ -0.21,\ 0.21,\ 0.01)_t E \tag{6.A.5a}$$

$$(V)_{over} = (1.03,\ 0.31,\ -0.24,\ -0.52,\ 0.58,\ -0.05)_t E \tag{6.A.5b}$$

where t is the transposed matrix.

From the above equations, the sheath voltages are given by:

Underground cable:

$$V_{sa} = -0.21 \text{ pu},\quad V_{sb} = 0.21 \text{ pu},\ V_{sc} = 0.01 \text{ pu}$$

Overhead cable:

$$V_{sa} = -0.52\ \text{pu},\quad V_{sb} = 0.58\ \text{pu},\ V_{sc} = 0.05\ \text{pu}$$

The above results indicate that the sheath voltages in the overhead cable are about 2.5 times higher than those in the underground cable. This comes from the difference of the refraction coefficients given in Table 6.A.2 which result from the difference of the surge impedances given in Table 6.A.1, calculated by Cable Constants of the EMTP.

Appendix 6.B

The relation of voltage V and current I of the double-circuit overhead line is given by:

$$(V) = [ZI](I) \quad \text{or} \quad \begin{pmatrix} (V_1) \\ (V_2) \end{pmatrix} = \begin{bmatrix} [Z_s] & [Z_m] \\ [Z_m] & [Z_s] \end{bmatrix} \begin{pmatrix} (I_1) \\ (I_2) \end{pmatrix} \tag{6.B.1}$$

where
$(V_1),\ (I_1)$: first circuit voltage and currents
$(V_2),\ (I_2)$: second circuit voltage and currents

In the case of double-circuit operation, for $(V_1) = (V_2) = (V_d)$ and $(I_1) = (I_2) = (I_d)/2$, the following relation is obtained.

$$(V_d) = (1/2)([Z_s] + [Z_m])(I_d) = [Z_d](I_d) \tag{6.B.2}$$

where
$[Z_d] = ([Z_s] + [Z_m])/2$

In the case of single-circuit operation, for $(I_2) = 0$, the following relation is obtained.

$$(V_s) = (V_1) = [Z_s](I_1) = [Z_s](I_s) \tag{6.B.3}$$

The impedance of the overhead line seen from node P in Figure 6.27 is obtained from (6.B.2) for double-circuit operation and from (6.B.3) for single-circuit operation, the refraction coefficient matrix is given by:

$$\text{Double-circuit operation:}\quad [\lambda_d] = 2[Z_c]([Z_d] + [Z_c])^{-1} \tag{6.B.4a}$$

$$\text{Single-circuit operation:}\quad [\lambda_s] = 2[Z_c]([Z_s] + [Z_c])^{-1} \tag{6.B.4b}$$

For Z_d being smaller than Z_s, λ_d is greater than λ_s.

References

[1] Shinozaki, H., Kawakado, T., Hamamoto, J. *et al.* (1971) Abnormal voltages of a core at a cross-bonding joint. *Journal of the Technology Laboratory of the Chugaku Electric Power Comapny*, **39**, 175–98.

[2] Nagaoka, N. (1992) Transient analysis of cable systems by means of a frequency-transform method. PhD thesis. Doshisha University.

[3] Nagaoka, N. and Ametani, A. (1983) Transient calculations on crossbonded cables. *IEEE Transactions on Power Apparatus and Systems*, **102** (4), 779–87.
[4] Nagaoka, N. and Ametani, A. (1988) A development of a generalized frequency-domain transient program – FTP. *IEEE Transactions on. Power Delivery*, **3** (4), 1996–2004.
[5] Dang, N.D. (1972) Transient performance of crossbonded cable systems. PhD thesis. University of Manchester.
[6] Wedepohl, L.M. and Indulkar, C.S. (1975) Switching overvoltages in short crossbonded cable systems using the Fourier Transform. *Proceedings of the IEE*, **122** (11), 1217–21.
[7] Wilcox, D.J. (1978) Transient phenomena in crossbonded cable systems; analytical results. *Proceedings of the IEE*, **125** (10), 999–1005.
[8] Humpage, W.D. and Nguyen, T.T. (1981) Z-transform electromagnetic transient analysis of crossbonded cable transmission systems. *Proceedings of the IEE*, **128** (2), 55–62.
[9] Nagaoka, N., Yoshida, K., Ametani, A., and Karasaki, T. (1991) Transient sheath voltage characteristic of a crossbonded cable installed within a tunnel. *IEE Japan Transactions on PE*, **B-111** (7), 784–90.
[10] Itoh, Y., Nagaoka, N., and Ametani, A. (1990) Transient analysis of a crossbonded cable system underneath a bridge. *IEEE Transactions on Power Delivery*, **5** (2), 527–32.
[11] Ametani, A. (1990) *Distributed-Parameter Circuit Theory*, Corona Publishing Co. (in Japanese).
[12] Ametani, A. (1980) Wave propagation characteristics of cables. *IEEE Transactions on Power Apparatus and Systems*, **99** (2), 499–505.
[13] Miyamoto, Y (2003) Transient analysis on cross-bonded cables. MSc thesis. Doshisha University.
[14] Ametani, A., Miyamoto, Y., and Nagaoka, N. (2003) An investigation of a wave propagation characteristic on a crossbonded cable. *IEE Japan Transactions on PE*, **B-123** (3), 395–401.
[15] Scott-Meyer, W. (1982) EMTP Rule Book, Bonneville Power Administration.
[16] Nagaoka, N., Miyamoto, M. and Ametani, A. (1985) Surge propagation characteristics of a POF cable. *Electrical Engineering in Japan*, **105** (5), 67–75.
[17] Ametani, A. (1972) The application of the fast Fourier transform to electrical transient phenomena. *International Journal of Electrical Engineering Education*, **10** (4), 277–87.
[18] Ametani, A. (1981) Surge propagation characteristics on gas-insulated cables. *IEE Japan Transactions on PE*, **B-101** (8), 491–7.
[19] Ametani, A., Nagaoka, N., Omura, H., and Goto, T. (2007) Induced surge characteristics on a control cable in a gas-insulated substation due to switching operation. *IEE Japan Transactions on PE*, **127** (12), 1306–12.
[20] Ametani, A., Motoyama, H., Ohkawara, K. *et al.* (2009) Electromagnetic disturbances of control circuits in power stations and substations experienced in Japan. *IET Proceedings – Generation, Transmission and Distribution*, **3** (9), 801–15.
[21] CIGRE WG C4.208 (2013) EMC within Power Plants and Substaions. *CIGRE Technical Brochure* 535.
[22] Japanese Electrotechnical Research Association (2002) Technologies of Countermeasures against Surges on Protection Relays and Control Systems. ETRA Report, Vol. 57 (3) (in Japanese).
[23] IEE Japan (2004) Standard of Low Voltage Control Circuits in Power Stations and Substations. *JEC-0103-2004*.

7

Examples of Cable System Transients

Teruo Ohno
Tokyo Electric Power Company, Tokyo, Japan

This chapter introduces examples of transients peculiar to cable systems [1]. Most of these transients are also studied with long ultra high voltage/extra high voltage (UHV/EHV) overhead lines (OHLs) but are considered more common to cable systems. These studies are performed as a part of feasibility studies, implementation studies, and forensic studies with a different degree of precision in a model setup. The implementation study is the stage where these studies are performed most frequently as they can affect the selection and the specification of related equipment of cable systems. For example, the leading current interruption and the zero-missing phenomenon can affect the selection and the specification of line breakers, and the cable discharge can affect those of voltage transformers.

Among the overvoltages discussed in this chapter, temporary overvoltages are most important for cable system transients since severe temporary overvoltages have been observed in past studies. Slow-front overvoltages related to cable systems are less severe compared with those related to OHLs. However, their differences are explained to help readers understand why slow-front overvoltages are less severe with cable systems.

7.1 Reactive Power Compensation

Cable lines become sources of reactive power like shunt capacitors. Especially, long EHV cable lines produce large reactive power and usually require reactive power compensation for the following reasons [1]:

- To suppress the steady-state overvoltage around the cable line.
- To suppress the steady-state overvoltage at the cable open terminal.

Cable System Transients: Theory, Modeling and Simulation, First Edition.
Akihiro Ametani, Teruo Ohno and Naoto Nagaoka.

- To prevent the reduction of the active power transmission capacity due to the large charging current.
- To reduce the leading current that flows through the line breaker so that it becomes lower than the leading current interruption capability of the line breaker.

When the compensation rate 100% is adopted, the installation of the cable line does not affect the reactive power balance around the cable line. Because of this, the compensation rate of 100 % is usually preferred in the planning of the cable line.

However, the compensation rate near 100 % cannot be achieved in some cases due to the unit size of shunt reactors for the compensation. For example, when a cable line has a charging capacity of 250 MVar, two units of 100 MVar shunt reactors may be installed for the compensation, which results in a compensation rate of 80%. In order to raise the compensation rate, the unit size needs to be increased to, for example, 120 MVar, but it is sometimes not a cost effective solution depending on the manufacturer.

When the compensation rate, as a result, becomes low, it leads to steady-state overvoltage on the cable line. In addition, the compensation rate becomes low when the cable line needs to be operated even if one unit of the shunt reactor is out of service. This requires careful consideration in the planning process.

The low compensation rate also leads to higher temporary overvoltages. It is highly recommended to study the temporary overvoltage in the feasibility study or at an earlier stage as it may affect the decision on the reactive power compensation in the planning process.

7.2 Temporary Overvoltages

7.2.1 Series Resonance Overvoltage

7.2.1.1 Theoretical Background

When inductance L and capacitance C are connected in series, the total impedance or series impedance becomes zero at the frequency $f_n = \frac{1}{2\pi\sqrt{LC}}$. When a voltage source of this frequency is connected to the series resonance circuit as shown in Figure 7.1, an infinitely large current flows into the series circuit in theory. Because of this large current, the voltages across the inductance V_L and the capacitance V_C become infinite if the resistance of the circuit is neglected. This is called series resonance, and V_L and V_C are the resonance overvoltage.

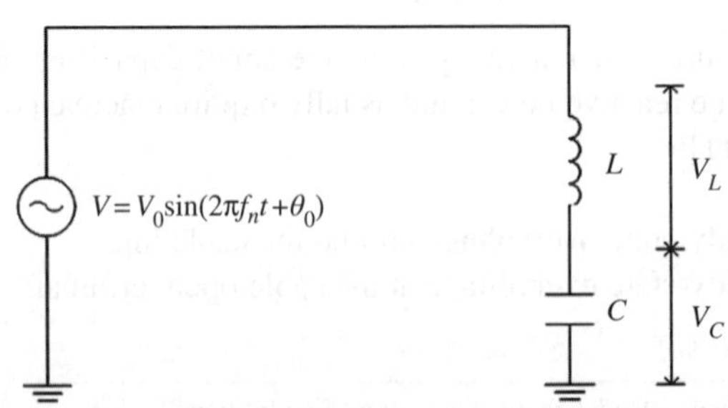

Figure 7.1 Simple series resonance circuit

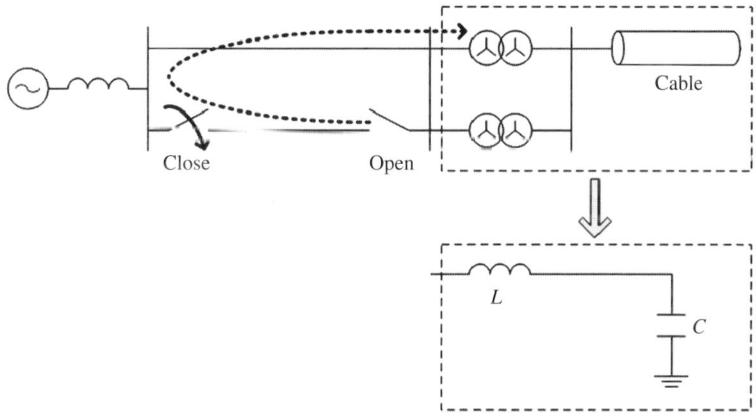

Figure 7.2 Example of series resonance circuits in actual power systems

The series resonance overvoltage can occur in an actual power system. Figure 7.2 shows an example of series resonance circuits. When one transmission line is energized, a part of the energization overvoltage travels into the series resonance circuit which is composed of transformer inductance L and cable capacitance C, as shown by the dotted arrow in the figure. If the energization overvoltage contains the natural frequency f_n of the series resonance circuit, the series resonance overvoltage can be caused at the secondary side of the transformer. Therefore, when studying the series resonance overvoltage, we need to know both the dominant frequency contained in the energization overvoltage and the natural frequency of the series resonance circuit.

In addition, when the transmission line energized in Figure 7.2 is a long EHV cable line, the dominant frequency contained in the energization overvoltage could be very low, depending on the source impedance. The low-frequency overvoltage is weakly damped, which can lead to a severe resonance overvoltage at a distant location.

The series resonance overvoltage is normally studied in the following procedure:

- Find the most severe switching scenarios.
- Simulate the most severe switching scenarios to find the dominant frequency contained in the energization overvoltage.
- Find the natural frequency of the series resonance circuit, for example, composed of transformer inductances and cable capacitances.
- Study a difference between two frequencies – dominant frequency and natural frequency.
- Simulate the most severe switching scenarios when the two frequencies are matched.

7.2.1.2 Analysis of Examples

Basic Example

Figure 7.3 shows an example of a series resonance overvoltage analysis. When the 400 kV cable line in Figure 7.3 is energized, a part of the overvoltage caused by the energization travels through 400 kV transformers to the secondary-side network. The series resonance overvoltage

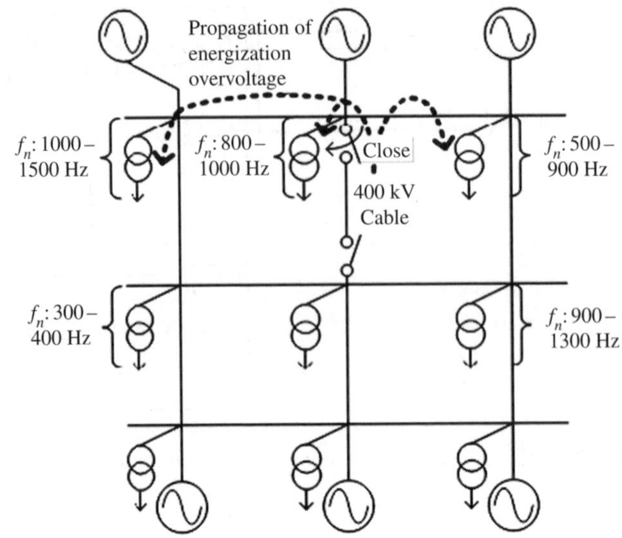

Figure 7.3 Example of a series resonance overvoltage analysis. *Source*: © 2012 IEEE. Reprinted, with permission, from [4]

can be caused when the dominant frequency component contained in the energization overvoltage matches the natural frequency of the network composed of the 400 kV transformers and the secondary-side network.

The natural frequency f_n can be calculated from the inductance between the primary and secondary windings of the 400 kV transformer and the capacitance of the secondary-side network, assuming that there is no load on the secondary-side network.

$$f_n = \frac{1}{2\pi}\sqrt{\frac{1}{CL_{ps}}} \tag{7.1}$$

where L_{ps} is the inductance of the 400 kV transformer and C is the capacitance of the secondary-side network.

When there is a load on the secondary-side network expressed by inductance L and resistance R connected in series, the natural frequency can be found as the frequency that minimizes the following magnitude of impedance X:

$$|X| = \left| R_{ps} + j\omega L_{ps} + \frac{1}{j\omega C + 1/(R + j\omega L)} \right| \tag{7.2}$$

In general practice, the natural frequency is found by frequency scans, but it is also possible to find the natural frequency from (7.2) using a mathematics software package.

Equations (7.1) and (7.2) indicate that the natural frequency changes depending on the conditions such as the network configuration and load. The resistive part of the load affects the damping of the series resonance overvoltage, and the overvoltage is severe in low-load conditions.

If it is possible to estimate the dominant frequency components contained in the energization overvoltage, it is possible to do without the series resonance overvoltage analysis caused by some lines. As shown in Figure 7.3, for example, it is possible to do without the series resonance overvoltage analysis caused by the 400 kV cable line if we can estimate that the dominant frequency in the energization overvoltage is 100 Hz since the natural frequency shown in Figure 7.3 is much higher than 100 Hz. In addition, the necessity of modeling the network composed of 400 kV transformers and the secondary-side network may be eliminated.

The theoretical formulas with the focus on long cables will be introduced in the next section. Since their energization overvoltage contains lower frequency components, it leads to severe conditions, namely low damping in the network.

Typical Example

This section shows an example of a series resonance overvoltage analysis with the network illustrated in Figure 7.4. The network shown in Figure 7.3 was simplified to Figure 7.4 in order to compare with the theoretical analysis. The loading condition was set to 142.6 MW + j67.19 MVar by setting $R = 100\ \Omega$ and $L = 150$ mH, which results in a natural frequency of 800 Hz.

Assumed physical and electrical characteristics of the 400 kV cable are shown in Table 7.1. The 400 kV cable is assumed to be laid in a flat formation at a depth of 1.3 m with a horizontal separation of 0.3 m.

First, it is necessary to find the dominant frequency contained in the energization overvoltage. Here, as in general practice, the time domain simulation of the cable energization was conducted to find the dominant frequency. The obtained cable energization overvoltage is shown in Figure 7.5.

Figure 7.6 shows the result of the Fourier analysis of the voltage waveform in Figure 7.5. As the cable energization overvoltage is almost damped by 40 ms, the time window of the Fourier

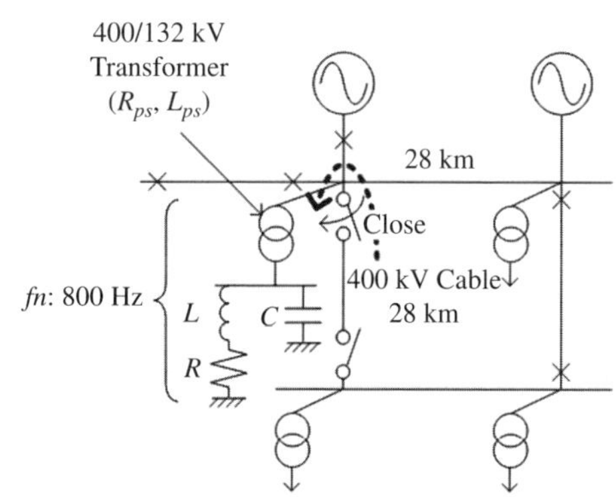

Figure 7.4 Simple simulation circuit for the series resonance overvoltage analysis. 400/132 kV transformer: $R_{ps} = 8\ \Omega$, $L_{ps} = 1916$ mH (at 400 kV); 132 kV network: $R = 100\ \Omega$, $L = 150$ mH, $C = 2.16\ \mu$F. *Source*: © 2012 IEEE. Reprinted, with permission, from [4]

Table 7.1 Physical and electrical characteristics of the 400 kV cable

Core (Aluminum)	Inner radius	0.0 cm
	Outer radius	2.6 cm
	Resistivity	2.84×10^{-8} Ωm
Insulation (XLPE)	Inner radius	2.80 cm
	Outer radius	5.50 cm
	Relative permittivity	2.4
Metallic sheath (Aluminum)	Inner radius	5.80 cm
	Outer radius	5.92 cm
	Resistivity	2.84×10^{-8} Ωm
Outer cover (PE)	Outer radius	6.35 cm
	Relative permittivity	2.4

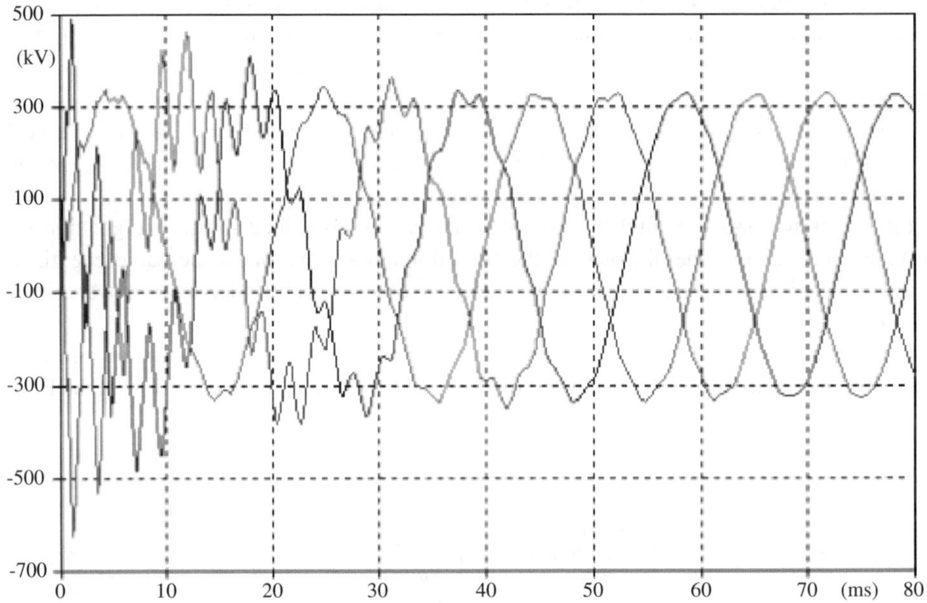

Figure 7.5 Results of the cable energization overvoltage analysis. *Source*: © 2012 IEEE. Reprinted, with permission, from [4]

transform is set to 0–40 ms. The figure shows that the dominant frequency contained in the cable energization overvoltage is 425 Hz.

Figure 7.7 shows the voltage waveform at the low voltage (LV) side of the transformer. As no series resonance occurs, the overvoltage is almost at the same level as that shown in Figure 7.5 (HV side).

In Figure 7.7, the natural frequency of the network composed of the 400 kV transformer and the 132 kV network was 800 Hz, which is far from the dominant frequency in the energization overvoltage (425 Hz). In order to excite the series resonance, the natural frequency was shifted

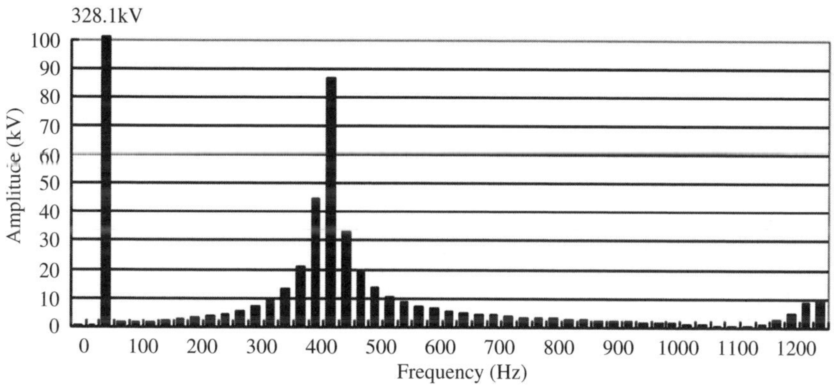

Figure 7.6 Results of the Fourier analysis of the cable energization overvoltage. *Source*: © 2012 IEEE. Reprinted, with permission, from [4]

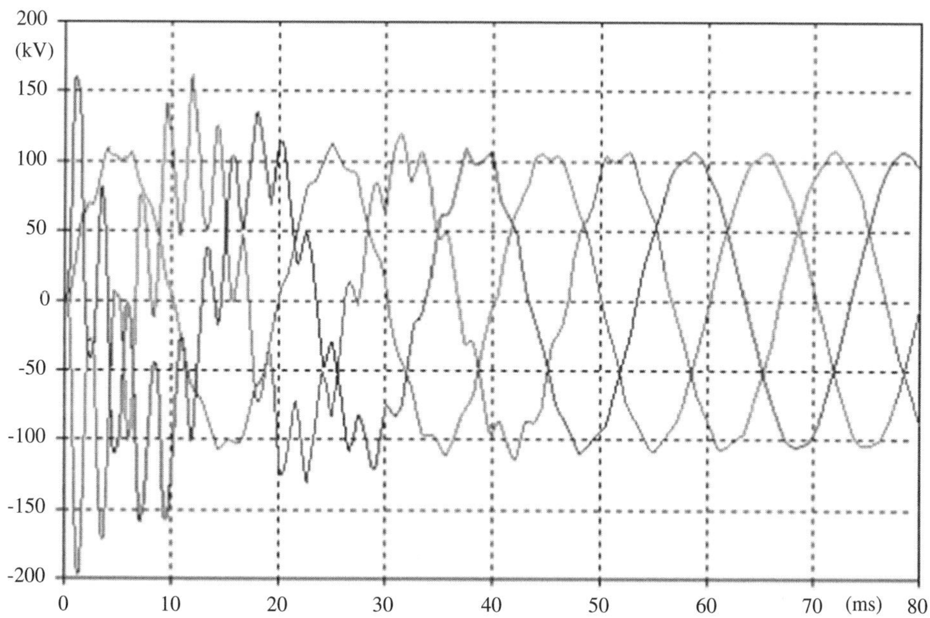

Figure 7.7 Results of the series resonance overvoltage analysis (f_n = 800 Hz). *Source*: © 2012 IEEE. Reprinted, with permission, from [4]

to 425 Hz by increasing the capacitance in the 132 kV network to 7.6 μF. The value of the capacitance was found by performing frequency scans.

Figure 7.8 shows the voltage waveform on the LV side of the transformer. When comparing this waveform with the one shown in Figure 7.5 (HV side), it becomes clear that the overvoltage is amplified on the LV side due to the series resonance since the natural frequency was set exactly at the dominant frequency. The Fourier analysis of the voltage waveform in Figure 7.8

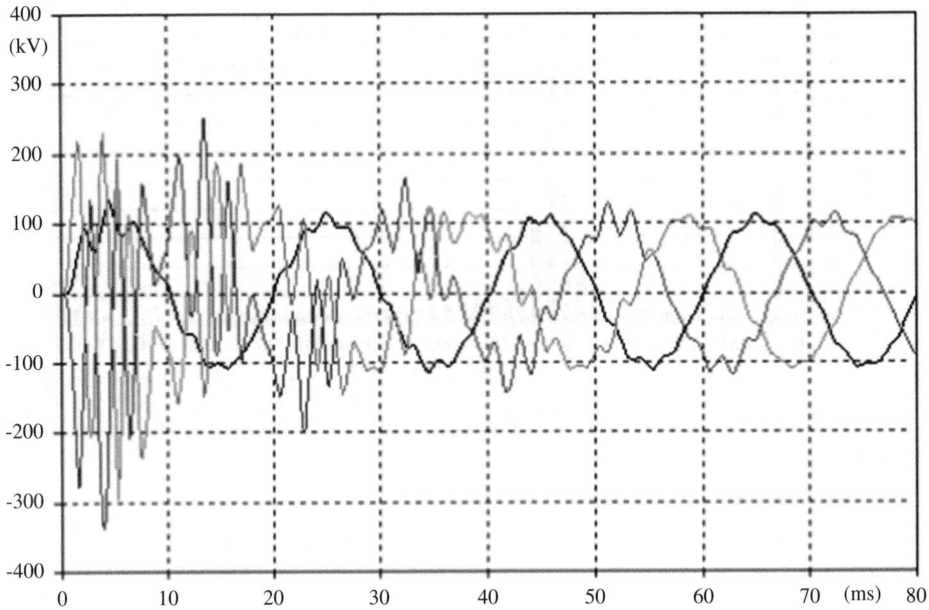

Figure 7.8 Results of the series resonance overvoltage analysis (natural frequency is adjusted to 425 Hz). *Source*: © 2012 IEEE. Reprinted, with permission, from [4]

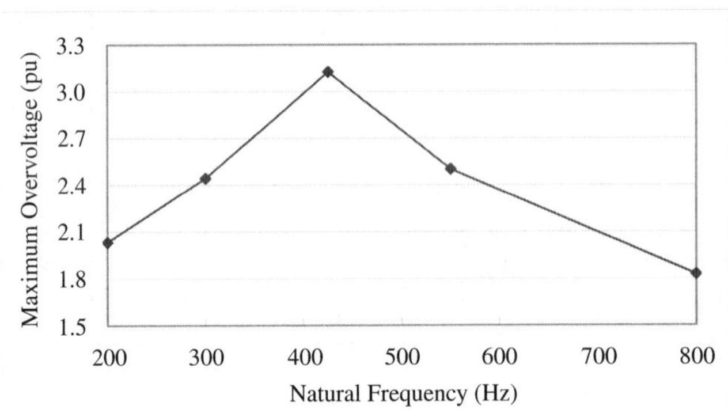

Figure 7.9 Effect of the natural frequency on the series resonance overvoltage. *Source*: © 2012 IEEE. Reprinted, with permission, from [4]

demonstrates that the dominant frequency contained in the series resonance overvoltage is 425 Hz and matches the cable energization overvoltage.

The time domain simulations were performed with different natural frequencies by changing the capacitance in the 132 kV network. Figure 7.9 shows that the series resonance overvoltage becomes more severe when the natural frequency becomes closer to 425 Hz.

The example given in this section illustrates the importance of understanding the natural frequency of the network and the dominant frequency in the energization overvoltage. Finding the possible range of the natural frequency of the network using (7.1) or (7.2) is fairly straightforward. The theoretical formulas of the dominant frequency in the energization overvoltage are derived in the following section.

7.2.1.3 Dominant Frequency in Energization Overvoltage

As discussed in the last section, the theoretical formulas are useful in order to find or estimate the dominant frequency before such simulations are implemented or to verify the dominant frequency found by these simulations.

As the modeling of a broader area is necessary for the temporary overvoltage analysis, it is time-consuming to set up network models and perform time domain simulations or frequency scans in EMT-type programs. As more long cables are being planned or installed, theoretical formulas to derive the dominant frequency are desirable for the efficient planning or operational planning since the theoretical analysis may eliminate the necessity to model some areas of the network.

The metallic sheath of a long cable is generally cross-bonded in order to reduce both sheath induced current and sheath voltage. Cross-bonding affects the cable impedance and admittance, and Nagaoka and Ametani found the impedance and admittance calculation of cross-bonded cables [2]. Based on the impedance and admittance calculation in References [2, 3], this section derives theoretical formulas to derive the dominant frequency contained in the switching overvoltage of long cables [1, 4].

Average Impedances and Admittances

Assuming that each minor section of a cable is of equal length, the average impedance of the cross-bonded cable for one major section is given in (3.131) as explained in Section 3.5.2.4.

$$[Z] = ([Z'] + [R][Z'][R]^{-1} + [R]^{-1}[Z'][R])/3 \tag{7.3}$$

Here, $[Z']$ is the impedance of the cable per unit length for one minor section, and $[R]$ is the rotation matrix as explained in Section 3.5.2.1.

$$[Z'] = \begin{bmatrix} [Z'_{CC}] & [Z'_{CS}] \\ [Z'_{CS}] & [Z'_{SS}] \end{bmatrix} \tag{7.4}$$

$$[R] = \begin{bmatrix} 1 & 0 & 0 & 0 & 0 & 0 \\ 0 & 1 & 0 & 0 & 0 & 0 \\ 0 & 0 & 1 & 0 & 0 & 0 \\ 0 & 0 & 0 & 0 & 0 & 1 \\ 0 & 0 & 0 & 1 & 0 & 0 \\ 0 & 0 & 0 & 0 & 1 & 0 \end{bmatrix} \tag{7.5}$$

where $[Z'_{CC}]$ is the impedance between conductors, $[Z'_{CS}]$ is the impedance between conductors and sheaths, and $[Z'_{SS}]$ is the impedance between sheaths.

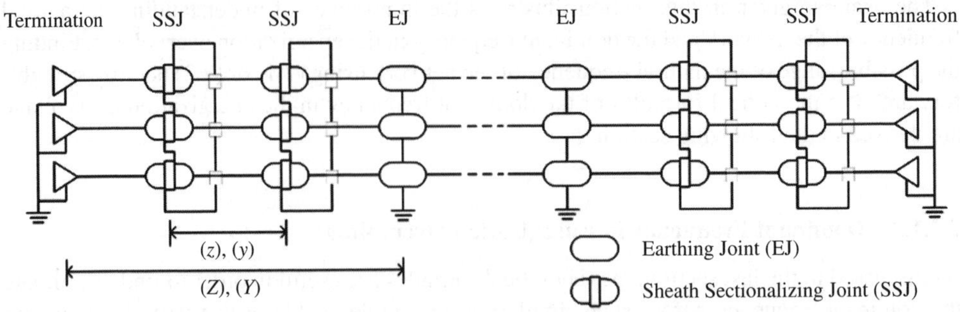

Figure 7.10 Cross-bonding diagram

The calculation of the average impedance of one major section, as shown in Figure 7.10, yields the following equations.

$$[Z] = \begin{bmatrix} [Z_{CC}] & [Z_{CS}] \\ [Z_{CS}]_t & [Z_{SS}] \end{bmatrix} \tag{7.6}$$

where

$$[Z_{CC}] = [Z'_{CC}] \tag{7.7a}$$

$$[Z_{CS}] = \begin{bmatrix} Z_{AS} & Z_{AS} & Z_{AS} \\ Z_{BS} & Z_{BS} & Z_{BS} \\ Z_{CS} & Z_{CS} & Z_{CS} \end{bmatrix} \tag{7.7b}$$

$$[Z_{SS}] = \begin{bmatrix} Z_{SS} & Z_{SM} & Z_{SM} \\ Z_{SM} & Z_{SS} & Z_{SM} \\ Z_{SM} & Z_{SM} & Z_{SS} \end{bmatrix} \tag{7.7c}$$

Impedances in (7.6) and (7.7a–7.7c) are given by:

$$Z_{AS} = (z_{Aa} + z_{Ab} + z_{Ac})/3 \tag{7.8a}$$

$$Z_{BS} = (z_{Ba} + z_{Bb} + z_{Bc})/3 \tag{7.8b}$$

$$Z_{CS} = (z_{Ca} + z_{Cb} + z_{Cc})/3 \tag{7.8c}$$

$$Z_{SS} = (z_{aa} + z_{bb} + z_{cc})/3 \tag{7.8d}$$

$$Z_{SM} = (z_{ab} + z_{bc} + z_{ca})/3 \tag{7.8e}$$

In subscripts of impedances, upper case letters show the phase of the conductor while lower case letters show the phase of the metallic sheath.

Similarly, the average admittance of the cross-bonded cable for one major section can be calculated as in (3.133):

$$[Y] = ([Y'] + [R][Y'][R]^{-1} + [R]^{-1}[Y'][R])/3$$
$$= \begin{bmatrix} [Y_{CC}] & [Y_{CS}] \\ [Y_{CS}] & [Y_{SS}] \end{bmatrix} \tag{7.9}$$

where

$$[Y_{CC}] = y_{AA} \cdot I \tag{7.10}$$

$$[Y_{CS}] = -\frac{y_{AA}}{3}\begin{bmatrix} 1 & 1 & 1 \\ 1 & 1 & 1 \\ 1 & 1 & 1 \end{bmatrix} \tag{7.11}$$

$$[Y_{SS}] = y_{SS} \cdot I \tag{7.12}$$

In (7.10) and (7.12), I is the identity matrix.

Since the metallic sheath of the cross-bonded cable is earthed at both ends of a major section, $[Z]$ and $[Y]$ can be reduced to 4×4 matrices. The reduction is, however, not performed as the dominant frequency can be found in a simpler form without the matrix reduction.

Derivation of Theoretical Formulas

Here, the main focus is to find the dominant frequency from $[Z]$ and $[Y]$ by means of simple theoretical formulas. For a given length of a line, one propagation velocity is linked to one dominant frequency by (7.13). The relationship can be found, remembering that the current is zero at the open terminal, as the fundamental resonance frequency from the general solution of the set of wave equations given by d'Alembert in the 1750s.

$$f_i = \frac{1}{4\tau} = \frac{v_i}{4 \times (\text{line length})} \tag{7.13}$$

Thus, we need to find the propagation velocity of the dominant frequency component in simple theoretical formulas.

It is known that the propagation velocity of each mode is found by [5]:

$$v_i = \frac{2\pi f_0}{\mathrm{Im}(\sqrt{z_i y_i})} \tag{7.14}$$

Here, z_i and y_i are ith diagonal entries of the modal impedance matrix $[z]$ and the admittance matrix $[y]$, respectively. The imaginary part of $\sqrt{z_i y_i}$ is denoted by $\mathrm{Im}(\sqrt{z_i y_i})$ and f_0 is the target frequency ($i = 1, 2, \ldots, 6$).

Equation (7.14) can also be expressed in a compact form as:

$$(v) = \mathrm{diag}\left(\frac{2\pi f_0}{\mathrm{Im}\left(\sqrt{[z]\,[y]}\right)}\right) \tag{7.15}$$

It is necessary to find $[z][y]$ from $[Z]$ and $[Y]$ without conducting diagonalization.

As already explained in detail in Chapter 3,

$$[z] = [A]^{-1}[Z][B], \quad [y] = [B]^{-1}[Y][A] \tag{7.16}$$

where $[A]$ and $[B]$ are the transformation (eigenvector) matrices for voltages and currents, respectively.

Also, $[Q] = [\Gamma]^2 = [z] \cdot [y]$ is calculated from $[P] = [Z] \cdot [Y]$ as explained in Chapter 3.

$$[z]\,[y] = [A]^{-1}[Z][Y][B] = [A]^{-1} \cdot [P] \cdot [B] = [Q] \tag{7.17}$$

where $[Q]$ is the eigenvalue matrix of $[P]$.

The above equation shows that the entries of $[Q]$ can be found from the eigenvalues of $[P] = [Z][Y]$.

Using (7.6)–(7.12), the matrix $[P]$ is given as:

$$[P] = \begin{bmatrix} [P_{11}] & [P_{12}] \\ [P_{21}] & [P_{22}] \end{bmatrix} \tag{7.18}$$

where

$$[P_{11}] = [Z_{CC}][Y_{CC}] + [Z_{CS}] \cdot [Y_{CS}] \tag{7.19a}$$

$$[P_{12}] = [Z_{CC}][Y_{CS}] + [Z_{CS}] \cdot [Y_{SS}] \tag{7.19b}$$

$$[P_{21}] = [Z_{CS}]_t[Y_{CC}] + [Z_{SS}] \cdot [Y_{CS}] \tag{7.19c}$$

$$[P_{22}] = [Z_{CS}]_t[Y_{CS}] + [Z_{SS}] \cdot [Y_{SS}] \tag{7.19d}$$

Thus,

$$[P_{11}] = \begin{bmatrix} F_{11}y_{AA} & F_{12}y_{AA} & F_{13}y_{AA} \\ F_{21}y_{AA} & F_{22}y_{AA} & F_{23}y_{AA} \\ F_{31}y_{AA} & F_{32}y_{AA} & F_{33}y_{AA} \end{bmatrix} \tag{7.20a}$$

$$[P_{12}] = \begin{bmatrix} F_{AX} & F_{AX} & F_{AX} \\ F_{BX} & F_{BX} & F_{BX} \\ F_{CX} & F_{CX} & F_{CX} \end{bmatrix} \tag{7.20b}$$

$$[P_{21}] = \begin{bmatrix} F_{AY} & F_{BY} & F_{CY} \\ F_{AY} & F_{BY} & F_{CY} \\ F_{AY} & F_{BY} & F_{CY} \end{bmatrix} \tag{7.20c}$$

$$[P_{22}] = \begin{bmatrix} F_{SS} & F_{SM} & F_{SM} \\ F_{SM} & F_{SS} & F_{SM} \\ F_{SM} & F_{SM} & F_{SS} \end{bmatrix} \tag{7.20d}$$

Here, variables $F_{11}, F_{12}, \ldots, F_{33}$ in $[P_{11}]$ can be calculated as:

$$F_{11} = z_{AA} - Z_{AS},\quad F_{12} = z_{AB} - Z_{AS},\quad F_{13} = z_{CA} - Z_{AS} \tag{7.21a}$$

$$F_{21} = z_{AB} - Z_{BS},\quad F_{22} = z_{AA} - Z_{BS},\quad F_{23} = z_{BC} - Z_{BS} \tag{7.21b}$$

$$F_{31} = z_{CA} - Z_{CS},\quad F_{32} = z_{BC} - Z_{CS},\quad F_{33} = z_{AA} - Z_{CS} \tag{7.21c}$$

where Z_{AS}, Z_{BS}, and Z_{CS} are given by (7.8).

In the submatrix $[P_{12}]$,

$$F_{AX} = -(z_{AA} + z_{AB} + z_{CA})y_{AA}/3 + Z_{AS}y_{SS} \tag{7.22}$$

F_{BX} and F_{CX} can be found similarly.

In the submatrix $[P_{21}]$,

$$F_{AY} = Z_{AS}y_{AA} - (Z_{SS} + 2Z_{SM})y_{AA}/3 \tag{7.23}$$

F_{BY} and F_{CY} can be found similarly.
In the submatrix $[P_{22}]$,

$$F_{SS} = -(Z_{AS} + Z_{BS} + Z_{CS})y_{AA}/3 + Z_{SS}y_{SS} \tag{7.24}$$

$$F_{SM} = -(Z_{AS} + Z_{BS} + Z_{CS})y_{AA}/3 + Z_{SM}y_{SS} \tag{7.25}$$

Instead of diagonalizing $[Z][Y]$, we assume that the overvoltage caused by the energization of a long cross-bonded cable is dominated by inter-phase modes. Under this assumption, the eigenvectors of inter-phase modes are found in an ideal form and it is possible to find the theoretical formulas in a simple form which does not require matrix calculation as explained in Chapter 3. In order for the coaxial mode to dominate the energization overvoltage, the dominant frequency must be higher than the critical frequency f_c given by (3.83) in Section 3.2.1 [6, 7].

$$f_c \approx \frac{\rho_s}{\pi \mu_s d^2} \tag{7.26}$$

where ρ_s is sheath resistivity, μ_s is sheath permeability, and d is sheath thickness.

The critical frequency calculated by (7.26) for a typical 400 kV XLPE cable ranges from 1000 to 1500 Hz. It is therefore reasonable to assume that inter-phase modes dominate the energization overvoltage, since the dominant frequency, which is higher than this critical frequency, cannot originate from long cross-bonded cables.

Since the ideal eigenvector of the first inter-phase mode is known as $A_1 = \begin{pmatrix}-1/3 & 2/3 & -1/3 & 0 & 0 & 0\end{pmatrix}_t$, the eigenvalue corresponding to this inter-phase mode can be found as:

$$\begin{aligned} Q_1 &= -(2z_{AB} - z_{AA} - z_{CA})y_{AA} \qquad \text{or} \\ &= (2z_{AA} - z_{AB} - z_{BC})y_{AA}/2 \end{aligned} \tag{7.27}$$

The eigenvalue corresponding to the second inter-phase mode $A_2 = \begin{pmatrix}1/2 & 0 & -1/2 & 0 & 0 & 0\end{pmatrix}_t$ is calculated as:

$$Q_2 = (z_{AA} - z_{CA})y_{AA} \tag{7.28}$$

Once the eigenvalues are found by (7.27) and (7.28), the propagation velocities are found by:

$$v_i = \frac{2\pi f_0}{\mathrm{Im}(\sqrt{Q_i})} \qquad (i = 1,\ 2) \tag{7.29}$$

Equations (7.27), (7.28) and (7.29) indicate that simple readily available data will provide the estimation of dominant frequency.

Effect of Source Impedance

Theoretical formulas have so far been derived assuming there is no source impedance. This may be justified in the slow-front overvoltage analysis as source impedance changes depending on network conditions and no source impedance is supposedly the most severe condition.

This section considers the effect of source impedance since it affects the dominant frequency. The amplitude of the overvoltage caused by cable line energization will be

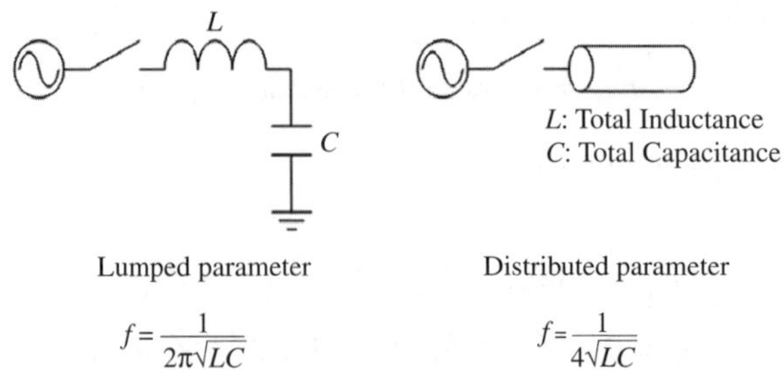

Figure 7.11 Comparison of dominant frequency. *Source*: © 2012 IEEE. Reprinted, with permission, from [4]

mitigated by the introduction of source impedance, but also the lowered dominant frequency may cause resonances and lead to higher resonance overvoltages.

First, when the source impedance is given as lumped parameter impedance Z_0, it has to be converted to distributed parameter source impedance z_0 by (7.30). Figure 7.11 illustrates the reason why the conversion is necessary. By the conversion, the lumped parameter source impedance can be considered a part of the distributed parameter cable line. The dominant frequency for the distributed parameter model can be derived from (7.13) and (7.14).

$$z_0 = Z_0 \times \left(\frac{2\pi}{4}\right)^2 \tag{7.30}$$

Assuming that the source impedance is a part of the cable line, the eigenvalues corresponding to the two inter-phase modes are found as:

$$\begin{aligned} Q_1 &= \{z_0 - (2z_{AB} - z_{AA} - z_{CA})\}y_{AA} \qquad \text{or} \\ &= (z_0 + 2z_{AA} - z_{AB} - z_{BC})y_{AA}/2 \end{aligned} \tag{7.31}$$

$$Q_2 = (z_0 + z_{AA} - z_{AC})y_{AA} \tag{7.32}$$

When the eigenvalues have been found, the propagation velocity and dominant frequency can be calculated by (7.29) and (7.13).

7.2.2 *Parallel Resonance Overvoltage*

7.2.2.1 Theory

When inductance L and capacitance C are connected in parallel, the total impedance or series impedance becomes infinitely large at the frequency $f_n = \frac{1}{2\pi\sqrt{LC}}$. When a current source of this frequency is connected to the parallel circuit as in Figure 7.12, infinitely large voltage might appear, in theory, on the source side of the inductance and the capacitance. Such a voltage is called parallel resonance overvoltage.

The parallel resonance overvoltage can occur in an actual power system. For example, transformer energization, cable energization, and HVDC can be a source of a harmonic current that excites a parallel resonance circuit. Among the potential harmonic current

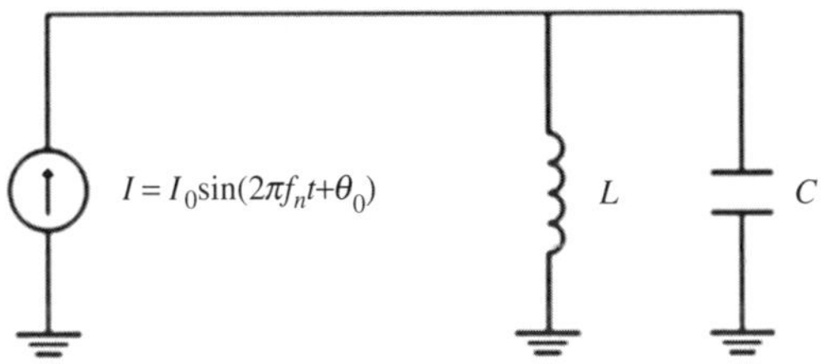

Figure 7.12 Simple parallel resonance circuit

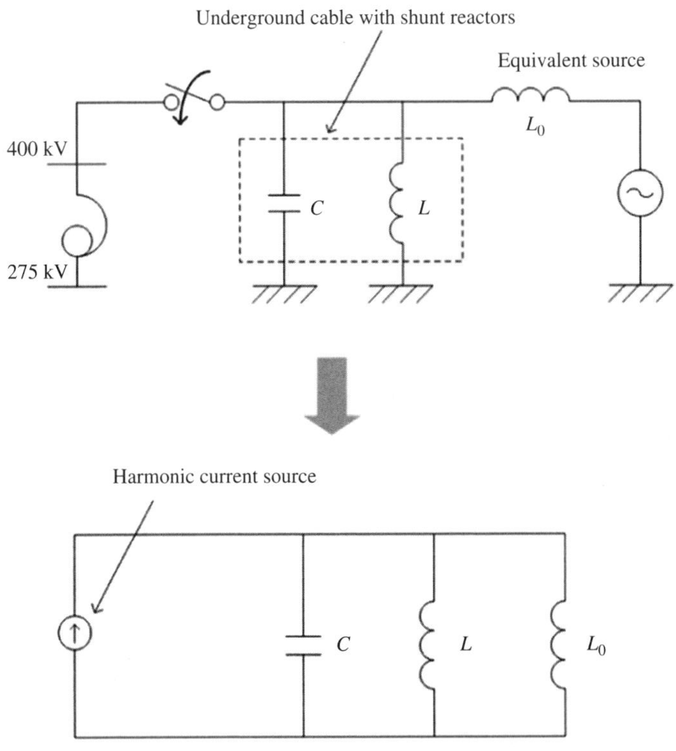

Figure 7.13 Equivalent circuit of parallel resonance caused by transformer inrush

sources, an inrush current caused by the transformer energization is considered to be the most onerous one because of its high harmonic contents, low frequency, low damping, and long duration.

Figure 7.13 shows a simplified circuit that can lead to a severe parallel resonance overvoltage caused by the transformer energization. When the 400 kV transformer is energized, harmonic current contained in the inrush current flows through parallel resonance circuit composed of charging capacity C, shunt reactor for reactive power compensation L, and equivalent source impedance L_0.

The harmonic equivalent circuit is also shown in Figure 7.13. The inrush current is expressed as harmonic current source. The voltage source behind the equivalent source impedance is ignored since the voltage source only generates fundamental frequency voltage.

The condition for the parallel resonance is:

$$\omega_n C = \frac{1}{\omega_n L} + \frac{1}{\omega_n L_0} \quad \therefore \quad f_n = \frac{1}{2\pi}\sqrt{\frac{1}{CL} + \frac{1}{CL_0}} \tag{7.33}$$

Here, f_n is the natural frequency of the parallel resonance circuit.

Since the second harmonic current is contained in the inrush current with a higher proportion compared with the other harmonic components, a parallel resonance overvoltage is most severe when the natural frequency of the parallel circuit is 100 Hz, that is:

$$f_n = \frac{1}{2\pi}\sqrt{\frac{1}{CL} + \frac{1}{CL_0}} = 100 \tag{7.34}$$

When the compensation rate of the cable is 100%, the natural frequency ignoring the source impedance is 50 Hz, that is:

$$f_n = \frac{1}{2\pi}\sqrt{\frac{1}{CL}} = 50 \tag{7.35}$$

The source impedance will raise the natural frequency, and it becomes 100 Hz when $L_0 = L/3$ as shown in (7.36).

$$\begin{aligned} f_n &= \frac{1}{2\pi}\sqrt{\frac{1}{CL} + \frac{1}{CL_0}} = \frac{1}{2\pi}\sqrt{\frac{1}{CL} + \frac{3}{CL}} \\ &= \frac{1}{2\pi}\sqrt{\frac{1}{CL}} \times 2 = 100 \end{aligned} \tag{7.36}$$

However, the source impedance is usually much smaller than $L/3$ in most networks in typical operating conditions. The natural frequency, therefore, becomes 100 Hz in weak networks such as in the black start operations. The load level is generally low in weak networks, and it contributes highly to a severe parallel resonance overvoltage.

The parallel resonance overvoltage is normally studied in the following procedure:

- Find the natural frequency of the parallel resonance circuit.
- Set the natural frequency to 100 Hz by adjusting source impedance.
- Simulate the most severe switching scenarios (transformer energization) with the natural frequency 100 Hz.

7.2.2.2 Example of the Analysis

In this section, the transformer energization in Figure 7.13 is simulated as an example of the parallel resonance overvoltage analysis. The transformer is energized through the same 400 kV cable as in Section 7.2.1.2. The 400 kV cable line is assumed to be a double-circuit line. Its charging capacity is compensated by four units of 120 MVar shunt reactors, located at both ends of the cable line. This realizes the compensation rate 100.6%.

Table 7.2 Hysteresis characteristic of the 400 kV transformer

I (A, peak)	Phi (Wb-T, peak)
−37.500	−1310.294
4.688	−1241.206
2.813	−1143.529
6.563	−889.412
12.656	770.294
19.688	1016.471
33.750	1167.353
62.344	1270.588
150.000	1350.000
206.250	1357.941

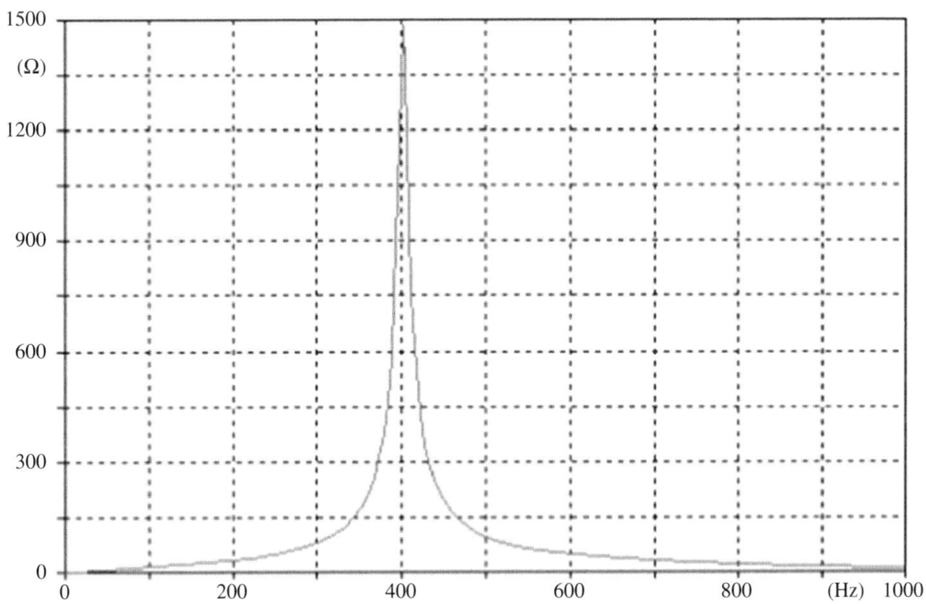

Figure 7.14 Results of frequency scan with source impedance 14.0 mH

One of the most important settings in the simulation is the hysteresis characteristic of the 400 kV transformer. It is set as shown in Table 7.2. The residual flux of 85% and −85 % is assumed in phases-a and b, respectively. The transformer is energized at voltage zero of phase-a in order to cause a large inrush current in phase-a.

First, the natural frequency of the network is set to 400 Hz by adjusting the source impedance L_0 to 14.0 mH. The natural frequency of the network is confirmed by the frequency scan as shown in Figure 7.14. Parallel resonance is not expected to occur in this condition since the magnitude of the impedance is low at 100 Hz.

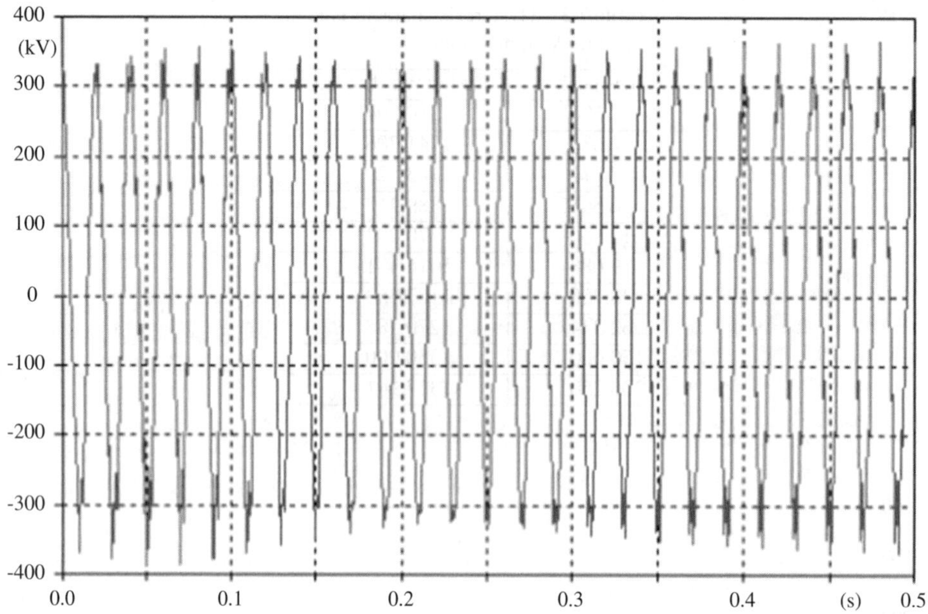

Figure 7.15 Results of parallel resonance overvoltage analysis ($f_n = 400$ Hz)

Figure 7.15 shows the results of the results of the parallel resonance overvoltage with the natural frequency 400 Hz. As expected from the results in Figure 7.14, a severe overvoltage is not observed under this condition.

Figure 7.16 and Figure 7.17 show the results of the parallel resonance overvoltage analysis with the natural frequency of 200 and 100 Hz, respectively. It is still difficult to observe the parallel resonance with the natural frequency 200 Hz. When the natural frequency goes down to 100 Hz, however, the voltage waveform is highly distorted due to the parallel resonance. The distortion of the voltage waveform results in the highest overvoltage 549.8 kV.

7.2.3 Overvoltage Caused by System Islanding

When one end of a long cable is opened, a part of a network can be separated from the main grid together with the long cable. The equivalent circuit expressing the situation is illustrated in Figure 7.18. Here, the assumed fault is a bus fault since a cable line fault will result in the removal of the cable line from the equivalent circuit.

From the equivalent circuit in Figure 7.18, the overvoltage caused by the system islanding can be expressed by the following equations [8]:

$$v(t) = V_m \sin \omega t - \frac{\omega}{\omega_0} V_m \sin \omega_0 t \tag{7.37}$$

$$V_m = \frac{E_m L}{L_0(1 - \omega^2 CL) + L} \tag{7.38a}$$

$$\omega_0 = \sqrt{\frac{1}{CL} + \frac{1}{CL_0}} \tag{7.38b}$$

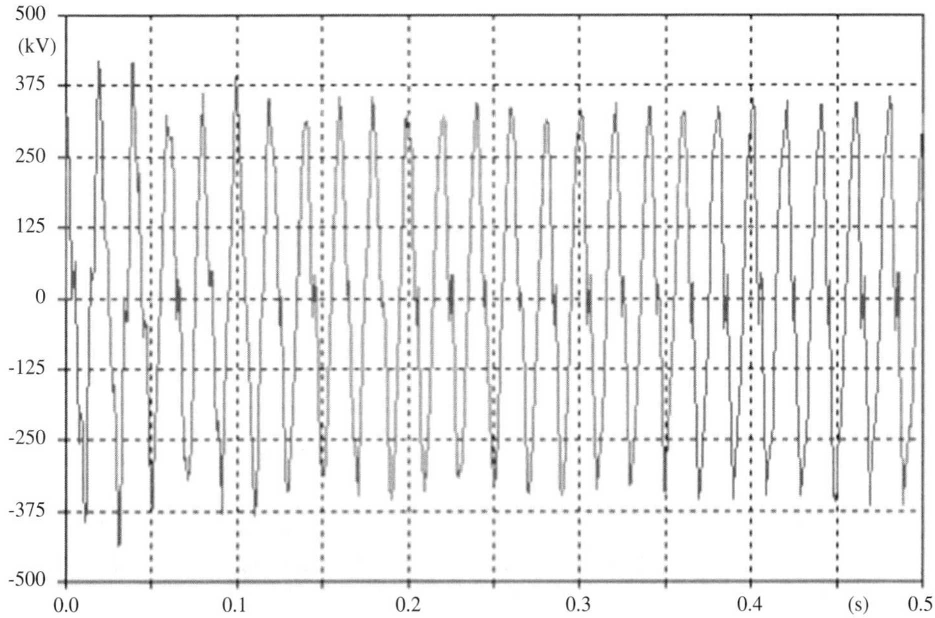

Figure 7.16 Results of parallel resonance overvoltage analysis (f_n = 200 Hz)

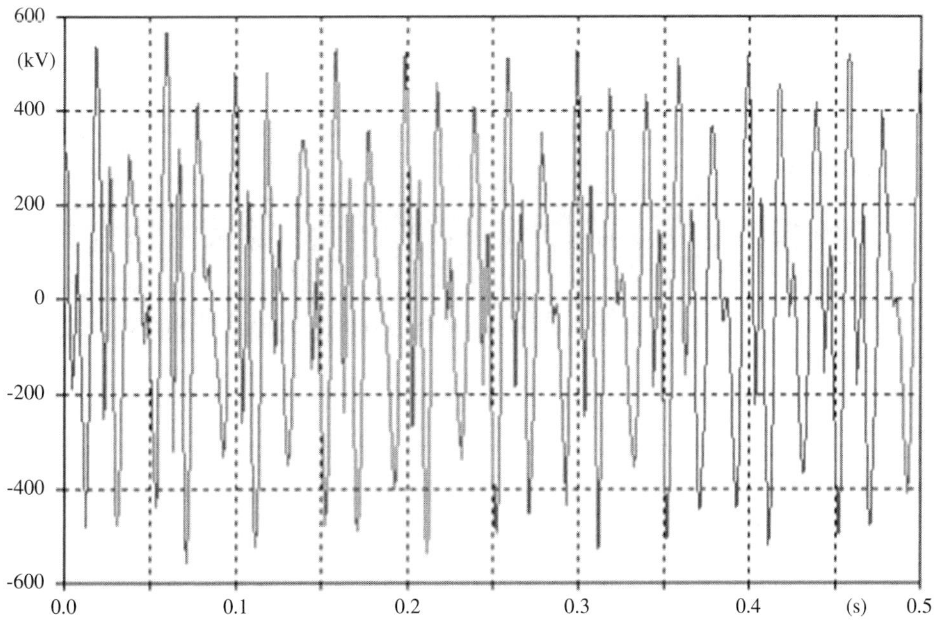

Figure 7.17 Results of parallel resonance overvoltage analysis (f_n = 100 Hz)

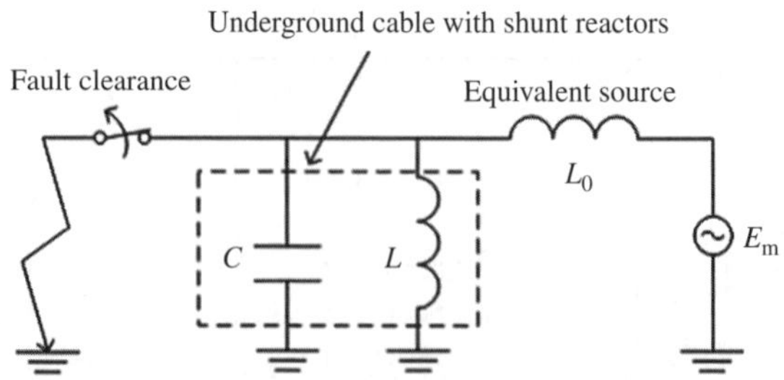

Figure 7.18 Equivalent circuit of the system islanding

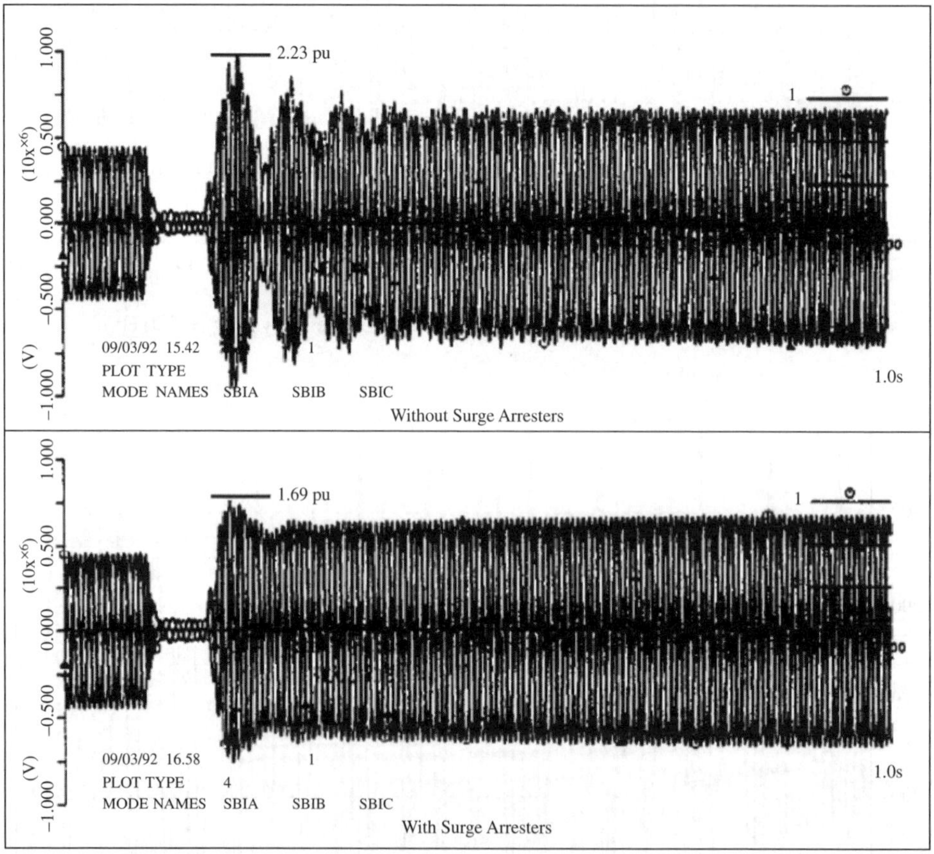

Figure 7.19 Example of the overvoltage caused by the system islanding: (a) without surge arresters; and (b) with surge arresters

where L_0 is source impedance of the weaker islanded system and E_m is the source voltage behind L_0. Charging capacity of the long cable and inductance of the shunt reactors directly connected to the cable are expressed by C and L, respectively.

Equation (7.37) shows that the overvoltage contains two frequency components, the nominal frequency ω and the resonance frequency ω_0. Since the overvoltage is caused by the superposition of two frequency components, the resulting overvoltage is oscillatory, and its level is often difficult to estimate before the simulation. The result of a simulation performed for the 500 kV Shin-Toyosu line is shown in Figure 7.19 [9].

The overvoltage level is sensitive to L_0, which expresses the short circuit level in the islanded system. In order to find the most severe overvoltage, it is necessary to study different fault current levels or network conditions.

7.3 Slow-front Overvoltages

It has never been reported that the slow-front overvoltage exceeds insulation strengths of equipment in cable systems. Until now, SIWV (switching impulse withstand voltage) determined with power systems predominantly composed of OHLs showed no problem with the power systems that contain the cable systems.

Presently, the slow-front overvoltage analysis is no longer conducted in the implementation study of OHLs in most cases. It is known from past studies that the slow-front overvoltage does not exceed SIWV as long as the typical one (generally, 1050 kV for 420 kV equipment) is selected. In contrast, it is typical practice to study the slow-front overvoltages for the installation of EHV cables, mainly due to the lack of past studies or experiences.

This section first introduces examples of energization overvoltages in cables, comparing them with those in OHLs [1, 10]. From a number of simulations, the statistical distribution of the energization overvoltage is derived and compared with the energization overvoltage of OHLs. The detailed analysis of differences in the statistical distribution is given in Section 7.3.3 [11]. Since energization overvoltages are affected by the modeling of feeding networks, the line energization both from a lumped source and from a complex source is discussed.

7.3.1 Line Energization Overvoltages from a Lumped Source

7.3.1.1 Overview

Statistical distributions of slow-front overvoltages have played an important role in determining insulation levels and considering insulation coordination of EHV systems. Especially, key values in the statistical distributions, such as 2%, maximum, and mean values, have been considered as indicators when assessing the insulation performance of EHV systems [12].

References [13–16] have studied the statistical distributions of line energization and reclosing overvoltages, gathering the simulation results by transient network analyzers and digital computers from all over the world. The results of the studies were used as the representative overvoltages and have formed the basis of today's insulation coordination.

When the study on statistical distributions was conducted in the 1970s, it focused on the overvoltages in EHV OHLs. EHV cables were not included in the study since their installed amount was limited, though some EHV cables were already in service [17–21]. Therefore, the statistical distribution of switching overvoltages of EHV cables is not available to date. Most of the insulation levels determined with OHLs have been applied to cable systems, which may not be an appropriate approach.

Since the study in the 1970s, the installed amount of EHV cables has grown to a level at which statistical evaluations are possible. CIGRE WG B1.07 reported that 5555 and 1586 circuit km of underground cables were installed, respectively, at 220–314 and 315–500 kV in

2007 [22, 23]. Furthermore, longer cables are planned or installed, including the 82 km 220 kV cable to the Anholt Offshore Wind Farm in Denmark. Thanks to this increase of cable installations, it is now possible to obtain various data which contain physical and electrical parameters of installed cables as well as cable layouts.

Based on the obtained cable data, this section derives the statistical distribution of energization overvoltages of EHV cables. Characteristics of the statistical distribution are found through a comparison between the statistical distributions of EHV cables and OHLs.

7.3.1.2 Study Conditions and Parameters

This section derives the statistical distributions of energization overvoltages of both EHV cables and OHLs. The distribution of overvoltages on OHLs has already been studied in References [13–16], but the energization overvoltages for OHLs are also studied in this section in order to compare the statistical distributions of EHV cables and OHLs under the same conditions and parameters. This will make it easier to compare the overvoltage distributions for cables and OHLs.

First, we discuss the common conditions and parameters for cables and OHLs. In this section, the statistical distributions of energization overvoltages were derived from the results of 200 line energization cases with statistical (random) switching. CIGRE Working Group 13.05 conducted 100 line energizations to obtain the distribution. The number of energizations can be increased due to the higher computer capability for higher accuracy in finding key values in the statistical distributions. With the 200 line energization cases, 2% value is the fourth highest value in the repeated simulations.

In the statistical switching, two kinds of randomness were considered. The first randomness is the phase angle (point-on-wave) when the line circuit breakers receive the command to close themselves. A uniform distribution from 0 to 360° was assumed for this randomness. The second randomness is the difference of close timings between three phases. A normal distribution with standard deviation of 1 ms was assumed for the randomness.

CIGRE WG 13.02 considered parameters such as line length, feeding network, shunt compensation, existence of closing resistors, trapped charge, and so on [13]. Among them, this section focuses on two parameters: line length and feeding network.

Table 7.3 shows the variation of the two parameters. The line length is increased up to 96 km, considering the fact that cable line as long as this level was studied [24]. The variation of the feeding network covers a weak source to a strong source within a reasonable range. Only the lumped parameter inductive source is considered as in the study by CIGRE WG 13.05. However, the source impedance 1 mH is also included in order to realize a steep initial voltage rise, which is expected in the line energization from a distributed parameter source [16].

Table 7.3 Study conditions and parameters for the cable energization overvoltage

Line length	24, 48, 72, and 96 km
Feeding network	1, 15, 30, and 100 mH (corresponds to fault current of 735, 49, 25, and 7 kA at 400 kV)

In all energization cases, charging capacity of EHV cables was compensated by shunt reactors directly connected to the cables. The compensation rate was set to 100% so that the power-frequency component of the overvoltage does not exist and only the transient component of the overvoltage can be observed. The power-frequency component of the overvoltage is not the focus of this section and can be calculated theoretically.

The shunt reactors were directly connected to the line at both ends when the line length was 24, 48, or 72 km. With the line length of 96 km, the shunt reactors were additionally connected to the center of the line. The necessary capacity of shunt reactors for 100% compensation was shared by shunt reactors at both line ends and at the center with the proportions of 1/4, 1/4, and 1/2. Saturation characteristics of shunt reactors were not modeled, as the characteristics for the different sizes of shunt reactors were not available.

Charging capacity of OHLs was also compensated to 100% in order to find the statistical distributions of overvoltages in the same condition as cables. However, the distributions were also studied without shunt compensation since it is not common to compensate charging capacity of OHLs of this length.

Cables

In order to derive the statistical distributions, 3200 line energization simulations (4 line lengths × 4 feeding networks × 200 random switchings) were performed with 10 types of EHV cables. Cable types, physical and electrical parameters, and burial layouts were selected based on actual installations. Table 7.4 shows key parameters of 10 types of EHV cables.

All 10 types of EHV cables are represented using the frequency-dependent model in the phase domain. It was assumed that their metallic sheaths are cross-bonded, which is the typical practice for a cable of this length. Figure 7.20 illustrates the simulation model for line energizations. The highest overvoltages were observed at the open end of the line.

In order to assure a reasonable computational speed and numerical stability, the number of minor sections was reduced to 12. This means that the length of the minor section was increased to 8 km for the line length of 96 km. This minor section length has not been achievable in underground cables, but it has been confirmed by test simulations that the errors introduced by this modification were below 5%.

OHLs

The statistical distributions of OHLs are derived from energization simulations with 10 types of OHLs. Physical and electrical parameters of the OHLs and tower configurations were determined from actual installations. Table 7.5 shows key parameters of 10 types of OHLs.

Table 7.4 Key parameters of 10 types of cables

	UGC1	UGC2	UGC3	UGC4	UGC5	UGC6	UGC7	UGC8	UGC9	UGC10
Voltage (kV)	400	400	400	500	275	275	275	230	230	230
Core	Al	Cu	Cu	Cu	Al	Cu	Cu	Cu	Al	Cu
Size (mm^2)	1600	1000	2500	2500	1600	2500	1400	630	1000	2000
Insulation	XLPE	SCFF	XLPE	XLPE	XLPE	XLPE	SCFF	SCFF	SCFF	XLPE
Sheath	Al	Pb	Al	Al	Al	Al	Al	Al	Al	Al
Layout	Flat	Flat	Flat	Tre-foil	Flat	Tre-foil	Tre-foil	Flat	Flat	Tre-foil
Phase separation (m)	0.3	0.3	0.5	0.17	0.5	0.17	0.5	0.3	0.3	0.14

SCFF, self-contained fluid filled.

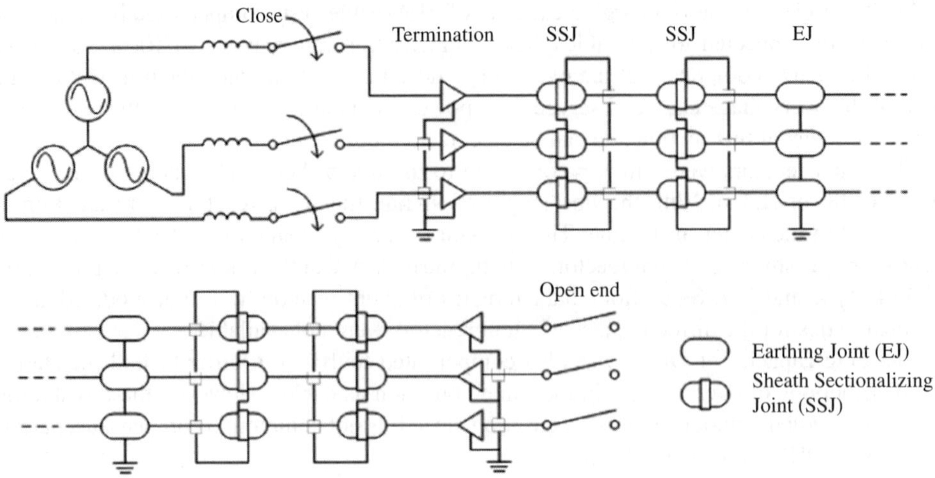

Figure 7.20 Simulation model for cable line energizations

Table 7.5 Key parameters of 10 types of overhead lines

	OHL1	OHL2	OHL3	OHL4	OHL5	OHL6	OHL7	OHL8	OHL9	OHL10
Voltage (kV)	400	400	400	500	275	275	275	500	400	400
Phase conductor	Martin	Finch	Curlew	TACSR 810	TACSR 610	Zebra	LGJ 240	LGJ 400	Dove	Cardinal
# of conductors in a bundle	2	2	2	4	4	2	6	4	4	4
Conductor separation (cm)	40	40	45	50	50	40	37.5	45	40	40
# of ground wires	2	2	2	2	2	2	2	2	2	2
Tower type	Barrel	Barrel	Single	Pine	Pine	Single	Two	Two	Danube	Danube
Height of lower conductor (m)	20	20	22	46	43.5	20	30.8	30.8	31.75	31.75

TACSR, ASCR with higher allowable temperature; Single, single level delta tower; Two, two-level delta tower.

7.3.1.3 Simulation Results and Statistical Distributions

This section discusses simulation results of the line energizations and obtained statistical distributions. The effects of studied parameters, that is, line length and feeding network, are also discussed.

The observed overvoltages are expressed in per unit, where 1 pu stands for the peak value of a phase-to-ground normal operating voltage.

Probability Distribution

Figure 7.21 shows cumulative probabilities by which the maximum overvoltage can exceed each overvoltage level (horizontal axis). First, each probability distribution was found from the results of 32 000 line energization cases. Comparing cumulative probabilities of cables and OHLs, the OHLs clearly generate higher overvoltages. The maximum overvoltage observed with cables and OHLs were, respectively, 2.54 and 2.91 pu.

In contrast, on the lower side, the differences are very small. For the overvoltage below 1.7 pu, the probabilities are virtually equal for the cables and the OHLs. It leads to the small difference in the mean values shown in Table 7.6. The standard deviation for the cables is therefore smaller than that for the OHLs.

As the charging capacity of the OHLs is relatively small, there exist only minor differences between the OHLs with 100% compensation and those without compensation.

Secondly, probability distributions are found from each 200 random energization case. With 32 000 line energization cases, 160 probability distributions are obtained, respectively, for cables and OHLs.

Probability distributions of random energization cases are often compared with the normal distribution [12]. For the parametric test, the skewness and kurtosis are obtained from 160 probability distributions. Table 7.7 shows average values of 160 skewness and kurtosis.

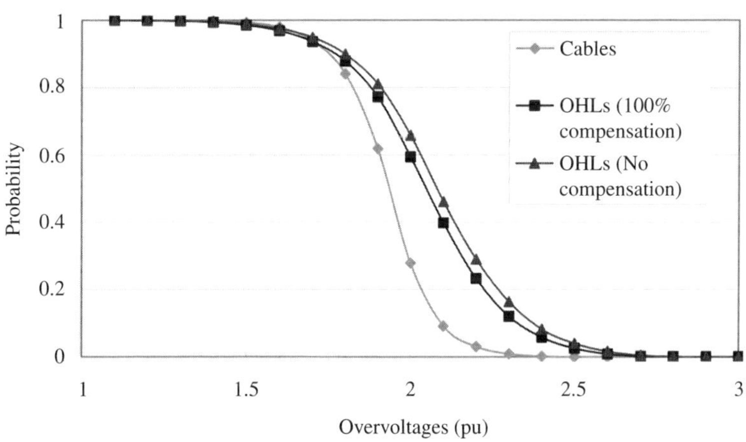

Figure 7.21 Cumulative probability distributions

Table 7.6 Means and standard deviations of probability distributions

	Mean (pu)	σ (pu)
Cable	1.93	0.14
OHL (100 % compensation)	2.05	0.23
OHL (no compensation)	2.08	0.23

σ, Standard deviation.

Table 7.7 Average skewness and kurtosis of probability distributions

	Skewness	Kurtosis
Cable	−0.65	4.22
OHL (100% compensation)	−0.52	3.62

The skewness (S) and kurtosis (K) are calculated as:

$$S = \frac{1}{200}\sum_{i=1}^{200}\left(\frac{x_i - \overline{x}}{\sigma}\right)^3 \tag{7.39}$$

$$K = \frac{1}{200}\sum_{i=1}^{200}\left(\frac{x_i - \overline{x}}{\sigma}\right)^4 \tag{7.40}$$

where x_i is sample value and $\overline{x}$ is the mean value.

Examples of probability distributions of cables and OHLs are, respectively, shown in Figure 7.22 and Figure 7.23 with fitted normal distribution curves. The skewness and kurtosis of the OHLs without compensation are not shown as the difference between with and without compensation is found to be negligible.

For both cables and OHLs, the average skewness takes negative values, which means that their probability distributions have longer tail to the left side than to the right side. As the average skewness of the cables is smaller than that of the OHLs, such a characteristic is more noticeable in the probability distributions of the cables. The characteristic can also be observed in Figure 7.22 and Figure 7.23.

Table 7.7 also shows the average kurtosis for both cables and OHLs. The kurtosis becomes equal to three when a probability distribution follows the normal distribution. The larger values of kurtosis mean probability distributions have more acute peaks than the normal distribution.

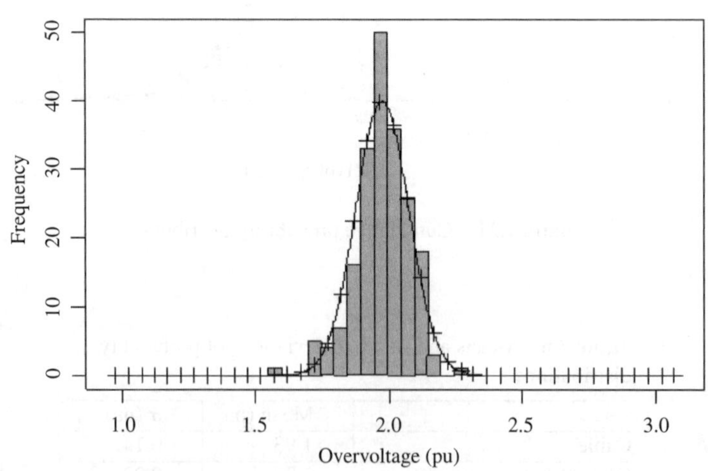

Figure 7.22 Example of probability distributions. UGC1: line length, 48 km; source impedance, 30 mH; mean, 1.98 pu; σ, 0.10 pu; skewness, −0.68; and kurtosis, 4.44

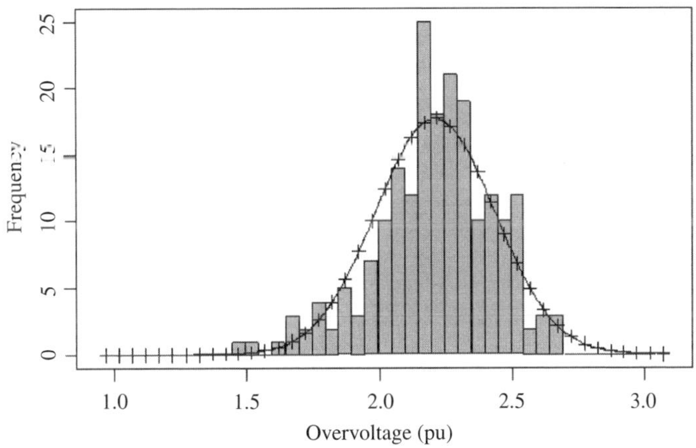

Figure 7.23 Example of probability distributions. OHL1 with 100 % compensation: line length, 48 km; source impedance, 30 mH; mean, 2.21 pu; σ, 0.23 pu; skewness, −0.54; and kurtosis, 3.35

The peaks of the probability distributions of the cables are more acute than those of the OHLs. The characteristic can also be observed in Figure 7.22 and Figure 7.23. In conclusion, both skewness and kurtosis show that the probability distributions of the OHLs are closer to the normal distribution than those of the cables.

In addition to the parametric test, a Kolmogorov–Smirnov test was performed on the probability distributions obtained [25]. In Figure 7.24, the horizontal axis is the significance level at which the hypothesis that the probability distributions follow the normal distribution is evaluated. The figure illustrates cumulative probabilities by which the probability distributions pass the Kolmogorov–Smirnov test at different significance levels. For example for cables, 16 probability distributions out of 160 (10%) pass the Kolmogorov–Smirnov test at the 20%

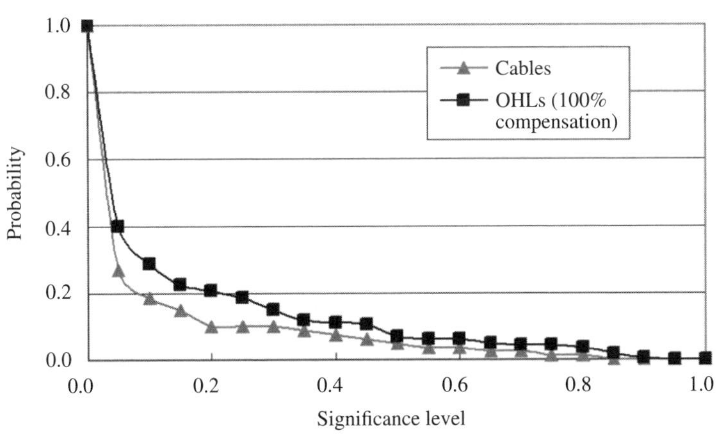

Figure 7.24 Cumulative probabilities by which probability distributions pass the Kolmogorov–Smirnov test

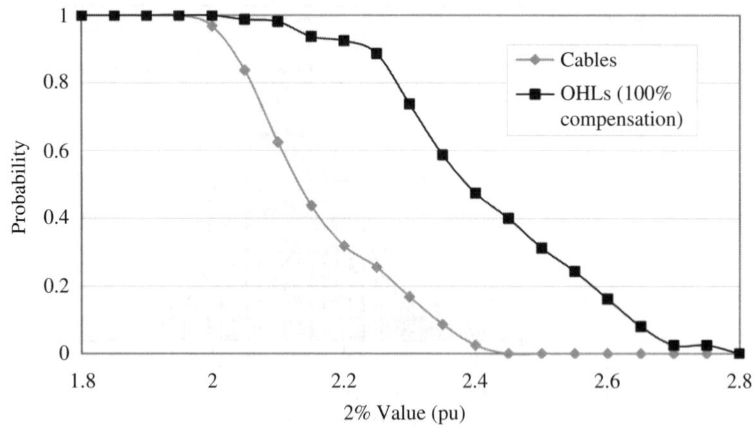

Figure 7.25 Cumulative probability distributions of 2% values

Table 7.8 Means and standard deviations of probability distributions of 2% values

	Mean (pu)	σ (pu)
Cable	2.16	0.12
OHL (100% compensation)	2.42	0.16

significance level. The Kolmogorov–Smirnov test, as well as the parametric test, demonstrates that probability distributions of the OHLs are closer to the normal distribution than those of the cables.

2% Value

The 2% values can be found as the fourth largest overvoltage in 200 random energization cases. With 32 000 line energization cases, one hundred and sixty 2% values were obtained. Figure 7.25 shows cumulative probabilities by which the 2% value can exceed each overvoltage level.

The characteristics and differences of the probability distributions observed in Figure 7.21 can also be observed in Figure 7.25. However, as shown in Table 7.8, the difference of the mean value between cables and OHLs becomes larger. In contrast, the difference of the standard deviations becomes smaller.

Comparison of Maximum Overvoltages with 2% Values

CIGRE WG 13.02 reported for OHLs that in 90% of all line energization cases, maximum overvoltages were not more than 6.5% higher than 2% values. In addition, in 50% of all the cases, maximum overvoltages were not more than 2.5% higher than 2% values.

The probabilities were also found from our simulation results for comparison. Table 7.9 summarizes the probabilities. The table shows that the differences between maximum overvoltages and 2% values are larger in cables, compared with OHLs, but the deviation between

Table 7.9 Comparison of maximum overvoltages with 2% values

	WG 13.02	Cables	OHLs
$P\ (r_{2m} \leq 6.5)$ (%)	90	90	96
$P\ (r_{2m} \leq 2.5)$ (%)	50	45	59

P, probability.

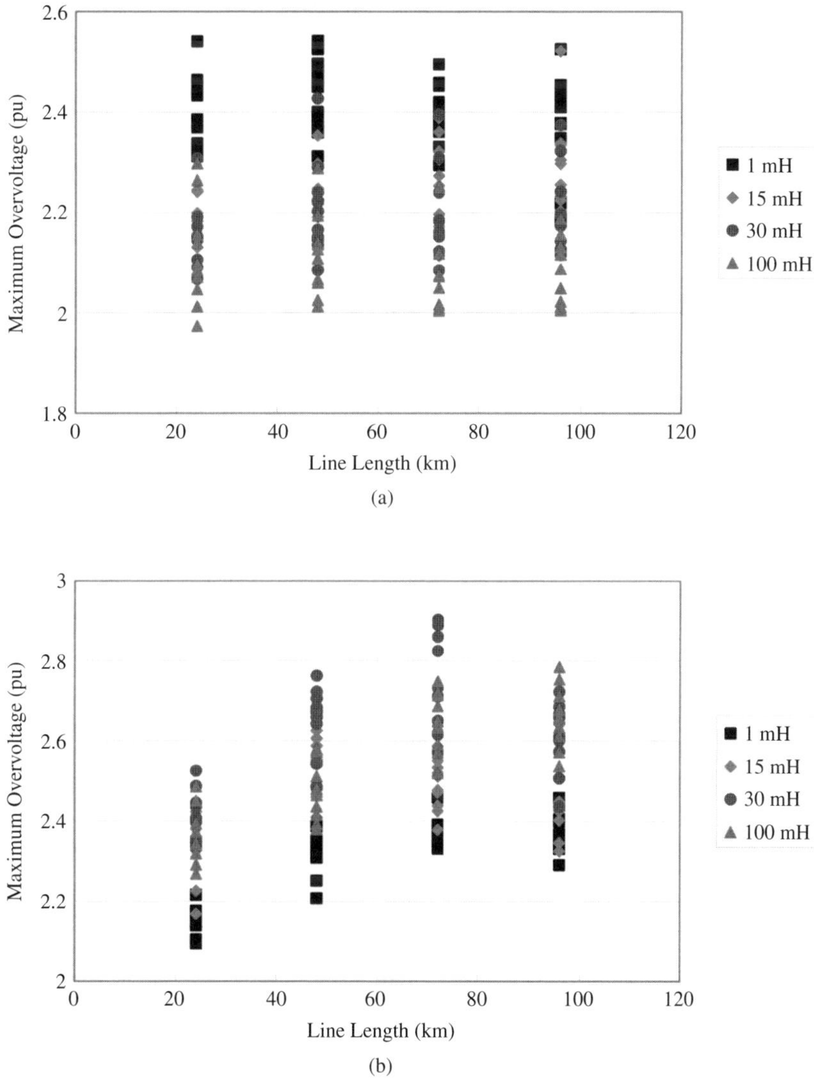

Figure 7.26 Effects of line length – maximum overvoltages: (a) cables; and (b) OHLs with 100% compensation

the cables and the OHLs is minor. In the table, r_{2m} is defined as:

$$r_{2m} = \frac{\text{Maximum overvoltage-2\% value}}{\text{2\% value}} \times 100 \tag{7.41}$$

Effects of Line Length

Figure 7.26 and Figure 7.27, respectively, show the effect of line length on the maximum overvoltages and 2% values in 200 line energizations. Different symbols in the figures show different types of line and different feeding networks (10 × 4 symbols).

The overvoltages in the OHLs have clear dependence on the line length. Regardless of line types or feeding networks, the highest overvoltages are observed when the line length is 72 km.

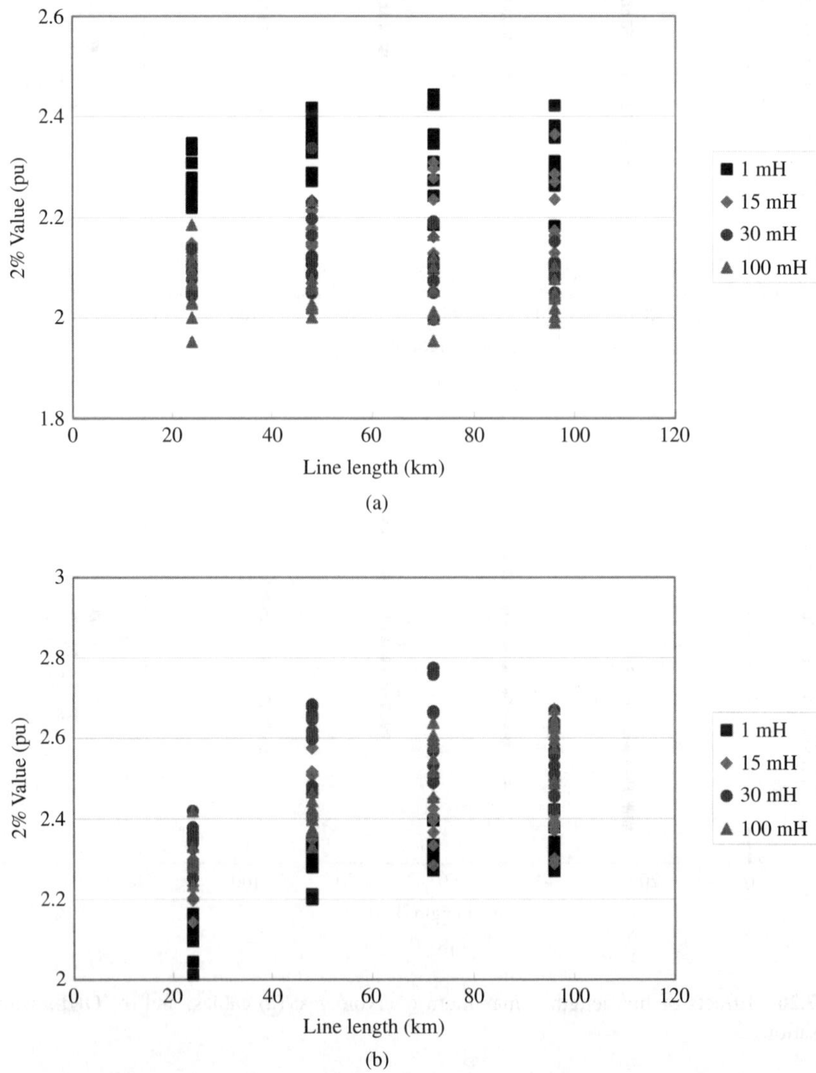

Figure 7.27 Effects of line length – 2 % values: (a) cables: and (b) OHLs with 100% compensation

The overvoltage level is raised with the increase of line length for short lines but is lowered after it reaches the highest with the line length 72 km.

In contrast, it is difficult to find any dependence of the overvoltages in cables on line length. The maximum overvoltages range around 2.0–2.6 pu regardless of line length and line types.

Effects of Feeding Network

Figure 7.28 and Figure 7.29, respectively, show the effect of a feeding network on the maximum overvoltages and 2% values in 200 line energizations. Different symbols in the figures indicate different types of the line and different line lengths (10 × 4 symbols).

It is observed in the figures that the overvoltages in the cables are dependent on a feeding network. The overvoltage level becomes lower for a larger source impedance (weaker feeding

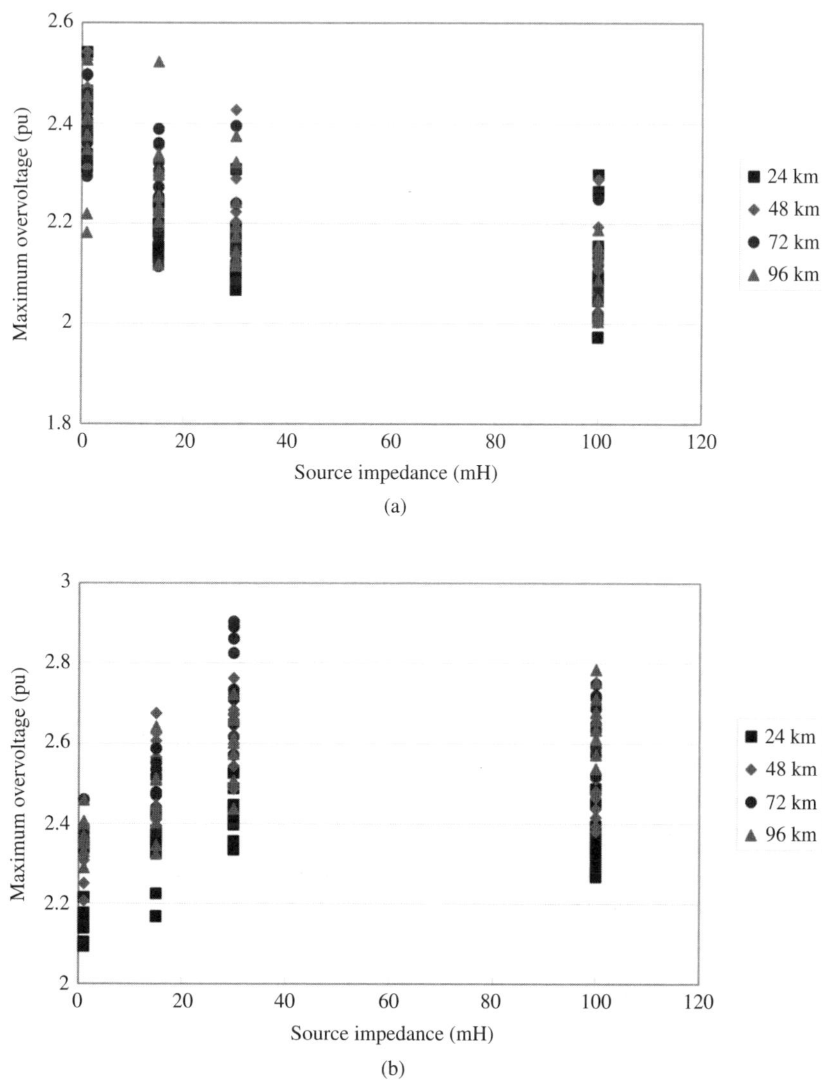

Figure 7.28 Effects of feeding network – maximum overvoltages: (a) cables; and (b) OHLs with 100% compensation

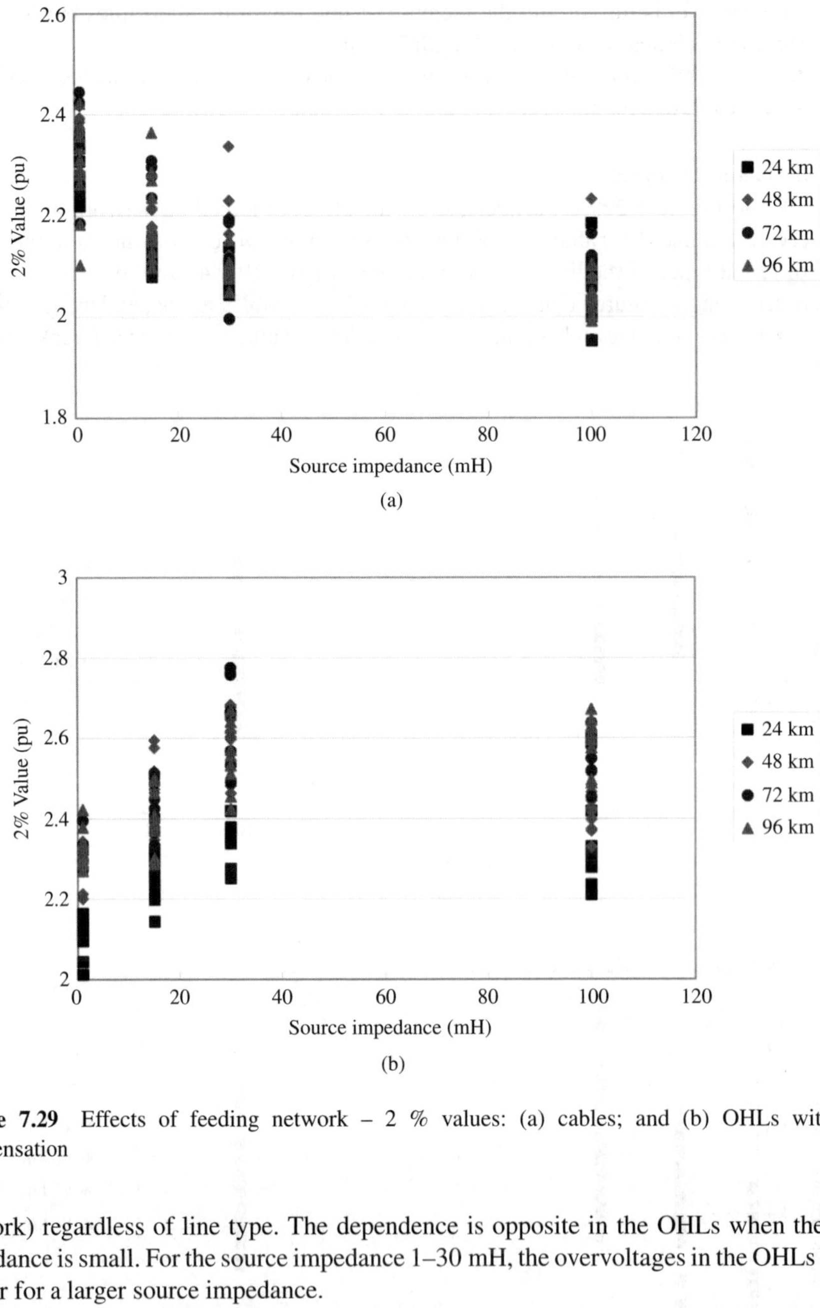

Figure 7.29 Effects of feeding network – 2 % values: (a) cables; and (b) OHLs with 100% compensation

network) regardless of line type. The dependence is opposite in the OHLs when the source impedance is small. For the source impedance 1–30 mH, the overvoltages in the OHLs become higher for a larger source impedance.

7.3.1.4 Concluding Remarks

The statistical distributions of energization overvoltages of EHV cables have been derived from repeated simulations. Based on a comparison with the statistical distributions of OHLs, it has been found out that line energization overvoltages in cables are lower than those in OHLs with respect to maximum, 2%, and mean values. In addition, standard deviations are smaller for cables.

From the statistical distributions introduced in this section, a power system with cables receives lower stress on its equipment compared with one with OHLs. As a result, if only energization overvoltages are considered, it is possible to apply lower withstand voltages to the power system with the cables. However, careful examinations are necessary as the possibility of a severe temporary overvoltage related to the cable systems has recently been reported [1, 8, 24, 26,27].

The effects of line length and a feeding network have also been studied. The energization overvoltages of cables become lower for a weaker feeding network regardless of line type. In contrast, it is difficult to find any dependence of the overvoltages in cables on line length.

7.3.2 Line Energization Overvoltages from a Complex Source

The previous section studied the cable/OHL energization from an inductive source. In this section, the energization from a complex source is studied as an example with a planned 400 kV cable line in Denmark. Figure 7.30 shows the planned 400 kV cable line from the substation ASV to KYV via TOR. The cable data is given in Table 7.1 in Section 7.2.1.2. The cable line is being planned by the Danish TSO, Energinet.dk.

According to the results of the past studies by CIGRE WG 13.02 and 13.05, it is obvious that the overvoltage level is lower when a line is energized from a complex source. It is estimated to be due to the fact that the overvoltage can propagate into adjacent lines when these lines are modeled in the simulations. In the previous section, the observed overvoltages were lower than SIWV. It is therefore expected that overvoltages observed in this section will be even lower than those in the previous section.

7.3.2.1 Study Conditions

As in the previous section, two kinds of randomness were considered in the statistical switching. The first randomness is the phase angle (point-on-wave) when the line circuit breakers receive the command to close themselves. A uniform distribution from 0 to 360° was assumed for this randomness. The second randomness is the difference of close timings between three phases. A normal distribution with standard deviation 1 ms was assumed for the randomness.

The cable line is energized from the ASV side in an assumed off-peak condition, and the energization overvoltages on the KYV side (open end terminal) were monitored. The profile of the overvoltage along with the line length was studied for the most severe case.

7.3.2.2 Simulation Results and Statistical Distributions

Figure 7.31 and Table 7.10 show the cumulative probability distributions and their characteristics. Comparing with the overvoltages caused by the line energization from lumped parameter inductive sources, it is clear that the overvoltage level is reduced by changing the lumped parameter sources to the distributed parameter source. This coincides with the results in

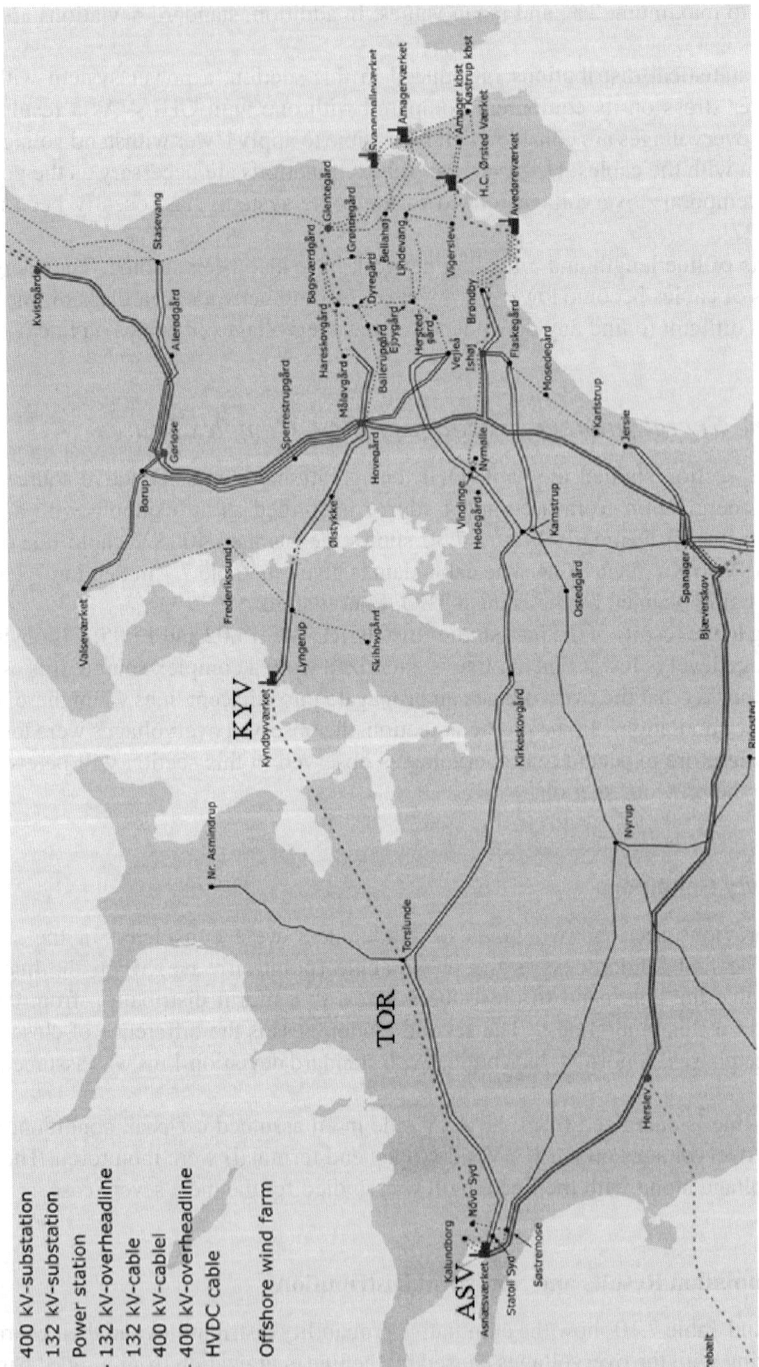

Figure 7.30 The planned 400 kV KYV–ASV cable line in the Eastern Danish grid. Courtesy of Energinet.dk

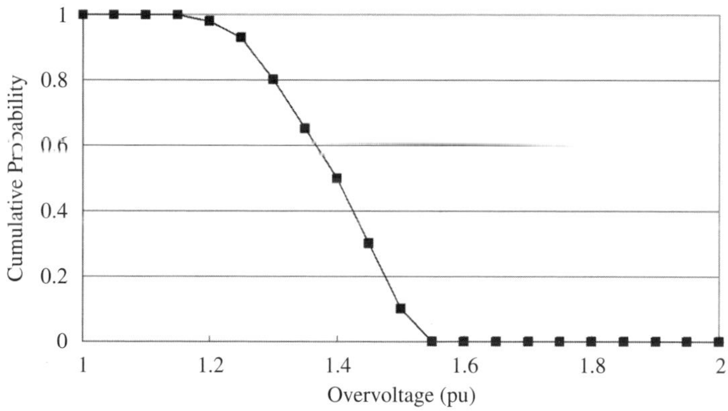

Figure 7.31 Cumulative probability distributions for the KYV–ASV cable line energization

Table 7.10 Characteristics of probability distributions for the KYV–ASV cable line energization

Maximum overvoltage (pu)	1.53
2% overvoltage (pu)	1.52
Average overvoltage (pu)	1.39
σ (pu)	0.0888
Skewness	−0.2813
Kurtosis	2.03

Reference [16]. The reduction of the overvoltage level is caused by multiple paths into which the reflected wave can propagate. Table 7.10 shows that the standard deviation is also lower with the distributed parameter source.

The most severe overvoltage was observed with the combination of the following switch timings:

Phase-a: 0.01831 s
Phase-b: 0.01972 s
Phase-c: 0.01967 s

Figure 7.32 shows the profile of the overvoltage along with the line when the most severe overvoltage was observed. The overvoltage is highest around the KYV side (open end terminal). Near the ASV side (sending end), the overvoltage is lower by approximately 0.08 pu. The overvoltage at TOR is slightly higher than both its sides since reflections occur at this point due to the separation of the sheath circuit. If the sheath circuits of the ASV side and the KYV side are connected at TOR, there is no local peak at TOR.

As the maximum overvoltage was observed at the KYV side (open end terminal), the waveforms of the energization overvoltage at the KYV side is shown in Figure 7.33. As discussed in

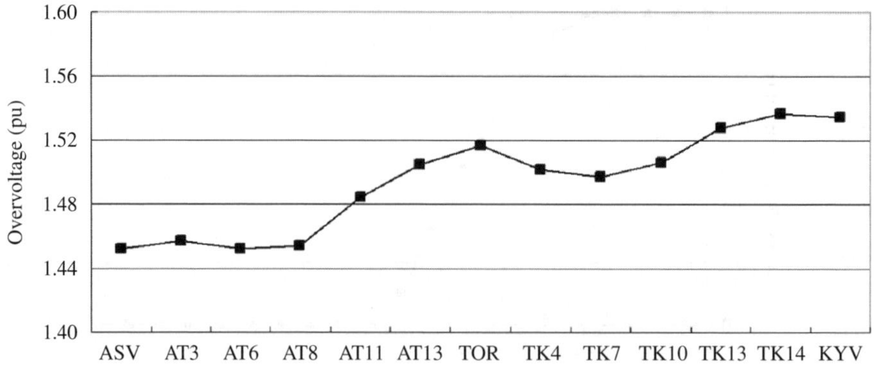

Figure 7.32 Profile of the energization overvoltage

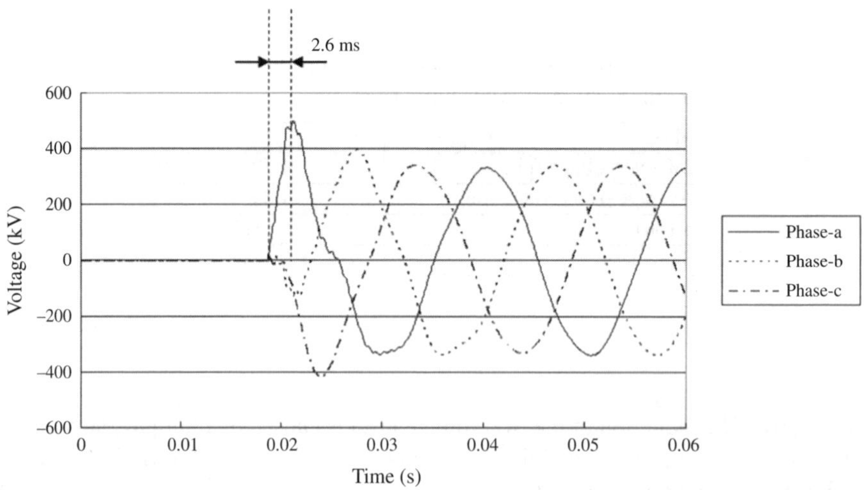

Figure 7.33 Voltage waveform of the energization overvoltage at the KYV side (open terminal)

Section 7.2, the dominant frequency contained in the energization overvoltage is very low due to the long length of the cable line and the weak source. The wavefront of the first overvoltage (2.6 ms) is approximately 10 times larger than the standard waveshape of the slow-front overvoltage (250/2500 μs). The overvoltage is highly damped and the voltage waveforms go back to the normal sinusoidal waveforms within one cycle.

7.3.3 Analysis of Statistical Distribution of Energization Overvoltages

Sections 7.3.1 and 7.3.2 have identified characteristics of statistical distributions of energization overvoltages of EHV cables, compared with those of OHLs. This section introduces causes and physical meanings of the characteristics.

Sections 7.3.1 and 7.3.2 studied 10 planned and existing EHV OHLs and cables, respectively. Four line lengths and four feeding networks (source impedances) were considered as parameters. The simulations were performed 200 times with random switches under each study condition. As a result, the study has found the following main characteristics of statistical distributions.

- Energization overvoltages for cables are lower than those for OHLs with respect to maximum, 2%, and mean values.
- Energization overvoltages for cables do not show any dependence on the line length while those for OHLs peak with the line length 72 km.
- Energization overvoltages for cables become lower for a larger source impedance while those for OHLs peak with the source impedance 30 mH.

This section focuses on these three main characteristics in the identification of causes and physical meanings. The analysis in this section shows only one type of OHLs and cables because the same characteristics and contributing factors were found in all OHL and cable types.

7.3.3.1 Analysis on the Highest Overvoltages

First, we study why energization overvoltages for cables are lower than those for OHLs with respect to maximum, 2%, and mean values. It is done by analyzing voltage waveforms of energization overvoltages. Here, we define the maximum overvoltage as the largest overvoltage found in 200 random simulations.

Figure 7.34 shows initial voltage waveforms of the energization overvoltage at the receiving end (V_r) and at the sending end (V_s) for one type of OHLs. The OHL was energized at the voltage peak of phase-a (5 ms), and the voltage waveforms of phase-a are shown in Figure 7.34. The line length and the source impedance were fixed to 48 km and 30 mH, respectively.

For the OHL, the overvoltage of the inter-phase mode travels at the speed of light. It arrives at the receiving end 0.16 ms after it departs from the sending end (1st wave):

$$48\ \text{km} \div 300\ \text{m}/\mu\text{s} = 0.16\ \text{ms}$$

The inter-phase mode is reflected back to the sending end, but V_r rises towards around 2 pu as the overvoltage of the earth-return mode arrives. The propagation time of the earth-return mode can be obtained by performing a frequency scan to the energized system or by performing Fourier transform to the voltage waveform. The result of the frequency scan shows the dominant frequency 980 Hz, which corresponds to the cycle time 1.02 ms. The result is reasonable as Figure 7.34 shows the voltage peak of the earth-return mode with an interval of 1.02 ms.

The initial voltage rise time at the receiving end is difficult to find theoretically. The receiving-end voltage starts to rise at 5.16 ms. Looking at the voltage amplitude of the earth-return mode, the voltage starts to rise from near the minimum. It suggests that it takes about 0.51 (= 1.02/2) ms to reach the maximum of the earth-return mode, but Figure 7.34 shows the earth-return mode reaches its maximum in approximately 0.32 ms after it starts to rise.

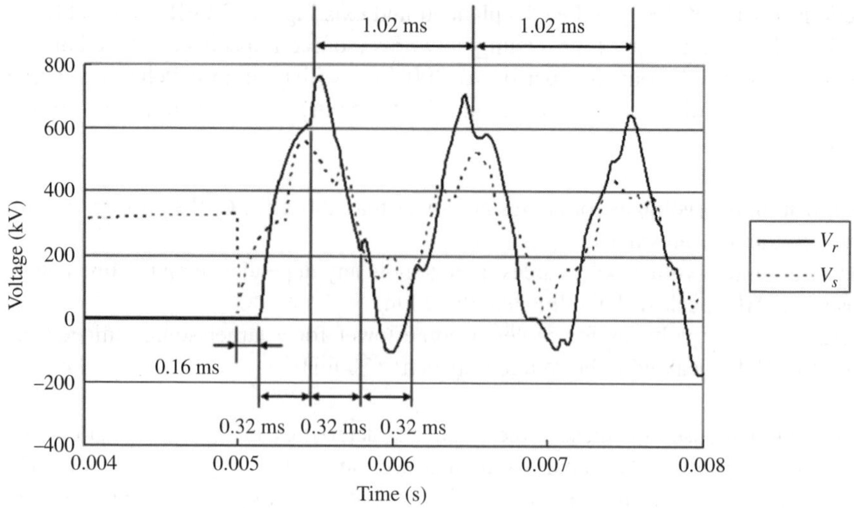

Figure 7.34 Initial voltage waveforms of the energization overvoltage for an OHL

This can be explained by the rate of voltage rise. When the voltage starts from the minimum, the rate of the voltage rise should increase in the first (1.02/4) ms and should decrease in the remaining (1.02/4) ms, assuming the receiving voltage follows the sinusoidal wave. However, in Figure 7.34, the rate of voltage rise is largest when the receiving voltage starts to rise. This makes the voltage rise time to 2 pu shorter, and the earth-return mode reaches its maximum in approximately 5.48 (= 5.16 + 0.32) ms.

This reduction of the voltage rise time affects the magnitude of the highest overvoltage since 5.48 ms is exactly when the inter-phase mode comes back to the receiving end (2nd wave). As shown in Figure 7.34, the inter-phase mode comes back to the receiving end with an interval of 0.32 ms as it travels at the speed of light. Since the peak of the earth-return mode (1st wave) and the peak of the inter-phase mode (2nd wave) reach their maximum almost at the same time around 5.48 ms, V_r goes above 2 pu and reaches 2.3 pu.

The same analysis on the initial voltage waveforms is performed for one type of underground cable. Figure 7.35 shows initial voltage waveforms of the energization overvoltage at the receiving end (V_r) and at the sending end (V_s). Study conditions, such as the switch timing (5 ms), line length (48 km), and source impedance (30 mH), are kept equal to the study on the OHL.

In the case of the underground cable, the overvoltage of the coaxial mode first arrives at the receiving end (1st wave). When a cable is not cross-bonded, the coaxial mode travels at:

$$\frac{300}{\sqrt{\varepsilon_{i1}}} = \frac{300}{\sqrt{2.852}} = 178\,\text{m}/\mu\text{s} \tag{7.42}$$

Here, ε_{i1} is the relative permittivity of the insulation corrected to account for semiconductive layers.

For a cross-bonded cable, it is known that the propagation velocity is reduced further, but it is not known how to analytically derive the reduced velocity. The propagation velocity is

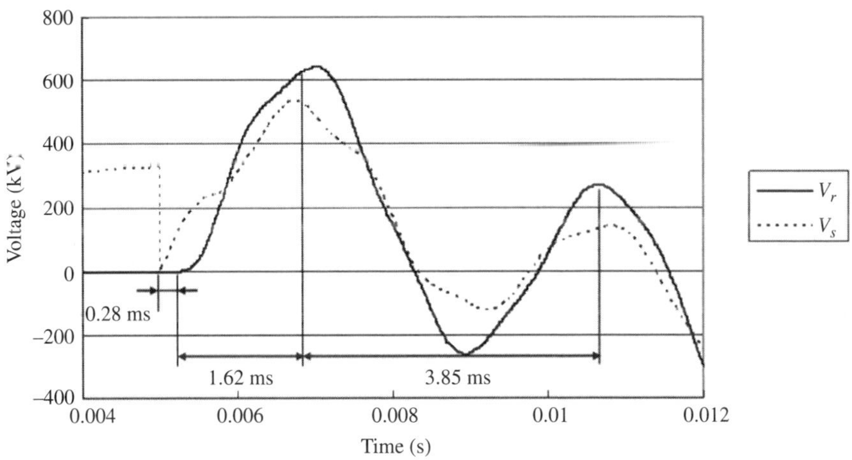

Figure 7.35 Initial voltage waveforms of the energization overvoltage for a cable (line length, 48 km; source impedance, 30 mH)

therefore derived from the propagation time. Figure 7.35 shows the coaxial mode (1st wave) arrives at the receiving end 0.28 ms after it leaves the sending end. The propagation velocity is calculated as 171 m/μs (= 48 km/0.28 ms). The reduction of 7 m/μs or 3.5% is a typical value.

Similarly to the OHL, the coaxial mode is reflected back to the sending end, but V_r rises towards around 2 pu as the overvoltage of the inter-phase mode arrives. The propagation time of the inter-phase mode can be obtained by performing a frequency scan to the energized system or by performing Fourier transform to the voltage waveform. The result of the frequency scan shows the dominant frequency 260 Hz, which corresponds to the cycle time 3.85 ms. The result is reasonable as Figure 7.35 shows the voltage peak of the inter-phase mode with an interval of 3.85 ms.

The modes found in the energization overvoltage of the OHL and the cable are summarized in Table 7.11.

Now, comparing the initial voltage waveforms of the OHL and the cable, the superimposition of two different modes found in the OHL cannot be observed clearly in the cable. There are some distortions of voltage waveforms on the inter-phase mode caused by the coaxial mode, but the distortions only have a small contribution on the highest overvoltage. This is because the overvoltage of the coaxial mode is smaller and more obtuse, compared with the inter-phase mode in the OHL.

Another factor affecting the difference in the highest overvoltage is the long cycle time of the inter-phase mode in the cable. About the time when the inter-phase mode reaches its

Table 7.11 Dominant mode and superimposed mode of the energization overvoltage

	OHL	Underground cable
Dominant mode (cycle time)	Earth-return mode (1.02 ms)	Inter-phase mode (3.85 ms)
Superimposed mode (propagation time)	Inter-phase mode (0.16 ms)	Coaxial mode (0.28 ms)

maximum (1st wave), the 4th wave of the coaxial mode arrives at the receiving end. At this time, the coaxial mode is highly damped and is difficult to observe. Even in the OHL, the inter-phase mode can have an impact on the highest overvoltage only until its 2nd wave. It is highly damped and is difficult to observe after the 3rd wave.

7.3.3.2 Analysis on the Effects of Line Length

Secondly, we study why energization overvoltages for cables do not show any dependence on the line length while those for OHLs peak with the line length 72 km.

Figure 7.36 shows initial voltage waveforms of energization overvoltages of the OHL at the receiving end (V_r). The switch timing is set to 5 ms as in the last section. The source impedance is set to 30 mH since the maximum overvoltage was observed with the source impedance 30 mH.

Figure 7.36 demonstrates that highest overvoltages are caused by the superimposition of the earth-return mode (1st wave) and the inter-phase mode (2nd wave), regardless of the line length. However, depending on the line length, the inter-phase mode (2nd wave) comes back to the receiving end at different points-on-wave of the earth-return mode (1st wave). For example, when the line length is 24 km, the inter-phase model (2nd wave) comes back to the receiving end before the earth-return mode (1st wave) reaches its maximum. As a result, the highest overvoltage becomes lower when the line length is 24 km. When the line length is 72 km, the inter-phase mode (2nd wave) comes back to the receiving end exactly when the earth-return mode (1st wave) reaches its maximum. This is why energization overvoltages of OHLs peak with the line length 72 km.

It has been shown that, for the source impedance 30 mH, the line length 72 km leads to the highest overvoltage. Considering how the highest overvoltages are caused, different line lengths should lead to the highest overvoltage for different source impedances. For example, when the source impedance is 100 mH, the cycle time of the dominant overvoltage becomes larger. Therefore, a longer line length, which requires longer propagation time, should lead to the highest overvoltage.

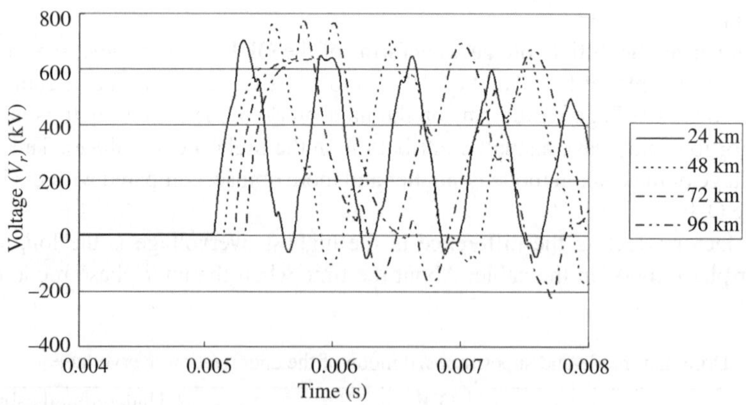

Figure 7.36 Initial voltage waveforms of the energization overvoltage for the OHL (source impedance, 30 mH)

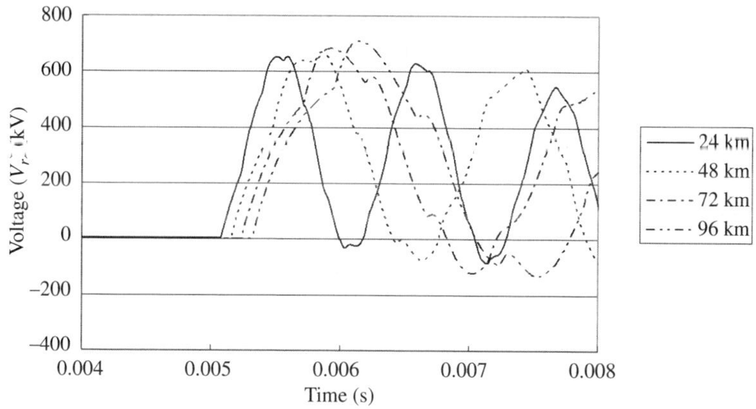

Figure 7.37 Initial voltage waveforms of the energization overvoltage for the OHL (source impedance, 100 mH)

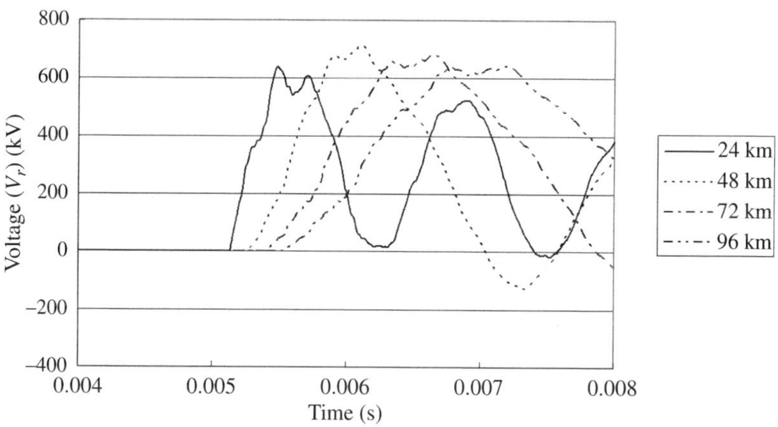

Figure 7.38 Initial voltage waveforms of the energization overvoltage for the cable (source impedance, 1 mH)

Figure 7.37 shows initial voltage waveforms of energization overvoltages of the OHL when the source impedance is 100 mH. It demonstrates that the line length 96 km leads to the highest overvoltage as expected. This was, however, not observed in Section 7.3.1.3 since the source impedance 100 mH was not the study condition that led to the maximum overvoltage. That is, the study condition 100 mH and 96 km caused a lower overvoltage than 30 mH and 96 km.

Figure 7.38 shows initial voltage waveforms of energization overvoltages of the cable at the receiving end (V_r). The source impedance is set to 1 mH since the maximum overvoltage was observed with the source impedance 1 mH. As the source impedance is set so small, not only the rate of voltage rise but also the magnitude of the overvoltage becomes larger, compared with other source impedance values.

Another significant difference from other source impedance values is how the coaxial mode is reflected at the sending end. With the source impedance 1 mH, the reflection at the sending end is negative reflection, whereas the reflection at the sending end is positive reflection with other source impedance values. This is because the voltage level at the sending end is fixed to the source voltage when the source impedance is 1 mH.

The difference of the reflection can be clearly observed in Figure 7.38. So far, the superimposition of the two modes raised the highest overvoltage. In contrast, in Figure 7.38, the negative coaxial mode (2nd wave) makes dents on the peaks of the inter-phase mode (1st wave). The inter-phase mode (1st wave) is larger for shorter line lengths, but the negative coaxial mode (2nd wave) is also larger for shorter line lengths. As a result, the highest overvoltage is observed in the case of the line length 48 km, and the effect of the line length on the highest overvoltage is reduced.

Figure 7.39 and Figure 7.40 compare initial voltage waveforms of energization overvoltages of the cable for source impedances 30 mH and 100 mH, respectively.

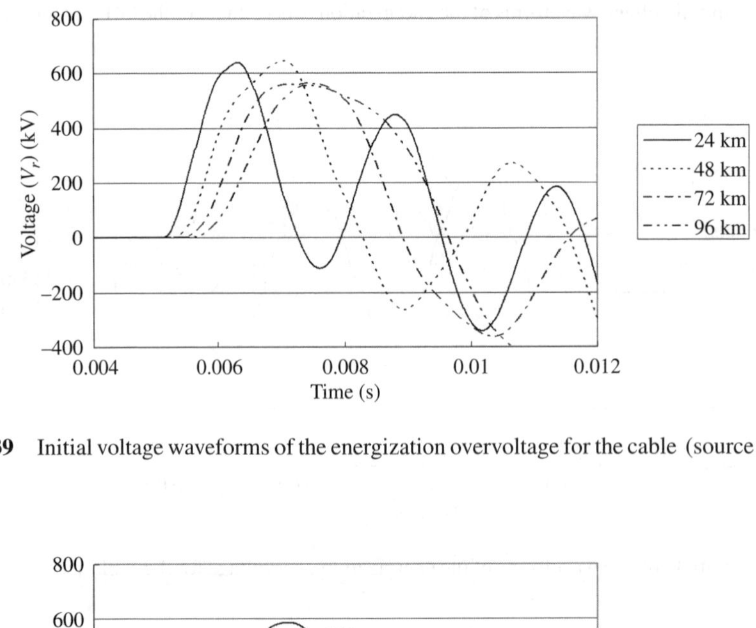

Figure 7.39 Initial voltage waveforms of the energization overvoltage for the cable (source impedance, 30 mH)

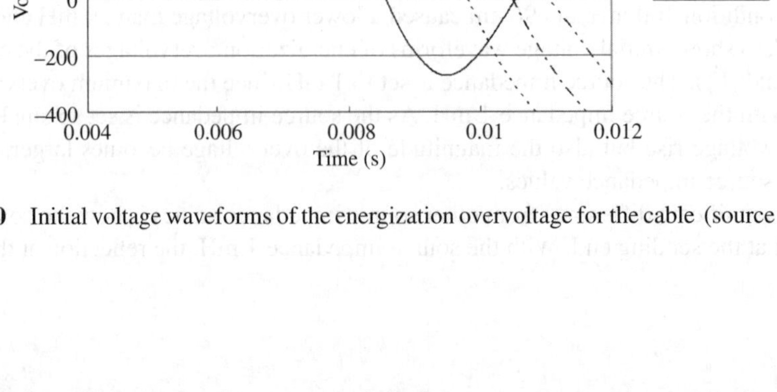

Figure 7.40 Initial voltage waveforms of the energization overvoltage for the cable (source impedance, 100 mH)

With the source impedances 30 mH and 100 mH, shorter line lengths lead to higher overvoltages as expected. This characteristic was, however, not observed in Section 7.3.1.3 since the source impedance 30 mH or 100 mH was not the study condition that led to the maximum overvoltage. The negative coaxial mode is not observed in these cases.

As we found in Section 7.3.3.1, the cycle time of the dominant mode of the cable is long (3.85 ms) with the line length 48 km and the source impedance 30 mH. It becomes even longer for longer line lengths or larger source impedances. As such, the horizontal axis of Figure 7.39 and Figure 7.40 is extended from 8 to 12 ms. The cable is energized at the voltage peak of phase-a (5 ms), but the source voltage goes down to zero at 10 ms. This also contributes to lower overvoltages in the case of longer line lengths or larger source impedances.

7.3.3.3 Analysis on the Effects of Feeding Network

Thirdly, we study why energization overvoltages for cables become lower for a larger source impedance while those for OHLs peak with the source impedance 30 mH.

Figure 7.41 shows initial voltage waveforms of energization overvoltages of the OHL at the receiving end (V_r). The switch timing is set to 5 ms as in the previous sections. The line length is set to 72 km since the maximum overvoltage was observed with the source impedance 72 km.

As discussed in Section 7.3.3.2, the highest overvoltage of the OHL is determined by the timing when the earth-return mode reaches its maximum and the inter-phase mode arrives at the receiving end. Figure 7.41 shows that source impedances 15 mH and 30 mH have the timing to cause high overvoltages. When the source impedance is 1 mH, the overvoltage of the inter-phase mode is smaller, which results in the lower overvoltage.

Figure 7.42 and Figure 7.43 show initial voltage waveforms of energization overvoltages of the cable for line lengths 24 km and 48 km, respectively. These two lengths were selected as we learned, in Section 7.3.3.2, that shorter line lengths lead to higher overvoltages except when the source impedance is 1 mH.

Figure 7.42 demonstrates that smaller source impedances lead to larger overvoltages as we found in Section 7.3.1.3. This is not true only when the source impedance is 1 mH.

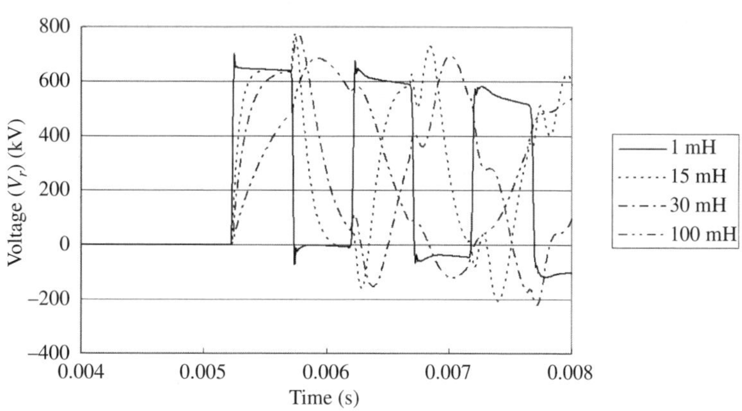

Figure 7.41 Initial voltage waveforms of the energization overvoltage for the OHL (line length, 72 km)

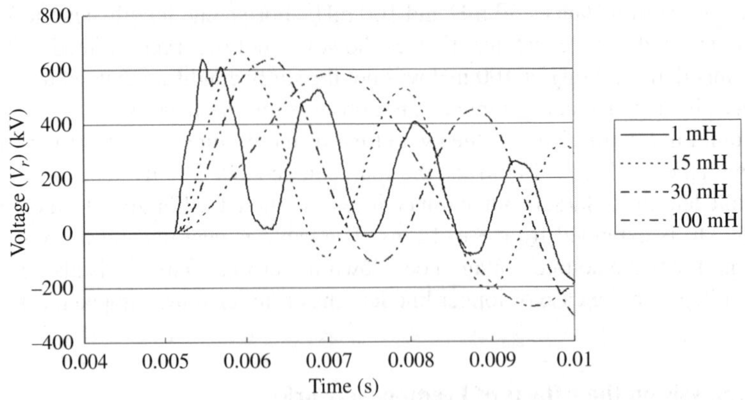

Figure 7.42 Initial voltage waveforms of the energization overvoltage for the cable (line length, 24 km)

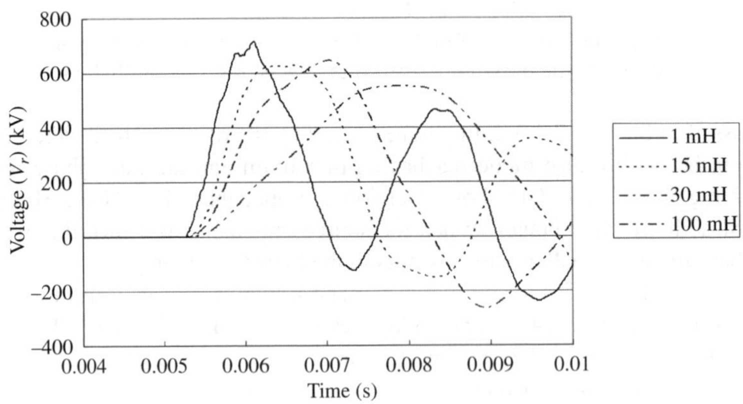

Figure 7.43 Initial voltage waveforms of the energization overvoltage for the cable (line length, 48 km)

When the source impedance is 1 mH, the highest overvoltage is caused when the line length is 48 km as we learned in the last section. Looking at the source impedance 1 mH in Figure 7.43 and other source impedances in Figure 7.42, it is clear that smaller source impedances lead to larger overvoltages.

It is reasonable to have larger overvoltages with smaller source impedances as the rate of voltage rise becomes larger for smaller source impedances. The larger rate of voltage rise is apparent as the smaller source impedances lead to a higher dominant frequency contained in the energization overvoltage. For larger source impedances, due to their slow rates of voltage rise, the fundamental component already starts to decrease from the voltage peak by the time when the inter-phase mode reaches its maximum.

7.3.3.4 Concluding Remarks

This section analyzed the characteristics of the energization overvoltage found in Sections 7.3.1 and 7.3.2. The analysis has found that these characteristics are not caused by random switching by accident, rather there are contributing factors and physical meanings behind the characteristics.

In summary, the analysis has identified contributing factors and has explained their physical meanings as follows:

- Energization overvoltages for cables are lower than those for OHLs because the overvoltage of the coaxial mode in cables is smaller and more obtuse, compared with the inter-phase mode in OHLs. The long cycle time of the inter-phase mode in cables also affects this characteristic.
- Energization overvoltages for cables do not show any dependence on the line length because the maximum overvoltages are observed with the source impedance 1 mH. The negative coaxial mode makes dents on the peaks of the maximum overvoltage when the source impedance is 1 mH. With the other source impedances, shorter line lengths lead to higher overvoltages.
- Energization overvoltages for cables become lower for larger source impedances, which is reasonable from the theoretical analysis.
- Highest energization overvoltages for OHLs are caused by the superimposition of the inter-phase mode on the earth-return mode. The timing of the superimposition determines the magnitude of the highest overvoltage, and the line length and the source impedance affect this timing. The energization overvoltage for OHLs peaks with the line length 72 km and the source impedance 30 mH because the combination provides the best timing for the superimposition within the studied conditions.

The obtained statistical distribution for cables, together with the contributing factors and physical meanings of the characteristics identified in this section, suggests a possible application of lower insulation levels to cable systems compared with OHLs. However, it may present a future challenge to define insulation levels for temporary overvoltages since a possibility of severe temporary overvoltages are reported in relation to cable systems [1, 8, 24, 26, 27].

7.4 Leading Current Interruption

When the leading current is interrupted at current zero, it occurs at a voltage peak assuming that the current waveform is leading the voltage waveform by 90°. After the interruption, the voltage on the source side of the circuit breaker changes according to the system voltage, whereas the voltage on the other side is fixed at the peak voltage E as shown in Figure 7.44. The most severe overvoltage occurs when a restrike occurs after half a cycle when the voltage on the source side becomes $-E$. As the voltage difference between the source side and the other side is $2E$, the overvoltage can go as high as $-3E$. The restrike can be repeated to cause a very severe overvoltage.

Considering the severe overvoltage that can be caused by the leading current interruption, preferred values of the leading current interruption capability of circuit breakers are defined in IEC 62271-100 [28]. Table 7.12 shows the leading current interruption capability of EHV line breakers according to IEC 62271-100.

When the charging capacity of a long EHV cable line is not compensated by shunt reactors directly connected to the line, the leading current interruption capability requires careful attention [29]. Considering typical capacitance 0.2 μF/km, the maximum line length for 400 kV

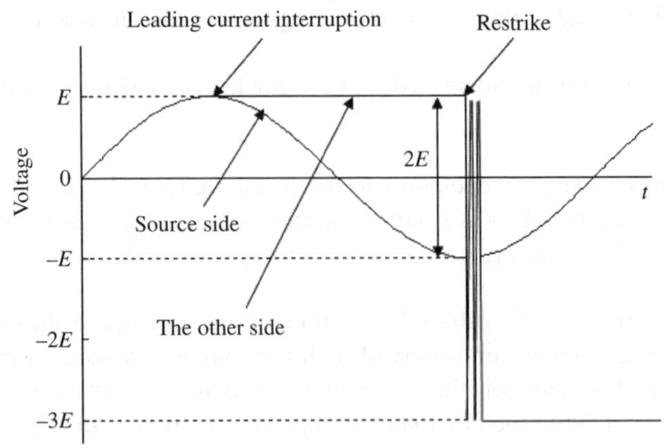

Figure 7.44 Overvoltage caused by leading current interruption and restrike

Table 7.12 Leading current interruption capability according to IEC 62271-100

Rated voltage (V)	Rated capacitive switching currents[a] (cable) (A)
420	400
550	500

[a]Preferred values, voltage factor: 1.4 pu.

cable line is limited approximately below 26 km. Here, it is assumed that the leading current is interrupted by one end, and the other end is open before the interruption.

Usually, the charging capacity of a long EHV cable line is compensated by shunt reactors directly connected to the line. When the compensation rate is high enough, the leading current interruption capability is not a concern. If the sequential switching is applied to a cable line as a countermeasure to the zero-missing phenomenon, however, the tripping of shunt reactors makes the compensation rate lower, sometimes even to 0%. This is the only occasion that requires careful attention and is discussed in more detail in Section 7.5.

7.5 Zero-missing Phenomenon

7.5.1 Zero-missing Phenomenon and Countermeasures

A d.c. offset current (zero-missing current) appears when an EHV underground cable is energized with its shunt reactors [30–32]. In this condition, an a.c. component of a charging current has the opposite phase angle to the a.c. component of a current flowing into the shunt reactors. If the compensation rate of the cable is 100%, the summation of these a.c. components becomes zero, and only the d.c. component remains. Since the d.c. component decays slowly with time, it can take more than 1 s, depending on the compensation rate, before a current that flows

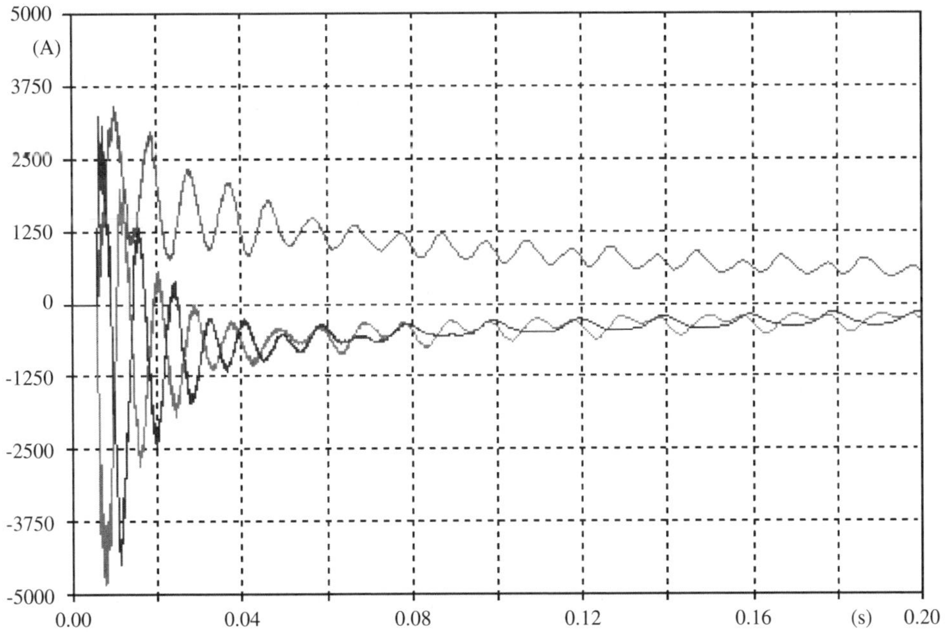

Figure 7.45 Zero-missing current in the underground cable energization

through the circuit breaker crosses the zero point. When the circuit breaker is to be operated to interrupt this current without the zero crossing, it may lead to a failure of this circuit breaker.

Figure 7.45 shows an example of current waveforms when an EHV underground cable is energized with its shunt reactors. It can be seen that the a.c. component of the energization current is very small since the compensation rate is close to 100%. The simulation was run for 0.2 s, but the energization current did not cross the zero point during this duration.

Since the zero-missing phenomenon is caused by a d.c. component of an energization current, it is most severe when the cable is energized when the maximum d.c. component is contained in the current. In order to realize this condition in phase-a, the cable was energized at the voltage zero point of phase-a in Figure 7.45.

Table 7.13 summarizes countermeasures of the zero-missing phenomenon. All of these countermeasures, except for Countermeasure (4), have already been applied to cable lines in operation. Countermeasure (1), which in particular has a number of application records to long EHV cable lines, is discussed in detail later in this section.

Countermeasure (2) is also a proven technology and more often applied to the transformer energization in order to minimize inrush current. When it is adopted as a countermeasure to the zero-missing phenomenon, a cable line will be energized at the voltage peak of each phase, which will lead to higher energization overvoltages on average. This requires careful attention even though cable energization overvoltages are lower than OHL energization overvoltages [11].

Countermeasure (3) can be implemented rather easier as long as a cable line is installed together with single-phase circuit breakers and current differential relays. For this reason, this countermeasure is more suited to EHV cable lines than HV cable lines.

Table 7.13 Countermeasures of zero-missing phenomenon

	Countermeasure	Notes
(1)	Sequential switching	• Requires higher leading current interruption capability • Requires single-phase circuit breaker and current differential relay
(2)	Point-on-wave switching (synchronized switching)	• May cause higher switching overvoltage • Requires single-phase circuit breaker
(3)	Delayed opening of healthy phases	• Requires single-phase circuit breaker and current differential relay • May not be possible to apply near a generator
(4)	Breaker with pre-insertion resistor	• May be necessary to develop a new circuit breaker (expensive)
(5)	Additional series resistance in shunt reactor for energization	• Requires special control to bypass series resistance after energization
(6)	Energize shunt reactor after the cable	• Causes higher steady-state overvoltage and voltage step

In this countermeasure, the faulted phase is opened instantly, but healthy phases are opened about 10 s later when the d.c. component is decayed enough. When this countermeasure is applied near a generator and especially when the cable line offers a radial path to the generator, it is necessary to evaluate the negative-sequence current capability of the generator as this countermeasure causes an unbalanced operation for a prolonged duration.

Countermeasure (4) is applicable in theory [33, 34], but the cost of the breaker may prohibit its application depending on the required size of pre-insertion resistors. Line breakers with pre-insertion resistors were, in the first place, proposed for OHLs in order to lower the switching overvoltage. However, it could not become mainstream mainly due to its cost. A breakthrough, especially on heat dissipation, would be necessary for it to be applied as a countermeasure.

In Countermeasure (5), a resistance is connected in series to shunt reactors when a cable line is energized. The resistance needs to be sized so that the d.c. component decays fast enough. After the cable line is energized, the resistance is bypassed in order to reduce losses. Considering the additional cost for the resistance, this countermeasure is more suited to HV cable lines than EHV cable lines.

Countermeasure (6) cannot always be applied; especially, it is difficult to apply it to long EHV cables due to the steady-state overvoltage.

7.5.2 Sequential Switching

When a cable line is switched onto a fault or when a fault occurs just after the energization before the zero-missing phenomenon disappears, the line breaker of healthy phases cannot interrupt the current, and it may result in a failure of the line breaker. Figure 7.46 shows an example of a zero-missing current with a single-line-to-ground (SLG) fault in phase-b.

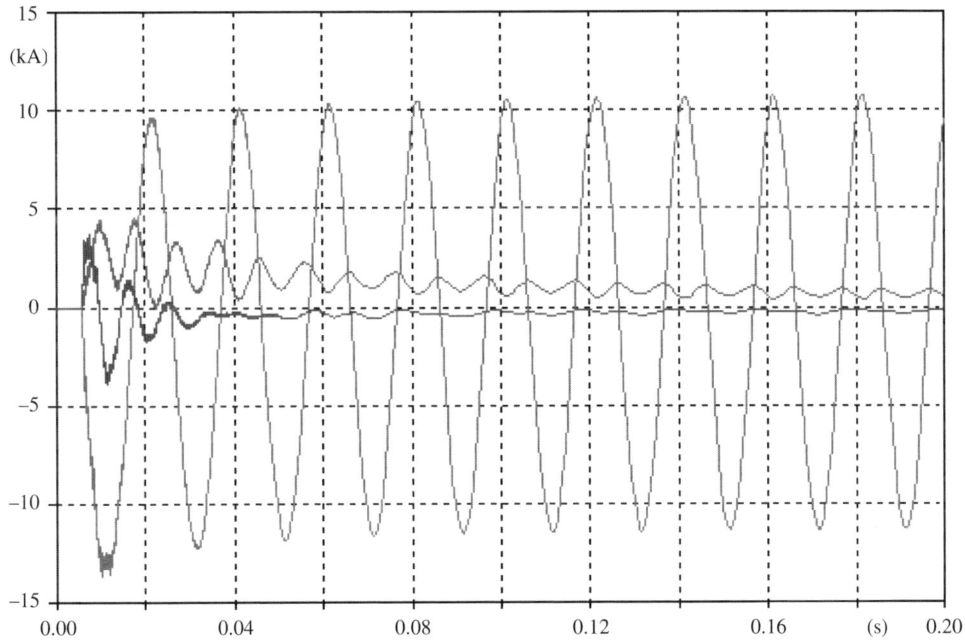

Figure 7.46 Zero-missing phenomenon with a SLG fault

In Figure 7.46, a SLG fault is applied in phase-b of an EHV underground cable before energization. The figure shows that the current in the faulted phase contains a large a.c. component due to the fault current. The line breaker of the faulted phase can interrupt this current since this fault current superimposed on the cable energization current crosses the zero point due to its large a.c. component.

Figure 7.47 shows the time sequence of sequential switching when the cable line is energized from Substation A. The line breaker of phase-b is opened, in Step 1, 60 ms after the fault, and the fault is cleared by this circuit breaker tripping. Since the fault is already cleared by the opening of the phase-b line breaker, there is no problem to take some time before opening line breakers of other healthy phases.

In Step 2, shunt reactors are tripped before the line breakers of healthy phases. It is necessary to trip shunt reactors of only healthy phases. At this time, it is not necessary and not recommended to trip shunt reactors of the faulted phase since the current through shunt-reactor breakers of the faulted phase does not cross the zero point.

It is now possible to open the line breakers of the healthy phases in Step 3. Figure 7.48 shows that the current in healthy phases contains the a.c. component and crosses the zero point after tripping the shunt reactors.

Since shunt reactors are tripped before line breakers of healthy phases, it is required to confirm that the leading current in the line breakers does not exceed its breaking capability in Table 7.12. For a very long cable, it is highly recommended to trip half of the shunt reactors before the line breakers of healthy phases in order to reduce the leading current in the line breakers of healthy phases.

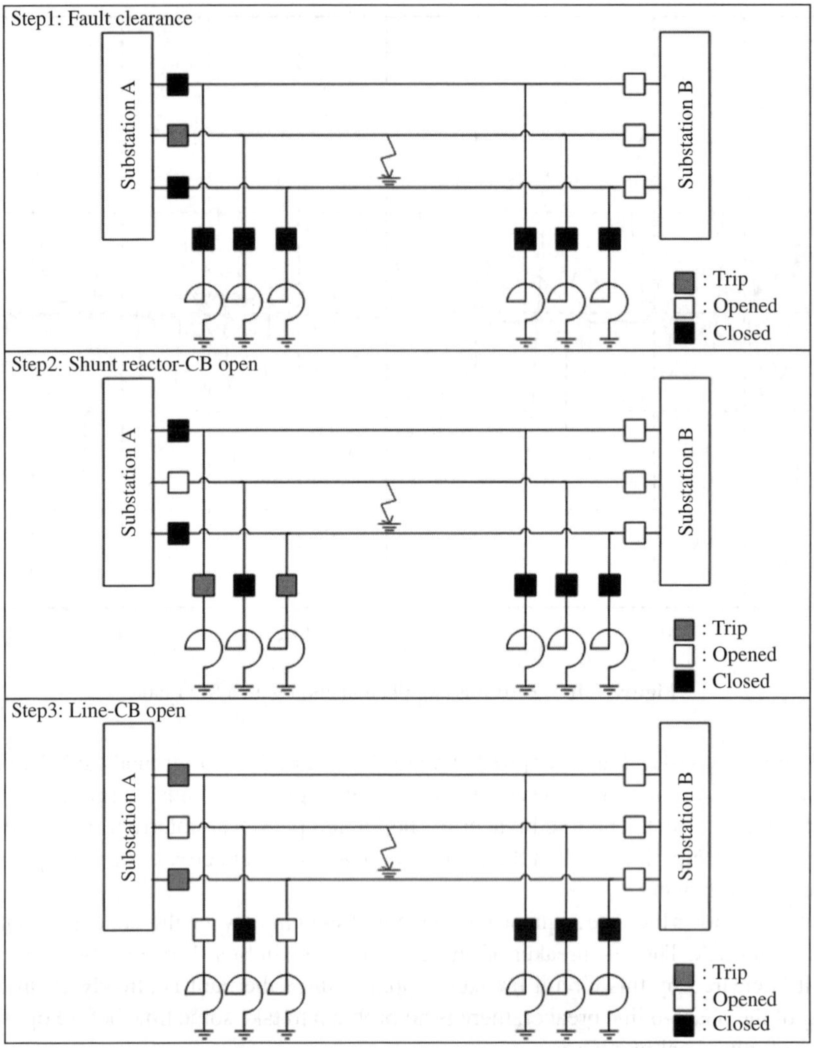

Figure 7.47 Time sequence of sequential switching

7.6 Cable Discharge

If a shunt reactor is directly connected to a cable line, the cable line is discharged through the shunt reactor when disconnected from the network. In this case, the time constant of the discharge process is determined by the quality factor (Q factor) of the shunt reactor. Since the Q factor ranges around 500 for EHV shunt reactors, the time constant of the discharge process is around 8 min.

If a cable line is disconnected from the network and energized again within a couple of minutes, there remains a residual charge in the cable line, and the residual charge can be highly dependent on the time separation between the disconnection and the re-energization.

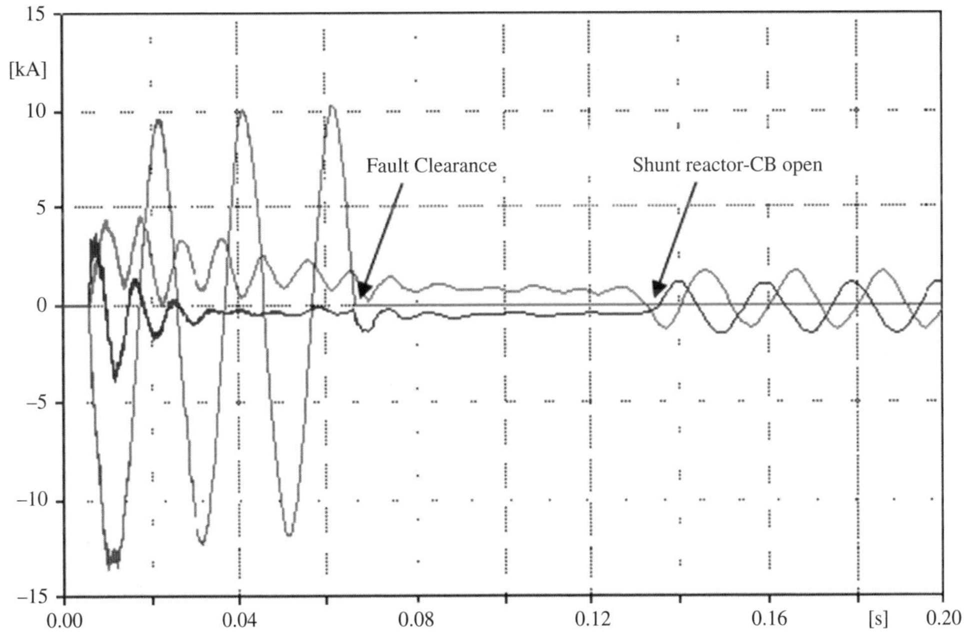

Figure 7.48 Zero-missing phenomenon with sequential switching

Under this condition, the re-energization overvoltage can exceed the SIWV, defined in IEC 60071-1 [35], when the residual voltage has an opposite sign to the source voltage at the time of re-energization.

This is usually an issue for OHLs since auto-reclose is applied to OHLs. For cable lines, it is not common to apply auto-reclose. If it is applied to the cable lines, they may experience higher overvoltages because of their higher residual voltage. System operators are required to know that they need to wait for about 10 min (maybe more to be cautious) before re-energizing a cable line though it would not be a typical operation to re-energize a cable line.

If a shunt reactor is not directly connected to the cable line, the cable line is discharged through inductive voltage transformers (VTs). In this case, the discharge process will be completed within several milliseconds. The inductive VTs need to have enough discharge capability if a cable line can be operated without a shunt reactor or all the shunt reactors are tripped by sequential switching. It takes several hours for the inductive VTs to dissipate heat after dissipating the cable charge. If the inductive VTs are required to dissipate the cable charge twice within several hours, the required discharge capability will be doubled.

References

[1] Ohno, T. (2012) Dynamic study on the 400 kV 60 km Kyndbyværket–Asnæsværket line. PhD thesis. Aalborg University.

[2] Nagaoka, N. and Ametani, A. (1983) Transient calculations on crossbonded cables. *IEEE Transactions on Power Apparatus and Systems*, **102** (4), 779–87.

[3] Ametani, A., Miyamoto, Y., and Nagaoka, N. (2003) An investigation on a wave propagation characteristic on a crossbonded cable. *IEEJ Transactions on PE*, **123-B** (3), 395–401.
[4] Ohno, T., Bak, C.L., Ametani, A. *et al.* (2012) Derivation of theoretical formulas of the frequency component contained in the overvoltage related to long EHV cables. *IEEE Transactions on Power Delivery*, **27** (2), 866–76.
[5] Hall, S.H. and Heck, H.L. (2009) *Advanced Signal Integrity for High-Speed Digital Designs*, Wiley-IEEE Press.
[6] Ametani, A. (1990) *Distributed-Parameter Circuit Theory*, Corona Publishing. Co. (in Japanese).
[7] Ametani, A. (1983) A study of cable transient calculations, Part II. Transient calculations. *Science and Engineering Review of Doshisha University*, **24** (2), 119–27.
[8] Karasaki, T., Goto, T., and Ametani, A. (1995) An abnormal overvoltage due to load rejection on EHV underground transmission lines. International Conference on Power System Transients (IPST), Lisbon, Portugal.
[9] Momose, N., Suzuki, H., Tsuchiya, S., and Watanabe, T. (1998) Planning and development of 500 kV underground transmission system in Tokyo metropolitan area. *CIGRE Session*, 37–202.
[10] Ohno, T., Bak, C.L., Ametani, A. *et al.* (2013) Statistical distribution of energization overvoltages of EHV cables. *IEEE Transactions on Power Delivery*, **28** (3), 1423–32.
[11] Ohno, T., Ametani, A., Bak, C.L. *et al.* (2013) Analysis of the statistical distribution of energization overvoltages of EHV cables and OHLs. International Conference on Power Systems Transients (IPST), Vancouver, BC, Canada.
[12] Hileman, A.R. (1999) *Insulation Coordination for Power Systems*, CRC Press.
[13] CIGRE Working Group 13.02 (1973) Switching Overvoltages in EHV and UHV Systems with Special Reference to Closing and Reclosing Transmission Lines. *Electra No.* **30**, 70–122.
[14] CIGRE Working Group 13.05 (1971) The Calculation of Switching Surges. *Electra No.* **19**, 67–78.
[15] CIGRE Working Group 13.05 (1974) The Calculation of Switching Surges – II. Network Representation for Energization and Re-energization Studies on Lines Fed by an Inductive Source. *Electra No.* **32**, 17–42.
[16] CIGRE Working Group 13.05 (1979) The Calculation of Switching Surges – III. Transmission Line Representation for Energization and Re-energization Studies with Complex Feeding Networks. *Electra No.* **62**, 45–78.
[17] Sato, J., Tsuchiya, Y., Arai, A. *et al.* (2001) Technical development and view of underground and overhead transmission line. *SWCC Showa Review*, **51** (1), 59–74 (in Japanese).
[18] Elgh, L., Jacobsen, C.T., Bjurstrom, B. *et al.* (1974) The 420 kV AC submarine cable connection Denmark-Sweden. *CIGRE Session*, 21–02.
[19] Gubinsky, I.A., Khanukov, M.G., Kusnetsov, L.A. *et al.* (1976) 500 kV cable lines for hydro power stations. *CIGRE Session*, 21–03.
[20] Hahn, W., Muller, U., and Peschke, E. F. (1976) The first 380 kV bulk power transmission in Germany. *CIGRE Session*, 21–08.
[21] Gualtieri, G., Lanfranconi, G.M., and Cavalli, W. (1973) 330-kV Oil-filled cables laid in 1600-ft vertical shaft at Kafue Gorge Hydroelectric Plant. *IEEE Transactions on Power Apparatus and Systems*, **92** (6), 1992–8.
[22] CIGRE WG B1.07 (2007) Statistics of AC Underground Cables in Power Networks. *CIGRE Technical Brochure* **338**.
[23] CIGRE WG B1.10 (2009) Update of Service Experience of HV Underground and Cable Systems. *CIGRE Technical Brochure* 379.
[24] Tokyo Electric Power Company (2009) Assessment of the Technical Issues relating to Significant Amounts of EHV Underground Cable in the All-island Electricity Transmission System. http://www.eirgrid.com/media/Tepco%20Report.pdf (accessed 27 November 2013).
[25] Massey, F.J. (1951) The Kolmogorov-Smirnov test for goodness of fit. *Journal of the American Statistical Association*, **46** (253), 68–78.
[26] Rebolini, M., Colla, L., and Iliceto, F. (2008) 400 kV AC new submarine cable links between Sicily and the Italian mainland. Outline of project and special electrical studies. *CIGRE Session*, C4–116.
[27] Colla, L., Lauria, S., and Gatta, F.M. (2007) Temporary overvoltages due to harmonic resonance in long EHV cables. International Conference on Power System Transients (IPST), Paris, France.
[28] IEC 62271-100 ed2.0 (2008) High-voltage Switchgear and Controlgear - Part 100: Alternating Current Circuit-breakers.
[29] Ohno, T. (2010) Operation and Protection of HV Cable Systems in TEPCO. http://www.ieee.ch/assets/Uploads/pes/downloads/1004/10042ohnoexperiencetepco.pdf (accessed 27 November 2013).
[30] Kulicke, B. and Schramm, H.H. (1980) Clearance of short-circuits with delayed current zeros in the Itaipu 550 kV-substation. *IEEE Transactions on Power Apparatus and Systems*, **99** (4), 1406–14.
[31] Michigami, T., Imai, S., and Takahashi, O. (1997) Theoretical background for zero-miss phenomenon in the cable network and field measurements. IEEJ General Meeting, Kyoto, Japan (in Japanese).

[32] Hamada, H., Nakata, Y., and Maekawa, T. (2001) Measurement of delayed current zeros phenomena in 500 kV cable system. IEEJ General Meeting, Nagoya, Japan (in Japanese).
[33] Silva, F.F., Bak, C.L., Gudmundsdottir, U.S. *et al.* (2009) Use of a preinsertion resistor to minimize zero-missing phenomenon and switching overvoltages. IEEE PES General Meeting, Calgary, AB, Canada.
[34] Silva, F.F. (2011) Analysis and simulation of electromagnetic transients in HVAC cable transmission grids. PhD thesis. Aalborg University.
[35] IEC 60071-1 ed8.0 (2006) Insulation Co-ordination - Part 1: Definitions, Principles and Rules.

8

Cable Transient in Distributed Generation System

Naoto Nagaoka
Graduate School of Electrical Engineering, Doshisha University, Kyoto, Japan

Some examples of transient simulations of cable systems in distributed generation systems by means of the electromagnetic transients program (EMTP) [1] will be given in this chapter.

8.1 Transient Simulation of Wind Farm

8.1.1 Circuit Diagram

Figure 8.1 illustrates a model circuit of a wind farm. Each feeder involves five wind generators. The generator of capacity 3 MW and its step-up transformer are assumed to be disconnected to focus on the transient characteristic of the cable system in the simulation. Each feeder is connected to a 22 kV bus in Substation L (s/s L). The buses have a bus-tie circuit breaker (CB) (SWC). The capacity of each transformer installed in Substation L is 18 MVA (P_{Tr}). When the bus-tie CB is closed, the condition is identical to that where the feeders are connected to the 66 kV bus via a single transformer of capacity 36 MVA. The wind farm is connected with a grid at Substation K (s/s K). The backward impedance of the grid is represented by an *RL* series circuit. Node names used in the EMTP simulation are composed of the characters in dashed boxes in Figure 8.1 followed by a character indicating phase-A, B, or C.

Table 8.1 shows the cable size and length used in this investigation. The cross section of the cable core decreases toward to the end of the feeder. The cables are insulated by a cross-linked polyethylene (XPLE) whose relative permittivity ε_i is 2.3. The metallic sheathes are bundled and grounded at the joints through a resistor of 0.5 Ω for 66 kV and 1 Ω for 22 kV cable, respectively.

Cable System Transients: Theory, Modeling and Simulation, First Edition.
Akihiro Ametani, Teruo Ohno and Naoto Nagaoka.

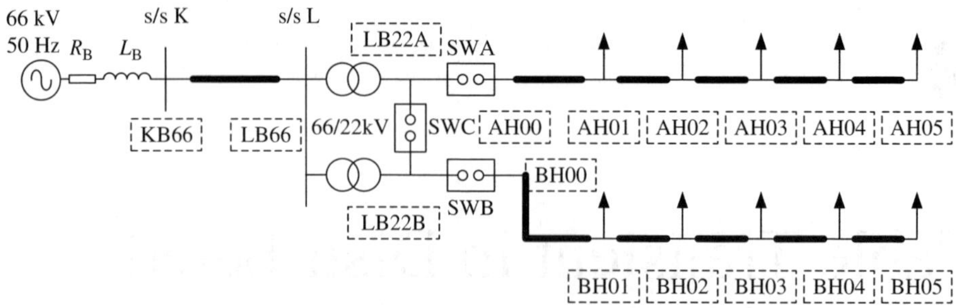

Figure 8.1 Circuit diagram of a wind farm

Table 8.1 Cable length

	Cable	Size (mm^2)	Length (km)
From	To		
KB66	KB66	600	10
AH00	AH01	250	2
AH01	AH02	150	1
AH02	AH03	150	1
AH03	AH04	60	1
AH04	AH05	60	1
BH00	BH01	250	4
BH01	BH02	150	1
BH02	BH03	150	1
BH03	BH04	60	1
BH04	BH05	60	1

8.1.2 *Cable Model and Dominant Frequency*

A choice of a cable model is important for a transient simulation of a cable system. The cable transient in a distributed generation system is determined by numerous elements because the system is generally composed of a number of various elements. To clarify the accuracy of the cable model, "pre-calculation" is important by a simplified circuit expressing the characteristic of the original cable circuit as closely as possible. Figure 8.2 shows a simplified circuit for the pre-calculation. A pulse current with 25 μs width [$i_p(t)$] is injected to a cable core (phase-A). The pulse width is determined from the total length of the cable (14 km) and the propagation velocity of the coaxial mode v_c defined by the relative permittivity ε_i(= 2.3) of XPLE. The traveling time within the cable becomes:

$$\tau = \frac{l}{v_c} = \frac{l}{c_0/\sqrt{\varepsilon_i}} = \frac{14 \times 10^3 \text{ (m)}}{300/\sqrt{2.3} \text{ [m/μs]}} \approx \frac{14 \times 10^3 \text{ (m)}}{200 \text{ [m/μs]}} = 70 \text{ (μs)} \tag{8.1}$$

If the pulse width is smaller than the traveling time, the reflections within the cable system will be easily recognized. The amplitude of the current is assumed to be 1 kA in this simulation.

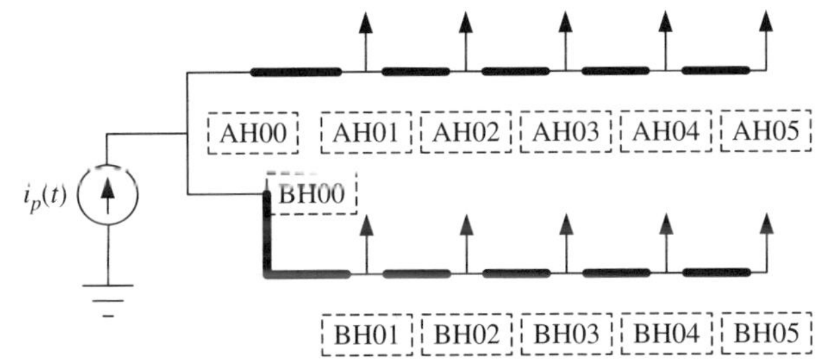

Figure 8.2 Circuit diagram for a pre-calculation

The cable models installed into the EMTP [1] are divided into three: pi equivalent circuit: constant-parameter line models (transposed and untransposed line); and frequency-dependent line models (Semlyen's, Marti's and Noda's models) [2–8].

The pi-equivalent circuit is the most simple and stable model of a cable, although the characteristic of traveling waves cannot be perfectly expressed. It is preferred that a cable is expressed as a distributed-parameter line model for a transient simulation.

In general, a transposed distributed-parameter line model cannot be applicable to represent a cable system, because the definition of the "transposed" in the EMTP is for a single-circuit overhead line, that is, a "perfectly transposed line." Even if the cables are perfectly twisted, a core and a metallic sheath of a cable cannot be transposed. An untransposed distributed line model (Dommel model) has to be used for expressing a cable by a constant-parameter line model.

The above two models (the pi and the untransposed distributed line model) neglect the frequency-dependent effects due to the cable conductors and the earth-return impedance. To improve the accuracy, Marti's, Semlyen's and Noda's line models have been installed into the EMTP. Because the Marti's and Semlyen's line models as well as the untransposed constant-parameter line model are based on modal theory [9], a transformation matrix is indispensable. The Noda's line model can automatically take into account the frequency-dependent effect of the transformation matrix because it is based on a phase domain analysis method [10–12]. However, a rich experience is required for the parameter calculation of the Noda's model. Marti's and Semlyen's models are widely used for taking into account the frequency-dependent effects. A constant transformation matrix is, however, assumed for a simplification of the models. The EMTP users have to select an appropriate frequency to calculate the transformation matrix.

One of the principles is the resonant frequency of a line. Figure 8.3 illustrates transient responses of a loss-less line whose characteristic impedance is Z_o and traveling time is τ. Both the sending-end current i_s and the receiving-end voltage v_r oscillate with a period of 4τ. This frequency of $1/4\tau$ is called the "dominant frequency of the distributed parameter line." In general, the transformation matrix is calculated at the dominant frequency. JMARTI SETUP and SEMLYEN SETUP [1] calculating the parameters of the line models have a special input column for the frequency for the transformation matrix calculation. If the other element, such as an inductor, is connected to the line, the basic oscillating frequency may be changed. The transformation matrix should be calculated at around the basic frequency if it is sited far from the dominant frequency.

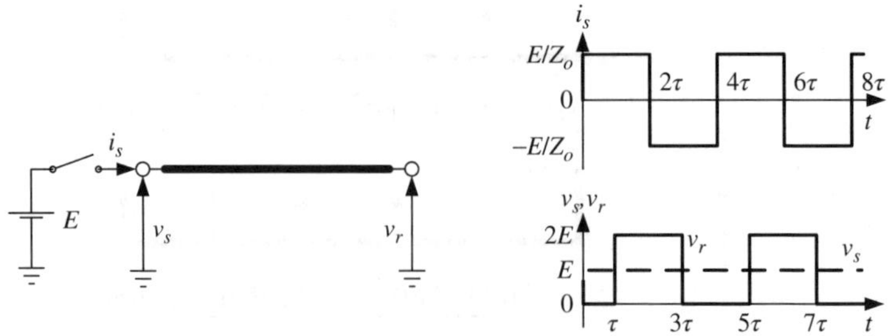

Figure 8.3 Transient response of an ideal line

8.1.3 Data for Cable Parameters

"CABLE PARAMETERS (CP)" is a subroutine installed in the EMTP-ATP (alternative transients program) for calculating cable parameters [13, 14]. It is a modified version of the original cable parameter calculation program, "CABLE CONSTANTS". In this section, the data structure is explained using a sample data for the 66 kV and the 22 kV cable shown in Data 8.1. Although a data editor using a graphical user interface has been recently used, input parameters are difficult to be shown by the interface. In this section, the input data are listed according to a conventional character based data format. The data file can be accepted by the EMTP and is easily converted to the data for the graphical data editor.

8.1.3.1 Declaration and Miscellaneous Data

Every EMTP data set starts from "BEGIN NEW DATA CASE" (Line 1) which separates data cases.

Keyword "CABLE PARAMETERS" (Line 2) is for invoking the modified version of the cable parameter calculation program.

A line keyed "C_" in columns one and two ("_" denotes a white space) as in Line 3 is a comment line. The line is ignored by the EMTP. There are two other types of comments (block comments and in-line comments) that can be used. The block comment called $DISABLE/ENABLE feature in the EMTP is explained in the *EMTP Rule Book* [1]. Any texts between braces ("{" and "}") or after an opening brace ("{") can be put as an in-line comment. The comment can be seen in Line 10 in Data 8.1.

Keyword "BRANCH" in Line 4 is a request for including the specified node names to output the branch data. The output data can be used in a transient simulation without any modifications.

Line 6 indicates a three-phase ([NPC]=3), single-core (ITYPE=[TY]=2) underground (ISYST=[SYS]=-1) cable system. The other miscellaneous data can be found in the *EMTP Rule Book* [1].

Value two in Line 8 shows the number of conductors of each cable, that is, a core and a metallic sheath. If a cable has an armor, the number becomes three.

Data 8.1 Data for CABLE PARAMETERS

```
1.  BEGIN NEW DATA CASE
2.  CABLE PARAMETERS
3.  C ANCH  [BUS1][BUS2][BUS3][BUS4][BUS5][BUS6][BUS7][BUS8][BUS9][BUSA][BUSB][BUSC]
4.  BRANCH  KB66_ALB66_AKB66_BLB66_BKB66_CLB66_CKB66_NLB66_NKB66_NLB66_NKB66_NLB66_N
5.  C TY][SYS][NPC][EAR][KMO][ZFL][YFL][NPP][NGD]
6.       2   -1    3    0    1    2    2         0
7.  C N1][ N2][ N3][ N4][ N4][ N5][ N6][ N7][ N8][ N9][N10][N11][N12][N13][N14][N15]
8.       2    2    2
9.  C   r1    ][   r2    ][   r3    ][   r4    ][   r5    ][   r6    ][   r7    ]
10.           0.0 1.6000E-2 2.6000E-2 2.6500E-2 3.0500E-2 {66kV XLPE  600sqmm
11. C  rohc  ][  muc    ][  muil  ][  epsil ][  rohs   ][   mus   ][  mui2  ][  epsi2 ]
12.    2.477E-8         1.0          1.0          2.3  1.80E-08         1.0         1.0         3.3
13. C   r1    ][   r2    ][   r3    ][   r4    ][   r5    ][   r6    ][   r7    ]
14.           0.0 1.6000E-2 2.6000E-2 2.6500E-2 3.0500E-2 {66kV XLPE  600sqmm
15. C  rohc  ][  muc    ][  muil  ][  epsil ][  rohs   ][   mus   ][  mui2  ][  epsi2 ]
16.    2.477E-8         1.0          1.0          2.3  1.80E-08         1.0         1.0         3.3
17. C   r1    ][   r2    ][   r3    ][   r4    ][   r5    ][   r6    ][   r7    ]
18.           0.0 1.6000E-2 2.6000E-2 2.6500E-2 3.0500E-2 {66kV XLPE  600sqmm
19. C  rohc  ][  muc    ][  muil  ][  epsil ][  rohs   ][   mus   ][  mui2  ][  epsi2 ]
20.    2.477E-8         1.0          1.0          2.3  1.80E-08         1.0         1.0         3.3
21. C  Vert1 ][ Horiz1 ][  Vert2 ][ Horiz2 ][  Vert3 ][ Horiz3 ][  Vert4 ][ Horiz4 ]
22.         0.50        -.15        0.50         0.0        0.50         .15
23. C      rohe     ][      Freq.    ][DEC][PNT][ DIST ][  IPUN  ]
24.              100.             5000.              10.E3          2
25. BLANK ending frequency data
26. BLANK ending CABLE PARAMETERS case
27. BEGIN NEW DATA CASE
28. CABLE PARAMETERS
```

```
C ANCH  [BUS1][BUS2][BUS3][BUS4][BUS5][BUS6][BUS7][BUS8][BUS9][BUSA][BUSB][BUSC]
BRANCH  AH00_AAH01_AAH00_BAH01_BAH00_CAH01_CAH00_NAH01_NAH00_NAH01_NAH00_NAH01_N
C TY][SYS][NPC][EAR][KMO][ZFL][YFL][NPP][NGD]
     2  -1   3   0   1   2   2       0
                                   :
                            Next Case
                                   :
C      rohe    ][      Freq.    ][DEC][PNT][ DIST ][  IPUN  ]
               100.          5000.               2.0E3          2
BLANK ending frequency data
BLANK ending CABLE PARAMETERS case
$PUNCH
BEGIN NEW DATA CASE
BLANK
```

8.1.3.2 Radii and Physical Constants

For an accurate simulation, cable data should be collected as accurate as possible. Figure 8.4 and Table 8.2 show an example of a configuration of XPLE cables provided by a cable manufacturer. To obtain cable impedance and admittance by CABLE PARAMETERS, further information and approximation is required to fulfill the assumption of the CABLE PARAMETERS [13]. The program is assuming that every conductor consists of a cylindrical conductor insulated by a single insulator between the conductors. For example, semiconducting layers, which are installed on the inner and outer surfaces of the insulator of a high-voltage cable, have to be approximately included into the input data. In this section, a simple approximation method is explained.

The semiconducting layer approximately acts as a conductor for the cable capacitance, and it can be negligible for a current conduction [15–18]. The electric permittivity of the main insulator ε_i is an important parameter because it determines the propagation velocity of the

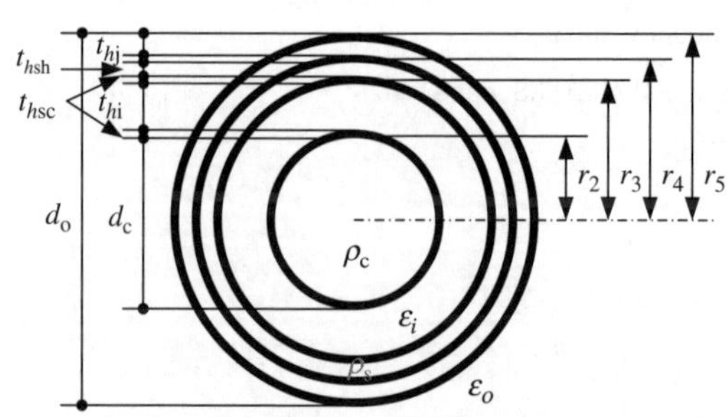

Figure 8.4 Sectional figure of a coaxial cable

Table 8.2 XPLE cable parameters

Voltage (kV)	66	22	22	22
Nominal cross section of the conductor, S (mm^2)	600	250	150	60
Outer diameter of the conductor, d_c (mm)	29.5	19	14.7	9.3
Thickness of insulator, t_{hi} (mm)	10	7	7	7
Thickness of semiconducting layer, t_{hsc} (mm)	1.25	—*	—*	—*
Thickness of metallic sheath (screen), t_{hsh} (mm)	0.5	0.3	0.3	0.3
Thickness of corrosion proof layer, t_{hj} (mm)	4	2.7	2.5	2.3
Outer diameter of cable, d_o (mm)	67	40	36	30
Conductor resistance, $R_{d.c.}$ (mΩ/km)	30.8	73.9	122	305
Corrosion proof layer	Vinyl	Vinyl	Vinyl	Vinyl

* Included in t_{hi}.

coaxial mode. The typical relative permittivity of XPLE is 2.3. The radii of the inner and outer surface of the insulator (r_2 and r_3) with the permittivity determine the characteristic impedance and capacitance of the coaxial mode. From the characteristics, the original permittivity should be used from a viewpoint of propagation velocity of the cable, so the radii have to be modified to express the coaxial-mode capacitance [19].

The inner and outer radii of the insulator (r_2 and r_3) for the CABLE PARAMETERS are determined to coincide with the coaxial capacitance. If the thickness of the semiconducting layer t_{hsc} for the 66 kV cable is 1.25 mm, the radii become:

$$r_2 = {}^{d_c}\!/_2 + t_{hsc} = {}^{29.5}\!/_2 + 1.25 = 16\ [\text{mm}] \tag{8.2}$$

$$\therefore r_3 = r_2 = t_{hi} = 16 + 10 = 26\ [\text{mm}] \tag{8.3}$$

In the above calculation the semiconducting layer on the inner surface of the metallic sheath, that is, on the outer surface of the main insulator, is neglected to coincide with the capacitance of the outside of the cable.

The modification of the inner radius of the insulator, that is, the radius of the conductor r_2 influences the conductor impedance. The effect is minimized by the modification of the conductor resistivity ρ_c.

$$R_{dc} = {}^{\rho_c}\!/_S = {}^{\rho_c}\!/_{\pi r_2^2}$$

$$\therefore \rho_c = \pi r_2^2 R_{dc} \tag{8.4}$$

Even if the radius is not modified for the semiconducting layer, the conductor resistivity should be modified due to the gap within the stranded conductor. The equivalent resistivity may be slightly larger than that of intrinsic resistivity of the material (copper = 1.8×10^{-8} Ωm).

The thickness of the metallic sheath (screen) t_{hsh} is another important factor of the cable transient characteristic. To keep the thickness, the outer radius of the metallic sheath is defined as:

$$r_4 = r_3 + t_{hsh} = 26 + 0.5 = 26.5\ (\text{mm}) \tag{8.5}$$

The outer radius of the cable becomes:

$$r_5 = r_4 + t_{hj} = 26.5 + 4 = 30.5\ (\text{mm}) \tag{8.6}$$

In the above calculation, some insulating materials on the outer surface of the metallic sheath is neglected to coincide with the original capacitance of the outside of the cable. The outer diameter ($2r_5 = 61$ mm) of the CABLE PARAMETERS data is smaller than that in the datasheet ($d_o = 67$ mm) shown in Table 8.2.

Table 8.3 shows the data for CABLE PARAMETERS for the cables. The relative permittivity of the insulator ε_i and the resistivity of the metallic sheath ρ_s are taken from typical values of 2.3 and 1.8×10^{-8} Ωm, respectively. An important frequency is derived from a relation between a penetration depth and the metallic sheath thickness t_{hsh} [20, 21].

$$
\begin{aligned}
t_{hsh} &= \sqrt{\frac{\rho_s}{\omega\mu_0}} = \sqrt{\frac{\rho_s}{2\pi f_c \mu_0}} \\
f_c &= \frac{\rho_s}{2\pi\mu_0 {t_{hsh}}^2}
\end{aligned}
\tag{8.7}
$$

where ρ_s denotes the resistivity of the metallic sheath.

A wave propagation of a cable core is determined by a coaxial mode, and a magnetic field is sealed within a metallic sheath in a frequency region above the critical frequency f_c. For the 22 kV cable used in this chapter, the critical frequency is:

$$
f_c = \frac{\rho_s}{2\pi\mu_0 {t_{hsh}}^2} = \frac{1.8 \times 10^{-8}}{2\pi \times 4\pi \times 10^{-7}(0.3 \times 10^{-3})^2} = \frac{5 \times 10^5}{2\pi^2} \approx 25(\text{kHz}) \tag{8.8}
$$

The relative permittivity of the corrosion proof layer is varied depending on its material. In this section, it is assumed to be 3.3.

The parameters are entered into Lines 10 and 12 in Data 8.1. The parameters of the other cable are entered after those of the first cable (Lines 13–20).

8.1.3.3 Position, Earth Resistivity and Frequency

Every cable is arranged in a flat formation with a phase distance of 150 mm and a depth of 0.5 m. The position is entered into Line 22.

Soil resistivity and a frequency at which the parameters are calculated are entered into Line 24. The resistivity of soil is assumed to be 100 Ωm. As already explained, the frequency for obtaining the constant-parameter line model is important. The frequency is temporally set by the dominant frequency determined by the cable length and its propagation velocity of the coaxial mode. The frequency for the cable of l km length becomes:

$$
f_d = \frac{1}{4\tau} = \frac{1}{4l}v = \frac{1}{4l}\frac{c_0}{\sqrt{\varepsilon_i}} = \frac{1}{4l}\frac{300 \times 10^3(\text{km/s})}{\sqrt{2.3}} \approx \frac{50}{l}(\text{kHz}) \tag{8.9}
$$

Table 8.3 Data for CABLE PARAMETERS

V (kV)	S (mm^2)	r_2 (mm)	r_3 (mm)	r_4 (mm)	r_5 (Mm)	ρ_c (Ωm)	ε_i	ρ_s (Ωm)	ε_c
66	600	16	26	26.5	30.5	2.47708×10^{-8}	2.3	1.8×10^{-8}	3.3
22	250	9.5	16.5	16.8	19.5	2.09528×10^{-8}	2.3	1.8×10^{-8}	3.3
22	150	7.35	14.35	14.65	17.15	2.07054×10^{-8}	2.3	1.8×10^{-8}	3.3
22	60	4.65	11.65	11.95	14.25	2.07184×10^{-8}	2.3	1.8×10^{-8}	3.3

The line length of the 66 kV cable, that of the 22 kV feeder, and the total length of the cable are 10, 6–8, and 14 km, respectively. The dominant frequency is in a range from 3.5 to 8.2 kHz. In this section, the cable parameters for a constant-parameter line model and the transformation matrix for a frequency-dependent line model (Semlyen's or Marti's model) are calculated at a frequency of 5 kHz.

The calculated cable parameters for the constant-parameter line model are independent of the cable length, that is, the results are obtained as per unit length. The column [DIST] is for the cable length for a branch data output.

Value two in the final column [IPUN] will result in the creation of EMTP branch data for a Dommel model of a length of DIST. Value one is alternatively for a pi-circuit model. For either non-zero value, the use of $PUNCH (Line 40) is required to extract such results from the EMTP buffer. For node names to be included, a BRANCH request (Line 4) is required.

8.1.4 EMTP Data Structure

In this section, EMTP data structure for a transient simulation will be explained taking the pre-calculation as an example. Data 8.2 shows the important part of the data.

8.1.4.1 Miscellaneous Data

Lines 3 and 5 in Data 8.2 are called "Misc. DATA." The time step (DT) of a time-domain transient simulation has a close relation to its accuracy. Strictly speaking, the physical behavior of a circuit to be simulated has to be previously known for the determination. It is, however, impractical. At least, the time step should be sufficiently smaller than the minimum traveling time of a distributed parameter line to enable a trace of traveling waves. In this case, the time step is determined as 0.1 μs from the minimum traveling time of 5 μs which is obtained from the minimum length of the cable of 1 km and the fastest modal propagation velocity of 200 m/μs.

The maximum observation time TMAX can be arbitrarily determined by the user if the number of calculation steps is reasonable.

The integer miscellaneous data (Line 5) are for a control of EMTP output. The first and second parameters (IOUT and IPLOT) specify the frequencies of calculated result to a disk file and to a graphic plotting, respectively.

The other parameters in Line 5 are usable in common.

8.1.4.2 Branch Data

Data of linear circuit elements as well as nonlinear elements are entered after the miscellaneous data. A metallic sheath grounding resistance of 0.5 Ω at Substation K (node: KB66_N) is specified at Line 8.

When there are two or more branches that have the same elements with the same values, the reference branch option can be used. The circuit values of the branch specified by the fields BUS3 and BUS4 are referred to the branch element connected between BUS1 and BUS2. This feature is useful for a change of the value at a time.

Data 8.2 Data for pre-calculation

```
BEGIN NEW DATA CASE
C    DT][ TMAX ][ XOPT ][ COPT ][EPSILN][TOLMAT]
  0.1E-6    .001
C  IOUT][ IPLOT][IDOUBL][KSSOUT][MAXOUT][ IPUN ][MEMSAV][ ICAT ][NENERG][IPRSUP]
    1000       1       1       0       1
C ------------------------------------------------------------ grounding resistance
C [BUS1][BUS2][BUS3][BUS4][  R ][  L ][  C ]                                      !
  KB66_N                        0.5
  LB66_N      KB66_N
C
  AH00_N                        1.0
  AH01_N      AH00_N
  AH02_N      AH00_N
  AH03_N      AH00_N
  AH04_N      AH00_N
  AH05_N      AH00_N
C
  BH00_N                        1.0
  BH01_N      BH00_N
  BH02_N      BH00_N
  BH03_N      BH00_N
  BH04_N      BH00_N
  BH05_N      BH00_N
$VINTAGE, 1
C  Branch data for constant-parameter distributed:
-1AH00_AAH01_A                5.31383E-03 2.04272E+01 1.38597E+07-2.00000E+03 1   6
-2AH00_BAH01_B                2.87094E-04 1.08288E+01 3.74862E+07-2.00000E+03 1   6
-3AH00_CAH01_C                3.76328E-04 1.16063E+01 4.55203E+07-2.00000E+03 1   6
                                   :
== Copy and paste the output of CABLE PARAMETERS ==
                                   :
   .34934649   .00000000   .65104393  -.65068821   .00000000  -.34917063
   .00070036   .00000000  -.00615022  -.03775591   .00000000  -.00307464
   .33714231  -.50010987  -.33731589   .33703352   .50016150  -.33698247
  -.00077768   .00278082   .00173212   .02101460   .02081577  -.00151437
```

```
$VINTAGE, -1,
-1AH02_AAH03_AAH01_AAH02_A
-2AH02_BAH03_B
-3AH02_CAH03_C
-4AH02_NAH03_N
-5AH02_NAH03_N
-6AH02_NAH03_N
-1AH04_AAH05_AAH03_AAH04_A
-2AH04_BAH05_B
-3AH04_CAH05_C
-4AH04_NAH05_N
-5AH04_NAH05_N
-6AH04_NAH05_N
-1BH01_ABH02_AAH01_AAH02_A
-2BH01_BBH02_B
-3BH01_CBH02_C
-4BH01_NBH02_N
-5BH01_NBH02_N
-6BH01_NBH02_N
 1BH02_ABH03_AAH02_AAH03_A
-2BH02_BBH03_B
-3BH02_CBH03_C
-4BH02_NBH03_N
-5BH02_NBH03_N
-6BH02_NBH03_N
-1BH03_ABH04_AAH03_AAH04_A
-2BH03_BBH04_B
-3BH03_CBH04_C
-4BH03_NBH04_N
-5BH03_NBH04_N
-6BH03_NBH04_N
-1BH04_ABH05_AAH04_AAH05_A
-2BH04_BBH05_B
-3BH04_CBH05_C
-4BH04_NBH05_N
```

```
71. -5BH04_NBH05_N
72. -6BH04_NBH05_N
73. BLANK ending BRANCH
74. C [BUS1][BUS2]                                            MEASURING             !
75.    LB22_AAH00_A                                           MEASURING             1
76.    LB22_ABH00_A                                           MEASURING             1
77. BLANK ending SWITCH
78. C [BUS1][][  AMP.  ]                                          [ TSTART ][  TSTOP ]
79. 11LB22_A-1        1.E3                                              0.0     25.E-6
80. BLANK ending SOURCE
81. C [BUS1][BUS2][BUS3][BUS4][BUS5][BUS6][BUS7][BUS8][BUS9][BUSA][BUSB][BUSC][BUSD]
82.    AH00_AAH00_BAH00_CAH05_AAH05_BAH05_CBH00_ABH00_BBH00_CBH05_ABH05_BBH05_C
83. BLANK ending OUTPUT Specification
84. C !![H][ST]END][MIN]MAX][BUS1][BUS2][BUS3][BUS4][ HEADING LABEL][ VERTICAL AXIS]
85.   144 .1  0.  1.0-4.E44.E4AH05_A                  Feeder A           VOLTAGE [V]
86.   144 .1  0.  1.0-4.E44.E4BH05_A                  Feeder B           VOLTAGE [V]
87.   144 .1  0.  1.0-4.E44.E4AH05_AAH05_B            Feeder A           VOLTAGE [V]
88.   144 .1  0.  1.0-4.E44.E4BH05_ABH05_B            Feeder B           VOLTAGE [V]
89. BLANK ending PLOT
90. BEGIN NEW DATA CASE
91. BLANK
```

The cable models generated by CABLE PARAMETERS are copied and pasted into this branch section (Lines 24–36). The feature of the reference branch can be also applicable to the constant-parameter line model (Lines 37–72).

The branch data are terminated by a BLANK line (Line 73).

8.1.4.3 Switch Data

A switch in the EMTP is widely used for representing various devices. For example, it expresses a switching device used in a power electronics circuit. In this chapter, the switch is used as a model of a CB and a current sensor, which are basic and common applications of the switch model.

The switch with a keyword "MEASURING" typed in columns 55–63 and a current output flag "1" in column 80 is a permanently closed switch, that is, a current sensor (Lines 75 and 76).

The switch data are terminated by a BLANK line (Line 77).

8.1.4.4 Source Data

The flag "11" in columns 1 and 2 in Line 79 indicates a step waveform (d.c.) source. The source is connected with a node of LB22_A. Columns 9 and 10 are used to select between a voltage and a current source. Negative integer ("−1") is a flag for a current source.

The amplitude of the current is specified by the field AMPLITUDE ([AMP.]).

The fields TSTART and TSTOP show times at which the function is activated and inactivated. The data in Line 79 show a pulse current source whose amplitude and width are 1 kA and 25 μs, respectively.

The source data are terminated by a BLANK line (Line 80).

8.1.4.5 Output Specification

The requested node voltages, and branch voltages and currents can be observed. The output of the node voltages can be only specified by these data. The other branch valuables can be also specified in column 80 of the branch data as shown in Lines 75 and 76.

The output specification data are terminated by a BLANK line (Line 83).

8.1.4.6 Plot Specification

Variables calculated by the EMTP can be plotted as a function of time by a batch mode plotting feature. The data are specified in Lines 85–88. An interactive plotting is also available after a simulation.

8.1.5 Results of Pre-calculation

Figure 8.5 shows the simulated results of the circuit illustrated in Figure 8.2. An energized-phase core voltage and an induced core voltage are shown in the figure. For reference, a calculated result obtained by a frequency transform method (FTP) [22] is shown in Figure 8.5a. The method accurately takes into account the frequency-dependent effect of a distributed-parameter line, such as a characteristic impedance, a wave propagation, and a transformation matrix.

From Figure 8.5b, the result obtained by the Dommel line model, whose parameter is obtained at a frequency of 5 kHz, involves spike voltages and high frequency oscillations. Excluding those in the result, the basic characteristics, such as the multiple reflections within the cable system and the attenuation agree with the calculated results by FTP. Figure 8.5c and d shows the results obtained by frequency-dependent Semlyen and Marti line models in the EMTP. The accuracy of the result obtained by Semlyen's line model decreases as time elapses. The attenuation of the result by Marti's model agrees with that by the frequency transform method. However, a spike voltage is observed at the wavefront and the induced voltage increases as time passes.

These line models are obtained by a certain least square error method for curve fitting. The accuracy depends on the parameters for the approximation, and trial and error is unavoidable

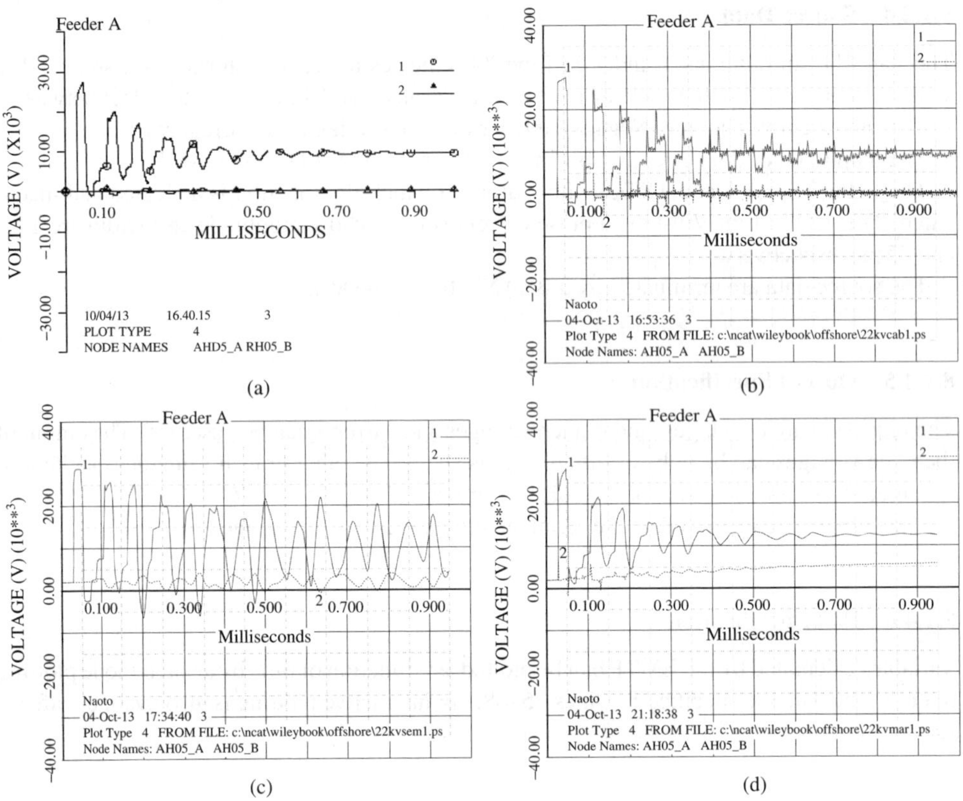

Figure 8.5 Simulated results of pre-calculation: (a) FTP; (b) Dommel (constant parameter); (c) Semlyen; and (d) Marti

to obtain a better model. The models used in the simulation are obtained by default or recommended parameters for the approximation explained in the *EMTP Rule Book* [1]. The frequency for a transformation matrix calculation can be determined by the dominant frequency. However, the frequency range for the approximation of the characteristic impedance and the wave propagation has to be modified according to the approximation accuracy by trial and error.

Because these models have been developed for a long line or cable system, there is a tendency that the accuracy for the short cable decreases compared with that for a long cable. A simulation using the stable model, that is, the constant-parameter Dommel line model, is valuable taking the time for the approximation into account.

8.1.6 Cable Energization

8.1.6.1 Data for Transient Simulation

Transients due to a cable energization in a wind farm illustrated in Figure 8.1 are simulated by EMTP. The data for the simulation is given in Data 8.3.

Data 8.3 Data for cable energization simulation

```
BEGIN NEW DATA CASE
C      DT][ TMAX ][ XOPT ][ COPT ][EPSILN][TOLMAT]
  0.1E-6  1.0E-3
C   IOUT][ IPLOT][IDOUBL][KSSOUT][MAXOUT][ IPUN ][MEMSAV][ ICAT ][NENERG][IPRSUP]
     1000        1       1       0       1
C ------------------------------------------------------------ backward impedance
$VINTAGE, 1
C [BUS1][BUS2][BUS3][BUS4][       R       ][       L       ][       C       ]     !
  SOUR_AJB66_A                     1.3068          41.597
  SOUR_BJB66_BSOUR_AJB66_A
  SOUR_CJB66_CSOUR_AJB66_A
$VINTAGE, 0
C ------------------------------------------------------------- Tr-A (66/22, D-Y)
C TRANSFORMER [BUS3]      [Iste][Pste][BTOP][Rmag]                               !
  TRANSFORMER                        TRAT_A
C       CUR     ][     FLUX     ]
                9999
C [BUS1][BUS2]            [ Rk ][ Lk ][Volt]                                     !
 1LB66_ALB66_B            6.534       66.
 2LB22AA                       7.703112.702
C TRANSFORMER [BUS3]      [Iste][Pste][BTOP][Rmag]                               !
  TRANSFORMER TRAT_A                 TRAT_B
C [BUS1][BUS2]            [ Rk ][ Lk ][Volt]                                     !
 1LB66_BLB66_C
 2LB22AB
C TRANSFORMER [BUS3]      [Iste][Pste][BTOP][Rmag]                               !
  TRANSFORMER TRAT_A                 TRAT_C
C [BUS1][BUS2]            [ Rk ][ Lk ][Volt]                                     !
 1LB66_CLB66_A
 2LB22AC
C ------------------------------------------------------------- Tr-B (66/22, D-Y)
                                       :
                          Transformer data for feeder B
                                       :
C ---------------------------------------------------------- grounding resistance
                                       :
```

```
                                          See Data 8.2
                                               :
$PREFIX, C:/XXX/YYY/
$INCLUDE, ZZZZ.PCH
-1AH02_AAH03_AAH01_AAH02_A
-2AH02_BAH03_B
-3AH02_CAH03_C
                                               :
                                          Reference
                                               :
-4BH04_NBH05_N
-5BH04_NBH05_N
-6BH04_NBH05_N
BLANK ending BRANCH
C [BUS1][BUS2]                                          MEASURING            !
  JB66_AKB66_A                                          MEASURING            1
  JB66_BKB66_B                                          MEASURING            1
  JB66_CKB66_C                                          MEASURING            1
C [BUS1][BUS2][ TCLOSE ][  TOPEN ][   Ie   ]                                 !
  LB22AAAH00_A       0.0 99999999.                                           1
  LB22ABAH00_B       0.0 99999999.                                           1
  LB22ACAH00_C       0.0 99999999.                                           1
C [BUS1][BUS2]                                          MEASURING            !
  LB22BABH00_A                                          MEASURING            1
  LB22BBBH00_B                                          MEASURING            1
  LB22BCBH00_C                                          MEASURING            1
C $DISABLE
C [BUS1][BUS2]                                          MEASURING            !
  LB22AALB22BA             { Bus tie CB }               MEASURING            1
  LB22ABLB22BB                                          MEASURING            1
  LB22ACLB22BC                                          MEASURING            1
C $ENABLE
BLANK ending SWITCH
```

```
70. C [BUS1][][  AMP.   ][  FREQ. ][  ANGLE ]                     [ TSTART ][  TSTOP ]
71. 14SOUR_A      53.89E3       50.0      -30.0                          -1.0
72. 14SOUR_B      53.89E3       50.0     -150.0                          -1.0
73. 14SOUR_C      53.89E3       50.0     -270.0                          -1.0
74. BLANK ending SOURCE
75. C [BUS1][BUS2][BUS3][BUS4][BUS5][BUS6][BUS7][BUS8][BUS9][BUSA][BUSB][BUSC][BUSD]
76.   KB66_AKB66_BKB66_CLB66_ALB66_BLB66_CLB22AALB22ABLB22ACLB22BALB22BBLB22BC
77.                                         :
78.                                         :
79. BLANK ending OUTPUT Specification
80. C !![H][ST]END][MIN]MAX][BUS1][BUS2][BUS3][BUS4][ HEADING LABEL][ VERTICAL AXIS]
81.  194 .1  0.  1.0 -80.  80.BRANCH                  S/S K                CURRENT [A]
82.                            JB66_AKB66_AJB66_BKB66_BJB66_CKB66_C
83.                                         :
84.                                         :
85. BLANK ending PLOT
86. BEGIN NEW DATA CASE
87. BLANK
```

Some components, which are not explained in the previous section, will be explained.

The reactance of the backward impedance $\%X_B$ and the ratio between the resistance and reactance is assumed to be 3% (10 MVA Base) and 10% ($R/X = 0.1$), respectively.

The backward resistance R_B and inductance L_B are obtained as follows:

$$X_B = \frac{\%X_B V^2}{100P} = \frac{3 \times 66^2}{100 \times 10} = 13.068\,(\Omega) \tag{8.10a}$$

$$R_B = \frac{X_B}{10} = 1.3068(\Omega) \tag{8.10b}$$

$$L_B = \frac{X_B}{\omega} = \frac{X_B}{2\pi f} = \frac{13.068}{2\pi 50} = 41.597\,(\text{mH}) \tag{8.10c}$$

These values are entered as branch data as shown in Lines 7–12 listed in Data 8.3.

The transformer model installed in the EMTP is expressed by its winding resistances, leakage inductances and magnetizing circuit. The magnetizing circuit (Y_m) is neglected in this case. In theory, either primary or secondary winding leakage inductance (L_1' or L_2') can be assumed to be zero for a two-winding transformer. The inductance of the secondary winding L_2, however, has to be non-zero in the EMTP. In this case, the winding resistances are combined and entered as a primary resistance R_1. The leakage inductances are entered into the secondary winding as shown in Figure 8.6.

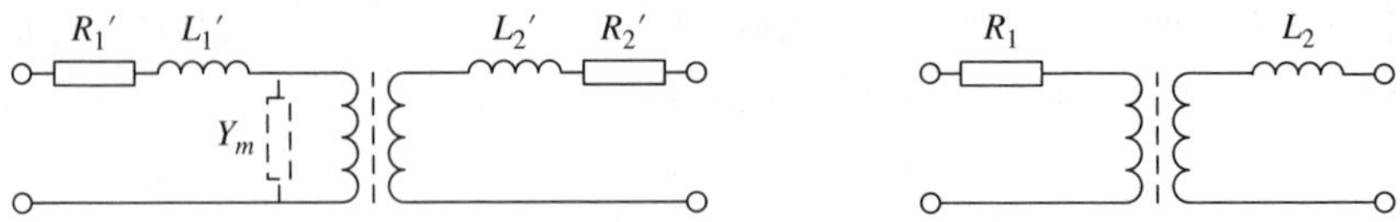

Figure 8.6 Single-phase transformer model

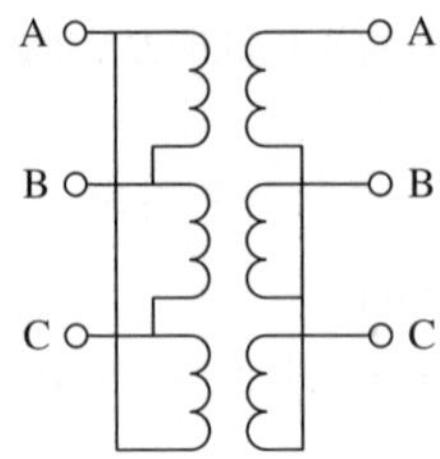

Figure 8.7 Three-phase Δ-Y transformer represented by three single-phase models

The three-phase transformer is modeled by three single-phase transformers in the EMTP (Figure 8.7). The delta-connected primary impedance has to be multiplied by three to the Y-connected impedance.

The reactance and resistance of the transformer of 18 MVA capacity (P_{Tr}) installed in Substation L are assumed to be 9% ($\%X_{Tr}$) and 0.9% ($\%R_{Tr}$), respectively. The loss-resistance referred to the high voltage side R_1 and leakage-inductance referred to the low voltage side L_2 are obtained as follows:

$$X_2 = \frac{\%X_{Tr} V^2}{100 P_{Tr}} = \frac{9 \times 22^2}{100 \times 18} = 2.42\,(\Omega) \tag{8.11a}$$

$$R_1 = \frac{\%R_{Tr} V^2}{100 P_{Tr}} \times 3 = \frac{0.9 \times 66^2}{100 \times 18} \times 3 = 6.534\,(\Omega) \tag{8.11b}$$

$$L_2 = \frac{X_2}{\omega} = \frac{X_2}{2\pi f} = \frac{2.42}{2\pi 50} = 7.7031\,(\text{mH}) \tag{8.11c}$$

The transformer is represented in Lines 15–30. The keyword TRANSFORMER with a name TRAT_A in Line 15 declares the beginning of the transformer data. A standalone terminator “9999” in Line 17 indicates to neglect a magnetizing inductance. If a magnetizing characteristic is included, a set of current versus flux characteristic is entered before the terminator with a magnetizing resistance in Line 15 ([Rmag]).

The primary and secondary windings are expressed in Lines 19 and 20 by the resistance and inductance with the following winding voltage ratio. It is entered into the column named [Volt].

$$66{:}\ 22/\sqrt{3} = 66{:}\ 12.702 \tag{8.12}$$

The transformers for other phases can be entered using a reference feature of the EMTP as shown in Lines 22–30.

The branch data of the cable are generated by CABLE PARAMETERS. The data can be copied and pasted into the file for a transient calculation. If the branch name is embedded into the result of the CABLE PARAMETERS by BRANCH request, the result can be directly used by the $INCLUDE feature of the EMTP. Lines 39 and 40 in Data 8.3 indicate the directory (folder) and file name of the output of the CABLE PARAMETERS.

After the branch data of the cables are completed, the cable data whose parameters are identical to the inputted cable can be entered using a reference feature of the EMTP without entering the numerous numbers.

After terminating the branch data by BLANK line (Line 50), the data of the CBs are entered. Lines 56–58 indicate that the three-phase CB SWA shown in Figure 8.1 is closed at $t = 0$. The switches in Lines 65–67 express the bus-tie CBs. In this case, the 22 kV buses are connected because the bus-tie CBs are expressed by measuring (permanently closed) switches. If the switches are commentouted by removing "C_" from the beginnings of Lines 63 and 68, the 22 kV buses are isolated. A $DISABLE line tells the EMTP to treat all following data lines as if they were comment lines, until the following appearance of a $ENABLE line. This feature is useful for a big block of comments, or for a temporary removal of components from a data case without deletion.

A three-phase source is expressed by three single-phase sources in the EMTP. The amplitude of the voltage source is given by the maximum phase-to-ground voltage rather than the root mean square phase-to-phase voltage. The backward system is expressed by three sources whose frequency and phase difference are 50 Hz and −120°, respectively. Their amplitudes are:

$$V_m = \sqrt{2} \times 66/\sqrt{3} = 53.89[\text{kV}] \tag{8.13}$$

The flag "14" in columns 1 and 2 in Line 71 indicates a sinusoidal source. The definition of the angle is based on a cosine function instead of a sine function. Because the EMTP is originally developed as a transient analysis program, the cosine voltage source is convenient since the maximum value is obtained with an angle of zero at the time origin ($t = 0$). The source with a negative TSTART flag indicates a steady-state calculation should be invoked before a transient calculation. If the flag is cleared, a transient calculation is only carried out.

8.1.6.2 Simulated Results

Figure 8.8a and b shows calculated results when Feeder-A is energized by closing SWA shown in Figure 8.1. The cables are modeled by Dommel line models. Feeder-B is assumed to be energized before the operation of SWA. The transformers operate in parallel because the bus-tie CB is closed. Pulse waveforms are observed both in the feeder currents and in the terminal voltages. The widths of 30 and 40 μs are determined by the traveling time of the coaxial mode of the Feeders-A and B. The oscillating period of 150 μs is generated by the reciprocal reflections within the 22 kV cable system. The reflections is due to the 22 kV buses being tied by SWC. The oscillations are attenuated by 1 ms.

Figure 8.8c and d illustrates simulated results when the bus-tie CB is opened. The cables in Feeder-A are energized through a transformer, the pulse current and voltage due to traveling waves are not observed. The sinusoidal oscillation can be expressed by a resonance of a lumped-parameter circuit. It indicates that the cable can be expressed by lumped-element equivalent circuits.

Figure 8.9 shows the calculated results using a Marti model. Although wave deformations are observed at wavefronts of the oscillations due to traveling waves, the difference between the results by the Dommel line model and the Marti model is small. If the 22 kV buses are separately operated, the difference is not observed as shown in Figure 8.9c and d.

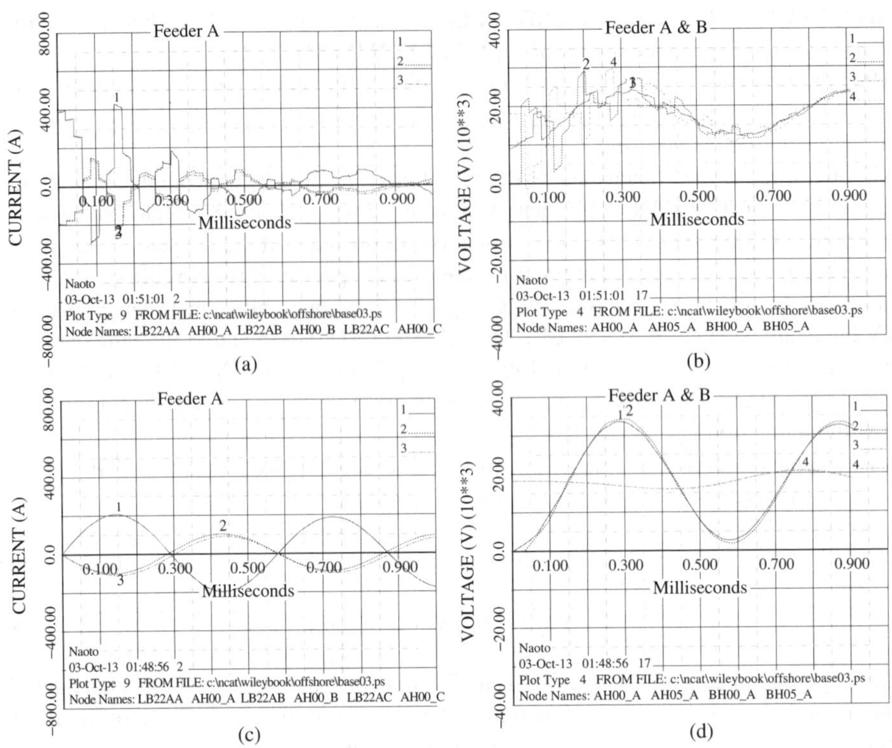

Figure 8.8 Results using the constant parameter (Dommel's) line model (SWB-On): (a) Feeder-A current (SWC-On); (b) sending- and receiving-end core voltage (SWC-On); (c) Feeder-A current (SWC-Off); and (d) sending- and receiving-end core voltage (SWC-Off)

Figure 8.10 illustrates an equivalent circuit for an analytical investigation. The system is expressed by lumped elements and the losses of transformers and cables are neglected. The 66 kV and the 22 kV cables are expressed by capacitors, C'_{66}, C_{22A} and C_{22B}. The backward inductance L_B and the capacitance of the 66 kV cable C_{66} are converted to the values for the 22 kV system.

$$L'_B = L_B(22/66)^2 \tag{8.14a}$$

$$C'_{66} = C_{66}(66/22)^2 \tag{8.14b}$$

The transient voltage at the 22 kV bus due to the operation of SWA can be explained by a superposition theorem. The transient consists of a steady-state and a transient term, which can be illustrated in Figure 8.10b. The transient term is obtained by:

$$V_{22}(s) = \frac{1}{sC_{22A}} I_{C22A}(s) = \frac{1}{sC_{22A}} \frac{V_s(s)}{Z_E(s)} \tag{8.15}$$

where $V_{22}(s)$ is Feeder A voltage, s is the Laplace operator, I_{C22A} is Feeder-A current, $V_s(s)$ is steady-state voltage across SWA, and $Z_E(s)$ is the driving point impedance seeing from

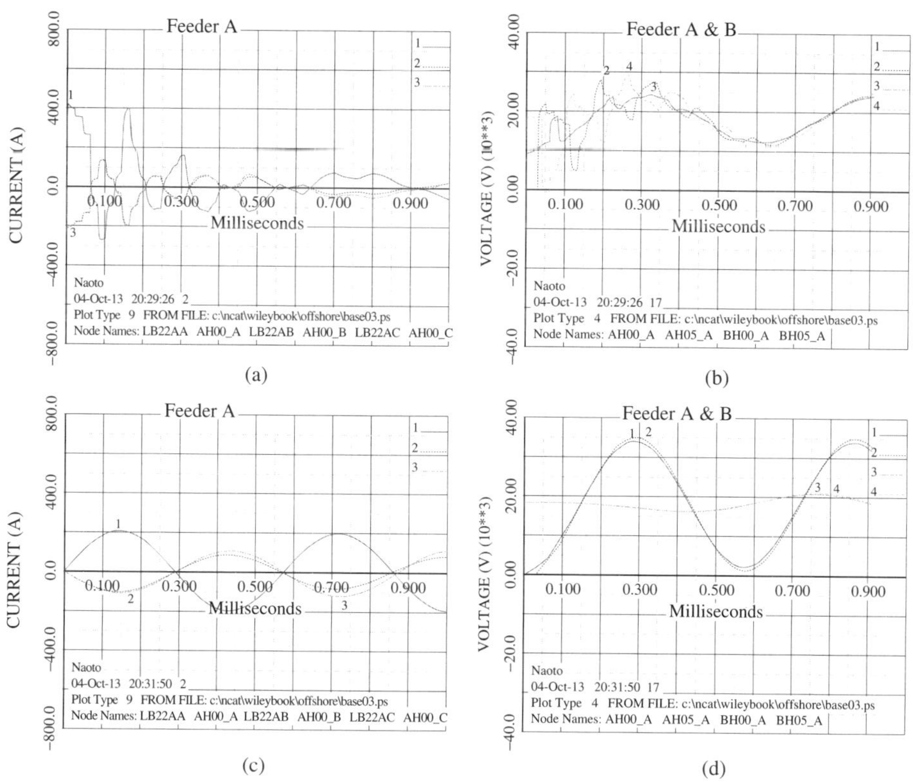

Figure 8.9 Results using Marti's line model (SWB-On): (a) Feeder-A current (SWC-On); (b) sending- and receiving-end core voltage (SWC-On); (c) Feeder-A current (SWC-Off); and (d) sending- and receiving-end core voltage (SWC-Off)

SWA terminals.

$$Z_E(s) = Z_{on}(s) = \left(\left(\left(sL'_B // \frac{1}{sC'_{66}}\right) + s\frac{L_{Tr}}{2}\right) // \frac{1}{sC_{22B}}\right) + \frac{1}{sC_{22A}}, \quad \text{SWC: on} \tag{8.16}$$

$$Z_E(s) = Z_{off}(s) = \left(sL'_B // \frac{1}{sC'_{66}} // \left(sL_{Tr} + \frac{1}{sC_{22B}}\right)\right) + sL_{Tr} + \frac{1}{sC_{22A}}, \quad \text{SWC: off} \tag{8.17}$$

where "//" denotes a parallel connection of impedances.

$$Z_{on}(s) = \frac{\begin{array}{c}((C_{22A}L_{Tr}C'_{66} + C_{22B}L_{Tr}C'_{66})s^4 + (2C_{22A} + 2C'_{66} + 2C_{22B})s^2)L'_B \\ +(C_{22A}L_{Tr} + C_{22B}L_{Tr})s^2 + 2\end{array}}{((C_{22B}s^4L_{Tr}C'_{66} + (2C'_{66} + 2C_{22B})s^2)L'_B + C_{22B}s^2L_{Tr} + 2)C_{22A}s} \tag{8.18}$$

$$Z_{off}(s) = \frac{\begin{array}{c}(L_{Tr}{}^2C'_{66}C_{22A}C_{22B}s^6 + L_{Tr}(2C_{22A}C_{22B} + C_{22A}C'_{66} + C_{22B}C'_{66})s^4 \\ +(C_{22A} + C'_{66} + C_{22B})s^2)L'_B + L_{Tr}{}^2C_{22A}C_{22B}s^4 + L_{Tr}(C_{22A} + C_{22B})s^2 + 1\end{array}}{((L_{Tr}C'_{66}C_{22B}s^4 + (C'_{66} + C_{22B})s^2)L'_B + s^2L_{Tr}C_{22B} + 1)C_{22A}s} \tag{8.19}$$

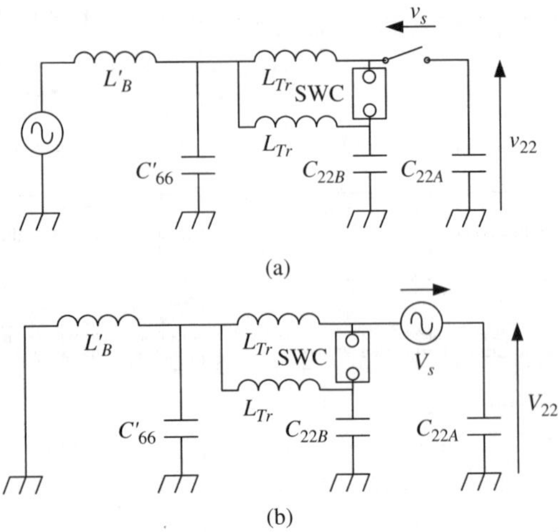

Figure 8.10 (a) Lumped equivalent circuit for the cable energization and (b) equivalent circuit for transient term

The above impedances are too complicated for an analytical investigation. In a high frequency region higher than the frequency of the resonance between L'_B and C'_{66}, the impedance of the backward inductance can be negligible. It corresponds to a time domain before 1 ms for the system. The above equations can be approximately expressed as follows:

$$Z_{on} \approx \left(\left(\frac{1}{sC'_{66}} + s\frac{L_{Tr}}{2}\right) // \frac{1}{sC_{22B}}\right) + \frac{1}{sC_{22A}}$$
$$= \frac{L_{Tr}C'_{66}(C_{22A} + C_{22B})s^2 + 2(C'_{66} + C_{22A} + C_{22B})}{(L_{Tr}C'_{66}C_{22B}s^2 + 2(C'_{66} + C_{22B}))C_{22A}s} \tag{8.20}$$

$$Z_{off} \approx \left(\frac{1}{sC'_{66}} // \left(sL_{Tr} + \frac{1}{sC_{22B}}\right)\right) + sL_{Tr} + \frac{1}{sC_{22A}}$$
$$= \frac{\begin{array}{l} L_{Tr}{}^2C'_{66}C_{22A}C_{22B}s^4 + L_{Tr}(C'_{66}C_{22A} + C'_{66}C_{22B} + 2C_{22A}C_{22B})s^2 \\ +C'_{66} + C_{22A} + C_{22B} \end{array}}{(L_{Tr}C'_{66}C_{22B}s^2 + C'_{66} + C_{22B})C_{22A}s} \tag{8.21}$$

The transient oscillating frequencies of the Feeder-A current $I_{C22A}(s)$ and 22 kV bus voltage $V_{22}(s)$ are determined by the poles of (8.15), that is, zeros of the impedance $Z_E(s)$. When the bus-tie CB (SWC) is closed, the poles are calculated from (8.20).

$$0 = L_{Tr}C'_{66}(C_{22A} + C_{22B})s_{zon}{}^2 + 2(C'_{66} + C_{22A} + C_{22B})$$
$$s_{zon} = \pm j\frac{1}{\sqrt{\dfrac{L_{Tr}}{2}\dfrac{C'_{66}(C_{22A} + C_{22B})}{C'_{66} + C_{22A} + C_{22B}}}} = \pm j\frac{1}{\sqrt{\dfrac{L_{Tr}}{2}\left(\dfrac{1}{\frac{1}{C_{22A}+C_{22B}} + \frac{1}{C'_{66}}}\right)}} \tag{8.22}$$

The above equation shows a resonance between an inductance of the parallel-connected transformer and a capacitance, which is determined by a series connected capacitance of the 66 kV and 22 kV cables.

When the bus-tie CB (SWC) is opened, the poles are calculated from (8.21).

$$0 = L_{Tr}{}^2 C'_{66} C_{22A} C_{22B} s_{zoff}{}^4 + L_{Tr}(C'_{66} C_{22A} + C'_{66} C_{22B} + 2C_{22A} C_{22B}) s_{zoff}{}^2 + C'_{66} + C_{22A} + C_{22B}$$

$$s_{zoff} = \pm j \sqrt{\frac{L_{Tr}(C'_{66} C_{22A} + C'_{66} C_{22B} + 2C_{22A} C_{22B}) \pm L_{Tr}\sqrt{(C'_{66} C_{22A} + C'_{66} C_{22B} + 2C_{22A} C_{22B})^2 - 4C'_{66} C_{22A} C_{22B}(C'_{66} + C_{22A} + C_{22B})}}{2L_{Tr}{}^2 C'_{66} C_{22A} C_{22B}}} \tag{8.23}$$

If a difference between the cable capacitances of the feeders is small ($C_{22A} \approx C_{22B}$), the poles can be approximately obtained by Taylor expansion.

$$s_{zoff} \approx j \frac{1}{\sqrt{2L_{Tr} \dfrac{C_{22A} C_{22B}}{C_{22A} + C_{22B}}}} = j \frac{1}{\sqrt{2L_{Tr} \dfrac{1}{\dfrac{1}{C_{22A}} + \dfrac{1}{C_{22B}}}}} \tag{8.24a}$$

$$s_{zoff} \approx j \frac{1}{\sqrt{2L_{Tr}\left(\dfrac{C_{22A} C_{22B}}{C_{22A} + C_{22B} + 4C_{22A} C_{22B}/C'_{66}}\right)}} = j \frac{1}{\sqrt{2L_{Tr} \dfrac{1}{\dfrac{1}{C_{22A}} + \dfrac{1}{C_{22B}} + \dfrac{4}{C'_{66}}}}} \tag{8.24b}$$

The first pole expresses a resonance within the feeders via the transformers. The second poles express the effect of the 66 kV cable capacitance.

The oscillating periods obtained from the poles are shown in Table 8.4. Although the physical behaviors of the oscillations explained by (8.22) and (8.24a,b) are different, the difference between the oscillating periods T_1 is small. The oscillations are observed in Figure 8.8 and Figure 8.9.

The longest oscillating period T_B is determined by a resonance between the backward inductance L'_B and cable capacitances $C'_{66} + C_{22A} + C_{22B}$.

Table 8.4 Oscillating periods obtained by analytical calculation

SWC	T_1 (ms)	T_2 (ms)	T_B (ms)	
On	0.605		2.21	Accurate
	0.608			Without LB
Off	0.568	0.674	2.21	Accurate
	0.569	0.676		Without LB
	0.600	0.632		By Taylor

$C'_{66} = 2.63 \times 9\,\mu F$, $C_{22A} = 1.12\,\mu F$, $C_{22B} = 1.59\,\mu F$, $L'_B = 41.6/9\,mH$,
$L_{Tr} = 7.70\,mH\ (= L_2)$.

From the above observation, the 66 kV cable between Substation K and Substation L has a minor effect on the transient of the cable energization.

A theoretical analysis can make a simulation-based investigation more valuable and effective, because the theoretical analysis explains the physical insight of a phenomenon.

8.2 Transients in a Solar Plant

A photovoltaic (PV) generation plant constitutes a d.c. power system. Transient characteristics of a.c. systems have been investigated from a broad perspective and stable operation is enabled by solutions of problems by collecting knowledge obtained from the investigations. For stable operation of a d.c. system, numerical simulation is a powerful tool to design a reliable d.c. system. For example, clarification of the current and voltage characteristics at a fault within a PV array contributes to an implementation of an appropriate countermeasure against it.

A fault simulation of a solar generating plant is explained in this section. The transient characteristic is determined by characteristics of a PV cell as well as a cable. An equivalent circuit of a PV module and the effect of an internal circuit of a power conditioning system (PCS) on the transient are explained.

8.2.1 Modeling of Solar Plant

Figure 8.11 illustrates a solar plant. It is a scaled model for a large-scale solar generation system. The plant consists of 21 PV modules with three cables connecting to a PCS.

8.2.1.1 Equivalent Circuit of a PV Module

In a PV array, some PV modules are connected both in series and in parallel. A set of series-connected solar modules is called a "string." A PV array is composed of plural strings connected in parallel. A current vs. voltage characteristic of a single module is required for a fault simulation. If the characteristic cannot be obtained, the single module characteristic has to be obtained from a measured array current. The current and voltage of the single module is obtained from those of the array divided by the number of parallel strings N_P and of series modules N_S, respectively.

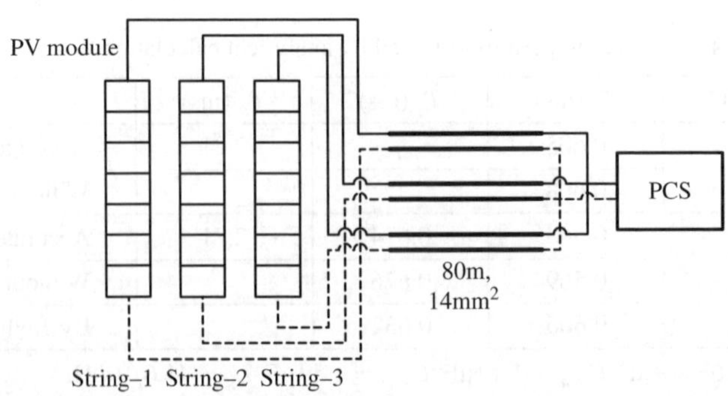

Figure 8.11 A solar plant

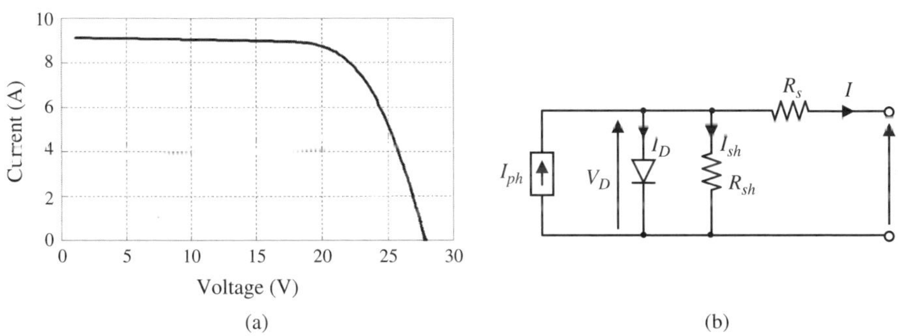

Figure 8.12 An equivalent circuit of a PV module: (a) current vs. voltage characteristic; and (b) circuit diagram

Figure 8.12a shows a current vs. voltage characteristic of a PV module. Although the current vs. voltage characteristic obtained by a measurement or a data sheet can be directly used for a numerical simulation, the roughness of the sampled characteristic reduces numerical stability of a nonlinear calculation. If the characteristic is expressed by a function, a smooth characteristic can be easily obtained. In addition, the effect of parameters, such as temperature, can be investigated.

The characteristic can be expressed by an equivalent circuit consisting of a d.c. source, a diode, a parallel resistor R_{sh} and a series resistor R_s as shown in Figure 8.12b. The module output current I is given in (8.25) as a function of a photoinduced current I_{ph}, a diode current I_D and a parallel resistor current I_{sh}.

$$I = I_{ph} - I_D - I_{sh} \tag{8.25}$$

A simple expression of the diode current I_D is:

$$I_D = I_0 \left\{ \exp\left(\frac{V_D}{nv_\tau}\right) - 1 \right\} = I_0 \left\{ \exp\left(\frac{V_D}{V_\tau}\right) - 1 \right\} \tag{8.26}$$

where V_D is the diode voltage, v_τ is the thermal voltage, and n is the correction factor.

The output current I is given as:

$$I = I_{ph} - I_D - I_{sh} = I_{ph} - I_0 \left\{ \exp\left(\frac{V + R_s I}{V_\tau}\right) - 1 \right\} - \frac{V + R_s I}{R_{sh}} \tag{8.27}$$

The parameters of (8.27) can be estimated from a measured characteristic or a data sheet shown in Figure 8.12a by a nonlinear parameter estimation program, such as Solver installed in Microsoft Excel. Table 8.5 and Figure 8.13 show the approximated parameters and the output characteristic reproduced from the estimated parameters. The maximum current difference between the measurement and approximation is 6.2% at 22.5 V. Although the voltage at the maximum power point is equal to the measured characteristic, the maximum power is smaller by 5.8%.

The diode current I_D is obtained by subtracting the photoinduced current I_{ph} from the output current I. Figure 8.14a shows the diode current calculated from the current vs. voltage characteristic shown in Figure 8.12a. Strictly speaking, Figure 8.14a shows that the characteristic should be expressed by at least two exponential functions.

Table 8.5 Estimated circuit parameters

I_{ph} (A)	I_0 (mA)	V_τ (V)	R_{sh} (MΩ)	R_s (mΩ)
9.10	0.466	2.82	2.65	101

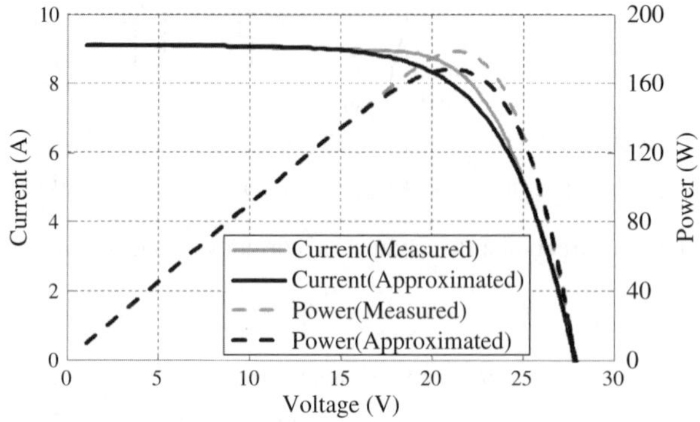

Figure 8.13 Approximated output characteristics of a PV module

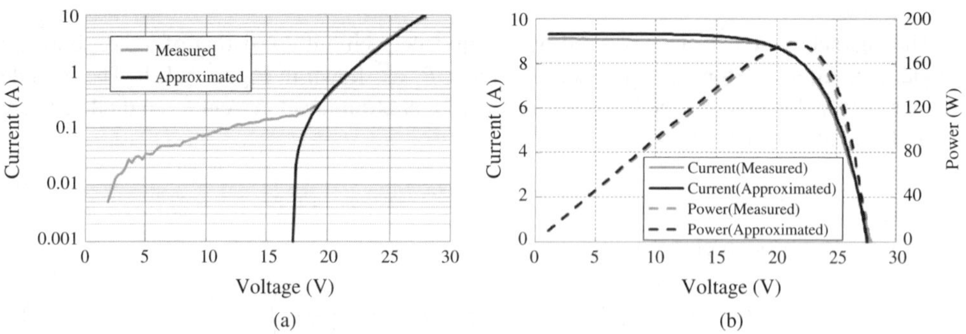

Figure 8.14 Estimation of nonlinear characteristic of the equivalent diode current: (a) nonlinear diode current; and (b) approximated current vs. voltage characteristic

A PV array mostly operates at around the maximum power point by the Maximum Power Point Tracking (MPPT) controller installed in a PCS. For a practical simulation, the approximation accuracy of the parameters of (8.27) can be improved using the characteristic at around the maximum power point instead of increasing the number of exponential functions. In addition, the current I_{sh} flowing through the parallel resistor R_{sh} can be ignored because the leak current I_{sh} is far smaller than the currents I and I_{ph}.

Table 8.6 shows the estimated circuit parameters for a single exponential function obtained from the output characteristic between the maximum power point and the open-circuit voltage. Figure 8.14 shows the estimated diode current using these parameters and the approximated output characteristic. Although a minor difference of 2.2% at the short-circuit current is observed, the important characteristic at around the maximum power point can be reproduced even if the single exponential function is employed to express the nonlinear characteristic.

Table 8.6 Estimated circuit parameters

I_{ph} (A)	I_0 (mA)	V_τ (V)	R_{sh} (MΩ)	R_s (mΩ)
9.32	0.238	2.61	1.50	51.3

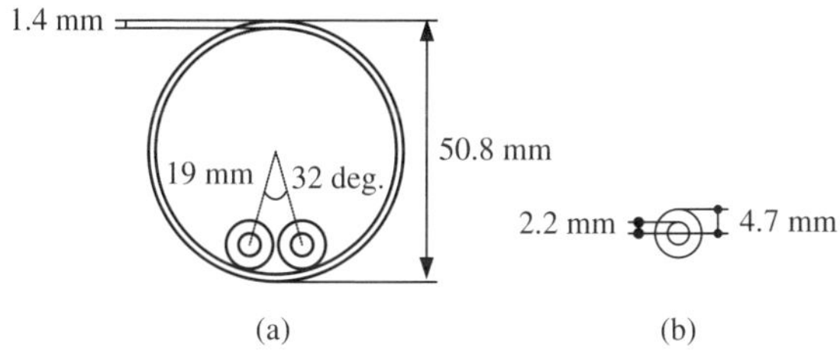

Figure 8.15 Cable configuration: (a) cable within a pipe; and (b) cross section of a cable

Although a PV array can be represented by assembling the plural single-module models, it becomes complicate to express a large-scale PV array. An array, which consists of N_S modules connected in series and N_P modules in parallel, can be expressed by the circuit shown in Figure 8.12b. The equivalent diode current I'_D, the parallel resistor R'_{sh} and the series resistor R'_s are given as follows:

$$I'_D = I_0 \left\{ \exp\left(\frac{V_D}{V_\tau \times N_s}\right) - 1 \right\} \times N_p \tag{8.28a}$$

$$R'_{sh} = R_{sh} \times \frac{N_s}{N_p} \tag{8.28b}$$

$$R'_s = R_s \times \frac{N_s}{N_p} \tag{8.28c}$$

8.2.1.2 Cable Model

Cables connecting a PV array with a PCS are assumed to be arranged within a steel pipe as shown in Figure 8.15. The cable parameters can be obtained as a pipe-type cable by the CABLE PARAMETERS.

Data 8.4 shows the data for a parameter calculation. In this section, differences from the data in Section 8.1 are explained. Line 6 indicates a two-phase ([NPC]=2), pipe-type (ITYPE=[TY]=3) overhead (ISYST=[SYS]=1) cable system. For a fault surge calculation of the PV system, the steel pipe voltage can be negligible. For simplifying a simulation, the pipe can be eliminated from the output of the CABLE PARAMETERS. For the simplification, the pipe thickness is assumed to be infinite by assigning NPP (columns 35–40) being zero, or the pipe is grounded by assigning NGD to one with NPP=1 (commetouted Line 7). The difference between the two models can be observed in the zero sequence (earth-return mode) impedance. There is no influence on the cable admittance.

Data 8.4 Data for calculating parameters of cables installed within a steel pipe

```
1. BEGIN NEW DATA CASE
2. CABLE PARAMETERS
3. C ANCH   [BUS1][BUS2][BUS3][BUS4][BUS5][BUS6][BUS7][BUS8][BUS9][BUSA][BUSB][BUSC]
4. BRANCH  PV_C1_PV21_PPV_N1_PV21_N
5. C TY][SYS][NPC][EAR][KMO][ZFL][YFL][NPP][NGD][DAT][IYG][PUN]
6.     3    1    2    0    1    0    0    0
7. C   3    1    2    0    1    0    0    1    1
8. C    Rp1  ][   Rp2  ][   Rp3  ][  rohp  ][  mur   ][  eps1  ][  eps2  ]
9.     24.0E-3    25.4E-3    26.0E-3  1.000E-7         1.          1.0         1.0
10. C  24.0E-3    25.4E-3    26.0E-3  1.000E-7       100.          1.0         1.0
11. C  Dist1 ][ Theta1 ][  Dist2 ][ Theta2 ][  Dist3 ][ Theta3 ][  Dist4 ][ Theta4 ]
12.     19.0E-3    -106.0  19.0E-3     -74.0
13. C N1][ N2][ N3][ N4][ N4][ N5][ N6][ N7][ N8][ N9][N10][N11][N12][N13][N14][N15]
14.     1    1
15. C    r1    ][   r2    ][   r3    ][   r4    ][   r5    ][   r6    ][   r7    ]
16.          0.0   2.20E-3   4.70E-3
17. C  rohc   ][  muc    ][  muI1   ][  epsI1 ][  rohs   ][   mus   ][  muI2   ][  epsI2 ]
18.   1.800E-8         1.0          1.0         1.0
19. C    r1    ][   r2    ][   r3    ][   r4    ][   r5    ][   r6    ][   r7    ]
20.          0.0   2.20E-3   4.70E-3
21. C  rohc   ][  muc    ][  muI1   ][  epsI1 ][  rohs   ][   mus   ][  muI2   ][  epsI2 ]
22.   1.800E-8         1.0          1.0         1.0
23. C center ]
24. C  30.0E-3
25. C      rohe      ][      Freq.     ][DEC][PNT][ DIST ][  IPUN  ]
26.                  100.          1.00E+3                 80.0          1
27. BLANK ENDING FREQUENCY DATA
28. $PUNCH
29. BLANK ENDING CABLE PARAMETERS CASE
30. BEGIN NEW DATA CASE
31. BLANK
```

The pipe parameters are entered into Line 9. The outer pipe radius R_{p2} has no meaning when NPP is specified as zero, because the thickness of the pipe is assumed to be infinite. When a bare steel pipe is assumed, the outer radius of the outer pipe insulator R_{p3} becomes arbitrary. The relative permeability μ_{rp} (mur) of the pipe can be taken into account. The data for $\mu_{rp} = 100$ are shown in Line 10 as a commentouted line. The relative permittivities of the inner and outer insulators of the pipe are assumed to be one.

The cable positions are specified by the distance from the pipe center and the angle as shown in Line 12.

The data structures of Lines 14–22 are identical to those of a single-core cable as shown in Section 8.1.

For the grounded pipe case (NGD = 1 with NPP = 1), the position of the pipe has to be entered. (The comment flag "C_" in Line 24 has to be removed.)

Table 8.7 shows the calculated elements of the impedance matrix. The thickness of the pipe is assumed to be infinite or the pipe is grounded, the impedance matrix size becomes 2 × 2 (two-conductor system). Although the matrix has four elements, the number of independent elements is two for the symmetry of the cable configuration. When the permeability of the pipe is 100, the effect of the simplification of the pipe becomes small. It can be explained by the critical frequency explained in Section 8.1.3, (8.8).

$$f_c = \frac{\rho_p}{2\pi\mu_0\mu_p t_p{}^2} = \frac{1\times10^{-7}}{2\pi\times4\pi\times10^{-7}\times\mu_p\times(1.4\times10^{-3})^2} = \frac{1\times10^6}{15.68\mu_p\pi^2} \approx \frac{6.5}{\mu_p}(\text{kHz}) \quad (8.29)$$

When the permeability is 100, the differences in the impedance between the simplifications are not observed at the frequency of 1 kHz, since the magnetic field within the pile is perfectly shielded. However, the effect on the simplification of the pipe on the positive sequence component, that is, the difference between the self and mutual impedances, is small.

8.2.2 *Simulated Results*

8.2.2.1 Pre-calculation of Current vs. Voltage Characteristic

The accuracy of the fault simulation heavily depends on a PV panel model. Before a simulation of a system, a pre-calculation is strongly recommended to confirm the accuracy of the current vs. voltage characteristic of the equivalent circuit of the PV module.

The diode model expressed by a single exponential function illustrated in Figure 8.16a is tested using a time-dependent load resistance shown in Figure 8.16b. The resistance is controlled by Transient Analysis of Control Systems (TACS). The load resistor is expressed by a TACS-controlled resistor (TYPE-91) and its resistance is sinusoidally changed using a TACS

Table 8.7 Calculated impedances at a frequency of 1 kHz (mΩ/m)

		$\mu_{pr} = 1$		$\mu_{pr} = 100$	
		NPP = 0, NGD = 0	NPP = 1, NGD = 1	NPP = 0, NGD = 0	NPP = 1, NGD = 1
Self	R_{cs}	1.578	1.918	2.940	2.938
	X_{cs}	2.614	2.526	5.099	5.100
Mutual	R_{cm}	0.231	0.571	1.586	1.584
	X_{cm}	0.599	0.511	2.507	2.519
Positive	$R_{cs} - R_{cm}$	1.347	1.347	1.354	1.419
sequence	$X_{cs} - X_{cm}$	2.015	2.015	2.592	2.581

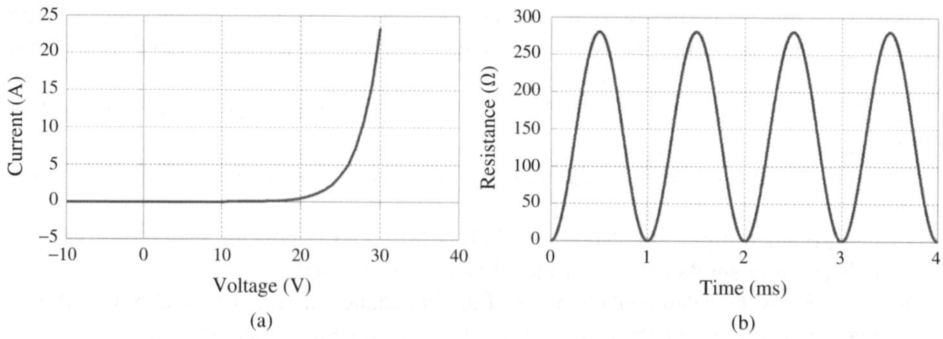

Figure 8.16 Characteristics of (a) a diode and (b) a load resistor

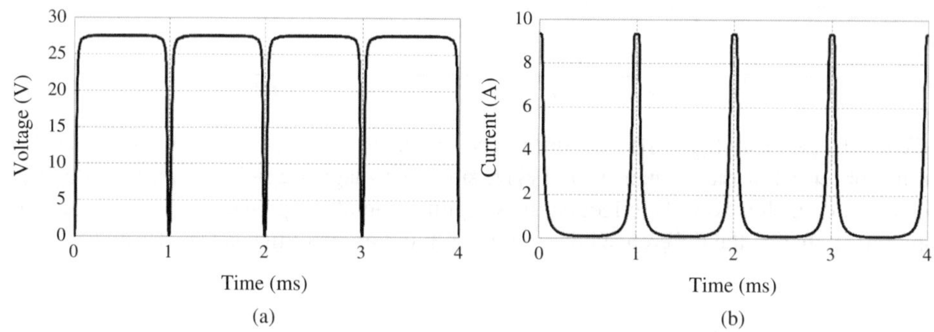

Figure 8.17 Waveforms of output (a) voltage and (b) current

signal source (TYPE-14). The diode characteristic is expressed by a nonlinear resistor model (TYPE-92) of the EMTP. Even if the nonlinear parameters are accurately estimated, the data conversion to the TYPE-92 model is one cause of errors. If the nonlinear voltage and current characteristic is estimated by Excel, the conversion can be easily carried out by formatting the EMTP data within the Excel sheet. The generated data can be used by pasting into an EMTP data file. Figure 8.17 shows the output voltage and current of the module, and Figure 8.18 shows a reproduced current vs. voltage characteristic. The simulated result agrees well with the approximated characteristic using (8.27). The number of current vs. voltage data for the TYPE-92 nonlinear resistor determines the accuracy of the simulation. In general, logarithmically sampled points give a good approximation. This test method is useful to confirm the accuracy of the exponential approximation and the number of sampled data for the TYPE-92 nonlinear resistor.

The accuracy of the simplification for a series connection, (8.28), is tested using the circuits shown in Figure 8.19a. The left string consisting of seven modules connected in series is expressed by an equivalent circuit [$N_S = 7, N_P = 1$ in (8.28)] and the right string is divided into two. The string is expressed by an equivalent circuit being $N_S = 6, N_P = 1$ with a single module model ($N_S = 1$, $N_P = 1$). From the simulated results shown in Figure 8.19b, it is clear that the model has a satisfactory accuracy for the simplification with respect to the series connection.

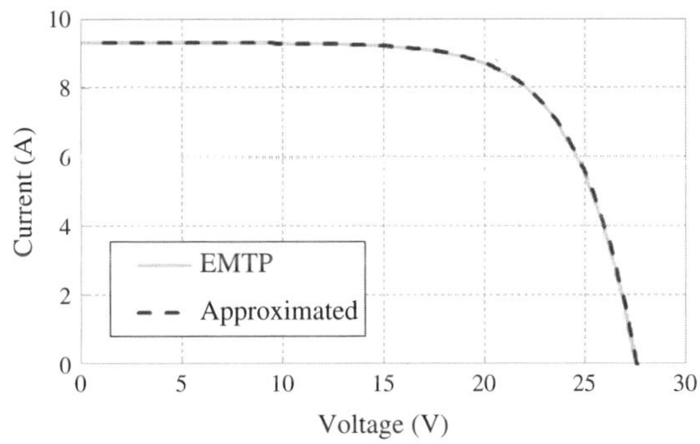

Figure 8.18 A reproduced current vs. voltage characteristic of a PV module by the EMTP

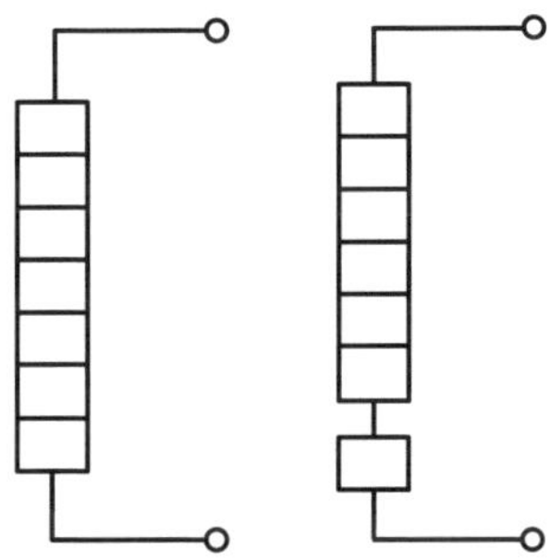

(a)

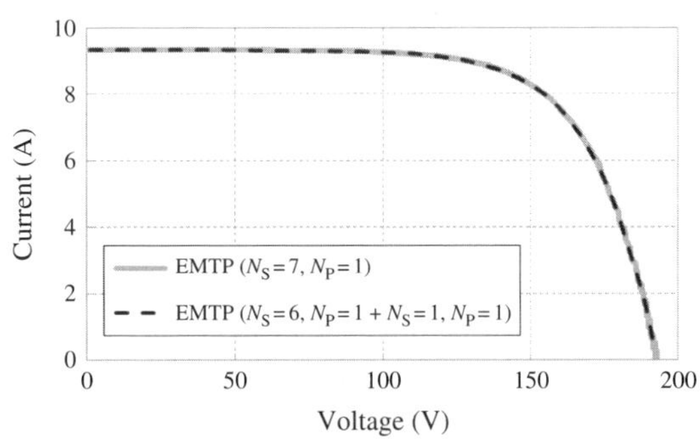

(b)

Figure 8.19 Series connection of equivalent circuits: (a) connection diagram; and (b) current vs. voltage characteristic

The simplification of the equivalent circuit with respect to a parallel connection is confirmed by the circuit shown in Figure 8.20a. The array consisting of 21 modules is expressed by an equivalent circuit [$N_S = 7, N_P = 3$ in (8.28)] in the first circuit. The right array is represented by three strings connecting in parallel, and each string is expressed by a reduced equivalent circuit ($N_S = 7,\ N_P = 1$). There is no difference in both simulated results shown in Figure 8.20b.

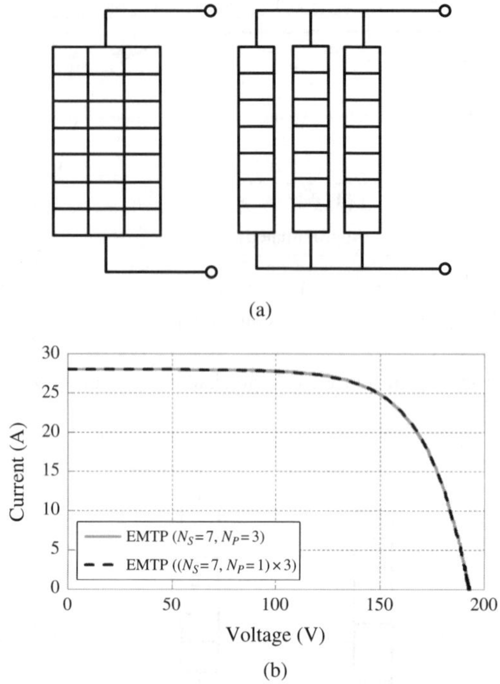

Figure 8.20 Parallel connection of equivalent circuits: (a) connection diagram; and (b) current vs. voltage characteristic

From the above, the simplification explained in the previous section, (8.28), is useful to simulate a large-scale PV generation by the EMTP.

8.2.2.2 Short-circuit Fault Simulation

Outline

A short-circuit fault within a PV array is simulated by the EMTP using the equivalent circuit of the PV module explained in Section 8.2.1. In this section, it is assumed that a positive terminal within a string is faulted to the negative terminal of the PV array.

Figure 8.21 shows simplified equivalent circuits to explain the fault characteristic. The resistance R_t and inductance L_t model a fault impedance.

The resistance R_k and inductance L_k express the cable impedance of the kth string.

$$R_k = \frac{1}{2}(R_{cs} - R_{cm})l_k$$
$$L_k = \frac{1}{2}\frac{(X_{cs} - X_{cm})l_k}{2\pi f_a} \qquad (8.30)$$

where R_{cs}, R_{cm}, X_{cs}, X_{cm} are the impedance elements shown in Table 8.7, l_k is the cable length, and f_a is the frequency at which the cable parameter is calculated.

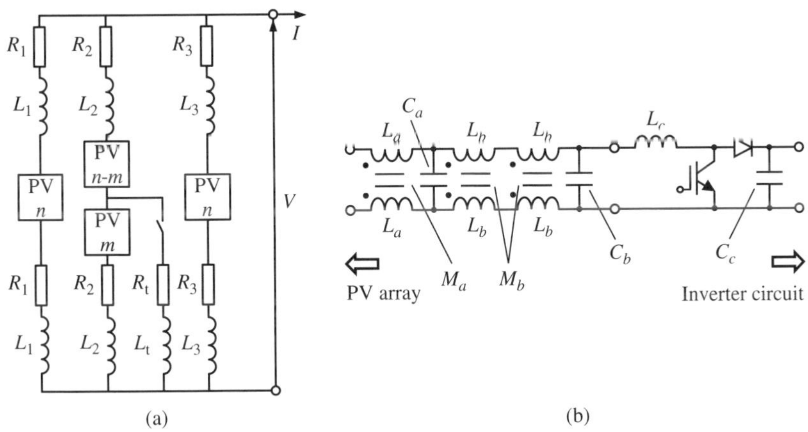

Figure 8.21 Simulation circuits: (a) PV array; and (b) input part filter and converter circuit of the PCS

The fault characteristic is determined by the nonlinear characteristic of the PV module and the line impedance. The cable can be modeled by a pi equivalent circuit because the multiple reflections within the cable can be neglected in the simulation. Furthermore, the cable can be expressed by a mutual coupled RL model (Type-51), since the effect of the cable capacitance is small.

The PCS consists of a filter, a d.c./d.c. converter and an inverter as shown in Figure 8.21b. Each coil in the filter circuit is wound on a toroidal core. The self and mutual inductances are $L_a = 11.8\,\mu\text{H}$, $M_a = 11.4\,\mu\text{H}$, $L_b = 1066\,\mu\text{H}$ and $M_b = 1064\,\mu\text{H}$. The inductors with a ceramic capacitor C_a (= 2.2 μF) and an electrolytic capacitor C_b (= 900 μF) make up a low-pass filter.

The d.c./d.c. converter boosts a voltage by the switching devices with an inductor L_c (= 1.6 mH) and an electrolytic capacitor C_c (= 3400 μF). An inverter circuit of the PCS is neglected in this simulation.

Single-module Short-circuit Fault

In this section, a transient characteristic is simulated when a positive terminal of a single module within the second string is faulted for 30 ms as shown in Figure 8.22. The photoinduced current I_{ph} of each string is assumed to be 6.6 A, and the operating voltage of the array is 143 V ($= V_a = 7 \times 20.4$ V), which is the voltage at the maximum power point. The operating point is kept after the fault, because the MPPT function of the PCS cannot trace the maximum power point. A d.c. source is connected to the output terminal of the d.c./d.c. converter as a load of the PV system for a simplification, because the voltage is kept by the converter operation. The duty ratio of the converter D is assumed to be 0.6, that is, the voltage of the d.c. source becomes 357 V [$= V_a/(1\text{-}D)$].

Figure 8.23a and b shows simulated results of the string-2 current I_2, the string-1 current I_1, the output voltage of the array V and the output current of the array I. The string-3 current I_3 is identical to I_1. When a module is short-circuited, the current vs. voltage characteristic of the string involving the fault module is nonlinearly changed as shown in Figure 8.23c. The current of the sound (no-fault) string is not affected by the fault and the current of the fault string decreases, because the voltage of the operating point is kept by the PCS. The frequencies

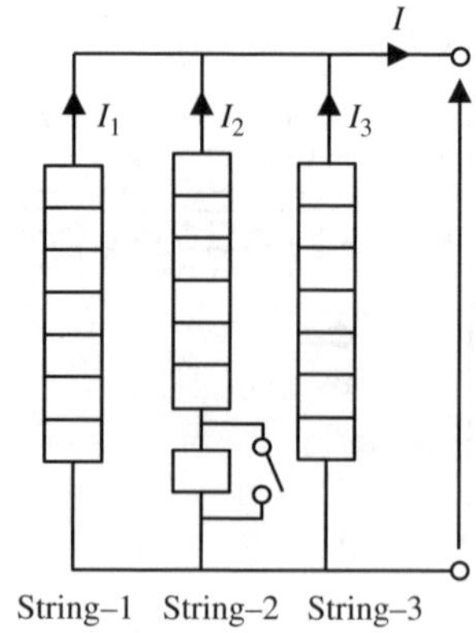

Figure 8.22 A simulation circuit for a single-module short-circuit fault

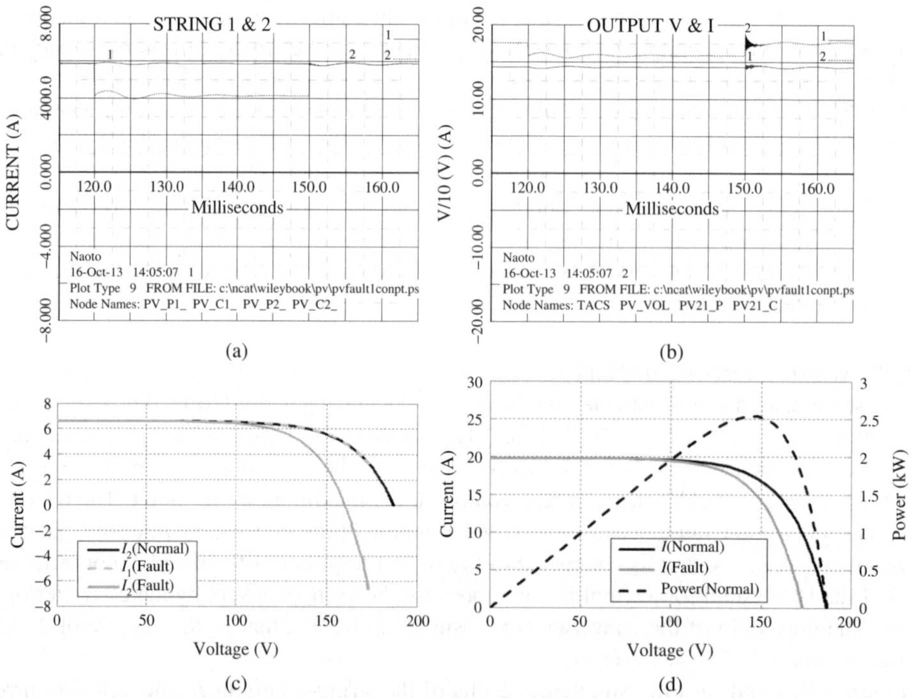

Figure 8.23 Simulated results at a single-module fault: (a) string currents; (b) output voltage and current; (c) current vs. voltage characteristic of a string (I_1 and I_2); and (d) current vs. voltage characteristic of a panel (I and power)

of the oscillations observed both at the fault and at the clearing are determined by the resonance between the capacitor C_b and the inductor L_c in the PCS.

To investigate the effect of the converter operation, the insulated-gate bipolar transistor in Figure 8.21 is removed from the circuit and a d.c. source of 143 V is connected in parallel with the capacitor C_b. Figure 8.24 shows the simulated results. When the converter circuit is neglected, the oscillations on the voltage and current waveforms shown in Figure 8.23a and b are not observed. The d.c. component of the current can be obtained with a satisfactory accuracy without the converter. The simplification is useful for the fault analysis.

Six-module Short-circuit Fault

Figure 8.25 illustrates a simulation circuit when six modules are short-circuited. The photoinduced current I_{ph} of each string is assumed to be 5.6 A. A resistor of 59.1 Ω (R_L) is connected to an output terminal of the converter as a load of the PV system.

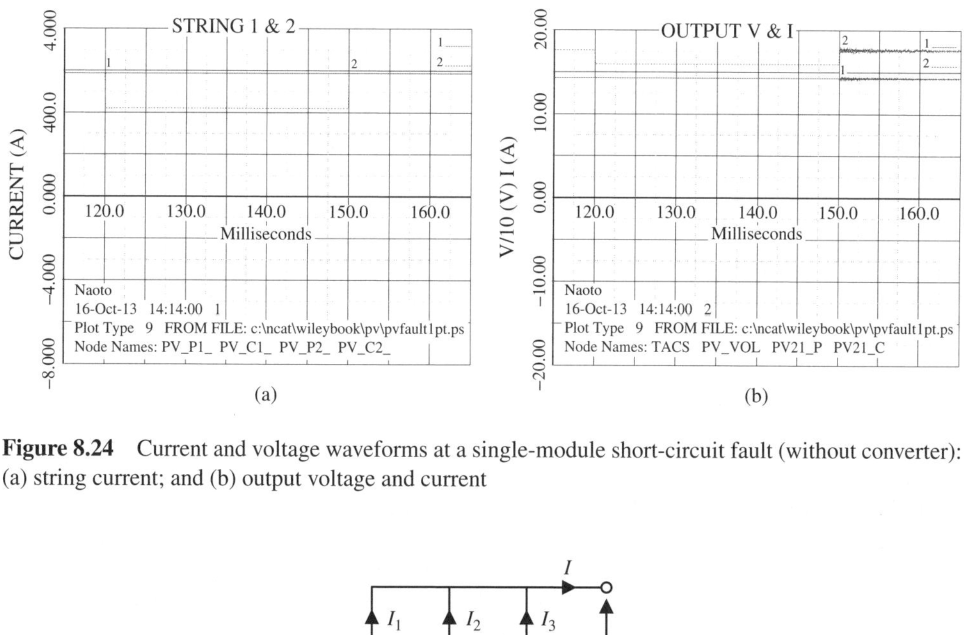

Figure 8.24 Current and voltage waveforms at a single-module short-circuit fault (without converter): (a) string current; and (b) output voltage and current

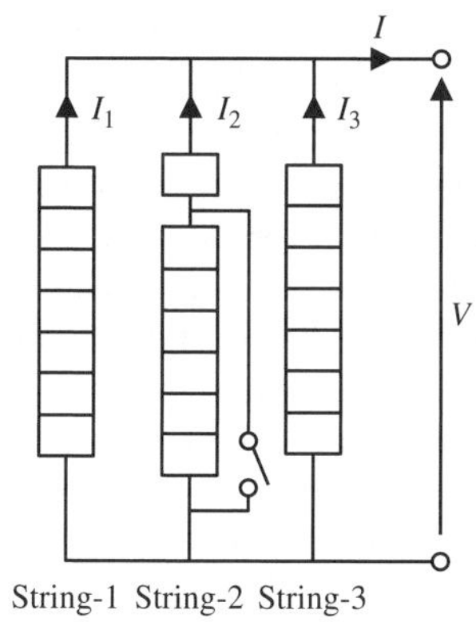

Figure 8.25 A simulation circuit for a six-module short-circuit fault

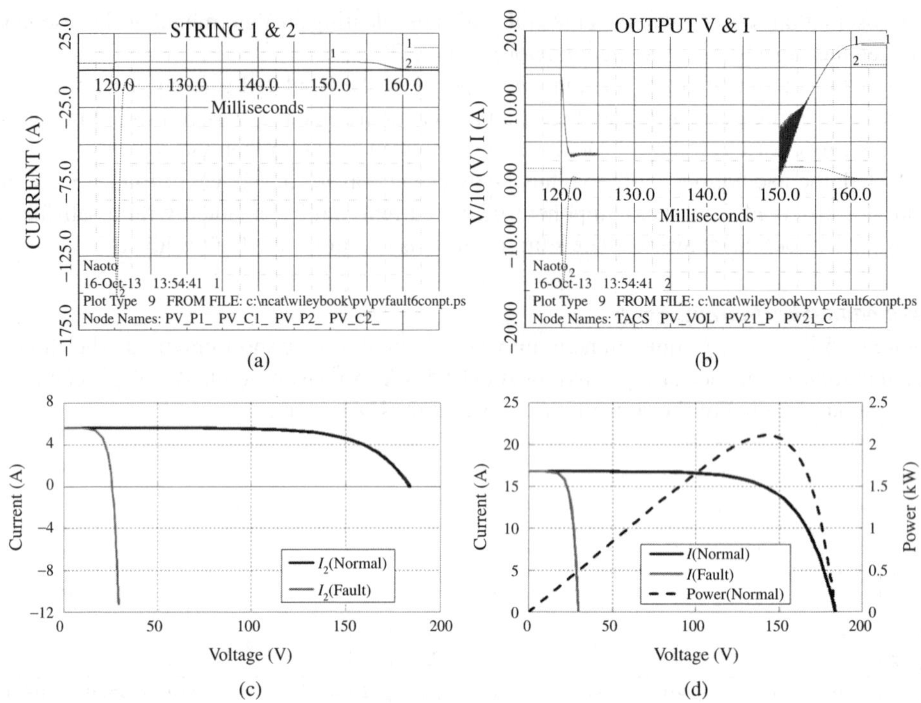

Figure 8.26 Simulated results at a six-module fault: (a) string current; (b) output voltage and current; (c) current vs. voltage characteristic of a string (I_2); and (d) current vs. voltage characteristic of a panel (I and power)

Figure 8.26 shows the simulated results. When the six modules are short-circuited, the current vs. voltage characteristics of the string is considerably changed compared with the single-module fault case as shown in Figure 8.23c. Therefore, the PCS stops after some tens of milliseconds from the fault due to the low voltage. In addition, the PCS cannot automatically recover after clearing the fault. In this simulation, the converter is stopped and the load is rejected 20 ms after the fault. The currents generated in string-1 and string-3 flow into string-2 as a reverse current. The operating voltage of the array is determined by the I-V characteristic at which the sum of the string currents becomes zero ($I_2 = -I_1 - I_3$). The fault string current becomes negative and the operating voltage decreases to around 30 V as shown in Figure 8.26c.

A spike current is superimposed on the reverse current when the modules are short-circuited. This spike current is the discharge current of the capacitors C_a and C_b. The charges in the capacitor C_c remain because a diode for the boost converter is connected to the capacitor. In addition, the discharge current does not flow into the no-fault strings because the impedances of the nonlinear diodes in the modules are far higher than that of the fault string.

Figure 8.27 shows simulated results when the converter circuit is neglected. A resistor of 9.5 Ω [$= R_L(1\text{-}D)^2$] is connected in parallel with the capacitor C_b. Minor differences of 3.2% on the peak value of the spike current and 3.7% on the reverse current are observed as shown in Figure 8.28. The converter can be negligible to simulate the multi-module fault with satisfactory accuracy as well as the single-module fault.

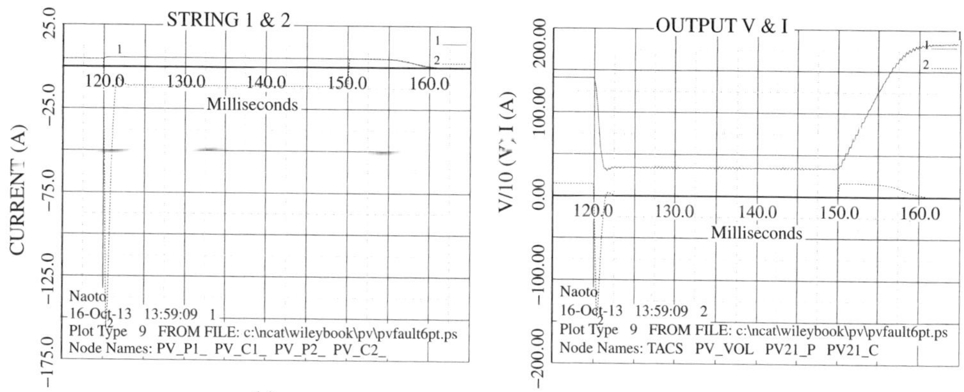

(a) (b)

Figure 8.27 Current and voltage waveforms at a six-module short-circuit fault (without converter): (a) string current; and (b) output voltage and current

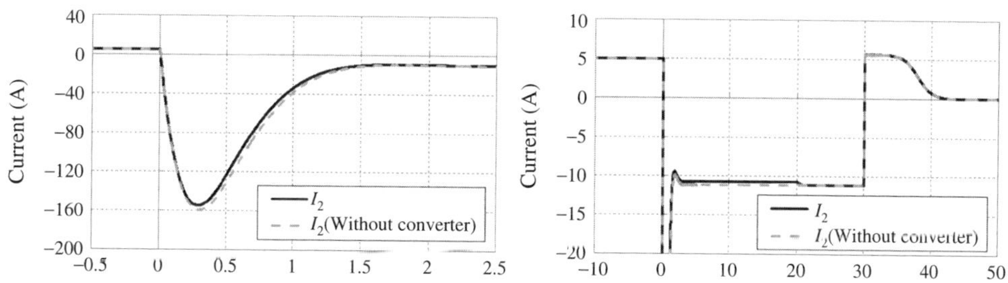

(a) (b)

Figure 8.28 Effect of the converter operation: (a) spike current (magnified); and (b) overall

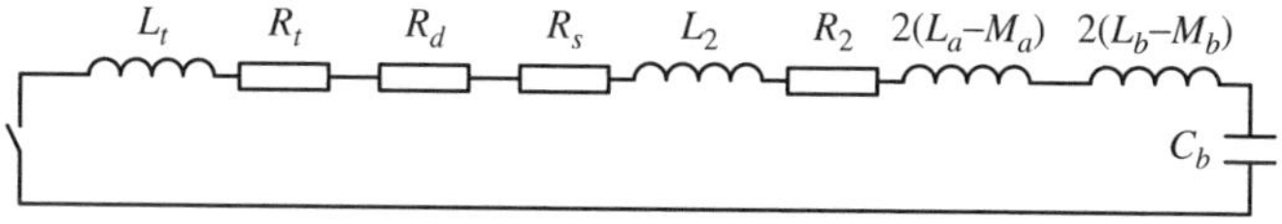

Figure 8.29 Simplified circuit for analytical calculation of the spike current

The spike current can be explained by an analytical calculation. Figure 8.29 shows an equivalent circuit for the analysis of the spike current. This circuit consists of the simplified filter circuit of the PCS (L_a, M_a, L_b, M_b, and $C_\mathbf{b}$), the cable impedance of string-2 (R_2 and L_2), the series resistor of the module (R_s), a differential resistance of the nonlinear resistor (diode) in the module equivalent circuit (R_d) and a fault impedance (R_t and L_t). The capacitor C_a is negligible because its capacitance is much smaller than that of C_b.

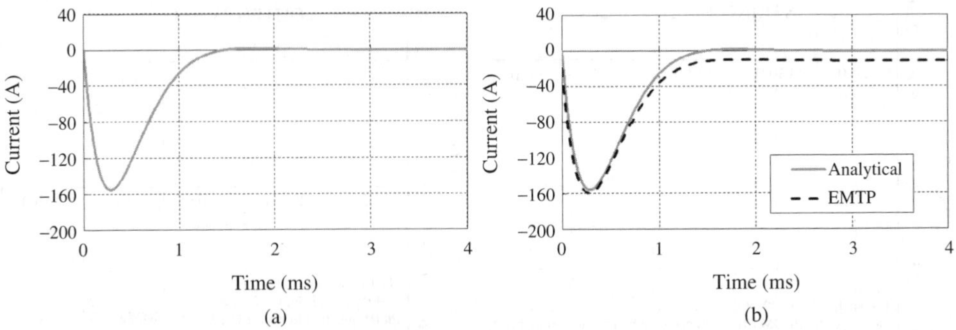

Figure 8.30 Analytical calculated result of the spike current: (a) analytical; and (b) EMTP

Figure 8.30a shows the current waveforms calculated by the analytical method. The initial voltage of C_b is assumed to be 141 V. For a comparison, the simulated result by the EMTP is given in Figure 8.30b, which is identical to Figure 8.28a. Although the d.c. component cannot be reproduced, the transient characteristic can be explained by the series *RLC* circuit. The oscillating period of the current waveform is determined by the capacitance C_b and the inductances of the cable $[L_2 >> 2(L_a\text{-}M_a) + 2(L_b\text{-}M_b)]$.

A nonlinearity of a current vs. voltage characteristic of a PV module should be modeled by at least two exponential functions for an accurate simulation. However, it can be expressed by a single exponential function for a fault simulation of a PV array. The characteristic can be obtained from a measured current vs. voltage characteristic of a PV module or array. A nonlinear curve fitting is useful for obtaining a smooth voltage vs. current characteristic, which is required for a stable numerical simulation. The curve fitting can be carried out using Microsoft Excel, and the data of the model for the EMTP are automatically generated by Excel. The data at around the maximum power point should be entered into the nonlinear approximation to keep the accuracy of the model. When a fault occurs in a PV array, a reverse d.c. flows into the fault string. A spike current due to a discharge of a capacitor in a PCS flows into the fault string when some modules are short-circuited, because the spike current depends on a capacitor of the input filter of the PCS and the cable impedance connecting the PV array with the PCS. However, the models of the converter and inverter circuit in the PCS are not required. The EMTP is useful for an accurate estimation of the reverse current and a design of countermeasures, such as a blocking diode.

References

[1] Scott-Meyer, W. (1982) EMTP Rule Book, Bonneville Power Administration.

[2] Dommel, H.W. (1968) Digital computer solution of electromagnetic transients in single and multiphase networks. *IEEE Transactions on Power Apparatus and Systems*, **PAS-88** (4), 388–99.

[3] Dommel, H.W. and Meyer, W.S. (1974) Computation of electromagnetic transients. *Proceedings of the IEEE*, **62** (7), 983–93.

[4] Meyer, W.S. and Dommel, H.W. (1974) Numerical modeling of frequency dependent transmission line parameters in an electromagnetic transients program. *IEEE Transactions on Power Apparatus and Systems*, **PAS-93** (5), 1401–9.

[5] Semlyen, A. and Dabuleanu, A. (1975) Fast and accurate switching transient, calculations on transmission lines with ground return using recursive convolutions. *IEEE Transactions on Power Apparatus and Systems*, **PAS-94** (2), 561–71.
[6] Ametani, A. (1976) A highly efficient method for calculating transmission line transients. *IEEE Transactions on Power Apparatus and Systems*, **PAS-95** (2), 1545–51.
[7] Marti, J. (1982) Accurate modeling of frequency-dependent transmission lines in electromagnetic transient simulations. *IEEE Transactions on Power Apparatus and Systems*, **PAS-101** (1), 117–57.
[8] Noda, T., Ono, T., Matsubara, H. *et al.* (2003) Charge-voltage curves of surge corona on transmission lines: Two measurement methods. *IEEE Transactions on Power Delivery*, **18** (1), 307–14.
[9] Wedepohl, L.M. (1963) Application of matrix method to the solution of travelling-wave phenomena in polyphase systems. *Proceedings of the IEE*, **110** (12), 2200–12.
[10] Ametani, A. (1973) Refraction coefficient method for switching-surge calculations on untransposed transmission lines – accurate and approximate inclusion of frequency-dependence. IEEE 1973 PES Summer Meeting, Vancouver, BC, Canada.
[11] Nakanishi, N. and Ametani, A. (1986) Transient calculation of a transmission line using superposition law. *IEE Proceedings*, **133** (5), 263–9.
[12] Marti, L. (1987) Simulation of transients in underground cables with frequency dependent modal transformation matrices. IEEE 1987 PES Winter Meeting, New Orleans, LA, USA.
[13] Ametani, A. (1994) Cable Parameters Rule Book, Bonneville Power Administration.
[14] Ametani, A. (1980) A general formulation of impedance and admittance of cables. *IEEE Transactions on Power Apparatus and Systems*, **PAS-99** (3), 902–10.
[15] Amekawa, N., Nagaoka, N., Baba, Y., and Ametani, A. (2003) Derivation of a semiconducting layer impedance and its effect on wave propagation characteristics on a cable. *IET Generation, Transmission and Distribution*, **150** (4), 434–40.
[16] Ametani, A., Miyamoto, Y., and Nagaoka, N. (2004) Semiconducting layer impedance and its effect on cable wave-propagation and transient characteristics. *IEEE Transactions on Power Delivery*, **19** (4), 1523–31.
[17] Steinbrich, K. (2005) Influence of semiconducting layers on the attenuation behavior of single-core power cables. *IEE Proceedings – Generation, Transmission and Distribution*, **152** (2), 271–6.
[18] Hasheminezhad, M., Vakilian, M., Blackburn, T.R., and Phung, B. T. (2006) Direct Introduction of Semicon Layers in XLPE Cable Model. POWERCON 2006, Chongqing, China, 1–7.
[19] IEE Japan Working Group (1991) Surge Phenomena in Cable Systems and Cuntermeasures. IEEJ Technical Report, 366.
[20] Ametani, A. (1980) Wave propagation characteristics of cables. *IEEE Transactions on Power Apparatus and Systems*, **PAS-99** (2), 499–505.
[21] Ametani, A. (1983) A study of cable transient calculations. *The Science and Engineering Review of Doshisha University*, **24** (2), 110–27.
[22] Nagaoka, N. and Ametani, A. (1988) A development of a generalized frequency-domain transient program – FTP. *IEEE Transactions on Power Delivery*, **3** (4), 1996–2004.

Index

Cable System Transients: Theory, Modeling and Simulation, First Edition.
Akihiro Ametani, Teruo Ohno and Naoto Nagaoka.